Muskeg and the Northern Environment in Canada provides
wide range of specialists – scientists, engineers, foresters, agriculturalists, and environmentalists – who deal with this unusual and remarkable resource. Continental water, northern forest resources, and petroleum exploitation are now pressing concerns of the northern
development complex and all involve wise analysis of the nature of muskeg.

The muskeg environment comprises ecosystems characterizing at least 180,000 square miles
of our country, an area bigger than Quebec and the Maritime Provinces combined. Muskeg is
found from the toe of the Ellesmere Ice Cap to the border with the United States, and it
reaches all three sea coasts. It occurs to considerable depth, sometimes to about 100 feet.
Whether it is in permafrost country or not, interference with it calls for the attention of
specialists if natural conditions are to be respected and wisely monitored.

The papers in this book, written by experts in various aspects of muskeg studies relating to
the northern environment, respond to the need for an intelligent analysis of the muskeg
question. The book reviews what is known and draws attention to limitations in current
knowledge of muskeg.

The volume also includes a valuable list of terms and definitions. *Muskeg and the Northern
Environment in Canada* demonstrates the wisdom of an interdisciplinary approach to the
subject and provides a welcome addition to the growing body of muskeg literature.

The papers gathered in this volume were originally presented at the 15th Muskeg Conference
held in 1973 and sponsored by the Muskeg Subcommittee of the National Research Council
of Canada's Associate Committee on Geotechnical Research.

N.W. Radforth is a member of the Department of Biology and Geology at the University of
New Brunswick and a consultant with Radforth and Associates, Fredericton. He served for
many years as chairman of the Muskeg Subcommittee of the Associate Committee on
Geotechnical Research, National Research Council of Canada.

C.O. Brawner is past chairman of the Muskeg Subcommittee of the Associate Committee on
Geotechnical Research of the National Research Council of Canada and a principal of the
firm Golder, Brawner & Associates Ltd., Vancouver.

National Research Council of Canada

ASSOCIATE COMMITTEE ON GEOTECHNICAL RESEARCH

CARL B. CRAWFORD, *Chairman*

MUSKEG SUBCOMMITTEE

C.O. BRAWNER	Golder, Brawner & Associates Ltd., Vancouver, British Columbia, *Chairman*
L.V. BRANDON	Department of Indian Affairs and Northern Development, Whitehorse, Yukon Territory
J.H. DAY	Soil Research Institute, Ottawa, Ontario, *Ex officio*
R.A. HEMSTOCK	Canadian Arctic Gas Study Limited, Calgary, Alberta
G.R. PELLETIER	Ministère de la Voirie, Province de Québec
N.W. RADFORTH	Muskeg Research Institute, Fredericton, New Brunswick
G.P. RAYMOND	Queen's University, Kingston, Ontario
P.J. RENNIE	Department of Fisheries and Forestry, Ottawa, Ontario
W. STANEK	Department of Fisheries and Forestry, Forest Research Laboratory, Sault Ste Marie, Ontario
J. TERASMAE	Brock University, St Catharines, Ontario
T.E. TIBBETTS	Department of Energy, Mines and Resources, Ottawa, Ontario
I.C. MACFARLANE	National Research Council of Canada, Ottawa, Ontario, *Research Advisor*
J. CURRAN (Mrs)	National Research Council of Canada, Ottawa, Ontario, *Secretary*

Muskeg and
the Northern Environment
in Canada

By the Muskeg Subcommittee
of the NRC Associate Committee
on Geotechnical Research

EDITED BY
N.W. Radforth and C.O. Brawner

University of Toronto Press
TORONTO AND BUFFALO

© University of Toronto Press 1977
Reprinted in 2018
Toronto and Buffalo
Printed in Canada

Library of Congress Cataloging in Publication Data

Muskeg Research Conference, 15th, Edmonton, Alta., 1973.
Muskeg and the northern environment in Canada.

Includes index.
1. Muskeg – Canada – Congresses. 2. Land use –
Canada – Congresses. I. Radforth, Norman W. II. Braw-
ner, C.O. III. National Research Council of Canada.
Associate Committee on Geotechnical Research. Muskeg
Subcommittee. IV. Title.
GB628.15.M86   1973   557.1   76-54734
ISBN 0-8020-2244-8
ISBN 978-1-4875-7308-9 (paper)

This book has been published during the
Sesquicentennial year of the University of Toronto

Contents

Foreword

Muskeg occurs, to some degree, in every Canadian province and territory. It covers a great expanse of the mid-Canada region, overlapping the northern forest/tundra zone, the boreal forest, and even occurs sporadically in various formations of limited size in the southern, more densely populated regions. Although no accurate figure is available on the extent of muskeg occurrence in Canada, it has been conservatively estimated to be in the order of 500,000 square miles (1,295,000 square kilometres).

The National Research Council of Canada has long been interested in all aspects of the terrain of Canada, including soil, rock, snow, ice, and muskeg. Canadian muskeg studies have developed slowly but steadily. Since the late 1940s they have been consistently supported by the Council through its own laboratory research program, and especially through the Muskeg Subcommittee of the Associate Committee on Geotechnical Research (formerly the Associate Committee on Soil and Snow Mechanics). The Subcommittee has been a major force in stimulating, encouraging, and coordinating muskeg research, as well as in information transfer. In 1955, the Subcommittee sponsored the first of its annual muskeg conferences; the proceedings of these conferences record the development in the sophistication of muskeg studies in Canada. These studies came of age in 1968 when the International Peat Congress, co-sponsored by the Muskeg Subcommittee, was held in Quebec and Canadian muskeg research came under the scrutiny of international workers in the field. In 1969, the *Muskeg Engineering Handbook*, another subcommittee undertaking, was published. This document, the first book of its type in the English language, has proved to be unusually valuable.

The present book, which is a compendium of papers presented at the 15th Muskeg Conference held in 1973, represents another major milestone in the history of muskeg studies in Canada. The papers, written by experts in various aspects of muskeg studies relating to the northern environment, have been edited by C.O. Brawner and N.W. Radforth, who were careful to retain the style of each author. The scope of this multidisciplinary book is extensive, and it should be of great interest to scientists, engineers, foresters, agriculturalists, environmentalists, and many others who are concerned with the development or use of the unusual and remarkable resource known as muskeg. The book reviews what is known and draws attention to limitations in knowledge pertaining to muskeg.

This book demonstrates the wisdom of a coordinative approach to muskeg studies and provides a welcome addition to the growing body of muskeg literature.

C.B. CRAWFORD
Chairman, Associate Committee on Geotechnical Research
National Research Council of Canada

Preface

Trees reflect, by their form, the effect of prevailing wind in coping with time and season. This is a biological response. Other plants, either singly or together, may also display the environmental effect in cosmic expressions of pattern, and so it is with muskeg.

In Canada, muskeg is sometimes called organic terrain or peatland. It develops as a multipatterned feature of the natural landscape. Each pattern arises from a particular biotic-abiotic interplay that invites understanding and categorization.

The muskeg environment comprises ecosystems characterizing at least 180,000 square miles of our country, an area bigger than Quebec and the Maritime Provinces combined. Muskeg is found from the toe of the Ellesmere Ice Cap to the border with the United States, and it reaches all three sea coasts. It occurs at any depth, sometimes up to about 100 feet. Whether it is in permafrost country or not, interference with it calls for the attention of specialists if natural conditions are to be respected and wisely monitored.

In recognition of this need, the Muskeg Subcommittee which reports to the Associate Committee on Geotechnical Research of the National Research Council of Canada has promoted fresh enquiry and understanding concerning the muskeg environment. Mr C.O. Brawner, the subcommittee chairman and co-editor of this book, conceived the idea that this topic should be discussed at a national symposium and the papers published in book form. He and the contributing authors are to be commended. Their response is timely: continental water, northern forest resources, and petroleum exploitation are now pressing factors of the northern development complex and all involve wise analysis of the muskeg question.

The editors are grateful for the cooperation of the authors, who have achieved a harmonious overview of the diverse approaches that the subcommittee has nurtured and wished to see expanded for the benefit of interdisciplinary activity.

A word of thanks is also due to the members of the Editorial Committee for their assistance, and particularly to Dr J.R. Radforth, who had the task of compiling the subject index, and to Dr W. Stanek, who compiled the List of Terms and Definitions.

The editors are grateful for the continued interest of Miss L. Ourom, University of Toronto Press, and to Mr R.H. Chow, Golder Associates Ltd., for editorial and author coordination.

Special appreciation must also be expressed for the support, moral and otherwise, of this project on the part of Carl B. Crawford, Director, Division of Building Research, National Research Council of Canada, and Chairman of the Associate Committee on Geotechnical Research, and for a subsidy from the Canadian Geotechnical Fund which has made it possible to publish the volume.

N.W. RADFORTH

Contributors

P. AITCIN, Department of Civil Engineering, University of Sherbrooke, Sherbrooke

R.J.E. BROWN, Geotechnical Section, Division of Building Research, National Research Council, Ottawa

A.L. BURWASH, Muskeg Research Institute, University of New Brunswick, Fredericton

S. CHORNET, Department of Chemical Engineering, University of Sherbrooke, Sherbrooke

M. COSSETTE, QAMA Laboratory, University of Sherbrooke, Sherbrooke

B. COUPAL, Department of Chemical Engineering, University of Sherbrooke, Sherbrooke

A.E. FEE, J.L. Richards & Associates Ltd., Ottawa

D.A. GOODE, The Nature Conservancy, Edinburgh, Scotland

ERKKI O. KORPIJAAKKO, Muskeg Research Institute, University of New Brunswick, Fredericton

N.A. LAWRENCE, Associated Engineering Services Ltd., Edmonton, Alberta

A.A. MARSAN, Department of the Environment, Province of Quebec, and Centre de Recherches écologiques de Montréal, Montreal

R.D. MEERES, Banister Pipelines, Edmonton, Alberta

J.-R. MICHAUD, Centre de Recherches écologiques de Montréal, Montreal

A.C. MOFFATT, Reid Crowther & Partners Ltd., Edmonton, Alberta

R.D. MUIR, Canadian Wildlife Service, Ottawa, Ontario

J.R. RADFORTH, Muskeg Research Institute, University of New Brunswick, Fredericton

N.W. RADFORTH, Muskeg Research Institute, University of New Brunswick, Fredericton

P.J. RENNIE, Canadian Forestry Service, Department of the Environment, Ottawa

M. RUEL, Environmental Emergency Branch, Environmental Protection Service, Department of the Environment, Ottawa

WALTER STANEK, Department of the Environment, Canadian Forestry Service, Great Lakes Forest Research Centre, Sault Ste Marie

J.M. STEWART, Department of Botany, University of Manitoba, Winnipeg

J. TERASMAE, Department of Geological Sciences, Brock University, St Catharines

M.E. WALMSLEY, Department of Soil Science, University of British Columbia, Vancouver

DAVID F. WOOLNOUGH, Land Registration and Information Service, Council of Maritime Premiers, Fredericton

Muskeg and the Northern Environment in Canada

Introduction

N.W. RADFORTH

Though climate plays an important role in influencing resource development in northern North America, terrain, another environmental factor, has recently become prominent as well. Muskeg (organic terrain) sometimes limits developmental activity, creating conditions that call for adjustments in normal engineering design and economics.

What makes muskeg important as an environmental entity is its response following impact. When interfered with, it becomes altered to present fresh physiographic conditions and its materials undergo change. After impact it becomes endowed with new sets of thermal, hydrological, structural and chemical characteristics. Often the changes are ir-reversible. The direct or indirect effects may be on a large scale and involve hundreds of square miles, or, if local, they may be intensive in implication.

The organic material, peat, of which muskeg is largely composed is a resource in its own right. As a primary product and as a base for certain processed or manufactured secondary commodities its value is already being realized. In terms of potential, these aspects of use are now attracting more attention than ever before among resource planners across the North American continent. In Europe, peat, both mined and in situ, is being exploited not only for use in agriculture and forestry but as an energy source as well. Most northern European countries have already taken steps to re-evaluate peatlands with a view to optimizing peat-use programs and establishing fresh priorities precipitated because of energy needs.

Clearly, the value of muskeg is increasing. South of the permafrost zone, if it is now untouched, the chances are that it will be brought into use soon either for agriculture or forestry or be controlled as a natural agent important in the manipulation of continental water. Meantime it deserves protection in anticipation that the best use will be made of it. Where peat is mined, concern for the exploited lands and their rehabilitation is no less important than it is for the waste conditions characterizing mining practice generally.

To foster good management it is reasonable to coordinate information on the various technologies associated with muskeg exploitation and peat utilization. Generally, this prin-ciple has been respected, as, for example, in Finland, the USSR, and Ireland. Elsewhere in Europe, as in Sweden, Norway, and Germany, certain technologies associated with peat utilization are now quite sophisticated but the accrued knowledge has little dependence on the direction that ecologically oriented studies give. In Canada, with little exception, the development of technological knowledge has always been independent of the ecological inference, as it has been in the United States and Britain. This book considers subject matter from both these areas of investigation. It is hoped that this gesture will not only stimulate further adoption of the coordinative approach for research in muskeg studies but will also provide guidance for resource planners and users of muskeg.

In addition to the present need in North America for integrative attention to methodology certain other areas are of more than potential importance. For instance, in the United States and especially in Canada, competition is strong for markets demanding peat products. Basic and applied research is needed to improve processing and to broaden the range of products offered to world markets. Furthermore, the exploitive developments now imminent where muskeg abounds in northern Canada and Alaska also call for specialized attention.

The present volume has been compiled in order to disseminate, among a wide spectrum of scientists, engineers, and economists, available information on various aspects of muskeg. It is hoped that it will assist them in dealing with the current situation, and that, as the need arises, it, like any broad review of endeavour, may be kept up to date by periodic revision.

In this review volume, consideration was given first to the developmental history of the muskeg phenomenon. Indeed, the book opens with a reasoned inference that knowledge of organic terrain as an environmental component is incomplete and inadequate without information on the sequential environmental events represented in the muskeg record.

A historical accounting sheds light on how the terrain was initiated. It also reveals the major physiographic changes that have occurred during development and discloses reasons for them. It suggests why the culminating environmental conditions including the existing vegetation have arisen. Finally, it forces attention on a conclusion that investigators must regard muskeg, and the materials constituting the condition, as embracing a great variety of types each consistently ordered by its particular set of physiochemical attributes and mechanical properties and by its propensity for recurrence at a particular frequency which depends on the region.

Peat chronology and history of climate do much to explain both transient and stable geobotanical states, and an understanding of these states contributes to developing a reassuring approach for good terrain management or the manipulation of the existing muskeg environment.

The ordered trends that muskeg reveals, as biotic and abiotic factors 'trade-off' through time, lend themselves to classification. Classification is needed, too, if the resource is to be categorized for best use. But classification, considered at length in Chapter 2, is often fraught with difficulties for the user who lacks sufficient background training in muskeg studies. Also misunderstandings arise because of the varied requirements to which classification systems must adapt. Definitions often refer to inappropriate scales of exactness, and sometimes to parameters which are not particularly suitable for the immediate purpose. Frequently, the characterizing expressions are expected to accommodate categories of performance when either the terrain or its products are under test.

Professional needs differ. Treed muskeg, perhaps classified satisfactorily for foresters, is usually avoided by peat producers, who in turn require information to distinguish non-treed conditions. The manufacturer of peat products is interested in the categories of peat which contain admixtures of fibre having the special physiochemical qualities he hopes to incorporate into commodities best suited to market demands. The designer of off-road vehicles, who is interested in both forested and non-forested muskeg, requires mechanical and hydrological evaluations in classification. Those who must know the thermal or drainage characteristics of peat and peatland in anticipation of pipeline installation or road-building, etc., require classifications of predictive conditions. The aim of the future is not to discard one system of classification for another or to expect the adoption of any single system, but to find a suitable model for which all of the classification systems are relevant and which would contain a cross-reference mechanism for those wishing to equate conditions defined in different systems.

In those countries where developments affecting muskeg are increasing rapidly, it is natural, with classification systems at hand, to create informative inventories of both the peat resources and the conditions characterizing the muskeg. The importance of taking an inventory is stressed in Chapter 3, and an example of how one might be initiated is provided. In Canada, where resource management is usually a provincial matter, it is quite possible that inventories will be made based on designs which differ from province to province. If this happens, it is hoped that the difficulties arising for the users will not be as confusing as those of classification which have already confronted developers and environmentalists.

Peat, water, and the chemical constitution of that water play significant roles in effecting organizational patterns reflected in the cover and topography of muskeg. These patterns are basic in formulating methods for undertaking inventories. The nutriments available in muskeg waters have a controlling or limiting effect on the productivity of the peaty environment. The importance of certain nutrient relations is expressed in Chapter 4. The contribution here, however, is perhaps best considered against the background developed in the following two chapters, on hydrology and permafrost, rather than in relation to the principles of inventory. It is the processes implicit in these two subjects that help establish and control the chemical differentials and distributional phenomena that occur in muskeg.

With the understanding and perspective developed in the book thus far, the reader will be better fitted to deal with the special aspects of muskeg utilization considered in Part II. General guidance on this subject is offered in Chapter 7, and Chapters 8 and 9 discuss two of the more important economic examples, one agricultural (the growing of wild rice) and the other relating to secondary industry (peat moss and the production of other products).

Unless the reader appreciates the indirect environmental impact of agricultural activities and product manufacturing on the lands bearing the virtually non-renewable muskeg the implications of Chapters 8 and 9 will not be fully realized. Production of wild rice unquestionably improves the productivity of muskeg for man but the peaty medium may suffer change and loss which in the long term may be significant on a national scale, as has happened with other agricultural applications where the peat is left in situ. The critical consequences arise when either natural or the managed hydrological conditions deteriorate or when the structural conditions of the peaty medium break down. The manufacturing of peat products, on the other hand, results in environmental loss which is immediate and devastating. Restoration in the sense of land rehabilitation is possible, but the cost is high and it is this cost that is seldom if ever considered in peatland management at times when benefits are being appraised.

The emphasis on evaluation of the special environmental effects of muskeg just now is in regard to off-road transportation, pipeline construction, improvement or conservation of continental water resources, and the care of wildlife resources. Each of these topics is considered in a separate chapter in Part III.

It is left to the reader to integrate and collate advice expressed throughout the book. It is hoped that this exercise will assist in formulating the management mechanisms appropriate for particular applications. Use of the book may also help save unharmed from ultimate loss the many kinds and natural conditions of muskeg which now abound.

It is now becoming appreciated that solutions to muskeg problems are unlikely to be reached without some measure of multi- or inter-disciplinary enquiry. In this context the contributing disciplines might be helped by an analysis of the muskeg terminology, some of which dates back to times when 'ecology' was a little-used expression and probably unfamiliar to engineers. Perhaps the list of terms and definitions provided at the end of the book will be most useful to engineers, who, these days, must account for their actions and claims wherever they refashion the muskeg landscapes.

The inclusion within the terminology of certain expressions which do not appear elsewhere in the text is deliberate. The chapter is meant to be more than a glossary. It may also contain definitions which differ somewhat from those expressed in the body of the text. There is perhaps no better illustration of this point than that involved in accounting for the expression 'bog.'

As with other terminology so with 'muskeg vocabulary' meanings change with time, often at rates and in directions which vary across the world. The terminology is rich and large and stems from a variety of unrelated endeavours. Interestingly, it contains no single term to mark the engineering science which has emerged as 'Muskeg Studies,' of which the content of this book is a significant expression.

Description of Muskeg

1
Postglacial History of Canadian Muskeg

J. TERASMAE

In 1732 the young Carl Linnaeus, who became one of the world's most eminent men in the natural sciences, made a memorable field excursion from Uppsala to Lapland in northern Sweden (Sjörs, 1950). He entered the following notes about muskeg in his diary, Iter Lapponicum: 'Strax efter begynte myrar, som mest stodo under vatten, dem måste vi gå en hel mil, tänk med vad möda; vart steg stod upp till knäs ... Hela denna lappens land var mest myra, *hinc vocavi Styx*. Aldrig kan prästen så beskriva helvete, som detta är ej värre. Aldrig har poeterna kunnat avmåla Styx så fult, där detta ej är fulare.' (Shortly afterwards began the muskegs, which mostly stood under water; these we had to cross for miles; think with what misery, every step up to our knees. The whole of this land of the Lapps was mostly muskeg, *hinc vocavi Styx*. Never can the priest so describe hell, because it is no worse. Never have poets been able to picture Styx so foul, since that is no fouler.)

Throughout the ages, muskeg (including bogs, fens, swamps, and marshes, although we are mainly concerned with the boreal peatland in this review) has been described with distaste and associated with something akin to mysticism — people were known to have disappeared without a trace in the swirling mists and dark gurgling waters of muskeg. It was assumed to be the dwelling place of bad spirits, and dangerous outlaws and witches (Glob, 1969). Rarely, indeed, is muskeg described as a thing of beauty or referred to with affection. Tom Alderman (1965), looking at muskeg from the viewpoint of industrial exploration in northern Canada, did not sound too enthusiastic when he described muskeg. 'It just lies there, smeared across Canada like leprosy. The machine hasn't been invented that can get through it consistently. It slurps up roads and railways, gobbles buildings and airfields, swallows the least trace of humanity. In summer, it's a rotting mushland of blackflies and mosquitoes, and the odor akin to backed-up septic tanks. In winter it's an eerie half-world, a frozen lifeless wasteland defying civilization.' However, Alderman also stated: 'Muskeg. Like it or not, more and more Canadians must learn to live on, with and alongside it. Beneath this matter could lie a wealth of crude oil, natural gas and minerals. Even more fascinating, locked within its slime is a fortune to be made from chemicals and synthetic materials, rich farmland and high quality woodland. Muskeg is both prize and plague, a crazy mixed-up quagmire with a split personality. It isn't all bad, say muskeg experts. And the sooner we realize it the sooner we'll reap its riches.'

No matter what has been said about muskeg, and whether we like it or hate it, the fact remains that muskeg is an important feature of the Canadian landscape. We may have more of it than any other country in the world, and yet we have spent a disproportionately small amount of money and effort to study it. We do not even know with reasonable accuracy how much muskeg we have in Canada, and we know less about the total volume of peat related to muskeg, the geographical extent of the different types of muskeg, and the biology and energetics of this important ecosystem. In spite of the great efforts made by a few people in muskeg research in Canada, the general lack of knowledge about the muskeg environment has cost us dearly — the oil business alone spent more than $150 million before 1965 in 'nuisance charges' arising from extra costs of operation in muskeg-covered regions. With increasing activity in northern Canada these 'nuisance charges' are also increasing, and frequently we have difficulty in substantiating the concern about environmental disturbance due to building of pipelines or other structures across muskeg because we simply do not understand this environment in the first place.

To gain an understanding of the muskeg environment we must first deal with the question of how, when, and from where all this muskeg arose. It is the purpose of this chapter to seek answers to this question.

DEGLACIATION AND MUSKEG DEVELOPMENT

Some 20,000 years ago, or even 15,000 years ago, there was no muskeg in Canada, except for relatively small areas that were not covered by continental ice sheets during the last glaciation (Figure 1). The initial development of muskeg in Canada was closely related to the process of deglaciation, which, in turn, was brought on by a significant change in climate.

Progress and Chronology of Deglaciation

The progress of deglaciation following the last glaciation (Wisconsin) in Canada has been summarized by Prest (1969), and some selected ice-marginal positions are shown in Figure 1. It is important to note that the retreat of the continental ice did not simply proceed from south to north, but from the periphery of the maximum extent of ice towards the last remnants of the ice sheets in the Cordillera region and in Labrador-Quebec. Deglaciation occurred during the time period from about 15,000 years B.P. (before present) to 7,000 years B.P. It did not proceed at a uniform rate, and there were several halts, readvances, and possible surges of the ice sheet margin, which in many areas had a lobate configuration. In addition, the development of a complex sequence of large ice-dammed lakes was a characteristic feature of deglaciation. These glacial lakes covered areas of substantial size (Prest et al., 1968) and formed barriers to plant migration during the late-glacial episode. When these lakes drained, the exposed large areas of lacustrine sediments provided suitable substrates for muskeg expansion.

In coastal areas, particularly along the eastern seaboard and around Hudson Bay the relative changes in sea level were an integral and important part of deglaciation. During the last glacial maximum, some 20,000 to 18,000 years ago, the sea level was about 130 metres lower than at present and substantial areas of the continental shelf were exposed, providing refuges for plants and animals to survive the glaciation.

The weight of the continental ice depressed the earth's crust and when deglaciation

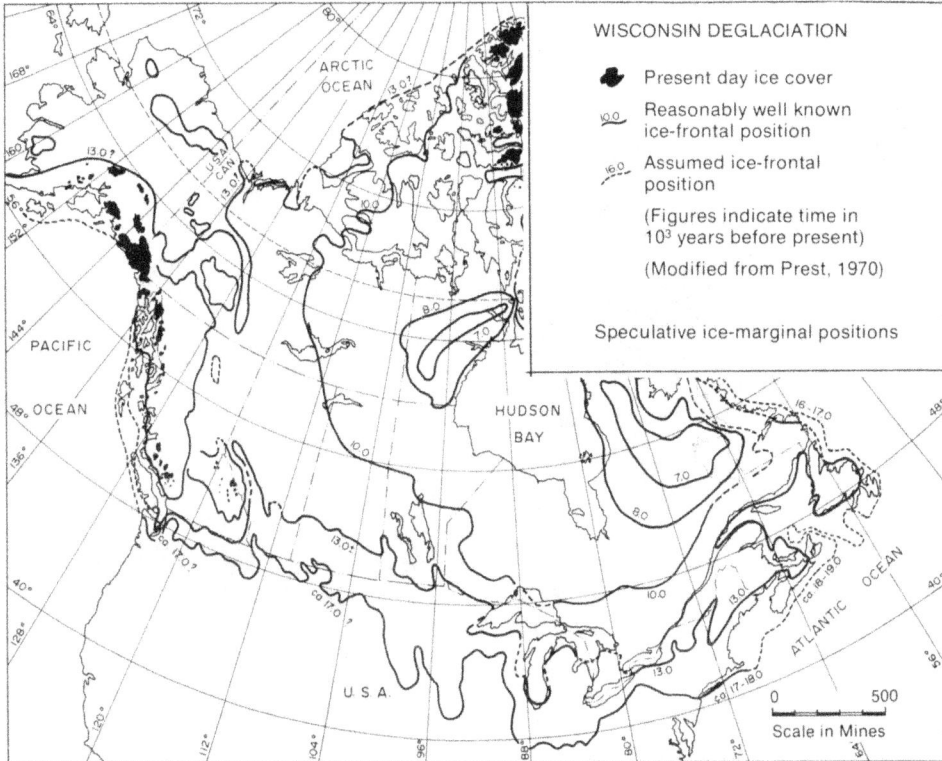

FIGURE 1   Outline of deglaciation. The contours refer to the marginal positions of ice, and the ages of these positions are given in 10^3 years B.P.

removed this load, crustal rebound occurred. However, the level of the sea was rising at a greater rate than the isostatic uplift of land (for example in the Hudson Bay area) and hence submergence of coastal areas was a characteristic feature of deglaciation. As a result of this process, large coastal lowland areas were covered by marine sediments and later as the uplift of land continued many of these areas, where poor drainage caused paludification, were covered by extensive muskeg.

Postglacial Migration of Muskeg

There is sufficient evidence (palynological and palaeobotanical) to indicate that muskeg vegetation survived the last glaciation south of the maximum limit of continental ice sheets in North America. In addition, muskeg survived the glaciation in unglaciated Alaska and Yukon. The presence of muskeg vegetation in the glacial refugia that existed along the east coast and possibly along the west coast is, however, only partially confirmed and further studies are required to settle this question.

From these glacial refugia muskeg vegetation recolonized the areas that were deglaciated during the melting of the continental ice sheets, and eventually occupied the region where muskeg is the predominant landscape component today. The migration of muskeg, as well as deglaciation, occurred in response to the changing climate. However, one must keep in mind

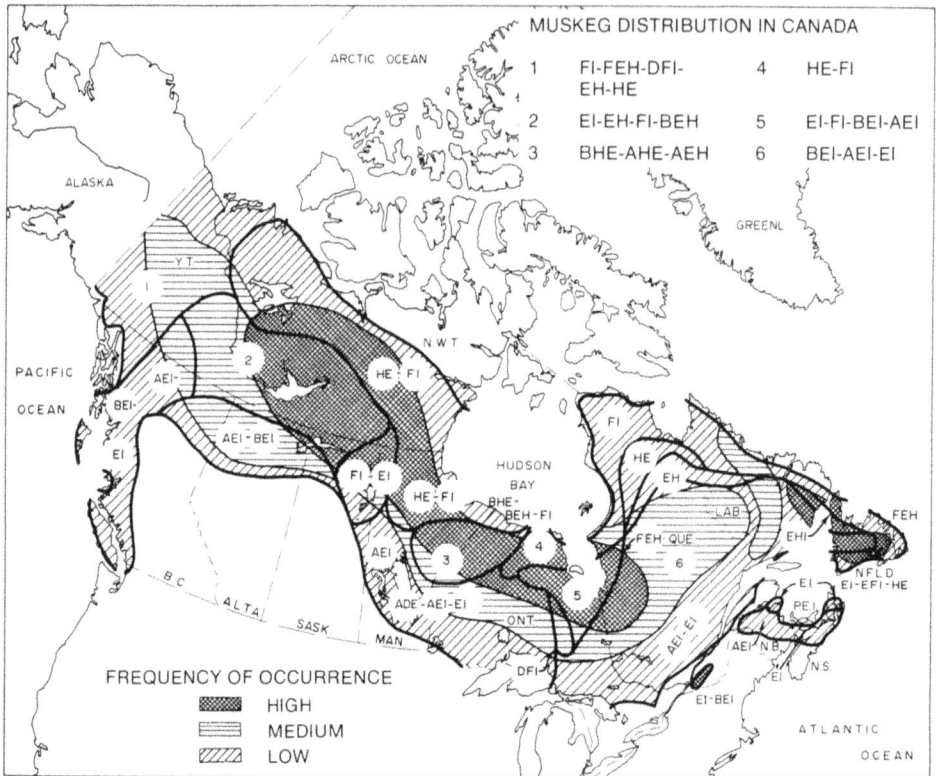

FIGURE 2   Muskeg distribution in Canada. Letter codes refer to the Radforth Classification (after Radforth).

the edaphic conditions which existed, because as the land emerged from the ice cover the soil-forming processes had not had time to modify or change the parent materials in terms of available nutrients, pH, and permeability. In some areas at least, the unfavourable soil conditions delayed development of muskeg, although apparently such delays were of short duration when climatic conditions were favourable for the growth of muskeg vegetation. In many areas, therefore, radiocarbon dating of the oldest postglacial peat has been used for estimating the minimum age of deglaciation.

Regional studies of palynostratigraphy, supported by radiocarbon dating, indicate that the migration of muskeg followed deglaciation rather closely and when the last remnants of continental ice disappeared some 7,000 years ago, the muskeg vegetation had already recolonized the region where it occurs now. However, a marked expansion of muskeg occurred somewhat later (beginning about 6,000 or 5,000 years ago) in response to a cooling trend in climate that was accompanied by an increase in available moisture.

Present Distribution of Muskeg

The present distribution of muskeg in Canada (Figure 2) coincides approximately with the boreal forest region, which is delineated by the range of distribution of black spruce (Figure 3), one of the characteristic species of Canadian muskeg. Studies of Canadian muskeg made

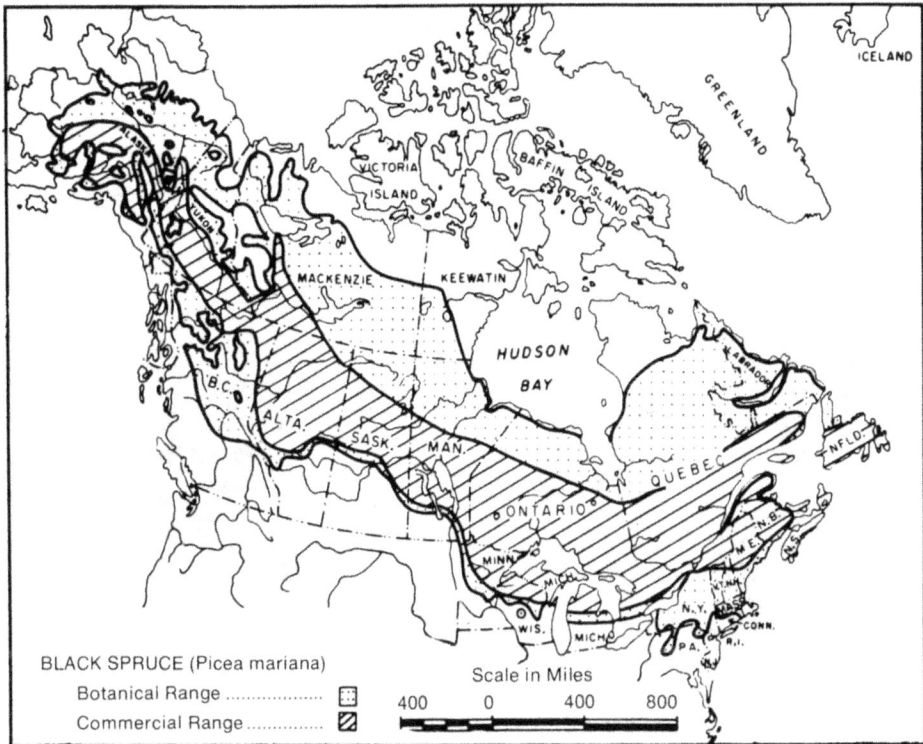

FIGURE 3   Distribution of black spruce *(Picea mariana)* in Canada.

by Radforth (1969a, 1969b) have indicated that muskeg varies from place to place, and that a limited number of consistently recognizable muskeg types can be established for the purpose of regional and local classification (Figure 2).

Palynological and palaeobotanical studies have demonstrated that muskeg was present for some length of time in areas outside of its present main distribution range in the late-glacial episode (Terasmae, 1968a) and also during some parts of the Holocene. It would seem that the distribution of muskeg is in a dynamic equilibrium with the existing environmental conditions and the limits of distribution can change in response to environmental changes.

Many small areas of muskeg (confined muskegs) exist outside the main distribution range and these can be considered to be relics of previously more extensive areas of muskeg. These confined muskegs outside the main limits of distribution are especially sensitive to disturbance and will be destroyed when burned or otherwise affected by man's activities, such as by drainage, cultivation, or commercial use of peat.

THE MUSKEG ECOSYSTEMS

There is no muskeg in Arizona, because the climate is too dry and warm. There is no muskeg in Peary Land in northern Greenland, or in the dry valleys of Antarctica, where the climate is cold but just as dry as in the southern deserts. There is relatively little muskeg in the subtropical region and the tropical rain forest region, where there is plenty of rainfall and the

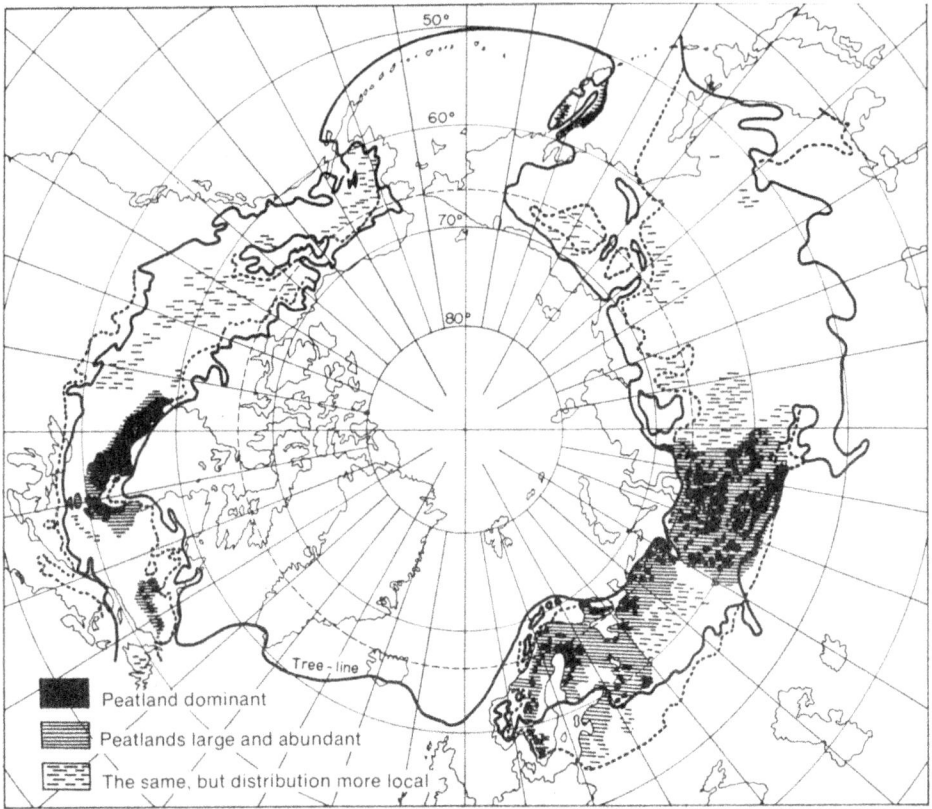

FIGURE 4   Muskeg distribution in the northern hemisphere (after Sjörs, 1961).

climate is warm and humid. Evidently dry climate is not favourable for muskeg development — no matter whether it is hot or cold. But neither is wetness of climate alone sufficient for muskeg growth. The presence of extensive muskeg is quite clearly associated with the boreal lifezone (Figure 4) of both hemispheres (Sjörs, 1961), and its subalpine equivalent in mountainous regions of the temperate and tropical zones. It is a reasonable assumption that the existence of a cool and moist climate and the occurrence of muskeg are related matters.

What about the effect of geology in this context? Does geology have any control over the occurrence of muskeg? Besides mountains (which are geological phenomena) in the temperate and tropical zones, which have an influence on climate through the altitudinal effect, it appears that muskeg has little, if any, respect for the different kinds of geological substrates (including rock types and soils). It is only in the transitional zone, mostly along the southern boundary of the muskeg region in the northern hemisphere, that geology exerts a definite influence on the occurrence of muskeg. In this case muskeg is frequently confined to depressions related to features of the geological landscape, and the chemical make-up of rocks and soils becomes an important factor in muskeg development.

In general, the boundaries of the muskeg region do not coincide with any of the boundaries of the major geological regions. One cannot say that muskeg is clearly related to igneous rocks, metamorphic rocks, or sedimentary rocks.

CLIMATE

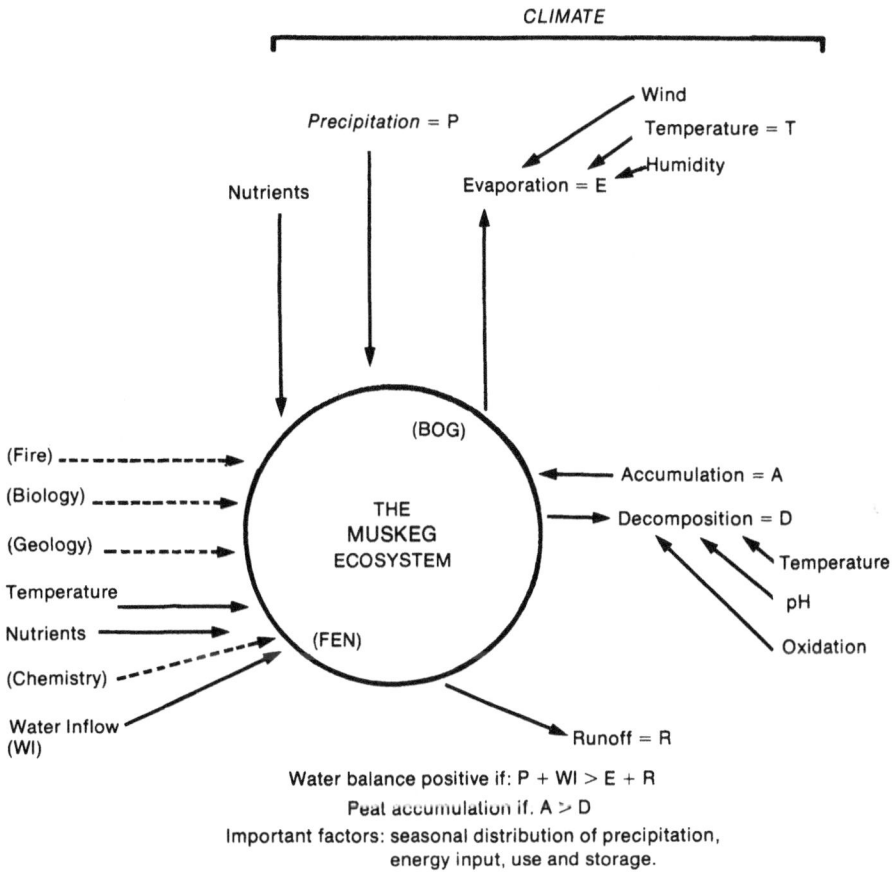

FIGURE 5 Diagram of the muskeg ecosystem.

Description and Definition of the Ecosystem

The factors that have, or can have, an influence on the muskeg ecosystem are shown in diagrammatic form in Figure 5. It is evident from an examination of this diagram that the development, survival, and 'health' of muskeg are governed by climatic factors. A positive water balance is essential and hence $P + WI$ must be greater than $E + R$. The presence of muskeg requires peat accumulation, and the rate of accumulation must be greater than that of decomposition. This condition $(A > D)$ prevents muskeg development in most parts of the tropics because of the very high decomposition rates. In desert areas (including the arctic desert, where the annual total precipitation is no more than a few inches of rain equivalent) the water balance is negative $(P + WI < E + R)$ and no muskeg develops. The seasonal distribution of rainfall (and snow) is important, because the humidity requirement for muskeg development cannot be met if large amounts of rainfall occur in very short periods of time only at specific times of the year. If all precipitation falls as snow, then the temperature factor will favour glacier formation and not muskeg development. Provided that the water balance and humidity and temperature requirements are met, and that $A > D$, then the muskeg ecosystem is set into motion, with a 'health potential' depending on the ratio of $P +$

OMBROGENIC PEATLANDS

A

1. Blanket Bog   2. Raised Bog   3. Forested Bog

SOLIGENIC PEATLANDS

Cross sections

Cross section

B

1. Basin muskeg   2. Hanging muskeg   3. Slope muskeg

TOPOGENIC PEATLANDS

lake

C

1. Infilled muskeg   2. Fen   3. Flat muskeg

MIXED PEATLANDS

D

1. Ombro-soligenic bog   2. Soli-topogenic fen

3. Spring fen

☐ Bog peat   ▦ Gyttja   ↘ Groundwater flow
■ Fen peat   ⇓ Precipitation
                                    --- Water table

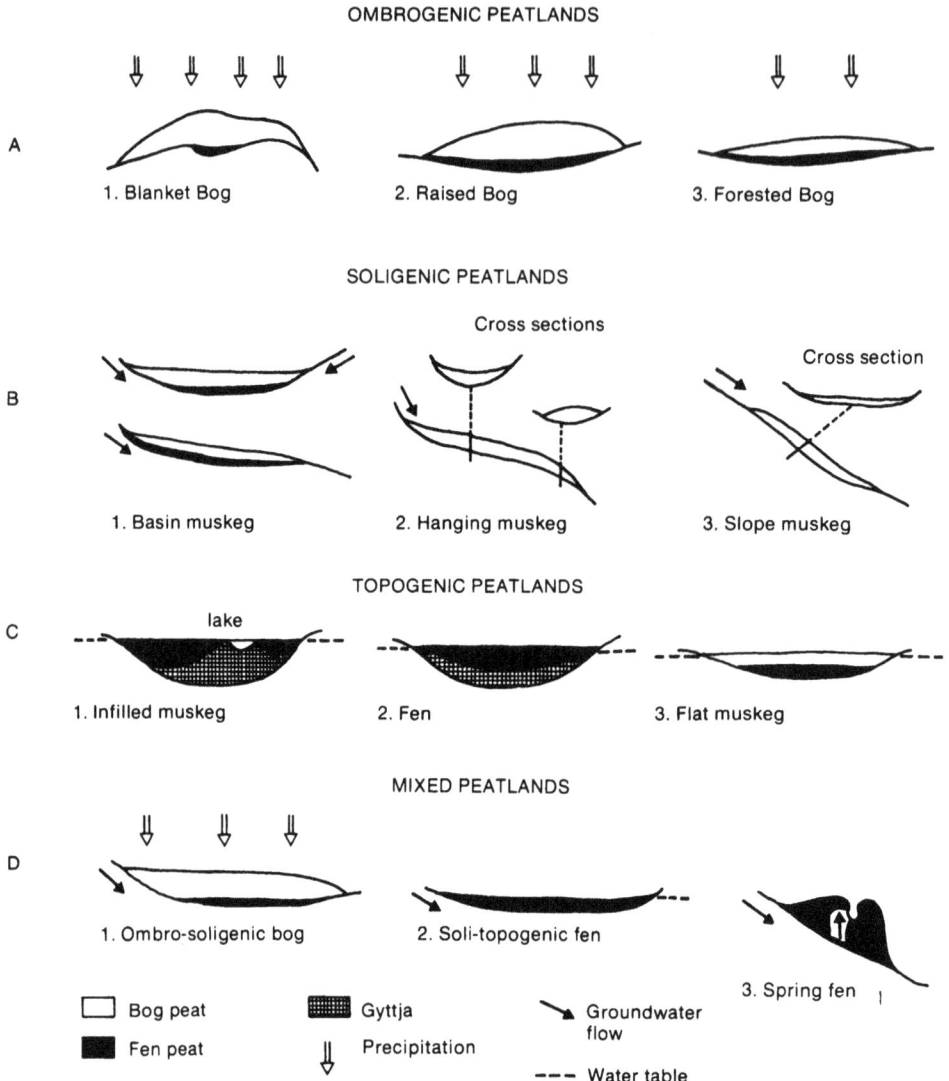

FIGURE 6   Diagram of different muskeg types (after Magnusson, Lundqvist and Granlund, 1957).

WI to $E + R$ and of A to D. For example, in a cool-moist maritime climate a muskeg cover can develop on almost any substrate and even on hill slopes (it is not restricted to depressions or lowland areas). As the $(P + WI)/(E + R)$ and A/D ratios decrease, the muskeg gradually becomes restricted and confined to lowland areas and local depressions. Under these conditions, the muskeg ecosystem resorts to its 'residual powers' of survival. These consist of the particular environmental 'resistance factors' that the muskeg ecosystem has, such as low pH (in the 3.5 to 5 range), the 'sponge effect' (water-holding capacity of peat), and microclimate. The muskeg 'outliers' (confined muskegs outside of the main muskeg region) survive because of these built-in resistance factors. If such a muskeg outlier is disturbed by natural or man-made causes it cannot recover and will be destroyed because of its sensitivity and marginal survival level relative to even small changes in the environment. Under optimum conditions 'healthy' muskeg is rather more resistant to disturbances, such as fire and several kinds of human activity.

The point should be made here that muskeg under stress in the southern and northern transitional zone of the main muskeg region is an extremely sensitive and delicate ecosystem that can be easily, and irrecoverably, destroyed by relatively little disturbance. This matter has been emphasized for the subarctic environment, but it applies equally well to the southern borderline of the muskeg region.

The different types of muskeg are illustrated in diagrammatic form in Figure 6 (after Magnusson, Lundqvist and Granlund, 1957). This simplified classification is based on the interaction between precipitation, landscape configuration, and surface groundwater flow patterns.

The differentiation within the muskeg ecosystem into bogs and fens depends largely on the nutrients that the plants which make up the muskeg vegetation have available to them. A true *bog* has its surface higher than the surrounding groundwater level and receives its nutrients only from rain water — hence the low pH values and poverty of nutrients. The various types of *fen* are influenced by groundwater and, hence, the pH values, as well as the levels of available nutrients, are generally higher.

Muskeg is always associated with peat — the product of accumulation of organic matter resulting from incomplete decomposition of plants that live and die in the muskeg ecosystem. One might wonder why this is so in the muskeg ecosystem and not in many other terrestrial ecosystems, such as grasslands, deciduous forests, rivers, and tropical rain forests. Peat accumulation in muskeg regions is accepted as a matter of fact and no further questions are asked. If, however, we wish to understand the muskeg ecosystem dynamics, it seems to me that this is one of the very fundamental aspects that must be explored.

Energy Flow in a Muskeg Ecosystem

All ecosystems differ in the matter of energy flow within them or through them. In most ecosystems there is no residual storage of energy. They are just like our affluent society — very active, with a rapid flow of energy within it and through it, but with no residual 'savings account.' In effect, we seem to be using the deficit budgeting approach in the human ecosystem. The affluence is associated with efficiency, which means that we are literally burning up our energy resources at an ever increasing rate. Any projected end result, based on this course of action, looks very dismal indeed. We could learn a lesson from even the most active natural ecosystems (such as the tropical rain forest ecosystem), where the energy budget is balanced so that they can maintain themselves over long periods of time.

The muskeg ecosystem is fundamentally very different. We might say that it is inefficient in the sense of our modern society because it accumulates its 'savings' in the form of peat. Figure 7 (data from Kormondy, 1969) illustrates the energy flow system in one particular type of muskeg environment. It is inefficient, like other ecosystems, in that it utilizes only a very small amount of the total available energy resource — the incoming solar radiation. However, it is very 'economy minded' since it uses only a small amount of energy. In effect, about 70 per cent of the energy used is stored in peat accumulation.

If the energy flow system of muskeg is changed in any substantial way (through warming of

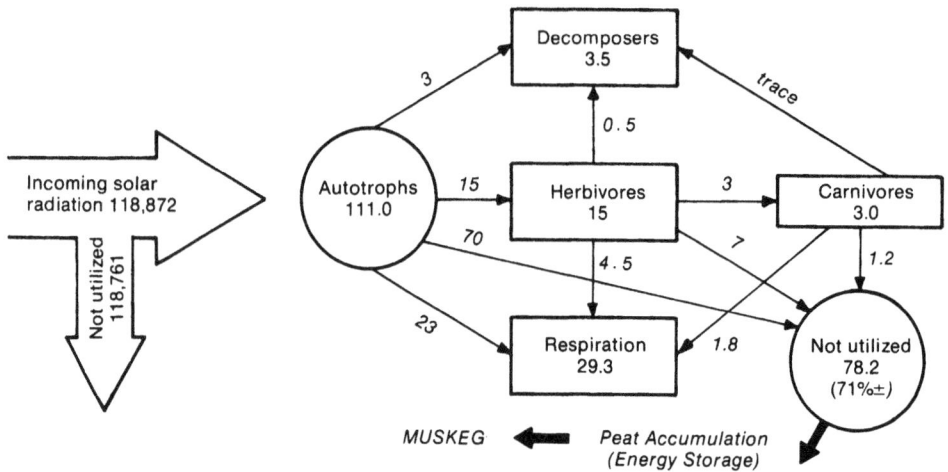

FIGURE 7   Diagram of energy flow in a muskeg ecosystem (Cedar Bog Lake, Minnesota). The figures are in gram-calories per square centimetre per year.

the climate or decrease in precipitation), the functioning of the ecosystem will also change. No matter how one looks at it, the muskeg ecosystem is responsive to, and influenced and *controlled* by the climate.

The acceptance of this conclusion helps one to understand the present muskeg ecosystem dynamics, which is essential in interpreting the history of muskeg development during postglacial time.

POSTGLACIAL CLIMATIC CHANGES AND MUSKEG

The termination of the last glaciation was caused by a significant climatic change that has been substantiated by many different kinds of evidence. A generalized temperature graph for late-Quaternary time is shown on Figure 8. It is also well known that several changes in climate have occurred during the Holocene (Sawyer, 1967). The most important of these changes in terms of muskeg development were those that led to cooler and moister conditions and resulted in an expansion of peatland at the expense of other types of vegetation. One such change occurred some 6,000 to 5,000 years ago and terminated the hypsithermal episode, which was characterized by a generally warmer and drier climate than the present. Another change to cooler and moister conditions occurred about 2,500 to 3,000 years ago, and several other changes of smaller magnitude have occurred since, some of which are documented by historical records.

Correlation between Climate and Muskeg

Figure 9 illustrates the response of vegetation and glaciers to climatic changes. It is evident from this illustration that vegetation responds relatively rapidly to a climatic change and hence it should not be surprising that muskeg migration was able to follow deglaciation closely, as is indicated by available field evidence.

Assuming that muskeg is a climate-controlled ecosystem, the present distribution limits of the muskeg region are best explained on the basis of the climatological reasoning and

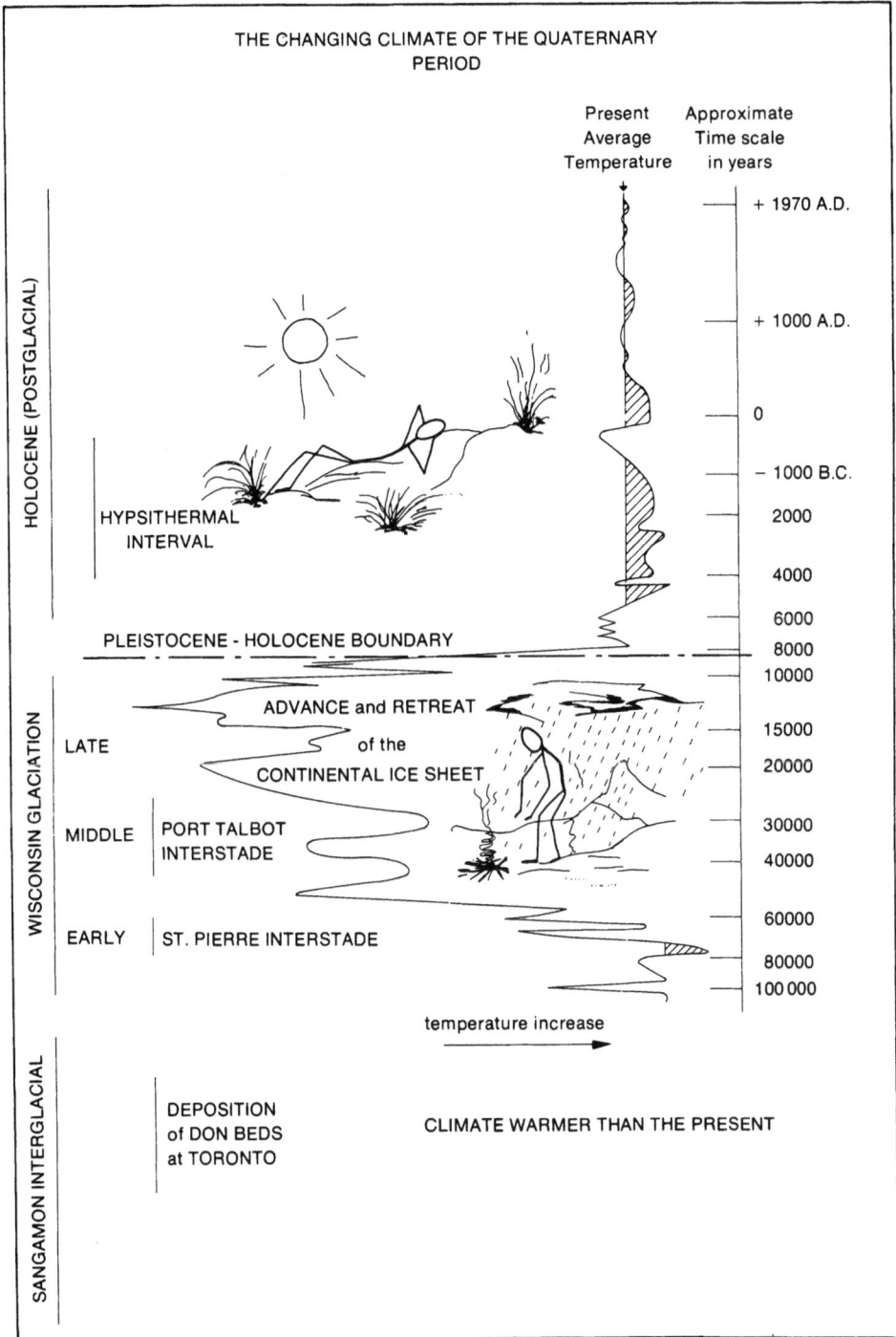

THE CHANGING CLIMATE OF THE QUATERNARY PERIOD

FIGURE 8 Estimated late-Quaternary temperature changes.

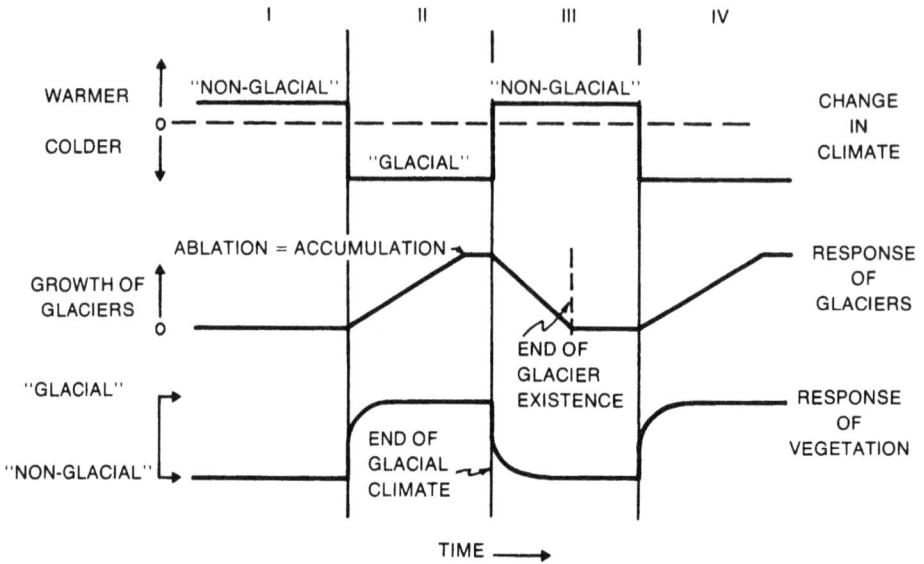

FIGURE 9   Response of vegetation and glaciers to climatic changes (after Bryson, 1967).

palaeoclimatological reconstructions presented by Bryson and Wendland (1967). These authors base their hypotheses on the assumption that the North American climate is controlled by the interaction of major air masses as shown in Figure 10. When the positions of interaction between these air masses are plotted for the summer and winter (Figure 11), and superimposed on a map of vegetation regions, it becomes strikingly evident that these climatic 'interaction zones' of air masses coincide very closely with the boundaries of major vegetation regions, such as the boreal forest. It is difficult to disregard this relationship as a mere concidence. A more reasonable conclusion is that the boundaries of vegetation zones are, in fact, controlled by climate.

   If we accept the above reasoning, then it follows that the distribution limits of muskeg can be defined in terms of climatological parameters.

   A climatological reconstruction (Figure 12) relating to the late-glacial episode (some 13,000 to 12,000 years ago) indicates the probable displacement and position of the boreal forest at that time (Bryson and Wendland, 1967). It is evident from this reconstruction that muskeg species could have survived the last glaciation south of the continental ice sheets.

Fossil Records in Peat

As the growth of muskeg leads to peat accumulation (Figure 13) the resulting deposit can be compared with a history book. Every year a small increment of peat is added, and this increment contains a record of plants that grew on the site, as well as pollen and spores from regional vegetation because these plant microfossils are transported by wind and will be deposited and preserved in peat. This process leads to the build-up of a fossil record that spans the time of peat deposition.

   Fossil records recovered from peat bogs have been used to reconstruct past vegetation and environmental conditions for a long time. Pollen analysis, or palynostratigraphy, was officially established as a research field by L. von Post in 1916, in Sweden, and this method

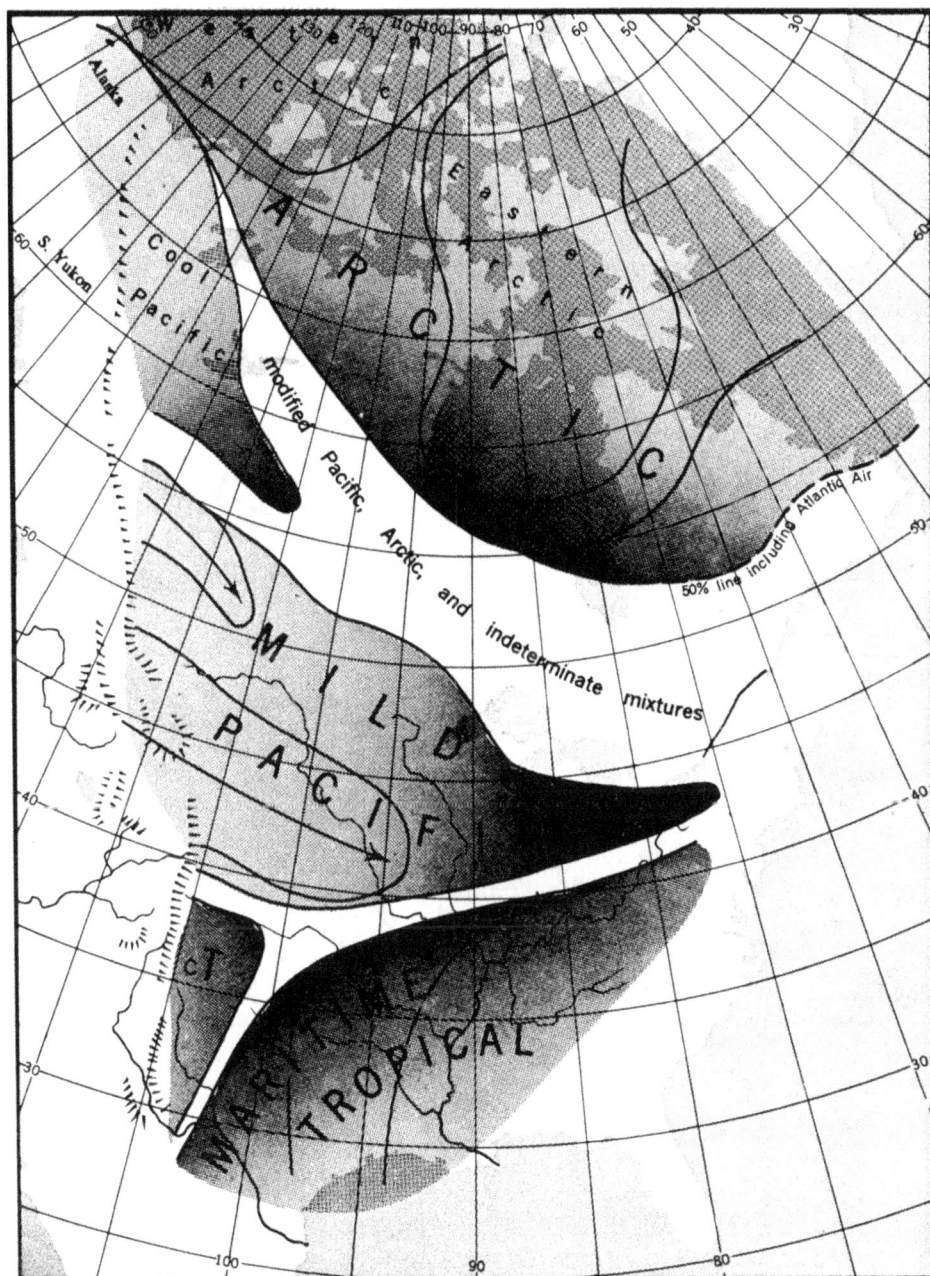

FIGURE 10 Predominant air masses that control the North American climate (after Bryson and Wendland, 1967).

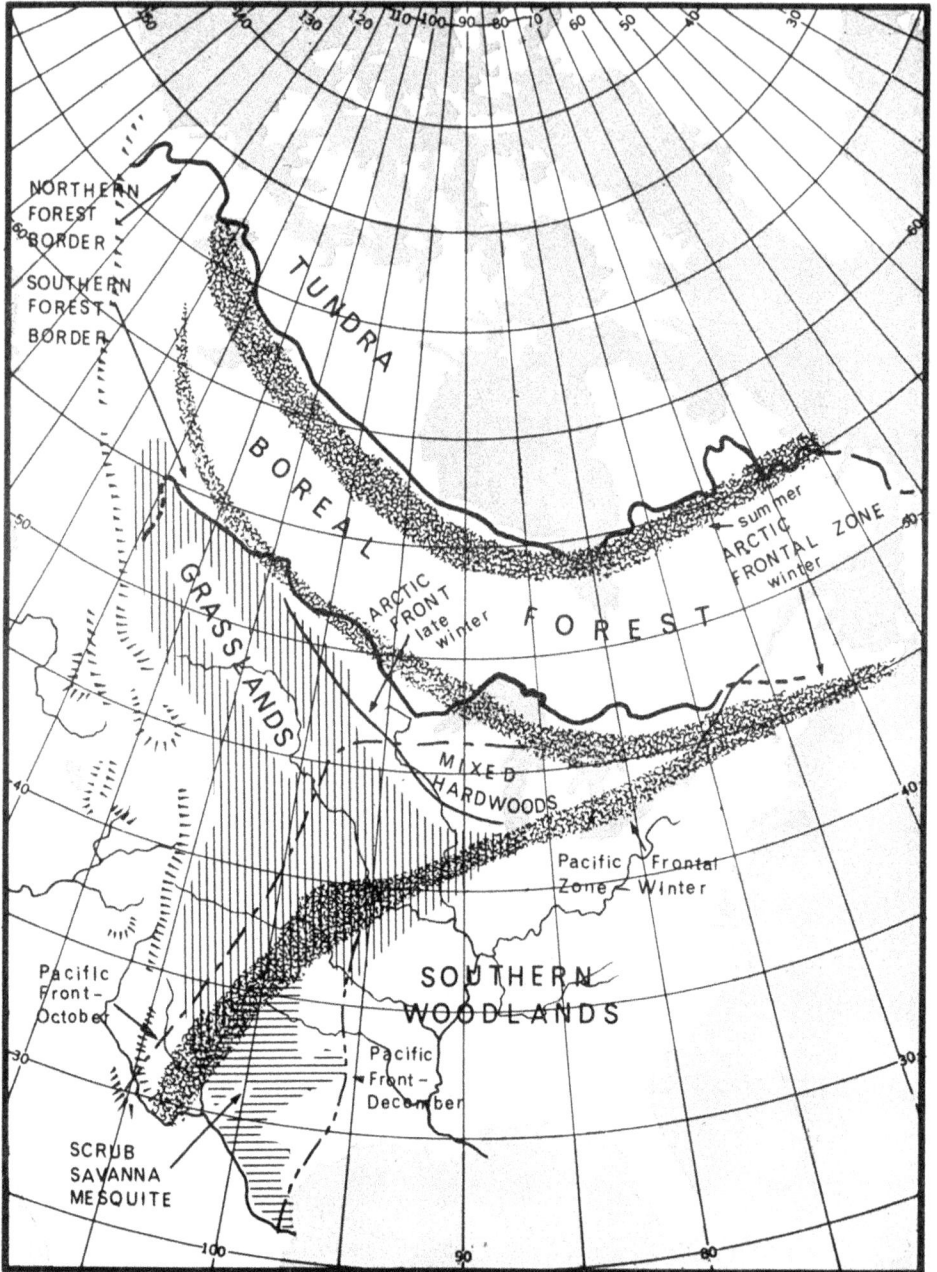

FIGURE 11   Relationships between boundaries of vegetation regions and air mass interaction zones (after Bryson and Wendland, 1967).

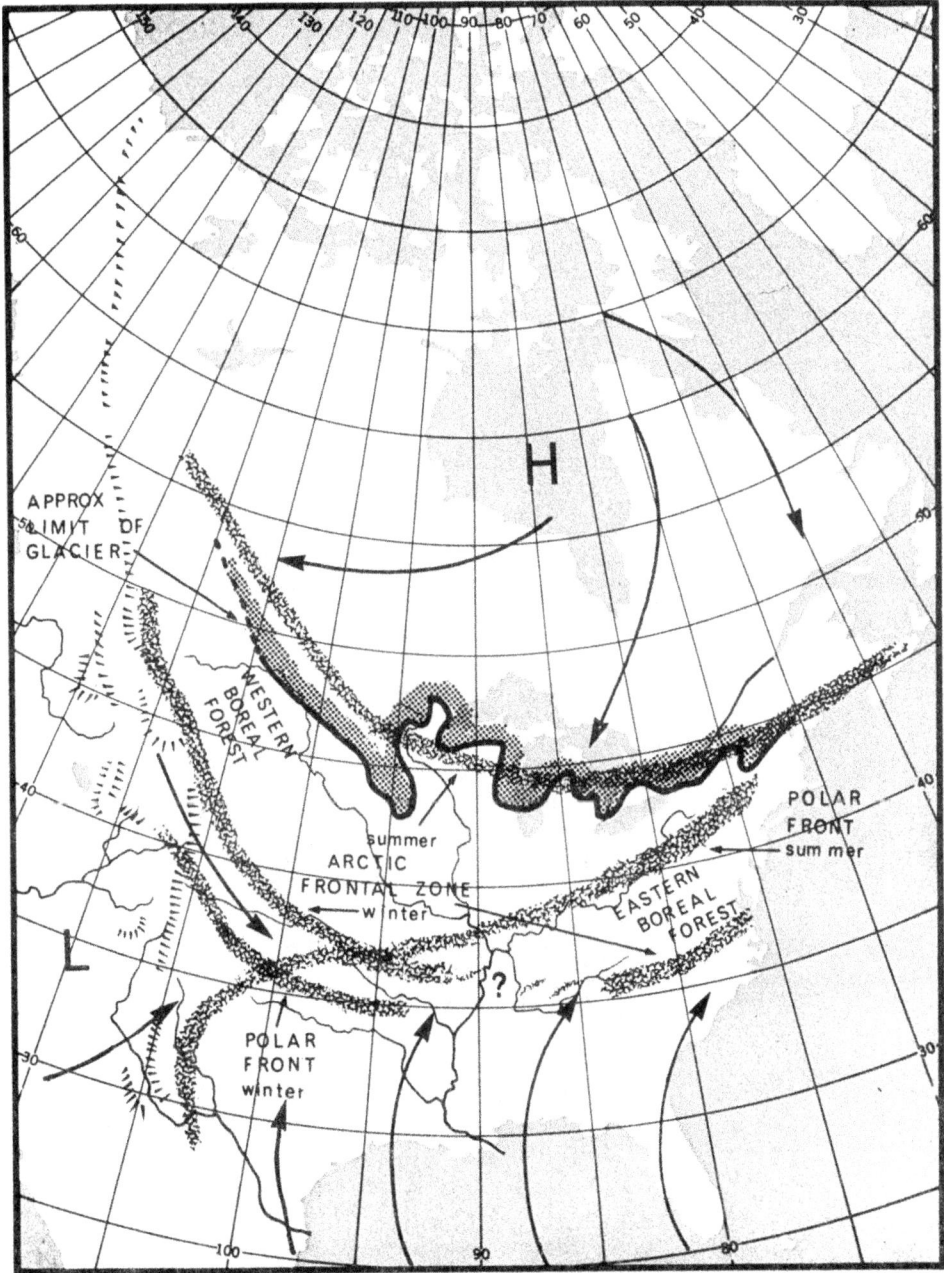

FIGURE 12   Vegetation regions and climatological boundaries in late-glacial time, about 13,000 to 12,000 years B.P. (after Bryson and Wendland, 1967).

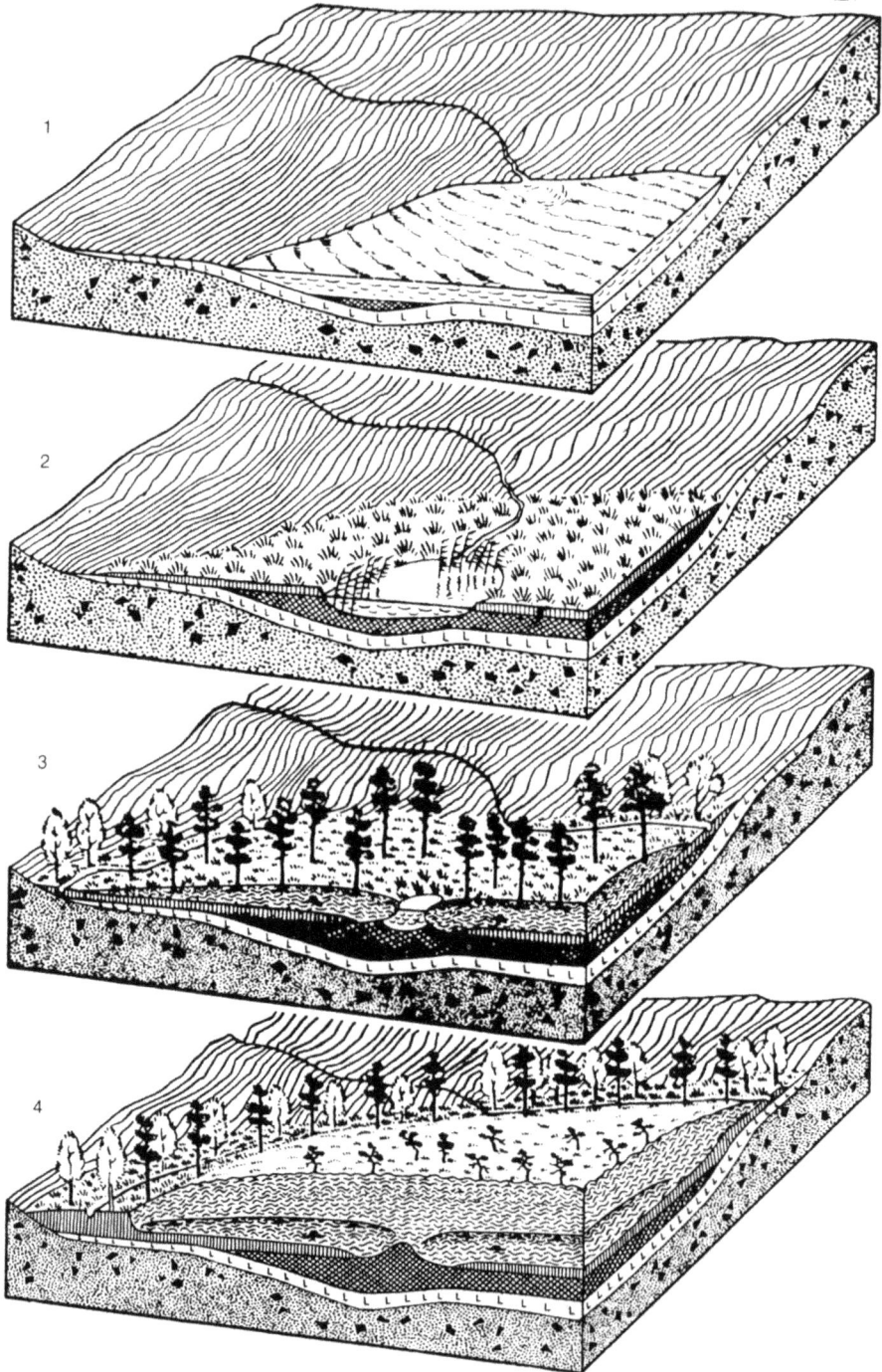

FIGURE 13   Diagram of muskeg development (after Kivinen, 1948).

has been used extensively for studies of Quaternary climatic changes. In addition to plant microfossils and and macrofossils (seeds, leaves, wood, etc.), the physical and chemical characteristics of peat have also provided much useful information about the past environmental conditions. In effect, a peat deposit is a record book containing information about past environments. The potential usefulness of this record book depends on our ability to 'read' and interpret the records accurately and reliably. In order that this can be done we must have a detailed knowledge of the present muskeg environment and its relationship to climate and other physical, chemical, and biological factors that have an influence on it. Without this knowledge we have little hope of interpreting correctly the fossil and other records that are stored in our peat deposits.

REGIONAL ASPECTS OF MUSKEG HISTORY

Studies of muskeg and peat deposits across Canada have indicated the presence of definite regional differences in both muskeg types and the characteristics of peat accumulation. Some aspects of these differences will be outlined in the following paragraphs.

Maritime Canada

The maritime climate (relatively cool and moist, without extremely cold winters and hot, dry summers) of eastern Canada is reflected in the types of muskeg and peat deposits which occur. Raised bogs, composed predominantly of *Sphagnum* peat ('peat moss'), are common in maritime Canada. This type of muskeg (ombrogenic peatland) requires high precipitation and abundant moisture throughout the year. Stratigraphic studies of bogs in this region have indicated that environmental conditions for muskeg development were favourable in maritime Canada throughout the Holocene and, because of this, peat has accumulated to a considerable thickness in many bogs. Therefore, peat bogs in maritime Canada provide better and more continuous fossil records than bogs in some other parts of Canada where peat deposition has been interrupted by frequent fires or high rates of decomposition, or by a lack of accumulation during some periods of the Holocene.

The Great Lakes and Hudson Bay Regions

In terms of muskeg distribution the Great Lakes region lies in the transition zone between the areas of extensive muskeg to the north and the isolated (confined, or relict type) muskegs to the south. Figures 14, 15, and 16 show the characteristic patterns of unconfined muskeg in the James Bay lowland. Figure 14 shows the relatively recent muskeg development at the mouth of Moose River near Moosonee, where the land is rising relatively rapidly (owing to crustal uplift following the last glaciation) and where muskeg under the *present* environmental conditions develops without delay on the emerged land. Figures 15 and 16 illustrate flow patterns (black spruce islands and string bogs) in the extensive muskeg area north of Smoky Falls and west of Wawa Lakes. The postglacial history of muskeg in northern Ontario and the Great Lakes region has been discussed by the writer in previous publications (Terasmae, 1967, 1968a, 1970). These studies have indicated that in northern Ontario large-scale muskeg expansion was generally delayed until some 6,000 to 5,000 years ago, after the hypsithermal interval, when the climate began to grow moister and colder. In southern Ontario extensive muskeg developed in late-glacial time on emerged lake sediment

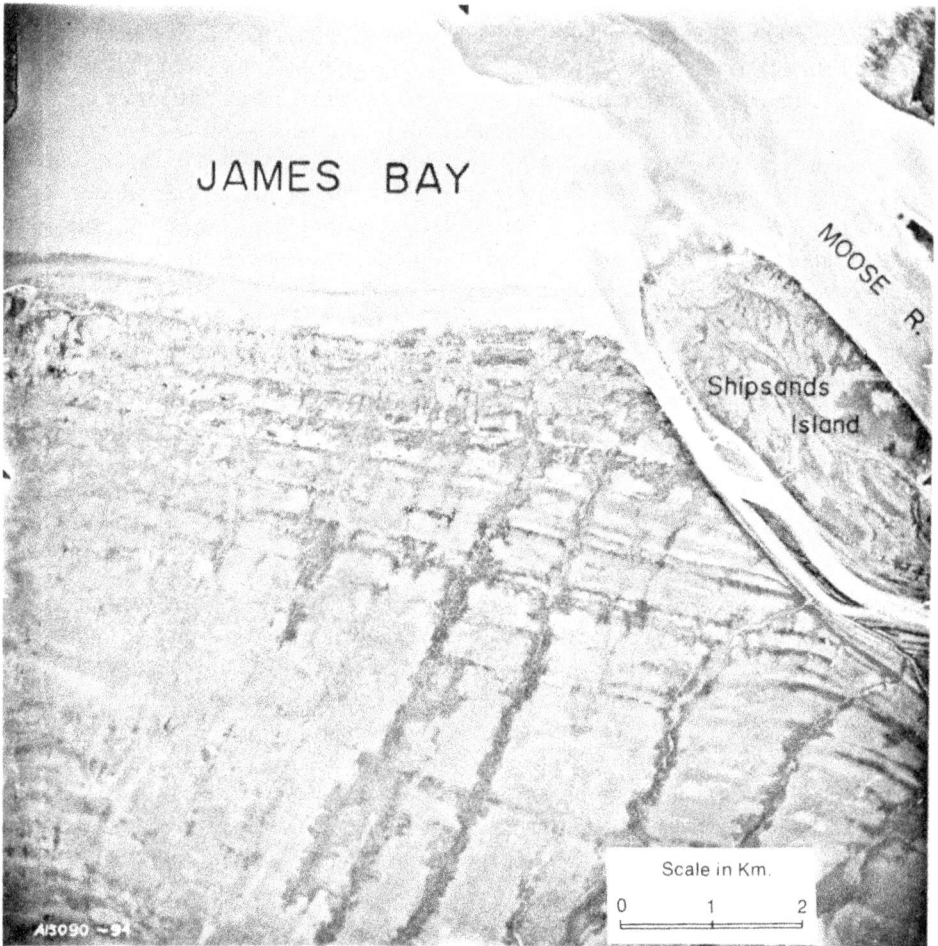

FIGURE 14 Muskeg patterns at the mouth of Moose River near Moosonee, Ontario.

plains, but as the climate warmed the rates of peat deposition and decomposition became about equal over many areas so that no further peat accumulation has occurred during the last 8,000 to 9,000 years, and in other areas the rate of decomposition of peat was greater than the deposition so that the whole peat deposit (formed in late-glacial time) has disappeared.

Mid-Western Canada

In late-glacial time muskeg appears to have been relatively abundant in southern Alberta and Saskatchewan, as indicated by palynological and palaeobotanical studies. However, as the climate became drier and warmer the muskeg vegetation was replaced by grassland. Further studies should be made to investigate this interesting phytogeographical and palaeoecological problem in greater detail, incorporating geochronometric age determinations of environmental changes wherever these can be established.

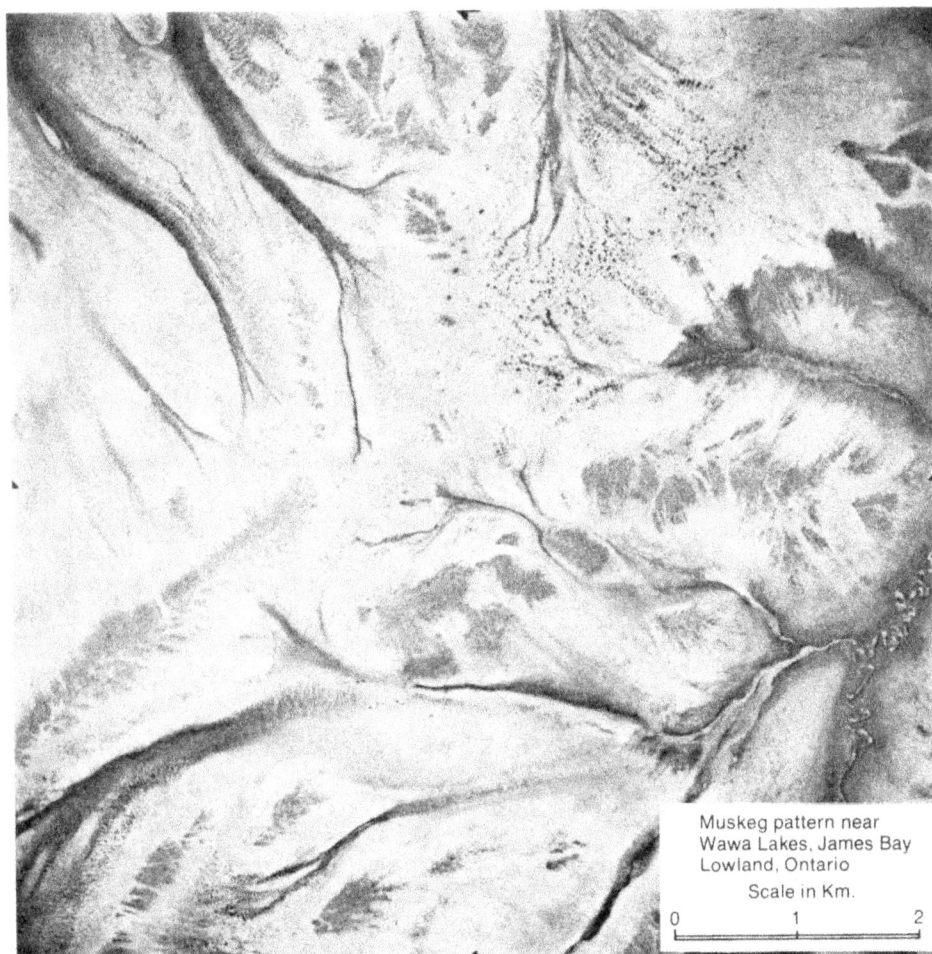

Muskeg pattern near
Wawa Lakes, James Bay
Lowland, Ontario
Scale in Km.

FIGURE 15   Muskeg pattern reflects groundwater flow near Wawa Lakes, James Bay lowland, Ontario.

The West Coast

Muskeg is relatively abundant in lowland and alpine areas along the west coast because of the high precipitation and maritime climate in general. Exceptions are interior valleys and eastern slopes that are in the rain shadow of mountains relative to the airflow from the Pacific Ocean, the source of moisture. In areas of exceptionally high precipitation a great thickness of peat accumulated during the Holocene (Heusser, 1960), and these deposits provide excellent fossil records for studies of palaeo-environmental changes.

The Yukon and Northwest Territories

The history of muskeg in the Yukon extends beyond the beginning of the Holocene in areas that were not glaciated (Rampton, 1971), but the rather extensive peat 'blanket' appears to have developed mostly during the Holocene. It is possible that the glacial climate was

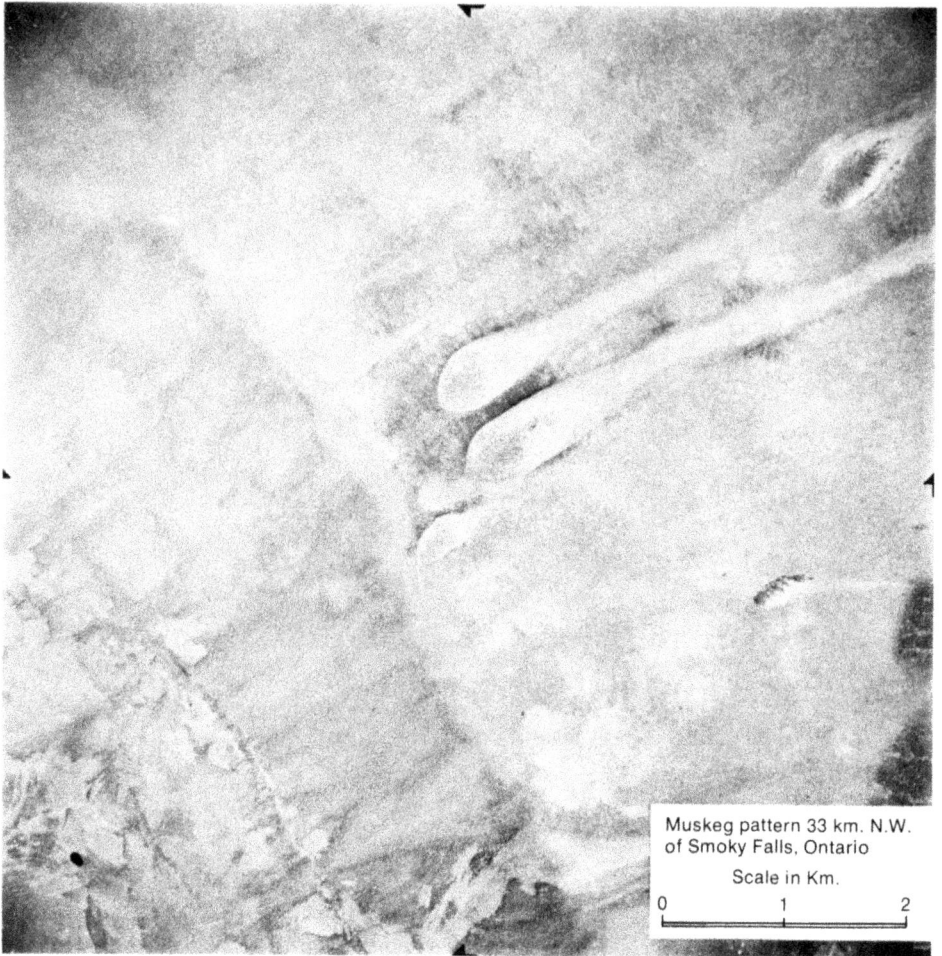

Muskeg pattern 33 km. N.W.
of Smoky Falls, Ontario

Scale in Km.

0   1   2

FIGURE 16   Muskeg pattern showing black spruce islands and groundwater flow direction north of Smoky Falls, Ontario.

generally too dry and cold for optimum development of muskeg (Terasmae, 1968b; Hopkins, 1972).

Muskeg is a common landscape feature in the Northwest Territories. The development of muskeg in arctic Canada (Ritchie and Hare, 1971) is something of a 'chicken and egg' type of problem. It is assumed that muskeg is abundant because permafrost forms an impervious layer in the soil that keeps up the water table and, therefore, surface water is present in abundance in spite of the low annual precipitation. On the other hand, peat provides the thermal insulation that helps to maintain the permafrost. How this interesting system originated after deglaciation is not fully understood in detail. At the present time, when the muskeg layer is destroyed the result is melting of permafrost and development of thermokarst features, which comprise a troublesome problem for arctic construction. Clearly a better understanding of the muskeg-permafrost system is desirable.

Palynological and palaeobotanical studies, supported by radiocarbon dating, have indicated that the muskeg layer in the arctic seldom represents accumulation throughout post-glacial time. This may indicate that permafrost, in fact, did develop first and provided the suitable moisture conditions for muskeg vegetation. However, it is also possible that peat accumulation was extremely slow and was interrupted by frost disturbance of soil, fires, or various types of wind erosion. Fossil records obtained from arctic peat deposits have been generally poor, discontinuous, and span a relatively short time.

SUMMARY AND CONCLUSIONS

Although we have a general outline of postglacial muskeg history in Canada, many details require confirmation or further investigation. Our knowledge of the present muskeg vegetation in a regional sense, and of the muskeg ecosystem in general, is totally inadequate for a satisfactory understanding of this important component of the Canadian landscape. Present and anticipated activities in northern Canada require that we acquire a vastly improved understanding of the muskeg region. Problems related to construction of northern pipelines point very clearly to this need, as does our concern about the northern environment.

Muskeg is important in terms of our water resources and, yet, studies related to this aspect of muskeg research are few.

We need to learn considerably more about the postglacial history of muskeg because changes in this environment have certainly occurred and will occur again in response to climatic changes in the future. Unless we know what has happened to the muskeg ecosystem in the past, and how it has responded to environmental changes, our discussions of changes that can be caused by natural events or human activities will be founded only on speculation.

REFERENCES

Alderman, T. 1965. It's a nuisance. Imperial Oil Rev. 49 (3): 6-10.

Bryson, R.A., and Wendland, W.M. 1967. Tentative climatic patterns for some late-glacial and post-glacial episodes in central North America. In Life, Land and Water, ed. W.J. Mayer-Oakes (Univ. Manitoba Press, Winnipeg), pp. 271-298.

Glob, P.V. 1969. The Bog People (Faber and Faber, London).

Heusser, C.J. 1960. Late-Pleistocene environments of North Pacific North America. Am. Geog. Soc., Spec. Publ. 35.

Hopkins, D.M. 1972. The paleogeography and climatic history of Beringia during late Cenozoic time. Inter-Nord, no. 12, pp. 121-150.

Kivinen, E. 1948. Suotiede (Werner Söderström, Helsinki).

Kormondy, E.J. 1969. Concepts of Ecology (Prentice-Hall, Toronto).

Magnusson, N.H., Lundqvist, G., and Granlund, E. 1957. Sveriges Geologi (Svenska Bokförlaget, Stockholm).

Prest, V.K. 1969. Retreat of Wisconsin and Recent Ice in North America. Geol. Survey Canada, Map 1257A.

Prest, V.K., Grant, D.R., and Ramptom, V.N. 1968. Glacial Map of Canada. Geol. Survey Canada, Map 1253A.

Radforth, N.W. 1969a. Classification of muskeg. In Muskeg Engineering Handbook, ed. I.C. MacFarlane (Univ. Toronto Press), pp. 31-52.

— 1969b. Airphoto interpretation of muskeg. In Muskeg Engineering Handbook, ed. I.C. MacFarlane (Univ. Toronto Press), pp. 53-77.

Rampton, V.N. 1971. Late Quaternary vegetational and climatic history of the Snag-Klutlan area, southwestern Yukon Territory, Canada. Geol. Soc. Am. Bull. 82: 959-978.

Ritchie, J.C., and Hare, F.K. 1971. Late Quaternary vegetation and climate near the Arctic tree line of northwestern North America. Quaternary Research, 1: 331-342.

Sawyer, J.S. (ed.) 1967. World Climate from 8,000 to 0 B.C. (Roy. Meteorol. Soc., London).

Sjörs, H. 1950. Myren och dess växtvärld (Muskeg and its plant life). Studentfören. Verdandi Småskrifter, no. 508, Stockholm.

— 1961. Surface patterns of boreal peatland. Endeavour, 20 (80): 217-224.

Terasmae, J. 1967. Postglacial chronology and forest history in the northern Lake Huron and Lake Superior regions. In Quaternary Paleoecology, ed. E.J. Cushing and H.E. Wright (Yale Univ. Press, New Haven, Conn.), pp. 45-58.

— 1968a. A discussion of deglaciation and the boreal forest history in the northern Great Lakes region. Proc. Entomol. Soc. Ontario, 99: 31-43.

— 1968b. Some problems of the Quaternary palynology in the western mainland region of the Canadian Arctic. Geol. Survey Canada, Paper 68-23.

— 1970. Postglacial muskeg development in northern Ontario. Natl. Res. Council, Assoc. Comm. Geotech. Research, Tech. Memo. 99, pp. 73-90.

2
Classification of Muskeg

WALTER STANEK

Muskeg is a North American term frequently employed for peatland (organic terrain) (37, 52, 154, 210). In nature it applies to every unit of peat-forming vegetation on organic soils and includes all peat originating from that vegetation. In the sense in which it is used here the term is synonymous with *mire* (British), *suo* (Finnish), *tourbière* (French), *das Moor* (German), *boloto* (Russian), *myr* (Swedish), etc. For convenience the following limitations are restated.

In muskeg peat must be more than 30 cm (12 in.) thick when drained or 45 cm (18 in.) when undrained; the ash content of the peat must be not more than 80 per cent (22, 37, 40, 41, 70, 71, 73, 101, 292, 297). Muskeg encompasses the living organisms (biocoenosis, i.e. phyto-, zoo-, and micro-biocoenoses) (137), the human influence, and the environment (ecotope or geocoenosis) (263, 264). The environment is made up of the 'climatope' (the atmosphere, macro- and micro-climate) and the 'edaphotope' (the soil, geology, and hydrology) or, more specifically, the topology of a locality (267), or the character of a site defined by elevation, exposure, slope, soil profile, and geology, all of which determine air and water drainage, temperature, and so on. Together these constitute a balanced system — called 'ecosystem' (137, 184, 270, 271), 'holocoenosis' (78, 240), 'biogeocoenosis' (263), 'climax formation' (44, 45, 46, 295), and so on.

The term muskeg has been applied also in a narrow, more restricted sense (125, 202, 227, 301).

Muskeg investigations of significance in Canada were begun at the turn of the century by botanists, explorers, geologists, and individuals interested in the utilization of peat. There are many classifications, according to the purpose for which they are destined (217). Those which are of interest here are ecologically meaningful and express relationships to environmental factors.

This report is an attempt to synthesize the information given in publications available to me which is pertinent to the classification of the Canadian muskeg environment.

MUSKEG REGIONS AND THEIR CHARACTERISTICS

In Canada there are numerous vast areas of muskeg. Radforth (213) published a small-scale

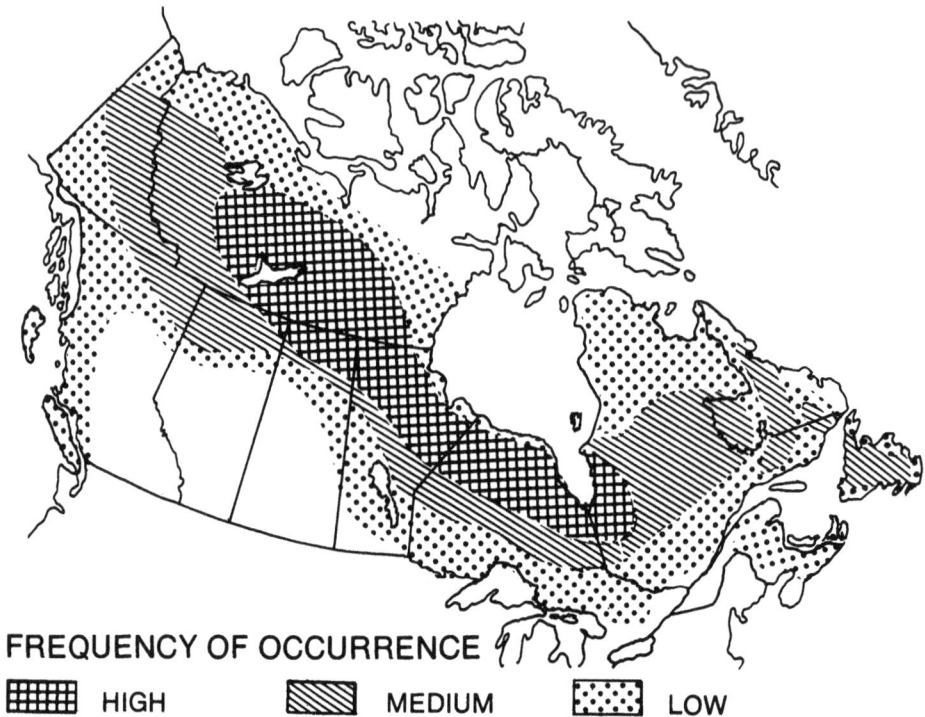

FREQUENCY OF OCCURRENCE

▦ HIGH   ▨ MEDIUM   ⬚ LOW

FIGURE 1   Frequency of muskeg occurrence in Canada according to Radforth (216).

map showing the frequency of muskeg occurrence in Canada (Figure 1). The highest concentration occurs in the boreal and subarctic regions and is affected by discontinuous permafrost (33-39, 214).

The boundaries of major forest regions (89, 236, 237) also appear to define regions with characteristic and prevalent muskeg landforms. They also correspond approximately to the positions of the isopleths of the mean annual number of degree days above 40°F (5.6°C), the potential evapotranspiration, and the mean annual length of the growing season (Figure 2) (8, 43, 93, 95). Therefore, the following Canadian muskeg regions, shown in Figure 3, are recognized: (1) the arctic, which consists of the high-arctic, the middle-arctic, and the low-arctic; (2) the subarctic, with parts of Ungava and Labrador in the eastern section, parts of northern Manitoba, Saskatchewan, and the Northwest Territories in the western section, and the Hudson Bay Lowland in between; (3) the boreal forest from the east coast to the foothills of the Rocky Mountains, with the clay belt in northern Ontario; (4) Newfoundland; (5) the Cordillera, consisting of the northern, southern, and coastal sections; (6) the Great Lakes-St Lawrence-Acadia region; and (7) the Prairies, with the Prairie Transition and Prairie Grassland.

The Arctic

This region is also called the tundra (4, 171, 235). It is treeless (10, 182, 183, 205) and is affected by continuous permafrost (16, 37, 38, 205, 259).

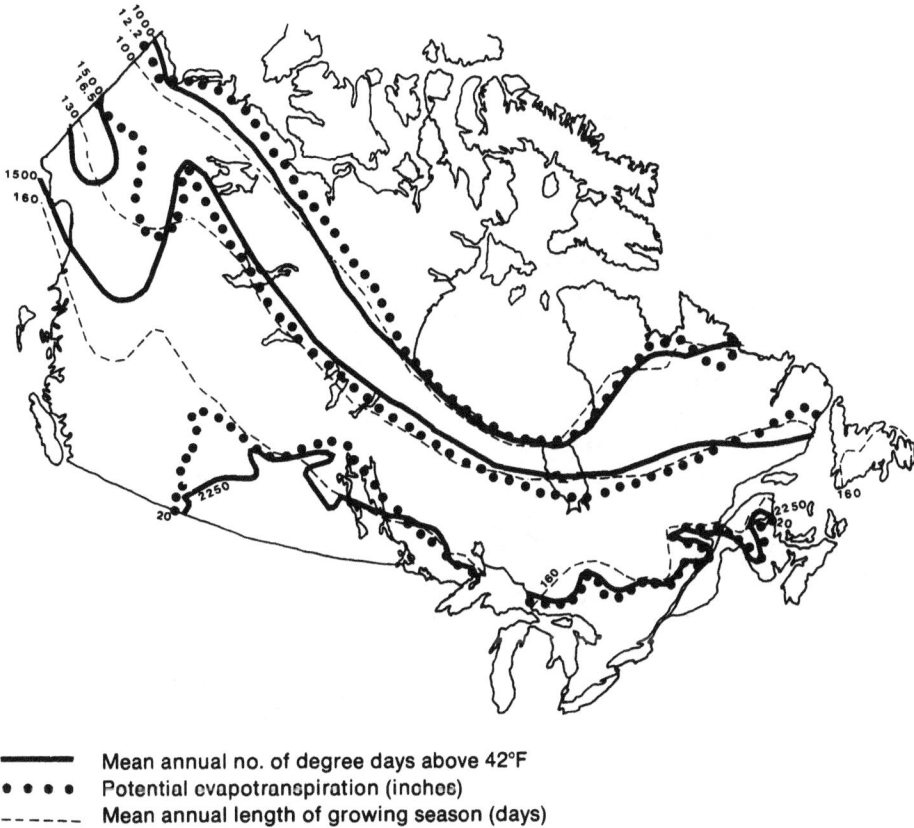

Mean annual no. of degree days above 42°F
• • • • Potential evapotranspiration (inches)
– – – – – Mean annual length of growing season (days)

FIGURE 2   Isopleths of three climatological variables pertaining to the muskeg region boundaries shown in Figure 3. Data compiled from (8), (43), (93), and (95).

Three vegetation belts (dashed lines in Figure 3) are recognized (27, 39, 200, 207). Nearest the treeline is the low-arctic. The vegetation here is essentially continuous and consists of sedges, mosses, lichens, ericaceous shrubs, and arctic willow. Muskegs develop entirely by the process of filling in of water-filled or wet depressions in the 'low tundra' (171, 259, 282, 299) and may consist largely of *Sphagnum*, although in the middle- and high-arctic they are formed by other plants (86).

The farther north one goes the less frequently one encounters muskeg (28, 280, 298). The number of plant species decreases drastically (27). In the middle-arctic, open, plantless areas are increasingly common and a stony, sedge-moss-lichen tundra predominates (39, 199, 279). In the high-arctic, rock desert and ice desert are most common (205, 279, 283), organic matter accumulation being minimal even in wet depressions (57, 280, 293).

Peats up to 2 metres (7 feet) deep (with permafrost) have been reported (28, 159). Wet marshes and wet meadows, with the characteristic tussocks or *têtes des femmes* of *Eriophorum* (151, 152, 199, 205, 259) and polygon formations (64, 65), have also been reported. However, similar communities and landforms also develop on mineral soil with an

FIGURE 3   Canadian muskeg regions: (1) arctic, (2) subarctic, (3) boreal, (4) Newfoundland, (5) Cordillera, (6) Great Lakes-St Lawrence-Acadia, and (7) prairies and aspen grove.

organic layer (H-layer) too shallow to be significant for certain applied purposes of reference and classification (63, 150, 151, 152, 159, 199, 205, 278, 279, 280, 282, 299).

The Subarctic

This region is almost entirely within the discontinuous permafrost zone (19, 36, 37, 156, 214, 229, 276). It consists mainly of what several authors refer to as the forest tundra (91, 96, 111, 112, 114, 168).

    The northeastern part has been frequently studied (94, 96, 110, 111, 112, 114, 115, 119, 161, 201, 202, 235), but the northwestern part needs much survey work although several valuable investigations have been carried out in specific localities (205, 221, 222, 223, 224, 230, 232, 233, 288). Raised and string muskegs occur in both parts (6, 37, 38, 111, 116, 117, 281, 299). They are confined to rock basins and are intermixed with barren rock and areas of mainly open stands of dwarfed trees. However, muskegs prevail in the Hudson Bay Lowlands (49, 230, 231, 232, 233). This vast area is nearly flat (estimated over-all slope 1:1,300) (9, 15, 49,

249, 284), and is recognized as a physiographic unit (18). The clay soils as well as permafrost add to the over-all poor drainage conditions (38, 249).

Virtually all of the 30 million hectares (110,000 square miles) of the Lowlands (249) are composed of muskeg except the banks of major rivers, rock outcrops, a narrow coastal strip near Churchill, and some beach ridges (37, 49, 119, 173, 213, 216, 232, 246). Over most of the area 'islands' with trees and extensive flat sedge areas, called water tracks (101, 245), and extensive string muskegs with pools, hummocks, hollows, and seepages of numerous shapes and sizes are common (21, 31, 37, 66, 90, 92, 97, 102, 103, 119, 133, 139, 176, 201, 231, 241, 246, 247, 249, 269, 274, 299).

North of the Nelson River in Manitoba, treeless, deep peats with ice-wedge polygons, palsas, and plateaus in all stages of development abound (39, 50, 100, 231, 232, 249, 275, 302, 303, 304). Adjacent to the boreal closed forest muskegs with tree cover become prominent (4, 17, 110, 201, 227, 228, 230, 232, 237, 249).

The Boreal Closed Forest

This region stretches in a broad belt from the Atlantic coast to the foothills of the Canadian Cordillera. In the eastern section, covering a large portion of the provinces of Ontario and Quebec, the terrain is rolling and in part mountainous with areas of raised muskegs filling the depressions and lowlands (62, 237).

In the northern clay belt, with clay soils and a nearly level topography (20, 167, 277), seemingly endless stands of black spruce alternate with extensive string muskegs, or muskegs with alder shrubs and a maze of tamarack, cedar, balsam fir, and spruce (24, 30, 104, 106, 109, 133, 170, 172, 237, 289, 290). Toward the west, particularly in former basins of Lake Agassiz and Lake Hyper-Churchill (9, 237), string bogs occur as well as peat landforms indicating the presence of permafrost (275, 302). The influence of limestone parent materials can be seen in the minerotrophic muskegs in central Saskatchewan (125, 127, 237).

It is principally in the boreal closed forest region that 'unmerchantable' stands, mainly of black spruce, grow on substantial peat layers (25, 61, 134, 148, 157, 163, 251, 253, 254, 256). However, poor drainage affects yield to a great extent (106, 164, 190, 254), and frequently beaver dams cause flooding (24, 246).

Newfoundland

The island is essentially a tilted plateau rising to the west, where elevations of over 750 metres (approximately 2,500 feet) can be found. Calcareous and some serpentine soils occur along the west coast, but acidic bedrock and glacial till prevail over most of the remainder of the island (54).

Approximately 17 per cent of the land area is covered by open muskeg. If forested types were also included, the coverage by muskeg would be much higher (100, 195, 196, 197, 225). The muskegs are mainly of the raised bog type, but not more than 2 metres (approximately 6 feet) in depth and generally small in size, or of the blanket bog type (195, 196, 197).

The forests are similar to the boreal closed forest of the immediate mainland (53, 54, 74), but only the central and western ecoregions are heavily forested. Treeless barrens and peatlands occur on higher elevations and on the remainder of the island (3, 197).

The Cordillera

This region covers all of British Columbia and the Yukon Territory and a strip of southeastern Alberta (9, 18, 19, 237). The mountainous relief causes variation in the local climate, and along the coast there is much humidity and drifting precipitation caused by the cold sea currents (13). The occurrence of arctic conditions and permafrost varies directly with latitude and altitude (35, 37, 192). The alpine tundra is similar to the arctic tundra (27). Its lower boundary coincides approximately with the treeline. In the southern interior it is at an average elevation of 1,500 to 1,800 metres (approximately 5,000 to 6,000 feet) above sea level. Toward the north it lies at lower levels and in the boreal closed forest region it is between 330 and 850 metres (approximately 1,100 and 2,800 feet) (252). Muskegs are mostly confined to rock basins (35, 39, 226). They increase in frequency toward the north, where either the texture of the mineral soils or the frozen ground maintains a high water table (237).

In British Columbia there are an estimated 190,000 hectares (approximately 4,600,000 acres) of organic soils, of which 40,000 hectares (approximately 1,000,000 acres) are arable (238). In the dry interior muskegs also occur and some are associated with alkaline soils (120, 146, 250).

Along the Pacific coast muskegs develop mainly on impervious soil (105, 226); some are not more than 400 years old (291). The high air humidity on Graham Island leads to the formation of muskeg bogs (237). The muskegs are comparable to those in the east, but the vegetation includes coastal and Columbian elements.

Great Lakes-St Lawrence-Acadia

The topography of this region is irregular but frequently plains occur (237). The climate supports an abundance of tolerant hardwoods (95, 160, 237). The frequency of muskeg occurrence in this region is low (216). Many of the muskegs are oligotrophic of the type described by Jurdant (128) and of unusually homogeneous structure (22, 85, 140). In the western part, near the sea, low relief and poor drainage have favoured the development of extensive muskegs with black spruce, as well as those with tamarack, eastern white cedar, alder, and willow. However, muskegs decrease in frequency toward the south and remain conspicuous only in 'obliterated' lakes and depressions (42).

The Prairies

This region consists of grassland and a belt, 80 to 240 km (approximately 50 to 150 miles) wide, called the aspen grove (125, 237). The topography varies from gently rolling in the west (32) to flat in the east in the former Lake Agassiz Basin (241). In the grassland the occurrence of muskeg is rare, low-lying wet parts being occupied by salt marshes (241). In the aspen grove, muskegs at various stages of development are associated with the occurrence of lakes. They are relatively small and minerotrophic, and increase in size and frequency toward the zone of discontinuous permafrost (154, 176, 222) and at higher elevations (123).

COVER CLASSES, AIRFORM PATTERNS, AND LANDFORMS

Among the primary factors which control the formation and nature of muskegs, climate, topography, and geology (climatic and edaphic factors) occupy a prominent place (292). These are responsible for the development of the characteristic vegetation and surface relief features in muskeg. Stoeckeler (259) in Alaska investigated the value of vegetation cover and landforms as indicators of soil texture, drainage, and permafrost conditions. He described several airform patterns for the use of airphoto interpreters (engineers and soils scientists with little knowledge of botany). Radforth (210, 211, 212) pioneered this approach in Canada. He aimed at the recognition from the air of structural differences and physiographic conditions which give rise to trafficability problems in muskeg. He established a system of coverage classes derived from the dominant vegetation physiognomy and published a map of Canada showing the distribution of common cover formulae (213, 216). Radforth also described airform patterns for inspection of the coverage class organization and estimated the large-scale patterned landscape distribution in Canada (216). If the user has considerable field experience, both of his systems help in identifying and interpreting terrain features. However, much closer linkage with environmental and phytosociological data is required (6, 149, 249) before they can be applied to biological and ecological problems. Recently Crampton (50) used Radforth's airform patterns to elicit information on the genesis of some organic terrain patterns. The guidelines provided by Adams and Zoltai (1) for an open water and wetland classification for land inventory purposes should also be mentioned. Their objective is to recognize and group ecologically significant wetlands into classes that are meaningful to a variety of resource managers and users, serving the needs of several disciplines. They introduce and define many terms. Their basic approach appears to follow the Swedish school of thought originated by DuRietz (68) and introduced to Canada mainly by Sjörs (246, 247, 248).

EDAPHIC FACTORS

Nutrients

A total chemical analysis of peat will give a good indication of its nutrient status (40, 107, 166, 185, 268, 287). The required minimum nutrient contents as a percentage of dry weight are: nitrogen, 1.00; calcium, 0.14 to 0.19; phosphorus, 0.04 to 0.09; potassium, 0.08. Most peats are low in total content of potassium and phosphorus (26, 40, 75, 87, 131, 132, 136, 165, 286), and show a calcium deficiency when strongly acidic (i.e. a pH less than 3.5) (40, 76, 297). The chemical properties of peat have been used to define their nutrient status (trophic value) in a poor-rich (i.e. oligotrophic-eutrophic) series. According to Ramenskij (218) (Figure 4) the richest muskegs are subeutrophic in the over-all nutrient series. Several authors use eutrophic as the best muskeg nutrient level, which of course applies only in a relative muskeg scale and must be interpreted as such.

Bruene (40) published a table of total nutrient contents of three basic muskeg types in Germany for use in conversion of peatland to agricultural land (Table 1). The three basic muskeg types (*Nieder, Uebergangs-*, and *Hochmoor*) are still in use (23).

In Newfoundland, Heikurainen (100) and Pollett (198) published tables of nutrient contents of peat soils of several major associations. Pollett (198) arranged them in three main groups (Table 2) corresponding in a sense to Bruene's (40) moor types.

Gauthier (79), Blouin and Grandtner (29), and Gauthier and Grandtner (81) supplied chemical analyses of peat for each of their vegetation communities. Data for representative

P = 55, 60, 65, 70, 75, 80 (vertical axis)

Vacc. vitis-idaea
Arctostaphylos
angustifolia

P

P
P
A
Pter.
Maj. c.
Linnaea

Oxalis

P P P A
Kalmia Trient. Athyr. Al fil. fem.

T
Le P dum P L Al
Carex pauc. Carex B disp. Al

Cassandra
Eri P oph. vag. L Calamag. canad. Car. Al diand.
Androm. Car. lasioc. Erioph. ang. Al
B S Impatiens

Car. limosa Car. rostr. Car. vesicaria
Calamagr. neglecta Car. rostr.

Scheuchzeria pal. Cicuta mac.
Potent. pal.
Menyanthes

Oligotr. | Suboligotr. | Submesotr. | Mesotr. | Permesotr. | Subeutr. | Eutr.

Cladonia sp.

Dicranum scoparium

Pleurozium schreberi

Drepanocladus sp.

Pleurozium schreberi
Polytrichum commune

Hylocomium splendens
Climacium dendroides

Mnium affine
Plagicchila asplenoides

Aulacomnium palustre

Sphagnum recurvum

Sphagnum fuscum

Sphagnum subsecundum

Sphagnum squarrosum

Sphagnum cuspidatum
Sph. dusenii

Sphagnum girgensohnii

Rhytidiadelphus triquetrus

Calliergon cordifolium

Calliergon giganteum

Calliergonella cuspidata

Sphagnum warnstorfii
Helodium blandowii

P Picea mariana B Betula papyrifera A Abies balsamea Al Alnus rugosa S Salix sp. L Larix laricina T Thuya occidentalis

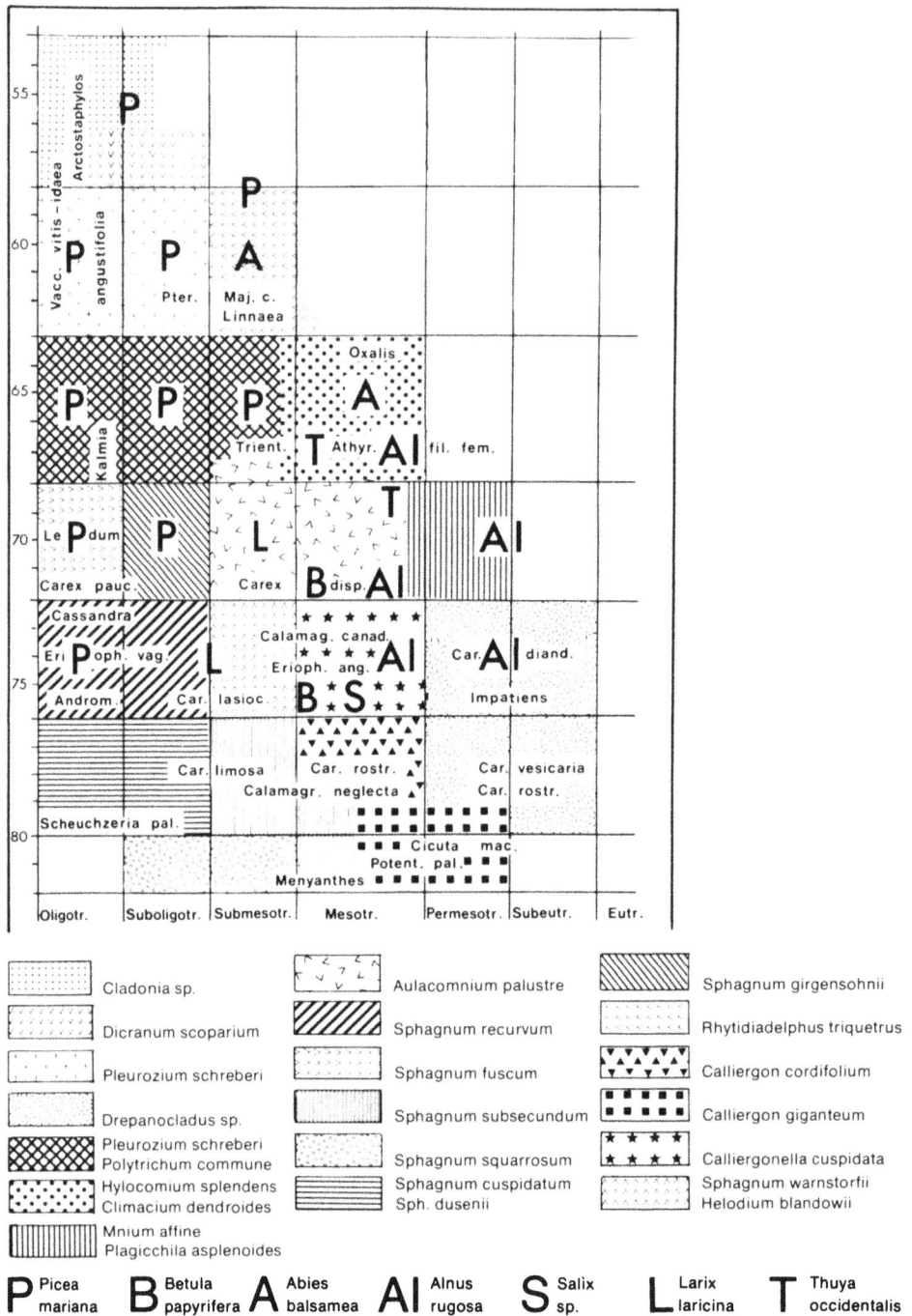

FIGURE 4 Ramenskij's scheme (232) of moisture and nutrient gradients (slightly modified).

TABLE 1
Nutrient content of the upper 20 inches (% dry weight) according to Bruene (40)

	'Hochmoor'	'Transition Moor'	'Niedermoor'
Nitrogen	1.20	2.00	2.50
Phosphorus	0.04	0.09	0.11
Potassium	0.05	0.10	0.10
Calcium	0.25	0.71	2.86

TABLE 2
Content of N, P, K, Ca, and Mg in % of dry weight and of Mn, Fe and Zn in ppm of dry weight according to Heikurainen (100) and Pollett (198)

	pH	N	P	K	Ca	Mg	Mn	Fe	Zn
Bog association	3.3- 4.0	0.67- 0.84	0.03- 0.04	0.04- 0.08	0.12- 0.31	0.12- 0.22	45.0- 67.0	740.0- 920.0	30.0- 36.0
Ksf-T Ksf-0 REn									
Weakly minerotrophic transition	3.8- 4.7	1.38	0.06	0.04	0.51	0.17	94.0	2,090.0	26.0
SSp									
Euminerotrophic lowmoor	4.8- 7.3	1.60- 2.09	0.05- 0.10	0.04- 0.10	1.61- 3.06	0.13- 0.32	122.0- 295.0	5,020.0- 10,070.0	44.0- 78.0
PCs BVu TPf									

TABLE 3
Chemical properties of peat from four associations according to Gauthier and Grandtner (81)

Associations	Layer	Depth	pH	Organic matter (%)	Total carbon (%)	Total nitrogen (%)	C/N ratio	Total saturation in ions (%)	CEC (me/ 100 g)	Exchangeable cations (ppm)			Available phosphorus (ppm)
										Ca^{++}	K^+	Mg^{++}	
Sphagnum cuspidatum	T₁	0-100	3.35	97.0	45.8	1.46	31	8	114	874	175	524	17
Eriophoretosum sphagnum variant	T₁	0-100	3.2	98.4	46.5	0.55	85	6	149	542	146	694	11
Ledetosum	T₁	10-120	3.8	98.1	51.7	0.75	69	20	143	2,748	1,224	1,309	21
Sphagno-Alnetum rugosae	T₁	15- 30	4.2	93.2	56.5	2.74	21	24	175	4,776	284	2,177	19
Thujetum occidentalis	T₁	30-105	5.7- 5.9	87.3	44.7	2.24	20	66	175	17,841	195	1,994	6

TABLE 4
Chemical properties of water samples according to Sjörs (247, 249)

	pH	Residue origin (mg/l)	Ca^{++}	Mg^{++}	K$^+$	HCO$_3^-$	Conductivity (mhos x 10^{-6})
			Ions, as ppm (mg/l)				
Poor bog pool	4.6	2	0.5	0.2	0.1	0.0	19
Rich fen flark	6.8	37	8.9	1.8	0.3	24.1	48
Very rich calcareous lake	7.9	—	32.1	6.8	0.6	131.0	207

associations are given in Table 3. Their associations could also be grouped into a trophic series.

Trophic types of muskegs of the Hudson Bay Lowlands were determined by Sjörs (247, 249) according to the chemical properties of water (Table 4). He considered the supply of mineral nutrients as extremely important among the factors affecting peatland vegetation (70, 249).

The pH of moist peat has been used to indicate the trophic level of peatlands (59, 60, 124, 126, 127), the significant gradients controlling distribution of species and communities.

Moisture

The vegetation communities have been further grouped according to the moisture status ('hygrotopy'; wet-dry, i.e. hydric-xeric series). This approach was first used in forest typology in Eurasia (142, 143, 144, 174, 194, 218, 261, 262). In North America the wettest 'hygrotopes' are called open hollows, ponds, moats, and pools (66, 249). At the relatively dry end are forested muskegs and the tops of elevated muskeg landforms (154, 176, 177, 178, 197, 249).

Depth to water level has been recognized as an important environmental factor correlating floristic and vegetational variations in peatlands (60, 124, 127). Dirschl and Coupland's (60) moisture regimes were determined according to the water depth in August in metres: above surface: (1) over 1.0; (2) 0.3 to 1.0; (3) 0.0 to 0.3; and below surface: (4) slightly below 0.0; (5) 0.0 to 0.3; (6) 0.3 to 1.0.

The 'trophotope' and 'hygrotope' are reflected in the vegetation community, and served as the basis for several types of classifications, particularly of the total ecosystem and biogeo-climate (265).

CLASSIFICATION OF VEGETATION COMMUNITIES OF MUSKEG

Environmental factors affect the development of muskeg (37, 40, 70, 71, 73, 215) and influence particularly the vegetation, which also produces its substratum, the peat (41). The nature of the vegetation is often used as a classifying criterion for muskeg.

I had hoped partly to overcome the difficulties of resolving inconsistencies arising as a result of choosing different bases of classification (141, 267, 300) by grouping the vegetation communities into those of open and forested muskegs below. These again I have arranged into oligotrophic, mesotrophic, and eutrophic muskeg nutrient classes, and, in each, the wettest of the communities has been mentioned first. The commonly quoted muskeg types

TABLE 5
Muskeg types arranged in a scheme of nutrient and moisture series

		Open (up to 25% tree cover)	Treed (more than 25% tree cover)
Nutrient regime	Very wet sites and transition to open waters	Wet sites and excessively wet sites, poorly aerated	Moist to wet sites, sufficiently aerated on seepage sites
Ombrotrophic muskeg (*Hochmoor*) oligotrophic	hollows, sinkholes, quaking, floating bog (*Verlandungs Weissmoor*)	raised (bog), blanket (bog), strings, palsas, mounds, plateaus, islands of muskeg complexes, low sedge muskeg *Weissmoor*)	black spruce muskeg, plateaus, islands, palsas of muskeg complexes (*Reisermoor*) lichen-rich spruce forest
Transition muskeg (*Uebergangsmoor*) suboligotrophic and submesotrophic	hollows, moats, ponds, marshes, floating mats, mud bottoms	transition bog, poor fen, etc., bowl bog, flat bog, sedge-, wet meadows	poor swamp, etc., black spruce-cedar swamp
Minerotrophic muskeg (*Niedermoor*) mesotrophic to subeutrophic	floating fen, spring fen, etc., fen pools and flarks (rimpis), mud bottoms (*Verlandungs Braunmoor*)	rich fen, minerotrophic fen, low moor, minerotrophic mire, etc. (*Braunmoor*)	alder-cedar-tamarack swamp, etc. (*Bruchmoor*)

are arranged according to the nutrient and moisture gradients given in Table 5. Frequent reference has been made to Cajander (41) because his types appear to fit Canadian conditions (66, 249).

Open Muskegs

Oligotrophic open muskegs have been called *Weissmoore* (41) and mud-bottoms (249) or oligotrophic bogs (55, 125). An example of the wettest community is one of *Sphagnum cuspidatum* (81) with floating, quaking mats of *Sphagna* (*S. cuspidatum*) and *Eriophorum virginicum* frequently supporting low shrubs of *Chamaedaphne calyculata, Carex limosa, C. oligosperma,* and *Scheuchzeria palustris,* the last frequently being dominant to the exclusion of all other species. Similar communities have been described in the arctic and in the southeastern Yukon (199, 206).

The low sedge muskeg (99, 100), with dominant *Eriophorum vaginatum, Trichophorum caespitosa, Carex pauciflora, Scheuchzeria palustris,* and ombrotrophic vegetation listed in Sjörs (249), is drier. The low sedge white-moor (41) and *Chamaedaphnetum calyculatae* or *Sphagno-Chamaedaphnetum calyculatae* (55, 79, 80, 83; 243, 258) are similar. Several subassociations have been recognized: *Sphagno-Chamaedaphnetum calyculatae eriophoretosum* (7, 83) and *Sphagno-Chamaedaphnetum calyculatae cassandretosum* (55), which are widespread and characteristic of the consolidation stage.

The last in the oligotrophic group is the less hydrophillous *Sphagnum fuscum* muskeg (41,

99, 100) or *Kalmio-Sphagnetum fusci* (197).

Toward the north a number of species which play a somewhat subordinate role farther south become dominant: they are *Betula glandulosa, Empetrum nigrum, Ledum groenlandicum, Vaccinium uliginosum, V. vitis-idaea* var. *minus,* and *Dicranum fuscescens* (55, 138). These indicate transitions to the treed muskegs (66, 229, 249), i.e., the *Sphagno-Picetum marianae* (83).

Mesotrophic open muskegs are transitional between the 'poor' and 'rich' muskegs. The wettest include floating mats of *Sphagna (S. squarrosum, S. subsecundum, S. warnstorfii,* etc.), and occasional *Calliergon giganteum, C. cordifolium* with *Menyanthes, Potentilla palustris* and *Carex rostrata, C. lasiocarpa,* and *C. chordorrhiza.* They are comparable with Cajander's (41) *Paludella* and *Hypnum trichoides* moors. Drury's (66) filling oxbows, Sjörs' (249) riparian-rich fen, the *Menyanthes trifoliae* of Dansereau and Segadas-Vianna (55), and the *Sphagnum* fen and herb-rich sedge bog in Newfoundland (100) appear to fit into this category.

The *Caricetum rostratae* with dominant sedges *Carex rostrata, C. lasiocarpa,* and *C. chordorrhiza* (55, 100) is less wet. This type is possibly comparable to the poor fen (249) in the Hudson Bay Lowlands, Ritchie's (229) sedge muskeg in Manitoba, Drury's (66) sedge meadow in Alaska, Allington's (6) sedge meadow in Labrador, and Cajander's (41) *Carex rostrata* muskegs in Finland.

The firmest, most lawnlike, and least waterlogged are the communities with dominant *Scirpus caespitosum* or *Eriophorum vaginatum.* Similarities occur between these and types described by Polunin (199) in the arctic, the *Scirpo-Sphagnetum papillosis* (197) and *Scirpetum elatum* (55). They occur in wet depressions or around pools and bodies of water (41, 249). These associations transgress on the forested mesophytic-eutrophic muskegs (66) (i.e. *Bruchmoore*) (41).

The *most eutrophic open muskegs,* also called minerotrophic (70, 249), are the open fens. These are the actual *Braunmoore* (41) dominated by 'brown mosses' such as *Paludella, Scorpidium, Drepanocladus, Calliergon, Aulacomnium* (99, 100, 119, 229), and frequently calciphilous vegetation (125, 155, 191, 197, 222, 245, 246, 247, 249). Sjörs (247) lists the following as typical of rich fen pools and flarks: *Carex limosa, C. livida, C. chordorrhiza, C. lasiocarpa, Menyanthes trifoliata, Equisetum fluviatile,* and others, as well as the mosses *Scorpidium scorpioides* and *Drepanocladus* species. Similar associations from southern Quebec are: *Eleocharetum smalii* (81), a spike rush community, where *Eleocharium* is an important species of the riparian flora associated with plants of the same biological type (i.e., *Equisetum fluviatile, Scirpus americanus*) (167). These may be compared with associations (84, 197), in which *Nuphar, Dulichium, Potentilla palustris, Sagittaria latifolia, Carex brunescens, C. stellata,* and *Calamagrostis canadensis* (167, 197) occur, and with the essentially floating, shallow-water vegetation of the rims of bog hollows and moats in Alaska (66), as well as with the *Scirpeto elatum* and *Calamagrosetum canadensis* (55) in Quebec and Ontario. The *Thalictro-Potentilletum fruticosae* and the drier, more heterogenous *Betulo-Vaccinetum uliginosi* (176, 197) and brown moss fen *(Campylium stellatum, Trichophorum)* (100) in Newfoundland should also be included here.

The saline meadows noted by Moss (176) in northwestern Alberta form a special class. There is a high salt concentration with halophytic vegetation prevailing (e.g., *Hordeum jubatum, Agropyron* species, *Distichlis, Elymus* species, *Muhlenbergia richardsonis, Puc-*

cinellia nuttalliana, Sueda species). Similar saline meadows and marshes were described in the Upper Liard River area, along the shores of Hudson Bay, and on Iles-de-la-Madeleine (84, 173, 222, 241).

Forested Muskegs

The most frequently encountered *oligotrophic forested muskegs* have parallel types in what Cajander (41) termed *Reisermoore* or what Ramenskij (218) called oligotrophic forests. They are drier than the open muskegs, and black spruce, often with vegetative propagation (253, 256), is their main tree.

In the boreal region, forested muskegs often contain merchantable trees (99, 113, 119, 134, 290) although, generally, only 'poorly productive' or 'unmerchantable' trees are classified as growing on muskegs (67). Jack pine sometimes takes the place of black spruce (29, 176), and in the western and eastern coastal area black and white spruce become ecologically interchangeable, and lodgepole pine and yellow cedar are common on the west coast (67, 146, 188, 226, 237, 257, 291).

Several authors have given excellent accounts of the individual associations. Although much of this work stems mainly from southern Quebec along the St Lawrence River, the *Sphagnum*-black spruce associations reappear in the boreal and forest tundra regions. *Sphagno-Picetum* (83) and *Picetum ericaceum* (55) are the main associations. *Sphagno-Picetum marianae chamaedaphnetosum* (98), *Sphagno-Picetum marianae ledetosum* (83, 229), and *Sphagno-Picetum marianae kalmietosum* (53, 79, 100) have been recognized as subassociations. The forest types on peat in Quebec (129, 148, 157), Manitoba (179, 180), Saskatchewan and Ontario (119, 125, 127, 158) appear to be very similar to this category.

The lichen-rich spruce forests are oligotrophic and relatively dry. They are common on droughty soils *(Pessière noir à Cladonie)* (84) and occur also on relatively 'dry' peats on tops of muskeg landforms such as palsas, plateaus, islands, and mounds (5, 114, 138, 218, 229, 246). Other types occur, however, in mixture with *Ledum, Pleurozium,* and *Cladonia rangiferina* (229, 246). The *Rubo-Empetrum nigri* (197) contains numerous lichens on blanket bogs on coastal areas of Newfoundland. In general, the stands consist of open-growing, small trees not higher than 10 metres (approximately 30 feet) (138).

Ahti (4) listed the following lichens as occurring on drier peat in arctic-to-temperate areas: *Cladonia mitis; C. rangiferina, C. alpestris* (mainly boreal), *C. pseudorangiferina* (which occurs in black spruce bog forests and is arctic to hemiboreal), *Cetraria nivalis* (146) (which occurs on palsas in the northern boreal zone and subarctic, and in the tundra frequently forms pure stands), and *Peltigera aphthosa* (which grow on hummocks in muskegs in the arctic-to-northern temperate climates).

Mesotrophic and eutrophic forested muskegs, called *Bruchmoore* (41) or swamps (wooded minerotrophic (103)), grow on 'nutrient-rich' sites (127, 138, 155, 218). The two nutrient regimes have been little differentiated in the literature. In the eastern boreal region they are made up of *Thuja occidentalis* and *Picea mariana* (29, 56, 118, 130, 138, 246, 249) and sometimes *Abies balsamea* and *Betula papyrifera.* Here belong muskegs with alder *(Alnus rugosae)* association, also called alder swamps (53, 83, 130, 218). They are frequently taken over by trees which overshade the alder (218). Most grow on thin, well-decomposed peat. However, the *Sphagno-Alnetum rugosae* (83, 130) with *Viburnum cassinoides,*

Nemopanthus mucronata, Ledum, Rhododendron, etc., grow on deep peat and are similar to the wet alder swamp in Newfoundland which grows on bog borders or where water moves into or out of muskegs (53, 100). The author has observed this type on peat 1 to 2 metres (approximately 3 to 6 feet) deep.

Muskegs with tamarack forests *(Laricetum laricinae)* (55), or tamarack swamps (138), occur in the boreal and northern deciduous forests as well as in the subarctic regions in aapa complexes, on elevated hummocks and ridges (5, 6, 100, 116, 117, 119, 138). Gauthier (130) described the association as *Sphagno-Laricetum laricinae*. The forest consisted of tamarack and black spruce, and the ground cover of ericaceous shrubs, *Carex*, and hardly any *Sphagna. Cladonia multiformis, Peltigera leucophlebia*, and *P. rufescens* grow on calcareous peats (4). After fire, a larch-birch community deleveoped into a birch-pine and *Hypnum* community (155).

Muskegs with cedar-tamarack-alder, usually called cedar swamps *(Thujo-Laricetum laricinae alnetosum* or *Thujo-Laricetum laricinae thujetosum)* (130), are frequently linked to the white cedar-balsam fir muskegs as well as to the black spruce-*Sphagnum* community with *Nemopanthus, Viburnum cassinoides*, and *Aronia mellanocarpa* (55, 83). The latter occurs mainly in the St Lawrence Lowlands but the dominant shrub species are also found in other boreal areas bordering oligotrophic muskegs. The herb-rich black spruce swamps *(Sphagnum-Cornus-Picea mariana)* (100) also belong here. The sedge tree swamp *(Carex-Picea mariana-Larix laricina)* (100) in Newfoundland is herb-rich, but without *Nemopanthus*.

THE VEGETATION COMMUNITIES OF MUSKEGS WITHIN THE ENVIRONMENTAL SYSTEM

Few muskegs are alike in every respect. Many are complexes resulting from a mosaic of plant communities. However, there would seem to be some parallelism between muskeg phytocoenoses occurring on comparable sites as can be concluded from North American and European literature (41, 54, 66, 111, 176, 194, 197, 201, 204, 218, 221, 222, 223, 246, 247, 249, 265).

Ramenskij's scheme (218) (Figure 4) demonstrates the range of phytosociological units in a system of edaphic factors. This ecological system is based on the coordinates of active richness and moisture content of the soil. The nutrient regimes are named and shown on the abscissa, and the moisture regimes are marked on the ordinate by numbers: 50-59 representing moist sites, sufficiently aerated in dry regions only on seepage sites; 60-69, wet sites, poorly aerated, in dry regions on sites with excessive wetness; 70-79, marshes and transitions to open waters. Also shown in the system are the cover distribution of dominant mosses (lichens) and the occurrence of some characteristic herbs, shrubs, and trees.

The original scheme has been only slightly modified here by updating plant names and either replacing European species names by their North American equivalents or deleting them. Most of the known major vegetation communities (plant associations) of the several existing schools can be fitted into the system. It serves, so to speak, as the conclusion of the theme of this chapter.

GLOSSARY

aapasuo   Finnish term signifying open muskegs of strings and flares (string muskeg, string bogs) (41).

blanket bog   A term used in the British Isles for blanket-like oligotrophic muskeg covering the whole landscape; usually the ombrogenous peat is fibrous and seldom more than 2 metres (approximately 6 feet) deep; common in cool, temperate, maritime regions at lower elevations having high rainfall, high humidity, and relatively low termperatures (48, 54, 195, 196, 273).

bog   All classes of wet, extremely nutrient-poor, acid, ombrotrophic muskeg with a vegetation in which *Sphagnum* species play a very important role usually supporting shrub and tree vegetation, and the remains of which make up a major part of the organic horizon (24, 53, 68, 111, 155, 176, 219, 247, 285).

collapse scar   Depression adjacent to the slumping edge of a palsa (peat plateau, or peat island), usually water-saturated, without permafrost, round in outline, treeless, minerotrophic; the melting palsa edge appears as a steep bank with leaning, mostly dead trees (302).

eutrophic   Refers to soils with high nutrient content and high biological activity (145, 194).

fen   English term for a meadow-like, minerotrophic muskeg which has a better nutritional status and a less acid condition than bog (246), signifying a low-lying, sedge-rich area with a substrate which has a high organic-matter content and an alkaline reaction (272); *Sphagnum* species are subordinate or absent, whereas *Campylium polygamum, C. stellatum, Scorpidium scorpioides*, and *Drepanocladus* species are abundant (53, 68, 246, 292). Synonyms: topogenous or soligenous muskegs (209), rheophilous peat-bogs (147), *Niedermoor* (40, 297), *Braunmoor* (41).

island   Same morphology as palsa; tree-covered; may gradually collapse from the edges (246, 247, 248, 302).

lagg   The zone where water collects at the edge of muskegs with the communities resembling those of a fen (47, 69, 70, 187, 297). Synonyms: sloughs (155), fen soaks, seepages (248).

landforms   In muskeg, surface relief features such as palsas, strings and flarks, islands, plateaus, collapse scars (1, 6, 50, 101, 275).

marsh   An open, flat or depressional area, with less than 25 per cent woody cover and subject to seasonal flooding and gravitational water tables; a rich site usually associated with alluvium, or with lakes and ponds. It is characterized by unconsolidated graminoid mats which are frequently interspersed with open water, or by a closed canopy of grasses, sedges, or reeds (shore vegetation). It is a stage of filling in of water bodies and progressive land consolidation. The substrate can vary from usually shallow, well-decomposed peat with high ash content to mineral soils. Waters are usually near neutral to alkaline in reaction (1, 2, 24, 53, 101, 176).

mesotrophic   Refers to soils with nutrient content intermediate between eutrophic and oligotrophic (194).

minerotrophic   'Nourished by mineral water'; adjective referring to muskeg areas which receive nutrients from mineral groundwater (66).

mire   1) An English word which is, in the general sense, a term embracing all kinds of peatlands and peatland vegetation (bog and fen) (82, 247). 2) A section of wet, swampy ground; bog; marsh; wet, slimy soil of some depth; deep mud, etc. (12). Synonym: muskeg, peatland (66, 125, 215).

moat   An area of open water several yards across, with pond lilies and other aquatics in it; in most bogs it is a broad zone of coarse sedges growing in deep water and closely resembling a lagg that is actively expanding by thawing of the surrounding frozen alluvium. It is associated with discontinuous permafrost in muskeg with palsas. A moat is wider than the marginal channel of most American bogs, and is formed by the break in vegetation resulting from the rise of the floating bog mat when flooded in the spring (66).

moor   A Germanic term for muskeg, mire, peatland; the English moor has a different meaning and is applied to high-lying country covered with mainly ericaceous dwarf shrubs (272).

muck   Peat soil containing more than 50 per cent ash, or in an advanced stage of decomposition; refers to soils which contain a high percentage of organic (vegetable) matter, in a well-decomposed condition. Peat signifies the rawer organic soils. It is evident that there is no line of demarcation. In general, the agricultural practices which are suited to muck are likewise adaptable to peat soil (292). Synonym: anmoor (145).

muskeg   1) A North American term frequently employed for peatland (organic terrain) (37, 52, 154, 210, 222); in nature applies to every unit of peat-forming vegetation on peat soils and includes all peat originating from that vegetation. The peat soil must be more than 30 cm (12 in.) thick when drained or 45 cm (18 in.) when undrained, the ash content not more than 80 per cent (60% and 70% have also been used as a limit) (22, 37, 40, 41, 66, 70, 71, 73, 96, 101, 258, 292, 297). The term muskeg has also been applied in a narrow, more restricted sense (49, 118, 125, 176, 202, 227, 301). 2) A term designating 'organic terrain,' the physical condition of which is governed by the structure of the peat and its related mineral sublayer, as considered in relation to topographic features and the surface vegetation with which the peat coexists (210, 215). 3) A short expression for the black spruce swamp forest with a *Sphagnum* moss cover and generally Labrador tea or leather leaf and similar shrubs 1/2 to 1 m (1.5 to 3 ft) high (111, 112, 155, 202, 228). 4) The word muskeg is of Indian (Algonquin) origin and is applied in ordinary speech to natural and undisturbed areas covered more or less with *Sphagnum* mosses, tussocky sedges, and an open growth of scrubby timber. The use of the word as an ecological term is limited generally to peat-forming vegetation in Alaska and northwestern Canada (52, 196). Synonyms: mire, peatland, organic terrain. Translations: Dutch, *veen;* French, *tourbière;* Finnish, *suo;* Swedish, *myr, myrmark, torvmark, sumpmark;* Russian, *boloto.*

muskeg complex   Huge muskeg areas originating from small primary muskegs, which grew or fused by horizontal expansion. They are composed of many muskeg types, sometimes interdependent, such as string muskeg (aapasuo) complex, palsa muskeg complex, raised muskeg (bog) complex (41).

oligotrophic   Refers to soils with low nutrient content and relatively low biological activity, generally formed on base-deficient parent rocks (219).

ombrogenous(ic)   'Produced by rain'; refers to peat deposited in ombrotrophic muskegs (297).

ombrotrophic   'Nourished by rain'; refers to muskeg areas dependent on nutrients from precipitation (68).

palsa   A peat mound with a permafrost core; usually ombrotrophic. Generally much less than 100 m (approximately 350 ft) across and from one to several metres high. In Fennoscandia palsas are generally treeless but in North America they commonly have a few stunted tamaracks or black spruce (wooded palsa). Palsas grow by the accumulation of water (derived from below or from the surroundings) on growing ice crystals (50, 77, 135, 162, 247, 299, 302).

palsa muskeg   A muskeg complex of the discontinuous permafrost region. Peat mounds (palsas), peat islands, peat plateaus, all with a permafrost core, are characteristic. They are protuberant beyond the adjacent muskeg without permafrost. Associated with palsas are collapse scars and moats. The protuberant landscape forms are usually ombrotrophic, the surroundings minerotrophic.

peat   An organic soil, exclusive of plant cover, consisting largely of organic residues formed in a water-saturated condition as a result of incomplete decomposition of the plant constituents due to the prevailing anaerobic conditions. The physical and chemical properties of the peat depend mainly upon the nature of the plants from which it has originated, the properties of the water in which the plants were growing, and the moisture relations during and following its formation and accumulation. It must have an organic matter content of not less than 20 per cent of the dry weight, and when it is used as fuel, more than 50 per cent of its dry weight must be combustible (22, 40, 41, 51, 71, 73, 101, 186, 292, 297). Several approaches to classification of peat in North America have been proposed (58, 71, 73, 122). See also under *muck*.

peat moss   The term is mainly a trade name for peat sold for horticultural uses and refers to peats composed generally of *Sphagnum*, such as cymbifolia peat and acutifolia peat, not humified or only slightly humified. It also includes peats which have a high percentage of other constituents such as *Carex*-moss peat, wood-moss peat, and moss-*Carex* peat (153). It is used chiefly as a mulch or seed bed, or for acidification (12).

peat mound   Usually ombrotrophic, in muskegs transitional between raised and blanket bogs (135, 247, 299); with permafrost core it is called a palsa.

peat plateau   Same morphology as palsa, but covers much larger areas, often over 1 km square (approximately 0.6 mile square) and seldom exceeds 1 m (3 ft) in height; surrounded by unfrozen, water-saturated peatland (37, 249, 266, 304); occurs in the continuous permafrost zone and shows frost polygons (247).

polygons   Patterned ground caused by permafrost with recognizable depressions along the polygonal circumference; in peat mainly with raised centres (mounds), but also with flat centres and raised periphery (65, 294).

pool   Depression characterized by permanent open water table in muskegs of any kind (249). Synonym: *Kolk* (41).

raised bog   See *raised muskeg*.

raised muskeg   A nutrient-poor (oligotrophic) muskeg which has grown above its site of origin, whose centre is higher than the margins and whose surface is convex. Growth is by *Sphagnum* proliferation and deposition of peat, water being supplied chiefly by rainfall (ombrotrophic). There are usually one or several very acid hummocks on its surface and wet depressions or ponds draining through soaks toward the lagg (66, 70, 72, 88, 103, 138, 220, 246, 296). Raised muskeg peats are nutrient-poor and their calcium oxide content does not exceed 0.5 per cent (40). Raised muskegs occur mainly in the southern part of the boreal closed forest region and south of it in the northern deciduous forests in the Atlantic and Pacific coastal areas (62, 90, 167, 237, 255, 291). Synonyms: raised bog, domed bog, high bog.

soaks   Broad troughs leading waters away from bogs or fens. Synonym: raised bog drains. Translations: German, *Ruellen* (296); Swedish, *drog* (70, 187).

soligenous(ic)   'Produced by soil'; refers to peat deposited in muskegs nourished by mineral water from higher surroundings (209).

string muskeg   A gently sloping muskeg complex consisting of long, low strings (ridges, ribs, lanières) up to 2 m (approximately 6 ft) high alternating with flarks (rimpis, flashets, mares, hollows), usually wet or water covered, both oriented across the slope of the peatland and perpendicular to the water movement. They may be parallel or occur in a webbed, sinuous, or net-like pattern. In flat watersheds, with no slope, the flarks become pools, ponds, and hollows of irregular size and shape. The ridges are frequently ombrotrophic and bear mainly *Carex* and *Sphagnum*, but also spruce, larch, dwarf birch, willow, and ericaceous shrubs. The intervening flarks are frequently minerotrophic. They contain open, shallow water or *Carex, Drepanocladus-Calliergon* communities. String muskegs usually appear as a coarse-textured expanse on an aerial photograph (vermiculoid airform pattern). The typical form occurs near the tree line (6, 21, 27, 41, 66, 90, 96, 101, 102, 103, 133, 196, 215, 230, 231, 246, 274, 282). Synonyms: patterned-muskeg, -bog, -fen; aapa-bog, -fen, -mire, -myr, -mosse, -moor. Translations: Finnish, *aapasuo;* German *Strangmoor.*

swamp   A type of wet, forested, minerotrophic muskeg, rich in herbs, with dominant tree species at least reaching pulpwood sizes (tamarack, white spruce, balsam fir, cedar). Black spruce, if present, is faster growing than on ordinary muskegs. A more luxuriant moss cover is dominated by brown mosses (101, 118, 125, 201); a minerotrophic site influenced by fluctuating or flowing waters, although the substrate usually is continually waterlogged, and is related to fens and marshes; includes wetlands having tree and/or shrub layers occupying more than 25 per cent of the area (215). Although most swamps are level, this is not a necessary condition, for hillside swamps are by no means uncommon, owing to a constant supply of percolating ground water which maintains the swampy condition (sloping fen) (275). Lake basins are occasionally filled with vegetation and sediment, thus becoming swamps. Swamps may be found on the floodplain of rivers as well as on their deltas; they are characteristic of the flat, ill-drained areas of the Atlantic coastal plain. Coastal salt-water swamps may develop in the zone between high and low tides or extend up river estuaries (15).

swamping   Process of peat formation (66). Synonyms: paludification (208); *Vertorfung* (299).

topogenous (ic)   'Produced by relief'; refers to peat deposited in so-called muskeg holes which started development in a water-filled depression (209).

transition muskeg   With a vegetation type and physical and chemical properties of the soil intermediate between oligotrophic and minerotrophic muskegs (bogs and fens); represents a stage in the succession from topogenous-soligenous to ombrogenous muskegs. In central Europe the pine and birch forest peat as well as the *Scheuchzeria palustris* peat belong to the transition-bog-peat. It has a mesotrophic vegetation type and physical and chemical properties of the soil intermediate between highmoor (bogs) and lowmoor (fens) types (40, 103). Synonym: poor fen (245). Translation: German, *Uebergangsmoor* (297).

tussock tundra   1) A vegetation type which predominates in the low-lying tundra regions of Alaska: Seward Peninsula (66, 244) and north of Brooks Range. It resembles the '*Eriophoretum*' of Tansley's (272) British bogs. (2) Arctic and subarctic terrain, sometimes muskeg, characterized by tussocks of *Carex-Eriophorum* frequently with frost ~olygons (203). Synonym: *têtes des femmes (234)*.

REFERENCES

1. Adams, G.D., and S.C. Zoltai. Proposed open water and wetland classification. In Guidelines for Bio-physical Land Classification, compiled by D.S. Lacate (Dept. Fisheries and Forestry, Publ. 1264, Ottawa, 1969), pp. 23-41.
2. Adams, G.D., et al. Contributions to a meeting of the Subcommittee on Organic Terrain Classification held April 11-12, 1972, Prairie Migratory Bird Research Centre, Univ. Saskatchewan, Saskatoon.
3. Ahti, T. Studies on the caribou lichen stands of Newfoundland. Arch. Soc. Vanamo, 30/4 (1959): 1-44.
4. — Macrolichens and their zonal distribution in boreal and arctic Ontario, Canada. Ann. Bot. Fenn. 1 (1964): 1-35.
5. Ahti, T., and R.L. Hepburn. Preliminary studies on woodland caribou range, especially on lichen stands, in Ontario. Ont. Dept. Lands and Forests Res. Rept. (Wildlife) 74 (1967).
6. Allington, K.R. The bogs of Central Labrador-Ungava; an examination of their physical characteristics. Geog. Ann. 43/3-4 (1961): 404-17.
7. Anctil, L. Reboisement des tourbières brûlées (unpublished thesis, Faculty of Forestry and Geodesy, Univ. Laval, Quebec, 1956).
8. Anonymous. Atlas of Canada (Dept. Mines and Technical Surveys, Geog. Branch, Ottawa, 1957), plates 11, 12, 24.
9. — Geological Map of Canada (Geol. Surv. Canada, Dept. Mines and Technical Surveys, 1962, Map 1045A).
10. — Climatic Maps (Canada Dept. Transport, 1967), ser. 1, sheet 7 T56-3667/1-7.
11. — Climatic Maps (Canada Dept. Transport, 1967), ser. 2, sheet 7 T56-3667/2-7.
12. — The Random House Dictionary of the English Language (Random House, New York, 1967).
13. — Climatic Maps (Canada Dept. Transport, 1968), ser. 4, sheet 2 T56-3667/4-2.
14. — Glacial Map of Canada (Geol. Surv. Canada, Dept. Energy, Mines, and Resources, 1968, Map 1253A).
25. — Van Nostrand's Scientific Encyclopedia (D. Van Nostrand (Canada) Ltd., Toronto, 1968).
16. — Permafrost in Canada (Joint Production of Geol. Surv. Canada, Dept. Energy, Mines and Resources, Map 1246A, and Div. Building Research, NRC, Ottawa, Pub. NRC 9769, 1969).
17. — Vegetation Patterns of the Hudson Bay Lowlands (Ontario Dept. Lands and Forests, 1969, Map).
18. — Physiographic Regions of Canada (Geol. Surv. Canada, 1970, Map Number 1254A).

19. — Newfoundland Forest Research Centre Program Review 1969-1972 (1972).
20. Antevs, E. Retreat of the last ice-sheet in eastern Canada. Geol. Surv. Canada, Mem. 146 (1925).
21. Auer, Y. Ueber die Entstehung der Straenge auf den Torfmooren, Acta Forest. Fenn. 12 (1920): 23-145.
22. — Peat Bogs in Southeastern Canada. Canada Dept. Mines, Geol. Surv. Mem. 162 (1930).
23. Baden, W. The types of peat deposits used in present-day agriculture and the methods of their utilization. Proc. 4th Internat. Peat Congr. 3 (June 1972): 7-20.
24. Baldwin, W.K.W. Plants of the clay belt of northern Ontario and Quebec. Nat. Museum of Canada, Bull. 156 (1958).
25. — Report on Botanical Excusion to the Boreal Forest Region in Northern Quebec and Ontario (Dept. Northern Affairs and National Resources, 1962).
26. Binns, W.O. Some aspects of peat as a substrate for tree growth. Irish Forestry 19/1 (1962): 32-55.
27. Bliss, L.C. Arctic and alpine plant life cycles. Ann. Rev. Ecol. and Systematics 2 (1971): 405-38.
28. Bliss, L.C., and R.W. Wein. Ecological problems associated with arctic oil and gas development. Proc. Can. Northern Pipeline Res. Conf. 2-4 (Feb. 1972): 65-77.
29. Blouin, J.L., and M.M. Grandtner. Etude écologique et cartographie de la végétation du comte de Rivière-du-Loup. Res. Service, Planning Branch, Quebec Dept. Lands and Forests, Forest Paper 6 (1971).
30. Bonner, E. Balsam Fir in the Clay Belt of Northern Ontario (MScF thesis, Univ. Toronto, 1941).
31. Bostock, H.S. Physiographic Regions of Canada (Canada Dept. Energy, Mines and Resources, Geol. Surv. Canada, 1967, Map 1254A).
32. Bowser, W.E. The Soils of the Prairies. In A Look at Canadian Soils. Agric. Inst. Rev. 15/2 (1960): 24-6.
33. Brown, R.J.E. The distribution of permafrost and its relation to air temperature in Canada and the USSR. Arctic 13/3 (1960): 163-77.
34. — Influence of vegetation on permafrost. Proc. Permafrost Internat. Conf., Nov. 1963 (1966): 20-5.
35. — Permafrost investigations in British Columbia and Yukon Territory. NRC Div. Building Research, Ottawa, Tech. Paper 253 (1967).
36. — Permafrost investigations in northern Ontario and northeastern Manitoba. NRC Div. Building Research, Ottawa, Paper 291 (1968).
37. — Occurrence of permafrost in Canadian peatlands. Proc. Third Internat. Peat Congr., 1968 (1970): 174-81.
38. — Permafrost as an ecological factor in the subarctic. Symp. Ecol. Subarctic Regions, Helsinki, 1966 (1970): 129-40.
39. — Permafrost in the Canadian Arctic Archipelago. Zeitschr. für Geomorphol., Suppl. 13 (July 1972): 102-30.
40. Bruene, F. Die Praxis der Moor- und Heidekultur (Paul Parey, Berlin and Hamburg, 1948).

41. Cajander, A.K. Studien ueber die Moore Finnlands. Acta Forest. Fenn. 2/3 (1913): 1-208.
42. Chalmers, R. Surface geology of southern New Brunswick. Geol. Surv. Canada, Ann. Rept. 1888-89 (1890): 70-1.
43. Chapman, L.J., and D.M. Brown. The climates of Canada for agriculture. The Canada Land Inventory Report no. 3 (1966).
44. Clements, F.E. Research Methods in Ecology (University Publishing Company, Lincoln, USA, 1905).
45. — Plant Succession. Carnegie Inst., Washington, Pub. 242 (1916).
46. — Plant Succession and Indicators (R.W. Wilson Co., New York, 1928).
47. Conway, V.M. The bogs of northern Minnesota. Ecol. Monogr. 19 (1949).
48. — Stratigraphy and pollen analysis of southern Pennine blanket peats. J. Ecol. 42 (1954).
49. Coombs, D.B. The physiographic subdivisions of the Hudson Bay Lowlands south of 60 degrees north. Geog. Bull. 6 (1954): 1-16.
50. Crampton, C.B. The distribution and possible genesis of some organic terrain patterns in the southern MacKenzie River valley. Can. J. Earth Sci. (Feb. 1973).
51. Dachnowski, A.P. Quality and value of important types of peat material. US Dept. Agric., Bur. Plant Industry Bull. 802 (1919): 1-40.
52. Dachnowski-Stokes, A.P. Peat resources in Alaska. US Dept. of Agric. Tech. Bull. 769 (1941).
53. Damman, A.W.H. Some forest types of central Newfoundland and their relation to environmental factors. Canada Dept. Forestry, Forest Res. Branch, Contrib. 596, Forest Sci. Monogr. 8 (1964).
54. — The distribution patterns of northern and southern elements in the flora of Newfoundland. Rhodora 67/772 (1965): 363-92.
55. Dansereau, P., and F. Segadas-Vianna. Ecological study of the peat bogs of easterm North America, 1. Structure and evolution of vegetation. Can. J. Bot. 30/4 (1952): 490-520.
56. Dansereau, P. Phytogeographica Laurentiana, II. The principal plant associations of the St. Lawrence Valley. Contrib. Inst. Bot., Univ. Montreal 75 (1959).
57. Day, J.H. Characteristics of Soils of the Hazen Camp Area, Northern Ellesmere Island, NWT (Defence Res. Board, Dept. National Defence, Canada, Operation Hazen (Hazen 24), Directorate of Physical Research (Geophysics), Ottawa, 1964).
58. — The classification of organic soils in Canada. Proc. Third Internat. Peat Congr., Quebec, Canada, 18-23 Aug. 1968, pp. 80-4.
59. Dirschl, H.J. Geobotanical processes in the Saskatchewan River Delta. Canadian Journal of Earth Sciences 9/11 (1972): 1529-49.
60. Dirschl, H.J., and Coupland. Vegetation patterns and site relationships in the Saskatchewan River Delta. Can. J. Bot. 50/3 (1972): 647-75.
61. Dixon, R.M. The Forest Resources of Ontario (Ontario Dept. Lands and Forests, 1963).
62. Dresser, J.A., and T.C. Denis. Geology of Quebec. Quebec Dept. Mines, Geol. Rept. 20 (1944).
63. Drew, J.V. A Pedologic Study of Arctic Coastal Plain Soils near Point Barrow,

Alaska (PHD thesis, Rutgers Univ. Library, New Brunswick, NJ, 1957). (Not seen, referred to by Drew and Tedrow, ref. 65.)

64. Drew, J.V., and J.C.F. Tedrow. Pedology of an arctic brown profile from Point Barrow, Alaska. Soil Sci. Am. Proc. 21 (1957): 336-9.

65. — Arctic soil classification and patterned ground. Arctic 15 (1962): 109-16.

66. Drury, W.H. Bog flats and physiographic processes in the upper Kuskokwim River region, Alaska. Harvard Univ. Gray Herbarium Contrib. 178 (1956).

67. Duffy, P.J.B. A forest land classification for the mixedwood section of Alberta. Canada Dept. Forestry, Pub. 1128 (1965).

68. DuRietz, G.E. Main units and main limits in Swedish mire vegetation. Svensk Bot. Tidskr. 43/2-3 (1949): 274-309.

69. — Phytogeographical mire excursion to the Billingen-Falbygden district in Vaestergoetland (southwestern Sweden). 7th Internat. Bot. Congr., Stockholm, 1950.

70. — Die Mineralbodenwasserzeigergrenze als Grundlage einer natürlichen Zweigliederung der Nord- und Mitteleuropaeischen Moore. Vegetatio 5-6 (1954): 571-85.

71. Ehrlich, W.A. Report on the classification of organic soils. Rept. on 6th Meeting Nat. Soil Survey Comm. Canada (1965): 68-75.

72. Eriksson, J.V. Balinge Mossars Utvecklingshistoria och Vegetation. Svensk Bot. Tidskr. 1912: 169-70.

73. Farnham, R.S., and H.R. Finney. Classification and properties of organic soils. Adv. Agron. 17 (1965): 115-62.

74. Fernald, M.L. A botanical expedition to Newfoundland and southern Labrador. Rhodora 13 (1911): 109-63.

75. Feustel, I.C., and H.G. Byers. The physical and chemical characteristics of certain American peat profiles. US Dept. of Agric. Tech. Bull. 214 (1930): 1-26.

76. Fleischer, M. Die Eigenschaften des Hochmoorbodens als Landwirtschaftliches Kulturmedium, III. Mitteilungen. Ueber die Arbeiten der Moorversuchsanstalt Bremen, in Landwirtschaftliches Jahrbuch XVII:2 (1891).

77. Forsgren, B. Studies of palsas in Finland, Norway and Sweden, 1964-1966. Biul. Peryclacjalny 17 (1968): 117-23.

78. Friedrichs, K. Okologie als Wissenschaft von der Natur oder biologische Raumforschung. Bios VII (Leipzig, 1937).

79. Gauthier, R. Étude écologique de cinq tourbières de Bas Saint-Laurent (Master's thesis, Univ. Laval, 1967).

80. — Study of five peat-bogs of the Lower St. Lawrence, 1. Ecology, II: Stable-litter peat. Quebec Dept. of Natural Resources, Mines Branch, Special Paper 10 (1971).

81. Gauthier, R., and M.M. Grandtner. Les tourbières de Bas Saint-Laurent étude phytosociologique (Univ. Laval, 1973).

82. Godwin, H., and V.M. Conway. The ecology of a raised bog near Tregaron, Cardiganshire. J. Ecol. 27 (1939): 313-63.

83. Grandtner, M.M. La forêt de Beauséjour, Comté de Levis, Québec. Laval Univ. Forest Res. Foundation Bull. 7 (1960).

84. — Premierès observations phytopédologiques sur les prés salés des Iles-de-la-Madeleine. Nat. Can. 93 (1966): 361, 366.

85. — Les ressources végétales des Iles-de-la-Madeleine. Laval Univ. Forest Res. Foundation Bull. 10 (1967).

86. Griggs, R.F. The problem of arctic vegetation. J. Washington Acad. Sci. 24/4 (1934): 153-75.

87. Gulley, E. Ueber die Beziehung zwischen Vegetation, chemischer Zusammensetzung und Duengerbeduerfnis der Moore, zugleich ein Beitrag zur Kenntnis der Moore Sued-Bayerns. Mitt. Kgl. Bayer. Moorkultur 3 (1909): 1-38.

88. Hagelund, E. Svenska Mosskulturfoereningens Torfgeologiska Undersoekningar (Svensk Mosskulturfoereningens Tidskrift, 1911).

89. Halliday, W.E.D. A forest classification for Canada. Can. Dept. Mines and Resources, Lands, Parks and Forests Branch, Forestry Service Bull. 89 (1937).

90. Hamelin, L.E. Les tourbières réticulées de Québec-Labrador subarctique. Cah. Géog. Québec 2/3 (1957): 87-106.

91. Hamet-Ahti, L. Notes on the vegetation zones of western Canada, with special reference to the forests of Wells Gray Park, British Columbia. Ann. Bot. 2 (1965): 274-300.

92. Hanson, H.C., and R.H. Smith. Canada geese of the Mississippi Flyway. Bull. Ill. Nat. Hist. Surv. 25 (1950): 67-210.

93. Hare, F.K. The Climate of the Eastern Canadian Arctic and Sub-Arctic and its Influence on Accessibility (PHD thesis, Univ. Montreal, 1950). (Not seen, referred to in 'The Climate of Canada,' Met. Branch, Air Services, Dept. Transport, 1962).

94. — Climate and zonal divisions of the boreal forest formation in eastern Canada. Geog. Rev. 40/4 (1950): 615-35.

95. — The boreal conifer zone. Geog. Studies 1 (1954): 4-18.

96. — A photo-reconnaissance survey of Labrador-Ungava. Canada Dept. Mines, Tech. Surv., Geog. Branch, Mem. 6 (1959): 1-64.

97. Hare, F.K., and R.G. Taylor. The position of certain forest boundaries in southern Labrador-Ungava. Geog. Bull. 8 (1956): 51-73.

98. Hatcher, R.J., and M.L.G. Jurdant. Chibougamou Research Forest, Quebec (Project Q-120). Forest Res. Branch Rept. 65-Q-5 (1965).

99. Heikurainen, L. Improvement of forest growth on poorly drained peat soils. In International Review of Forestry Research, vol. 1 (Academic Press, New York, London, 1964), pp. 39-113.

100. — Peatlands in Newfoundland and possibilities of utilizing them in forestry. Canada Dept. Fisheries and Forestry, Can. Forestry Service, Information Rept. N-X-16 (1968).

101. Heinselman, M.L. Forest sites, bog processes, and peatland types in the glacial Lake Agassiz Region, Minnesota. Ecol. Monogr. 33 (1963): 327-74.

102. — String bogs and other patterned organic terrain near Seney, Upper Michigan. Ecol. 46/1-2 (1965): 185-8.

103. — Landscape evolution, peatland types, and the environment in the Lake Agassiz peatlands natural area, Minnesota. Ecol. Monogr. 40/2 (1970): 235-61.

104. Henderson, A. Agricultural resources of Abitibi. Rept. Bureau Mines 14/1 (1905): 213-19.

105. Heusser, C.J. Radiocarbon dates of peats from North Pacific North America. Am. J. Sci. Radiocarbon Suppl. 1 (1959): 29-34.

106. Hills, G.A. Soil-vegetation relationships in the boreal clay belts of eastern Canada. In Baldwin, ref. 25, pp. 39-53.

107. Holmen, H. Skogsproduktion pa Torvmark, summary, afforestation of peatlands. Kgl. Skogs- och Lantbruksakad. Tidskr. 5 (1969): 216-35.

108. Hopkins, D.M., and R.S. Sigafoos. Frost action and vegetation patterns on Seward Peninsula, Alaska. us Geol. Surv. Bull. 974-C (1951): 51-100.

109. Hosie, R.C. Forest regeneration in Ontario. Univ. Toronto, Forestry Bull. 2 (Res. Council of Ontario, 1953).

110. Hustich, I. Notes on the coniferous forest and tree limit on the east coast of Newfoundland-Labrador, including a comparison between the coniferous forest limit in Labrador and in northern Europe. Acta Geog. 7/1 (1939).

111. — On the forest geography of the Labrador Peninsula. A preliminary synthesis. Acta Geog. 10/2 (1949): 1-63.

112. — Phytogeographical regions of Labrador. Arctic 2 (1949): 36-42.

113. — Notes on the forests on the east coast of Hudson Bay and James Bay. Acta Geog. 11/1 (1950).

114. — The lichen woodlands in Labrador and their importance as winter pastures for domesticated reindeer. Acta Geog. 12/1 (1951).

115. — Forest-botanical notes from Knob Lake area in the interior of Labrador Peninsula. Ann. Rept. Natl. Mus. (Canada) for the Fiscal Year 1949-1950 (1951), pp. 166-217.

116. — On forests and tree growth in the Knob Lake area, Quebec-Labrador Peninsula. Acta Geog. 13/1 (1954).

117. — Forest-botanical notes from the Moose River area, Ontario, Canada. Acta Geog. 13/2 (1954): 3-50.

118. — La forêt d'épinette noire à mousses du Québec septentrional et du Labrador. Nat. Can. 95 (1968): 413-21.

119. — On the phytogeography of the subarctic Hudson Bay Lowland. Acta Geog. 16/1 (1957): 1-48.

120. Illingworth, K., and J.W.C. Arlidge. Interim report on some forest site types in lodgepole pine and spruce-alpine fir stands. bc Forest Service Res. Notes 35 (1960).

121. Ilvessalo, Y. Notes on some forest site types in North America. Acta Forest. Fenn. 34/39 (1929): 1-111.

122. International Peat Congress. Report by Commission I and V on classification of peat. 4th Internat. Peat Congr., Helsinki, June 1972.

123. Jeffrey, W.W., L.A. Bayrock, L.E. Lutwick, and J.F. Dormaar. Land-vegetation typology in the upper Old Man River basin, Alberta. Forestry Branch Departmental Pub. 1202 (1968).

124. Jeglum, J.K., C.F. Wehrhann, and J.M.A. Swan. Comparisons of environmental ordinations with principal component vegetational ordinations for sets of data having different degrees of complexity. Can. J. Forest Res. 1/2 (1971): 99-112.

125. Jeglum, J.K. Boreal forest wetlands near Candle Lake, central Saskatchewan, I. Vegetation. The Musk-Ox 11 (1972): 41-58.

126. — Personal communication (1973).

127. — Boreal forest wetlands near Candle Lake, central Saskatchewan, II. Relationships of vegetational variation to major environmental gradients. The Musk-Ox 12 (1973): 32-48.

128. Jurdant, M. Carte phytosociologique et forestière de la forêt experimentale de Mont-

morency. Canada Dept. Forestry 1046F (1964).

129. Jurdant, M., and G.J. Frisque. The Nicauba Research Forest. A research area for black spruce in Quebec, Part I. Forest land survey. Information Rept. Q-X-18 (1970).

130. Jurdant, M., and M.R. Roberge. Etude écologique de la forêt de Watopeka. Ministère des Forêts, Pub. 1051F (1965).

131. Kaila, A. Viljelysmann orgaanisesta fosforista. Valtion Maatalouskoitoiminta Julkaisuja 129 (1948): 1-118.

132. — Phosphorus in various depths of some peat lands. Maataloustieteelimen Aikakauskirga 28 (1956): 90-104.

133. Kalela, A. Notes on the forest and peatland vegetation in the Canadian clay belt region and adjacent areas, I. Comm. Inst. Forest. Fenn. 55/33 (1962): 1-14.

134. Ketcheson, D.E., and J.K. Jeglum. Estimates of black spruce and peatland areas in Ontario. Can. Forestry Service, Dept. Environment, Information Rept. O-X-172 (1972).

135. Kihlman, A.O. Pflanzenbiologische Studien aus Russisch-Lappland. Acta Soc. Fauna et Flora Fenn. VI n:03 (1890).

136. Kivinen, E. Untersuchungen ueber den Gehalt an Pflanzennaehrstoffen in Moorpflanzen und ihren Standorten. Acta Agral. Fenn. 27 (1933): 1-140.

137. Klika, J. Nauka o Rostlinných Společenstvech (Fytocenologie) (Nakladatelství Československé Akademie Věd, Prague, 1955).

138. Knapp, R. Die Vegetation von Nord- und Mittelamerika (Gustav Fischer Verlag, Stuttgart, 1965).

139. Knollenberg, R. The distribution of string bogs in central Canada in relation to climate. Univ. Wisconsin, Dept. Meteorol. Tech. Rept. 14 (1964): 1-44.

140. Korpijaakko, M-L., and N.W. Radforth. On postglacial development of muskeg in the province of New Brunswick. Proc. 4th Internat. Peat Congr., Helsinki, 1973, pp. 341-60.

141. Krajina, V.J. Ecosystem classification of forests: summary. Recent Adv. Bot. 1961: 1599-1603.

142. Kruedener, Von A. 1916. Not seen, referred to by Svoboda, ref. 267.

143. — 1917. Not seen, referred to by Svoboda, ref. 267.

144. — Waldtypen, Klassifizierung und ihre volkswirtschaftliche Bedeutung (Forest Types, Classification and their Importance for the National Economy) (Verlag Neumann, Neudamm, 1927).

145. Kubiena, W.L. The Soils of Europe (Thomas Murby and Company, London, 1953).

146. Kujala, V. Waldvegetationsuntersuchungen in Kanada. Ann. Acad. Sci. Fenn. Ser. A, IV, Biol. 7 (1945).

147. Kulczynski, S. Peat bogs of Polesie. Mém. Acad. Polon. Sci. et Lettres, Ser. 15 (Cracovie, 1949).

148. Lafond, A. Notes pour l'identification des types forestières des concessions de la Québec (North Shore Paper Company, 1967).

149. Larsen, J.A. Major vegetation types of western Ontario and Manitoba from aerial photographs. Univ. Wisconsin Dept. Meteorol. Tech. Rept. 7 (1962).

150. — Vegetation of the Ennadai Lake area, NWT: Studies in subarctic and arctic bioclimatology, Ecol. Monogr. 35 (1965): 37-59.

151.   — Vegetation of Fort Reliance, Northwest Territories. Can. Field Nat. 85/2 (1971): 147-78.

152.   — The vegetation of northern Keewatin. Can. Field Nat. 86/1 (1972): 45-72.

153.   Leverin, H.A. Peat deposits in Canada. Ottawa Dept. Mines and Resources, Pub. 817 (1946).

154.   Lewis, F.J., and E.S. Dowding. The vegetation and retrogressive changes of peat areas in central Alberta. J. Ecol. 14/2 (1926): 317-41.

155.   Lewis, F.J., E.S. Dowding, and E.H. Moss. The swamp, moor and bog forest vegetation of central Alberta. J. Ecol. 16/1 (1928): 19-70.

156.   Lindsay, J.D., and W. Odynsky. Permafrost in organic soils of northern Alberta. Can. J. Soil Sci. 45 (1965): 265-9.

157.   Linteau, A. Forest site classification of the northeastern coniferous section boreal forest region Quebec. Canada Dept. Northern Affairs and National Resources, Forestry Branch Bull. 118 (1955).

158.   Losee, S.T.B. Site classification at Abitibi Woodlands Laboratory. Woodlands Index 1404 (F-2) (1955).

159.   Lotz, J.R. Soils and Agriculture Possibilities in the Knob Lake Area P.Q. (M.SC. thesis, McGill Univ., 1967).

160.   Loucks, O.L. A forest classification for the Maritime Provinces. Proc. Nova Scotian Inst. Sci. 25/2 (1962): 85-167.

161.   Low, A.P. Report on explorations in the Labrador Peninsula. Ann. Rept. Geol. Surv. Canada 8 (new ser.), Rept. L (1895).

162.   Lundqvist, G. En palsmyr sudost om kebnekaise. Stockholm, Geol. Föreningen Förhandl. 73/2 (1951): 209-25.

163.   MacLean, D.W. Note on the forests in the northern clay forest section. In Baldwin, ref. 25, pp. 54-67.

164.   MacLean, D.W., and G.H.D. Bedell. Northern claybelt growth and yield survey. Canada Dept. Northern Affairs and Natural Resources, Forest Res. Div. Tech. Note 20 (1955).

165.   Malmer, N., and H. Sjörs. Some determinations of elementary constituents in mire plants and peat. Lab. Plant Ecol., Bot. Mus. Univ. Lund (Medd. Fraon Lunds Bot. Mus. 109) Bot. Not. 108/Fasc. 1 (1955): 47-80.

166.   Malmstroem, C. Om moejlighterna att omfoera Myrmark till productiv Skogsmark. Saertryck ur Beten Vallar Mossar 9 (1956): 1-8.

167.   Marie-Victorin, F. Flore laurentienne (Les Presses de l'Univ. Montréal, 1964).

168.   Marr, J.W. Ecology of the forest-tundra ecotone on the east coast of Hudson Bay. Ecol. Monogr. 18 (1948): 117-44.

169.   Maycock, P.F., and B. Matthews. An arctic forest in the tundra of northern Ungava, Quebec. Arctic 19/2 (1966): 114-44.

170.   McMillan, J.G. Explorations in Abitibi. Rept. Bur. Mines 14/1 (1905): 184-212.

171.   Middendorf, A. Uebersicht ueber die natur Nord- und Ostibiriens, Vierte lieferung: Die gewaechse Sibiriens. Sibirische Reisen 4/1 (1864): 525-783.

172.   Millar, J.B. The Silvicultural Characteristics of Black Spruce in the Clay Belt of Northern Ontario (M.SC.F. thesis, Univ. Toronto, 1936).

173.   Moir, D.R. Beach ridges and vegetation in the Hudson Bay region. Proc. N. Dakota Acad. Sci. 8 (1954): 45-8.

174. Morozov, G.F. The significance for forestry of Dokuchaev's work. J. Pochvo-vedenie, 1904. Not seen, referred to by Svoboda, ref. 267.

175. — Die Lehre vom Walde (J. Neumann, Neudamm, 1928). Not seen, referred to by Svoboda, ref. 267.

176. Moss, E.H. Marsh and bog vegetation in northwestern Alberta. Can. J. Bot. 31 (1953): 448-70.

177. Moss, G.E. The vegetation of northern Cape Breton Island, Nova Scotia. Trans. Conn. Acad. Arts Sci. 22 (1918): 249-467.

178. Moss, M.S. Taxonomic and Ecological Studies of Sphagnum Species and Bogs of Alberta (unpublished thesis, Univ. Western Ontario, London, Ontario, 1949).

179. Mueller-Dombois, D. The forest habitat types of southeastern Manitoba and their application to forest management. J. Bot. 42 (1964): 1417-44.

180. — Eco-geographic criteria for mapping forest habitats in southeastern Manitoba. Forestry Chron. 41/2 (1965): 188-205.

181. Muller, P.E. Studien ueber die natuerlichen Humusformen (Berlin, 1887; Danish original 1879).

182. Nordenskjoeld, O. Le monde polaire (Librairie Armand Colin, Paris). Not seen, referred to by Rousseau, ref. 235.

183. Nordenskjoeld, O., and L. MacKing. The geography of the polar regions. Spec. Pub. Am. Geog. Soc. 8 (1928).

184. Odum, E.P. Fundamentals of Ecology (Saunders, Philadelphia, 1959).

185. Ogg, W.G. The soils of Scotland, Part I. Empire J. Exptl. Agric. 3 (1935): 174-88.

186. Olenin, A.S., M.I. Neistadt, and S.N. Tyuremnov. On the principles of classification of peat species and deposits in the USSR. Proc. 4th Internat. Peat Congr., Helsinki, June 1972, pp. 41-8.

187. Osvald, H. Die Vegetation des Hochmoores Komosse Svenska Vaextsoc. Saells-kapet Handlingar I (Uppsala, 1923).

188. — Vegetation and stratigraphy of peatlands in North America. Nova Acta Reg. Soc. Sci. Upsal. (Acta Univ. Upsal.), Ser. V:C, 1 (1970).

189. Paczowski, J.K. La vegetation de la bialowieza. Edition Ministry Agriculture (War-saw), Ser. E, V-me Excursion Phytogeographic International (1928).

190. Payandeh, B. Projection of stumpage charge and harvesting costs for northern Ontario black spruce pulpwood to the year 2010. Forestry Chron. 48 (1972): 1-5.

191. Persson, H., and H. Sjörs. Some bryophytes from the Hudson Bay Lowland of Ontario. Svensk Bot. Tidskr. 54/1 (1960): 247-68.

192. Pewe, T.L. Permafrost and Its Effect on Life in the North (Oregon State Univ. Press, 1966).

193. Pogrebniak, P.S. 1929. Not seen, referred to by Svoboda, ref. 267.

194. — Fundamentals in forest typology (Kiev, 1944). Not seen, referred to by Svoboda, ref. 267.

195. Pollett, F.C. Certain Ecological Aspects of Selected Bogs in Newfoundland (M.SC. thesis, Memorial Univ., Newfoundland, 1967).

196. — Ecology and utilization of peatlands in Newfoundland. Canada Dept. Fisheries and Forestry, Can. Forestry Service, Internal Rept. N-6 (1968).

197. — Classification of peatlands in Newfoundland. Proc. 4th Internat. Peat Congr.,

Helsinki, 1 (June 1972): 101-10.
198. — Nutrient contents of peat soils in Newfoundland. Proc. 4th Internat. Peat Congr., Helsinki, 3 (June 1972): 461-8.
199. Polunin, N. Botany of the Canadian eastern arctic, III. Vegetation and ecology. Can. Nat. Mus. Bull. 104 (1948): 1-304.
200. — Aspects of arctic botany. Am. Sci. 43 (1955): 307-22.
201. Pomerleau, R. Au sommet de l'Ungava. Rev. Univ. Laval 4/9 (1950): 1-16.
202. Porsild, A.E. Flora of the Northwest Territories. In Canada's Western Northland (1937), pp. 130-41.
203. — Earth mounds in unglaciated arctic northwestern America, forest and peatland at Hawley Lake, northern Ontario. Natl. Mus. Canada Bull. 171 (1938): 46-58.
204. — Notes from a Labrador peat bog. Can. Field Nat. 58 (1944): 4-6.
205. — Plant life in the arctic. Can. Geog. J. 42/3 (1951): 121-45.
206. — Botany of southeastern Yukon adjacent to the Canol Road. Nat. Mus. Canada Bull. 121 (1951).
207. — Illustrated flora of the Canadian Arctic Archipelago. Nat. Mus. Canada Bull. 146 (1957): 1-209.
208. Post, von L. The geographical survey of Irish bogs. Irish Nat. J. 6 (1937): 210-27.
209. Post, von L., and E. Granlund. Soedra Sveriges Torvtillgaengar. Sveriges Geol. Undersoekning, Pub. Ser. C, no. 335 (Yearbook), 19/2 (1926).
210. Radforth, N.W. Suggested classification of muskeg for the engineer. Eng. J. 35/11, Tech. Memo 24 (1952).
211. — Organic terrain organization from the air (altitudes less than 1000 feet), Handbook number 1. Dept. Natl. Defence, Canada, Rept. DR 95 (1955).
212. — Organic terrain organization from the air (altitudes 1000 to 5000 feet), Handbook number 2. Dept. Natl. Defence, Canada, Rept. DR 124 (1958).
213. — Organic terrain. In Soils in Canada, ed. R.F. Legget (Roy. Soc. Canada Special Publ. 3, rev. ed.; Univ. Toronto Press, 1961), pp. 115-39.
214. — Organic terrain and geomorphology. Can. Geog. 6/3-4 (1962): 166-71.
215. — Organic terrain as it affects the development of national objectives. In Proc. Ninth Muskeg Research Conf., ed. I.C. MacFarlane and J. Butler (NRC Tech. Memo. 81, Ottawa, 1964), pp. 1-6.
216. — Airphoto interpretation of muskeg. In Muskeg Engineering Handbook, ed. I.C. MacFarlane (Univ. Toronto Press, 1969), pp. 53-77.
217. — Relationship between artificial and natural classification of organic terrain. Proc. 4th Internat. Peat Cong., Helsinki, June 1972: 389-400.
218. Ramenskij, L.G. 1928. Not seen, referred to by Svoboda, ref. 267.
219. Ratcliffe, D.A. Mires and bogs. In The Vegetation of Scotland, ed. J.H. Burnett (Oliver and Boyd, Edinburgh and London, 1964), pp. 426-66.
220. Ratcliffe, D.A., and D. Walker. The Silver Flowe, Galloway, Scotland. J. Ecol. 1958: 407-45.
221. Raup, H.M. Phytogeographic Studies in the Peace and Upper Liard River Regions, Canada, with a Catalogue of the Vascular Plants (Contribution Arnold Arboretum, 1934).
222. — Botanical investigations in Wood Buffalo Park. Nat. Mus. Canada, Bull. 14 (1935).

223. — Phytogeographic studies in the Athabasca-Great Slave Lake region, II. J. Arnold Arboretum 27/1 (1946).

224. — The botany of the southwestern MacKenzie. Sargentia 6 (1947): 1-95.

225. Rayment, A.F., and H.W.R. Chancey. Peat soils in Newfoundland, reclamation and use. Agric. Inst. Rev. January-February 1966. Not seen, referred to by Heikurainen, ref. 100.

226. Riggs, G.B. Some sphagnum bogs of the North Pacific coast of America. Ecol. 6/3 (1925): 260-78.

227. Ritchie, J.C. The vegetation of northern Manitoba, II. A prisere on the Hudson Bay Lowlands. Ecol. 38 (1957): 429-35.

228. — A vegetation map from the southern spruce forest zone of Manitoba. Geog. Bull. 12 (1958): 39-46.

229. — The vegetation of northern Manitoba, III. Studies in the subarctic. Arctic Inst. N. Am. Tech. Paper 3 (1959).

230. — The vegetation of northern Manitoba, V. Establishing the major zonation. Arctic 13/4 (1960): 211-29.

231. — The vegetation of northern Manitoba, VI. The lower Hayes River region. Can. J. Bot. 38 (1960): 769-88.

232. — A geobotanical survey of northern Manitoba. Arctic Inst. N. Am. Tech. Paper 9 (1962).

233. — The vegetation of northern Manitoba, IV. The Caribou Lake region. Can. J. Bot. 38 (1968): 185-99.

234. Rousseau, J. Notes sur quelques lichens antibiotiques de la province de Québec. Aspects écologiques et phytogéographiques. Rev. Can. Biol. 10 (1951): 181-2.

235. — Les zones biologiques de la péninsula Québec-Labrador et l'hémiarctique. Can. J. Bot. 30 (1952): 436-74.

236. Rowe, J.S. Forest regions of Canada. Canada Dept. Natl. Affairs and Natural Resources, Forestry Branch, Bull. 123 (1959).

237. — Forest Regions of Canada (Canada Dept. Environment, 1972).

238. Rowles, C.A., L. Farstad, and D.G. Laird. Soil resources of British Columbia. Trans. 9th B.C. Natural Resources Conf., Victoria, B.C., (1956), pp. 84-112.

239. Sanderson, M.E., and D.W. Phillips: Average annual water surplus in Canada. Canada Dept. of Transport, Meteorol. Branch, Climatol. Studies 9 (1967).

240. Schmid, E. Vegetationsguertel und Biocoenose. Ber. Schweiz. Bot. Gesell. 51 (1941).

241. Schofield, W.B. The salt marsh vegetation of Churchill, Manitoba, and its phytogeographic implications. Nat. Mus. Can. Bull. 160 (1959): 107-32.

242. Scoggan, H.J. Flora of Manitoba. Nat. Mus. Can. Bull. 140 (1957).

243. Segadas-Vianna, S. Ecological study of peat bogs of eastern North America: the Chamaedaphne community in Quebec and Ontario. Can. J. Bot. 33 (1955): 647-84.

244. Sigafoos, R.S. Soil instability in tundra vegetation. Ohio J. Soil Sci. 51/6 (1951): 281-98.

245. Sjörs, H. Myrvegetation i Bergslagen. Acta Phytogeog. Suec. 21 (1948).

246. — Bogs and fens in the Hudson Bay Lowlands. Arctic 12/1 (1959): 1-19.

247. — Forest and peatland at Hawley Lake, northern Ontario. Nat. Mus. Can. Bull. 171 (1961): 1-31.

248. — Surface patterns in boreal peatland. Endeavor 20/80 (1961): 217-24.

249. — Bogs and fens on Attawapiskat River, northern Ontario. Nat. Mus. Can. Bull. 186 (1963): 45-133.

250. Sprout, P.N., D.S. Lacate, and J.W.C. Arlidge. Forest land classification survey and interpretation for management of a portion of the Niskonlith Provincial Forest, Kamloops District, B.C. Dept. Forestry Pub. 1159, B.C. Forest Service, Tech. Pub. T60 (1960).

251. Stanek, W. The Properties of Certain Peats in Northern Ontario (M.SC.F. thesis, Univ. Toronto, 1961).

252. — Occurrence, Growth and Relative Value of Lodgepole Pine and Engelmann Spruce in the Interior of British Columbia (PH.D. thesis, Univ. British Columbia, 1966).

253. — Development of black spruce layers in Quebec and Ontario. Forestry Chron. 44/2 (1968): 25-8.

254. — A forest drainage experiment in northern Ontario. Pulp and Paper Mag. Sept. 20, 1968: 58-62.

255. — Amelioration of water-logged terrain in Quebec, Part I. Description of the area, hydrology and drainage. Canada Dept. Fisheries and Forestry, Can. Forestry Service, Information Rept. Q-X-17 (1970).

256. — Natural layering of black spruce, *Picea mariana* (Mill.) B.S.P. in northern Ontario. Forestry Chron. 37/3 (1971): 245-58.

257. Stanek, W., and V.J. Krajina. Preliminary report on some ecosystems of western coast on Vancouver Island. *In* Ecology of the Forests of the Pacific Northwest, ed. V.J. Krajina (Progress Rept. Univ. British Columbia, 1964), pp. 57-66.

258. Stanek, W., and L. Orloci. A comparison of Braun-Blanquet's method with sum-of-squares agglomeration for vegetation classification. Vegetatio 27 (1973): 323-45.

259 Stoeckeler, E.G. Identification and Evaluation of Alaskan Vegetation from Airphotos with Reference to Soil, Moisture and Permafrost Conditions. A preliminary paper. (St. Paul, Minn.) U.S. Army, Corps of Engineers, St. Paul District, 1949. Mimeographed.

260. Sukachev, W.N. 1932. Not seen, referred to by Svoboda, ref. 267.

261. — An experimental investigation of the struggle for existence between biotypes of the same plant species. Trav. Inst. Biol. Peterhof 15 (1935): 69-88.

262. — 1928. Not seen, referred to by Svoboda, ref. 267.

263. — Biogeocoenology and phytocoenology. Dokl. Akad. Nauk USSR 47/6 (1945): 429-31.

264. — Relation between the concepts of geographical landscape and biogeocoenose. Vop. Geogr., coll., 16 (1949). Not seen, referred to by V. Sukachev and D. Dylis, ref. 265.

265. Sudachev, V., and N. Dylis. Fundamentals of Forest Biogeocoenology, tr. Dr J.M. MacLennan (Oliver and Boyd, Edinburgh and London, 1964).

266. Svenson, H. Tundra polygons. Photographic interpretation and field studies in the north-Norwegian polygon areas. Norg. Geol. Undersikelse 223 (1962): 298-327.

267. Svoboda, P. Přínos Sovětské Vědy k Lesní Typologii. Lesn. Práce 28/11-12 (1949): 453-535.

268. Tamm, C.O. Determination of Nutrient Requirements of Forest stands. *International Review of Forestry Research* 1(1974):115-70.

269. Tanner, V. Outline of the geography, life and customs of Newfoundland-Labrador. Acta Geog. 8/1 (1944).

270. Tansley, A.G. The classification of vegetation and the concept of development. J. Ecol. 8 (1920): 118-49.

271. — The use and abuse of vegetational concepts and terms. Ecol. 16/3 (1935): 284-307.

272. — The British Islands and Their Vegetation (Cambridge Univ. Press, London, 1939).

273. — The British Islands and Their Vegetation, 2 vols. (Cambridge, 1949).

274. Tantuu, A. Ueber die Entstehung der Buelten und Strange der Moore. Acta Forest. Fenn. 4/1 (1915).

275. Tarnocai, C. Classification of Peat Landforms in Manitoba (Canada Dept. Agric. Res. Station, Pedology Unit, Winnipeg, 1970).

276. — Some characteristics of cyric organic soils in northern Manitoba. Can. J. Soil Sci. 52 (1972): 485-96.

277. Taylor, G. Canada (London, 1947). Not seen, referred to by Baldwin, ref. 24.

278. Tedrow, J.C.F. Concerning genesis of buried organic matter in tundra soils. Soil Sci. Am. Proc. 29 (1965): 89-90.

279. — Polar desert soils. Soil Sci. Am. Proc. 30 (1966): 381-7.

280. — Pedogenic gradients of the polar regions. J. Soil Sci. 19 (1968): 197-204.

281. — Soils of the Subarctic region. Proc. Helsinki Symp. Ecol. Subarctic Regions (1970), pp. 189-205.

282. Tedrow, J.C.F., and J.E. Cantlon. Concepts of soil formation and classification in arctic regions. Arctic 11/3 (1958): 166-79,

283. Tedrow, J.C.F., and L.A. Douglas. Soil investigations on Banks Island. Soil Sci. 98 (1964): 53-65.

284. Terasmae, J. Geological notes and impressions. In Baldwin, ref. 25, pp. 36-8.

285. Thunmark, S. Orientierung ueber die Exkursionen des IX Internationalen Limnologenkongresses im Anebodagebiet (Limnologie, Stuttgart, 1940).

286. Vahtera, E. Metsaenkasvatusta Varten Ojitettujen Soitten Ravinnepitoisuuksista (Ueber die Naehrstoffgehalte der fuer Walderziehung entwaesserten Moore). German Referat, Comm. Inst. Forest. Fenn. 45 (1955): 1-108.

287. Valmari, J. Beitraege zur chemischen Bodenanalyse. Acta Forest. Fenn. 20 (1921) 1-67.

288. Villeneuve, G.O. Aperçu climatique du Québec. Québec Ministère des Terres et Forêts, Bull. 10 (1948).

289. Vincent, A.B. Growth and numbers of speckled alder following logging of black spruce peatlands. Forestry Chron. 40/4 (1964): 514-18.

290. — Growth of black spruce and balsam fir reproduction under speckled alder. Canada Dept. Forestry Pub. 1102 (1965).

291. Wade, L.K. Vegetation and History of the Sphagnum Bogs of the Tofino Area, Vancouver Island (M.SC. thesis, Univ. British Columbia, 1965).

292. Waksman, S.A. The peats of New Jersey and their utilization. New Jersey Dept. Conservation and Development, Geol. Ser., Bull. 55 (1942).

293. Warren Wilson, J. Observations on the temperatures of arctic plants and their environment. J. Ecol. 45 (1957): 499-531.

294. Washburn, A.L. Classification of patterned ground and review of suggested origins. Bull. Geol. Soc. Am. 67 (1956): 823-66.

295. Weaver, J.E., and F.E. Clements. Plant Ecology (McGraw-Hill, New York and London, 1938).

296. Weber, C.A. Ueber die Vegetation und Entstehung des Hochmoors von Augustmal im Memeldelta (Berlin, 1902).

297. — Die Entwicklung der Moorkultur in den letzten 25 Jahren (Die Wichtigsten Humus- und Torfarten und ihre Beteiligung an dem Aufbau Norddeutscher Moore). Festschrift zur Feierdes 25 Jährigen Bestehens des Vereins zur Förderung der Moor-kultur im Deutschen Reich (Berlin, 1908), pp. 80-101.

298. Wein, R.W. Report on a vegetation survey along the proposed gas pipeline route — Peel Plateau to Old Crow Mountains, Appendix III: Vegetation (1971).

299. Wenner, C.G. Pollen diagrams from Labrador (a contribution to the quartenary geology of Newfoundland-Labrador, with comparisons between North America and Europe). Geog. Ann. 29/3-4 (1947): 137-373.

300. Whittaker, R.H. Classification of natural communities. Bot. Rev. 28/1 (1962).

301. Zawitz, E.J. Report of James Bay Forest Survey, Moose River Lower Basin (Ontario Dept. Lands and Forests, Forestry Branch, 1923).

302. Zoltai, S.C. Southern limit of permafrost features in peat landforms, Manitoba and Saskatchewan. Geol. Assoc. Can. Special Paper 9 (1971): 305-10.

303. Zoltai, S.C., and C. Tarnocai. Permafrost in peat landforms in northern Manitoba. Papers presented at the 13th Annual Manitoba Soil Science Meeting, Dec. 10 and 11, 1969, pp. 3-16.

304. — Properties of a wooded palsa in northern Manitoba. Arctic and Alpine Res. 3/2 (1971): 115-29.

3
Peatland Survey and Inventory

ERKKI O. KORPIJAAKKO and DAVID F. WOOLNOUGH

There is a serious need for a comprehensive inventory and survey of Canadian peatlands. To assist those interested in carrying out this task, a brief description of the methods and techniques used is given in this paper. An example of a detailed ground survey method giving a list of the basic items which should be included in the survey is discussed briefly. References to various systems of classification that may be used in surveys and inventories are given but only the von Post system is described in detail. Also, the main features of remote sensing techniques and of computerized mapping and automatic remote sensing techniques which are now used for peatland inventories, or whose use is under study, are discussed briefly.

INTRODUCTION

The greatest concentrations of peatlands occur in the temperate climatic zones of the northern hemisphere according to various authors (cf. Bülov 1925 and Figure 1), but the statistics given for the surface areas and volumes of peat in different parts of the world vary greatly. According to Kivinen (1948) the following estimates may be given for a few countries: USSR, 110 million hectares (275 million acres); Finland, 10.2 million hectares (25.5 million acres); Great Britain and Ireland, 2.5 million hectares (6.25 million acres). According to some estimates, the surface area of peatland in Canada may be 130 million hectares (500,000 square miles minimum, Radforth, 1962), which would give Canada the largest peat resources in the whole world.

In most developed countries, peatlands have long been recognized as a useful resource and treated accordingly. In a number of countries with significant peat resources, surveys and inventories to determine the extent and quality of this resource have been carried out in considerable detail. In Canada, despite its vast peat resources, no concentrated effort has yet been put into a survey. According to estimates made by Coupal (1972) the areas of peatlands surveyed so far in Canada are as follows:

British Columbia	2,385,000 acres	(950,000 hectares)
Alberta	2,000,000 acres	(800,000 hectares)

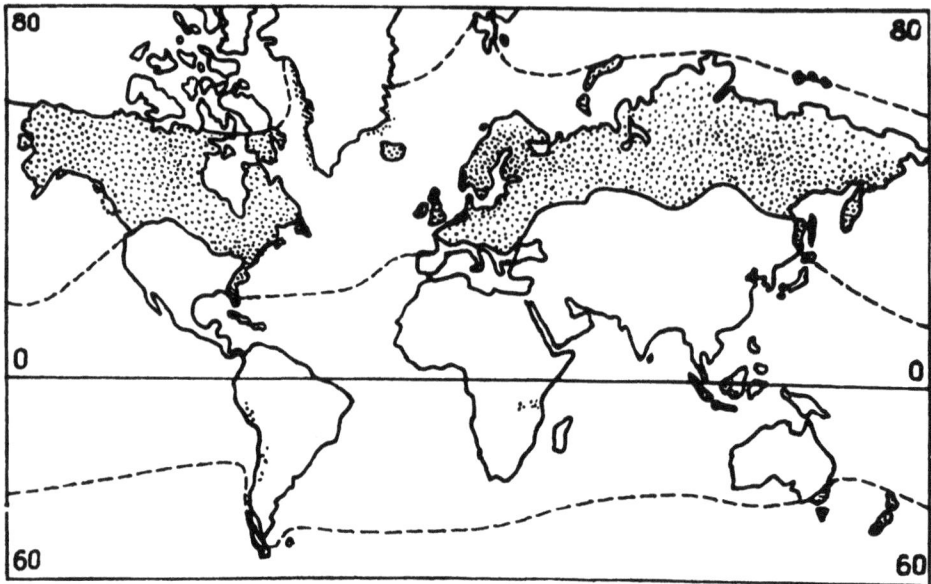

FIGURE 1   Distribution of peatland on the earth (Bülov 1925).

Saskatchewan	10,000,000 acres	(4,000,000 hectares)
Manitoba	92,500 acres	(37,000 hectares)
Quebec	255,000 acres	(100,000 hectares)
Nova Scotia	9,000 acres	(3,600 hectares)
Prince Edward Island	6,400 acres	(2,500 hectares)
Newfoundland	28,000 acres	(11,000 hectares)

No survey data are included for Ontario and New Brunswick in the above. However, the Muskeg Research Institute, with the financial support of the Department of Natural Resources, Province of New Brunswick, has carried out a preliminary peatland survey of the province of New Brunswick, which is the only one covering an entire province of Canada. According to this survey, about 10 per cent of the total land area of New Brunswick is covered by muskeg. This is equivalent to about 7,250 square kilometres (2,800 square miles) of peatland. If this figure is added to the above estimates of surveyed peatlands in Canada, it appears that only about 6.7 million hectares (about 16.5 million acres) have been surveyed, which represents only about 5 per cent of the estimated 130 million hectares (500,000 square miles, minimum, Radforth, 1962) of peatlands in Canada.

Before any proper planning can be carried out, it is essential that more of these peatlands be surveyed and the resources inventoried. Peat can be an engineering embarrassment, as the increased activity in the north has revealed, and even for this reason alone a survey should be done. Peat is not as homogeneous a material as it may seem to be to a casual and uninitiated observer. Its engineering and ecological properties vary greatly, depending on large-scale forces such as climate, topography, and the mineral soil type surrounding and underlying the deposits, and the uses indicated for the various types of peat differ. For proper planning of peatland utilization it is therefore necessary to consider in the surveys the different influences acting upon peat formation as well as the different peat types and their inherent properties.

The lack of surveys and inventories of peat and peatlands in Canada is fortunately being remedied by increased interest and activity in this field by various agencies. The following account is offered with the hope that it may help interested persons in deciding how and what to survey when they investigate peatlands.

OBJECTIVES OF SURVEY AND INVENTORY

The basic objectives of any survey of peatlands may be summarized as follows:

1. Peatlands constitute a natural resource with a great potential for a large variety of uses. For proper planning, this resource, like any other resource, should be properly surveyed and an inventory of its characteristics and distribution should be made. Depending on the location of a peatland area and on the type of peat it contains, it can be used for large-scale purposes such as agriculture, as a source of horticultural peat, for forestry or industrial purposes (for fuel, as a source of activated carbon, resins and waxes, etc.). In recent years peat moss has become a widely used commodity not only in the countries where it is found but also in countries with warmer climates, such as Israel and some African nations, to provide suitable replenishment of organic matter in their field and greenhouse soils. Recently also, the threatening shortage of fossil fuels has aroused an increased interest in peat as an alternative source of fuel, at least in some European countries.

Pollution control needs have recently escalated studies on new media, including peat, which has appeared as a potential source for an absorbing agent in fighting oil spills. It has proved to be promising as an agent in filters and as a good basic material for manufacturing activated carbon.

2. Peatlands form ecosystems with special characteristics, and a study of them may provide information for a number of scientific and applied purposes. Some unique features inherent in peatland ecosystems may be used to determine ecological changes in the environment brought about by changes in external factors, such as changes in climatic conditions during the last 10,000 to 12,000 years, whose effects have been preserved in the 'archives' formed by peat. Also, the effect of man on the environment is portrayed in these same peat formations, and an interpretation of these records can be used to predict possible changes in the environment due to man's future activities in such areas as agriculture and forestry, and to assist him in planning wisely.

Geological studies can detect past climatic changes by using pollen analysis, carbon-14 dating, and the ecological indicators of the peat record. This information may, in turn, be applied even for such an exotic purpose as climate forecasting.

These few examples indicate the potential applications of the study of peat deposits for other scientific and also practical purposes. A detailed survey and inventory of peatlands providing information on their unique features is obviously a good beginning.

3. Peatlands often impose unique problems for engineering activities. Frequently these problems can be solved only by using special methods. Because peat is not a homogeneous medium, as mentioned earlier, a survey of different peat types would be useful for engineering purposes to avoid unnecessary embarrassment. With the increased activity in the north, where peatland is often encountered, it is particularly important that a survey of its different

states be made to assist in construction of roads, pipelines, airstrips, and other structures.

Thus the objectives of a survey of peatland can be summarized in one statement: to obtain an intimate knowledge of different facets of the region through well-planned survey and inventory programs in order to make proper use of the resources and the land.

OBJECTS OF SURVEY AND INVENTORY

One of the main problems in survey and inventory work is deciding what to survey. This is particularly valid when dealing with peatland which is the cumulative result of biotic and abiotic influences, so that a total survey may involve both engineering and biological principles and a reconciliation of any possible conflict between the two. This means that a great number of factors such as climate, edaphic conditions, ecological conditions, botanical characteristics, and various engineering characteristics of peat, as well as its physical properties and characteristics, should all be accounted for in a detailed survey. The interaction between the living and non-living world is the unique feature that produces peatland, and they continue to interact and modify one another.

Keeping in mind this involved system, there are a number of steps that might be considered as suitable in carrying out surveys. A list of factors and aspects to be accounted for is given below. This list cannot, of course, satisfy the needs of investigators from all fields, and should be amended to suit the requirements of a particular worker or purpose.

In areas for which no previous survey data are available, a preliminary survey might satisfy the basic needs. This step should involve as little time-consuming ground work as possible and should give information about the general distribution of peatland as a percentage of land area and also, if needed, information about the distribution of the different kinds of peatlands as a percentage of the total amount of peatlands or of the total land area. Information can be given in the form of maps depicting the probabilities of peatland occurrence rather than the exact locations of individual peatland areas. Preliminary generalized information such as this can be used for determining priority areas for more detailed surveys.

The second step should be a survey revealing more details of particular features such as the location of individual peatland areas, their extent, their estimated depth, and the distribution of the main types of peatland (e.g. treed vs. open peatland). Part of this information can be obtained from existing topographic maps but some supplementary field work should be done to determine the depth of the deposits and also to check the extent as well as the frequency of various cover types. The information can be transcribed in the form of maps of varying scales from 1:10,000 to 1:250,000, and can be used for engineering planning or as a basis for planning detailed surveys.

The third step should consist of a detailed survey and inventory of some of the characteristics of peatlands. The aspects to be included may vary greatly depending on the end purpose of the survey, but the following features should form part of the standard information collected in the field.

1.   Extent of peat formation. This can be shown in the form of a large-scale map (e.g. 1:10,000).
2.   Airform patterns.
3.   The relationship of the peat deposit to the surrounding mineral terrain (topography).

4.   General genetic features (for example ombrogeny-minerogeny).
5.   Local or regional climate.
6.   Distribution of different cover types on the individual peatlands, using any of various classification systems. This should also include open water areas.
7.   Density and type of tree cover.
8.   Floristic and phytosociological analyses of the surface vegetation.
9.   Progress of paludification.
10.   Surface contour.
11.   The depth of peat, in the form of a contour map.
12.   Distribution of different peat types (maps and profiles).
13.   Various physical and chemical characteristics (pH, nutrient content, permeability, degree of humification, water content, ash content, trace elements, engineering properties such as shear strength).
14.   Distribution of woody erratics in the peat.
15.   Bottom soil type.
16.   Water regime. Origin of water.
17.   General drainage conditions.
18.   Possible existence of permafrost.
19.   Man-made features such as drainage ditches, peat mining, farming, etc.
20.   Wildlife.

METHODS

Equipment

For the preliminary survey airphotos, stereoglasses, and mirror stereoscopes form the basic tools. If computer analyses are indicated, access to computer facilities is also required.

For the detailed survey, suitable compasses, prisms (for obtaining 90° angles for survey lines), measuring tapes or cables, maps, and aerial photographs are needed.

For production of contour maps, the peatlands have to be levelled. For this purpose a self-levelling levelling machine is recommended to compensate for the inevitable slight swaying of the machine on the soft ground.

The sampling for a detailed survey may be carried out by using a variety of samplers. For visual field investigation, the most commonly used sampler is the Hiller sampler. It, and certain others, have been described in more detail in the *Muskeg Engineering Handbook*, chapter 5 (J.R. Radforth, 1969).

For obtaining undisturbed and uncontaminated samples, a piston sampler is required. The basic piston sampler has also been described in the above-mentioned publication. The Muskeg Research Institute has modified it to suit permeability and nutrient analyses sampling (M. Korpijaakko and Radforth, 1972). The modified version (Figure 2) has an inside diameter of 97 mm and can be dismantled to ensure the least possible disturbance of the sample. The sample in its original tube section can be placed in a permeability apparatus for permeability measurements.

Frozen peat can be sampled with various large non-portable conventional samplers. It can also be sampled to limited depths by a chain saw. A portable powered sampler for sampling frozen peat (Figure 3) has been developed by the Geological Survey of Finland. This sampler

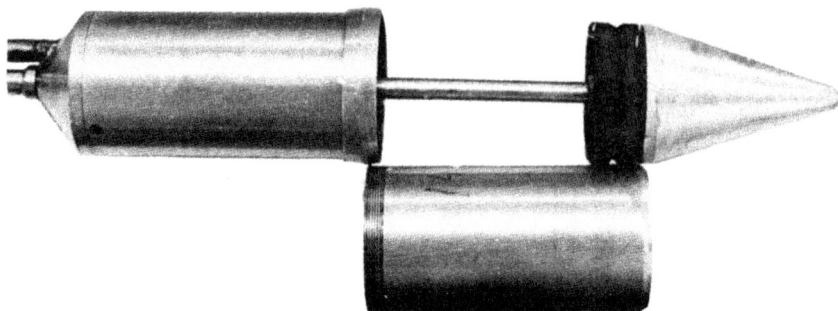

FIGURE 2   Piston sampler for permeability and nutrient content analyses.

has been used successfully in both Canada and Finland to extract cores of frozen peat to the depth of several metres.

Equipment which can be used in the field for conventional field testing of peat for engineering purposes has been described by J.R. Radforth in chapter 5 of the *Muskeg Engineering Handbook* (1969), to which the reader is referred. In addition, the Department of Civil Engineering of the University of New Brunswick in concert with the Muskeg Research Institute is in process of developing large-scale field equipment for field testing. This equipment is designed to overcome certain drawbacks of the conventional portable field equipment when applied to peat studies. This new equipment will be described in the future, after its testing has been completed. For laboratory testing, conventional soil engineering equipment can be used. The Muskeg Research Institute has developed a modified version of a constant head permeability apparatus (M. Korpijaakko, 1972) for which the principle has been described in the literature (Sarasto, 1961). The modified apparatus does not deform peat samples as would a triaxial machine. This is achieved by using a very low pressure created by a constant head of water of only 50-100 mm (Figure 4).

For laboratory analyses of peat samples, to finalize the results of the field investigation, conventional laboratory equipment is used.

Classification Systems used in Surveys

A great number of classification systems are available for peatland study. It is quite obvious that it is impossible to design a single classification system which would satisfy the needs of investigators from a variety of fields, such as agriculture, forestry, engineering, biology, and industry. Surveys designed for certain specific purposes may not be applicable to others through lack of the necessary information. In general purpose surveys the surveyor has therefore a difficult choice to make in deciding which aspects should be included and which excluded to satisfy as many people as possible. The data collected should be adaptable to as many fields as possible and this means that certain compromises must be made.

The most recent attempts at a comprehensive system in which all others are subsidiary and contributory lends itself to cross reference and computer treatment. It was first developed for presentation at the Third International Peat Congress, Helsinki and is now presented elsewhere in the literature.

One system will be described below, that of von Post, which has been used in its modified version perhaps most widely in peatland surveys in various parts of the world and perhaps is the best compromise solution, especially when supplemented by other systems in the field.

FIGURE 3   Powered sampler for sampling frozen peat.

Some other systems will be referred to, but they will not be described in any detail because descriptions are readily available in the literature cited. This does not imply that these systems would not be as good as the von Post system. On the contrary, for their specific purposes, they are superior, and they should be used to supplement field data whenever possible.

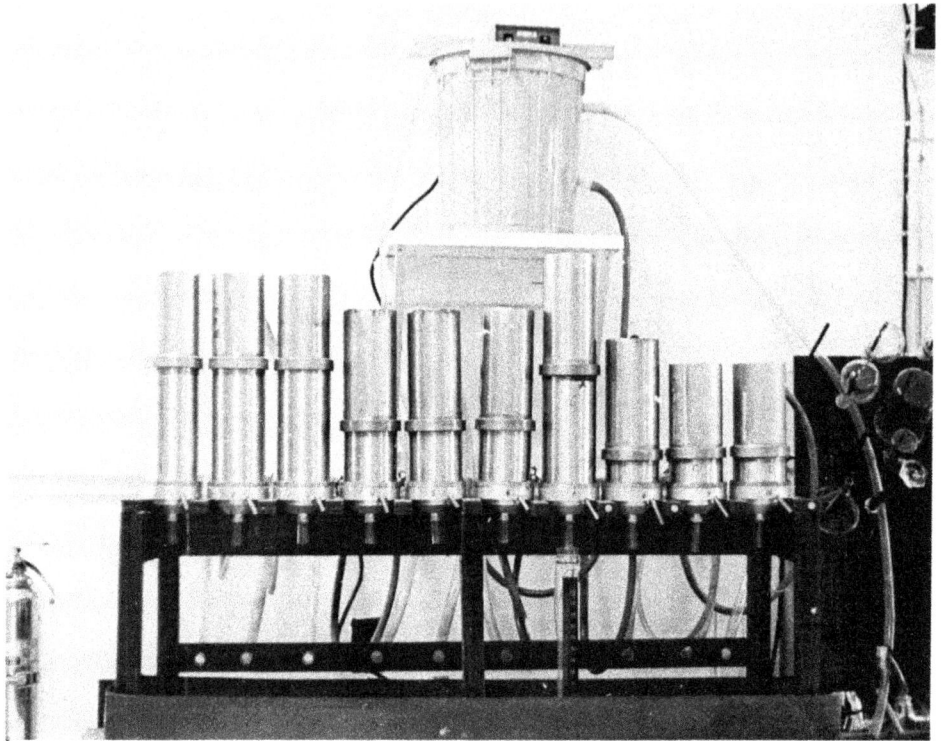

FIGURE 4   Permeability apparatus for measuring permeability of peat.

The von Post system was mainly designed originally to describe the degree of humification of peat. Later it was modified greatly to include numerous other properties of peat as well as surface characteristics of peat deposits. The following is a fairly detailed description of this system in its modified version. The application of the data extracted in the field will be described in a later chapter.

At each sampling site the first step is to make notes on the general features and the surface cover. These include the density of tree cover, the species which occur and the average height, and cover classifications. It is desirable to use several systems if possible, among which the Radforth (1952) system and the system used by foresters might be mentioned here (Heikurainen, 1960, 1968; Pollett, 1972). Also, the botanical description can be given following systems described in the literature (Ruuhijärvi, 1960). The list of features should also include reference to ponds, ridges, and rimpis and also a short description of the features of the surrounding mineral terrain.

After these general notes have been made, the peat profile is usually studied. For visual inspection, the Hiller borer can be used. Laboratory samples should be taken with a piston sampler and have to be taken only from a few selected locations depending on the homogeneity of the deposit as determined by inspection of the Hiller borer samples. For a detailed peat formula, according to the modified von Post system, the following items have to be recorded:

1. The depth (preferably in centimetres) from which the sample has been extracted.

2. The peat type, utilizing the recognizable features of the original plant constituents that formed the peat. The most common plants forming peat and the symbol used for them in the peat formula are: *Sphagnum* (S), *Carex* (C), *Eriophorum* (Er), *Equisetum* (Eq), *Phragmites* (Ph), *Scheuchzeria* (Sch), shrubs (N = nanolignidi), wood (L = lignidi), and mosses (other than Sphagna), i.e. Bryales (B). In some cases it is desirable, if possible, to specify the composition more accurately by direct reference to the species composition (e.g. *Sphagnum fuscum* peat).

Any other plant material recognizable as a significant contributor in the peat deposit under investigation should be recorded too.

Peat is usually a combination of remains of several plant groups, rather than of a single group. It has been customary in some European countries to mention the less significant constituents first, but for the sake of logic, it is recommended that they should be listed after the most significant ones. According to this principle, a peat composed mainly of *Carex* but with smaller amounts of *Sphagnum* would be called *Carex-Sphagnum* peat (CS peat).

3. The degree of humification, which is indicated by the letter 'H' and is divided into 10 categories described as follows according to Kivinen (1948), Scottish Peat Surveys (1968), and the author's own experience:

H_1 Completely undecomposed peat which, when squeezed, releases almost clear water. Plant remains easily identifiable. No amorphous material present.

H_2 Almost completely undecomposed peat which, when squeezed, releases clear or yellowish water. Plant remains still easily identifiable. No amorphous material present.

H_3 Very slightly decomposed peat which, when squeezed, releases muddy brown water, but for which no peat passes between the fingers. Plant remains still identifiable, and no amorphous material present.

H_4 Slightly decomposed peat which, when squeezed, releases very muddy dark water. No peat is passed between the fingers but the plant remains slightly pasty and has lost some of the identifiable features.

H_5 Moderately decomposed peat which, when squeezed, releases very 'muddy' water with also a very small amount of amorphous granular peat escaping between the fingers. The structure of plant remains is quite indistinct although it is still possible to recognize certain features. The residue is strongly pasty.

H_6 Moderately strongly decomposed peat with a very indistinct plant structure. When squeezed, about one-third of the peat escapes between the fingers. The residue is strongly pasty but shows the plant structure more distinctly than before squeezing.

H_7 Strongly decomposed peat. Contains a lot of amorphous material with very faintly recognizable plant structure. When squeezed, about one-half of the peat escapes between the fingers. The water, if any is released, is very dark and almost pasty.

H_8 Very strongly decomposed peat with a large quantity of amorphous material and very indistinct plant structure. When squeezed, about two-thirds of the peat escapes between the fingers. A small quantity of pasty water may be released. The plant material remaining in the hand consists of residues such as roots and fibres that resist decomposition.

H_9 Practically fully decomposed peat in which there is hardly any recognizable plant structure. When squeezed, almost all the peat escapes between the fingers as a fairly uniform paste.

H_{10} Completely decomposed peat with no discernible plant structure. When squeezed, all the wet peat escapes between the fingers.

4. The moisture regime of each peat sample, estimated by utilizing a scale of 1–5 and the symbol 'B' (derived from Swedish *blöthet* = wetness).

B_1 Dry peat
B_2 Low moisture content
B_3 Moderate moisture content
B_4 High moisture content
B_5 Very high moisture content

5. The fine fibre content, estimated and expressed by utilizing a scale of 0-3 and the symbol 'F.' The fibres are, in this case, mainly derived from *Eriophorum*. If any other fine fibres are recorded, they should be identified in the peat formula.

F_0 Nil
F_1 Low content
F_2 Moderate content
F_3 High content

The coarse fibres are also estimated by using a scale of 0-3 and the symbol 'R' in the manner tabulated above. In this case, mainly rootlets are recorded. Often the rootlets of *Carex* are referred to and also identified in the formula as such.

6. The presence of woody remnants in the peat, which may be a serious problem especially when peat is mined for peat moss production. For this purpose a scale of 0-3 and the symbol 'W' are used.

W_0 Nil
W_1 Low content
W_2 Moderate content
W_3 High content

7. Other information as pertinent, added to the end of the formula using suitable symbols and abbreviations. Thus layers of charcoal indicating old fires, seeds, and other recognizable plant remnants may be noted.

In the whole sample site, the last items to be recorded are the possible layers of detritus ooze, gyttja, and finally the type of bottom soil.

When all the information about the site has been recorded, a set of peat formulae for the site will have been obtained. One formula in the set might look something like the following:

Ste A1
Cover F1 (Radforth system), large sedge open white moor
0-30 CS H_{1-2} B_3 F_0 R_2 W_0 (seeds; Pot. palustr.)

This indicates that from the surface to the depth of 30 centimetres the peat is composed of

Carex-Sphagnum remnants, the decomposition varies from nil to very slight, the quantity of moisture is moderate, the quantity of fine fibres is nil and that of coarse fibres low, no woody material is present, and *Potentilla palustris* seeds were identified.

This information is quite readily convertible to other systems. Thus, from the above information, one can deduce that according to the Radforth (1955a) peat classification system the peat would be non-woody coarse fibrous. Supplemented with samples for laboratory analysis, this information can be processed to satisfy the needs of investigators from various fields.

The other classification systems already quite widely used in connection with surveys for special purposes such as agriculture, forestry and engineering activities, will not be described here in greater detail. This does not mean that they would be in any way inferior to the von Post system. On the contrary, when used for the purposes for which they were designed, they may be superior to a generalized system such as the one described here. However, it is quite a simple task to add the information that can be gleaned in the field by utilizing other systems in conjunction with the von Post system to achieve results that may satisfy a larger number of people and also may save time by avoiding successive surveys for future needs.

The most advanced systems used in Canada are the Radforth cover and peat classification systems (Radforth, 1952, 1955a). These systems are designed basically for engineering purposes, but the information can be interpreted and used with certain limitations in other fields.

Another rapidly developing system is one designed for use by soil scientists, especially for agricultural purposes. This system has been described by Day (1968), and the reader is referred to his paper for further information. Foresters tend to use their own systems, many of which are based on practices imported from northern Europe, especially from Finland. The reader may be referred in this context to various papers by Heikurainen (1960, 1968), some of which deal with Canadian conditions, and to papers by Pollett (1972) dealing with Newfoundland.

The remaining systems available are based on phytosociological methods and are suitable for studying surface vegetation and ecology of peatlands.

All these aspects can be and have been used for deciding the mode of peatland utilization. The more that can be included in the survey data, the better it will be for efficient planning.

Ground Survey

This section deals with the actual procedure of making a detailed ground survey of a peatland. For convenience, a commonly used system of notation has been adopted. The first step is to determine, on an aerial photograph, the main axes of the bog. Generally, the main survey line is run through the long axis of the deposit and through its centre, if possible. This line may be designated as the A-line and is marked with stakes so that A0 is at the edge of mineral terrain. From this point, the line is laid out with the aid of a compass, and stakes are driven into the peat at 100 metre intervals or closer if required. Each stake is marked with its appropriate designation so that the first stake, at a distance of 100 metres from A0, is marked AI and the following ones, using Roman numerals, AII, AIII, AIV, and so on, as long as they are spaced at regular 100 metre intervals. The final stake at the edge of the bog is marked with the distance, in metres (e.g. A 1,520m), from the A0 location.

A number of secondary lines, perpendicular to the main line, are staked out at 100 metre in-

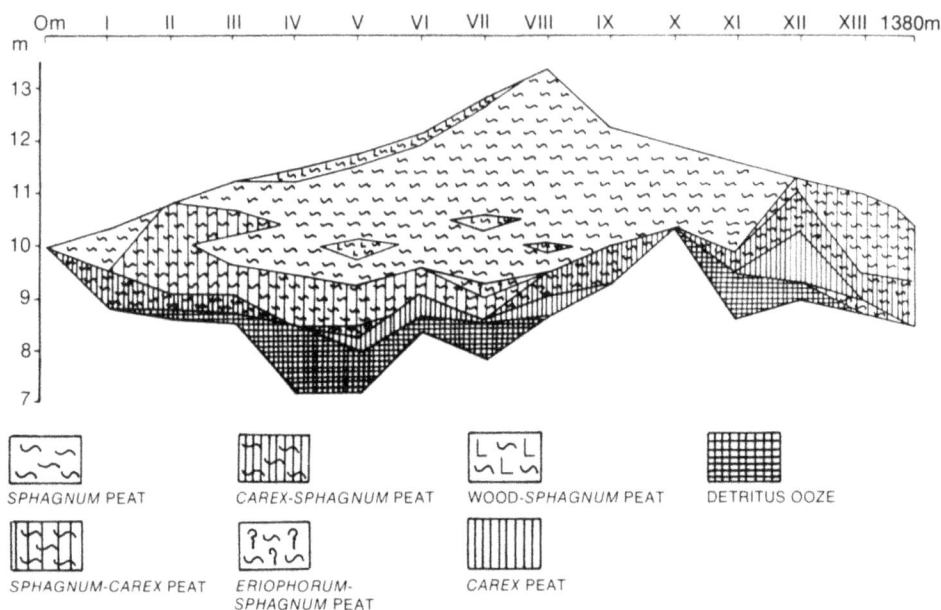

FIGURE 5   Peat profile of a raised bog (Bull Pasture Bog, New Brunswick). The vertical scale is exaggerated 80x over the horizontal scale.

tervals. These are marked by the location at which they intersect the main line and by an additional number, which bears a negative sign for the arm extending to the left of the main line and a positive sign for the right (as viewed along the line from the starting point A0). The stakes along the secondary line passing through AIV are thus marked with the identification AIV together with numbers indicating the distance from the main line in tens of metres and the appropriate sign. Thus the arm extending to the right of the main line would carry the designations AIV + 10, AIV + 20, etc., the last stake carrying the figure indicating the total distance from the main line (e.g. AIV + 1,250m).

At the end of each line, and at the intersection of the main line and the secondary lines, a sturdy stake is driven as deep into the peat as possible and left at ground level as a stable foundation for levelling purposes. The total line grid is levelled and tied to the absolute elevation at a suitable bench mark.

The surface characteristics and the peat profile are studied at each stake as described earlier, by using the von Post or some other suitable system. The record can be made at smaller or larger intervals than 100 metres, depending on the size of the deposit and on the homogeneity of the area. However, near the edge of the deposits it is advisable to record at least the depth of peat more frequently until the depth of one metre is reached, in order to facilitate later calculations of volumes and quantities of peat, and also to facilitate the planning of drainage operations.

For proper accuracy in areas of treed peatlands, it is also necessary to cut a narrow line through the trees; but one should try to avoid unnecessary cutting.

When these procedures have been completed, the next step is to analyse the data acquired and display it suitably.

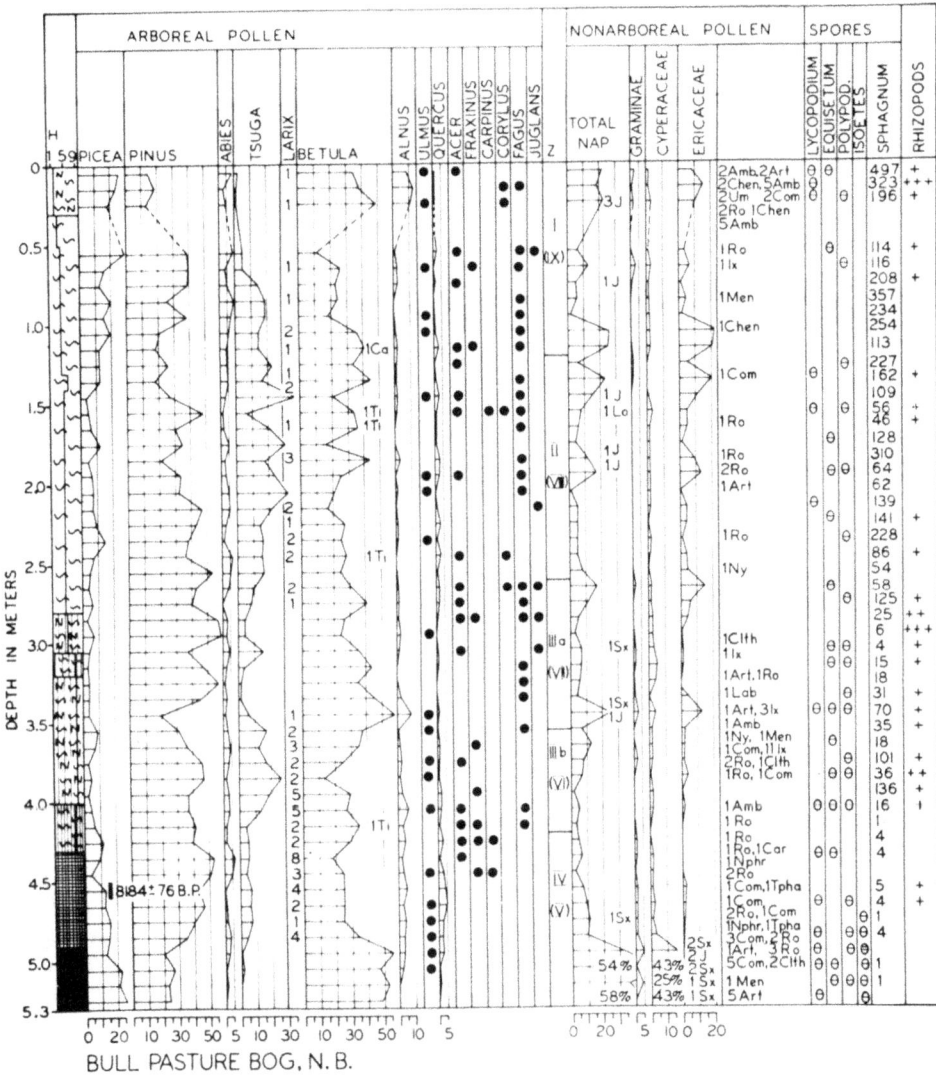

FIGURE 6 Pollen diagram from Bull Pasture Bog, New Brunswick (M.L. Korpijaakko and N.W. Radforth, 1972).

DISPLAY OF SURVEY RESULTS

The main modes of displaying the results of a survey are maps, diagrams, peat profiles, histograms, and tables.

Experience has shown that for a detailed survey a map scale of 1:10,000 is, in most cases, the most suitable. The map showing the extent of the peat deposit should also show the location of the survey lines and all the sampling sites. For each sampling site location, certain basic information, such as the depth of the peat deposit and the type of bottom soil, in addition to the sample site number, may be shown on the map.

Maps showing the distribution of different cover types (displayed by suitable shading or symbols), the peat depth contour map (for example at one metre intervals), the surface contour map, and all other maps that may be required (e.g. bottom soil type, drainage features, treed vs. open bog types, airform patterns) should display the survey line network.

The distribution of peat types and of the degree of humification can best be illustrated by using symbols. Figure 5 is an example of an actual peat profile from a bog in New Brunswick. Originally the horizontal scale was 1:4,000, and the vertical scale was exaggerated, being 1:50. This figure is somewhat reduced in size from the original. Such profiles may also display the types of cover and bottom soil. The symbols commonly used are also shown in the figure.

The information that can be calculated from the field data, including the average depth of the deposit, the volume of peat, the volumes of the various types of peat, the engineering properties, and the results of laboratory analyses, can be shown in tables and graphs.

The results of pollen analyses are shown in the form of diagrams which follow a fairly well-standardized international convention. Figure 6 is an example of a typical pollen diagram from New Brunswick (Korpijaakko and Radforth 1972).

REMOTE SENSING METHODS

Although the traditional method of inventory has been the ground survey, the large areas to be covered and the problems associated with travelling over muskeg have necessitated the use of a remote sensing method. Begun from the direct use of aerial photographs, this has now developed to the state that the characteristics of the underlying terrain are gathered by such sophisticated equipment as radar line scanners, magnetometers, infrared cameras, and thermal scanners.

Methods Based on Photography

At first, only the traditional methods of aerial photography were applied to muskeg inventory. Using various types of cameras, from 35mm to photogrammetric cameras with focal lengths up to 305mm on a 23 x 23 cm format, black and white (panchromatic) photographs of the terrain for the whole of Canada have been regularly available since World War II. All of Canada has also been mapped photographically by the Department of Energy, Mines and Resources, at scales varying from 1:5,000 to 1:100,000, and the photographs are obtainable from the National Airphoto Library. Figure 7 is an example of a muskeg area in New Brunswick at the scale 1:15,840.

For interpreting muskeg conditions from the air, a system of Airform Patterns has been developed (N.W. Radforth, 1955b, 1956a, 1956b, 1958). This system is based on the observation that certain conditions in peatland correspond to distinct patterns discernible from the air. When the conditions and smaller units contributing to their formation are known to the investigator, it is possible to interpret certain features of peatland from the air or aerial photographs.

It has been realized that panchromatic photographs are of limited use in detailed inventories, so the use of colour photography has become more common recently. The greatest breakthrough has come with the initiation of multispectral and colour infrared photography. It is possible, by using combinations of film types and filters, to restrict the record on the film to a narrow band of the electromagnetic spectrum. Different surface terrain types reflect at

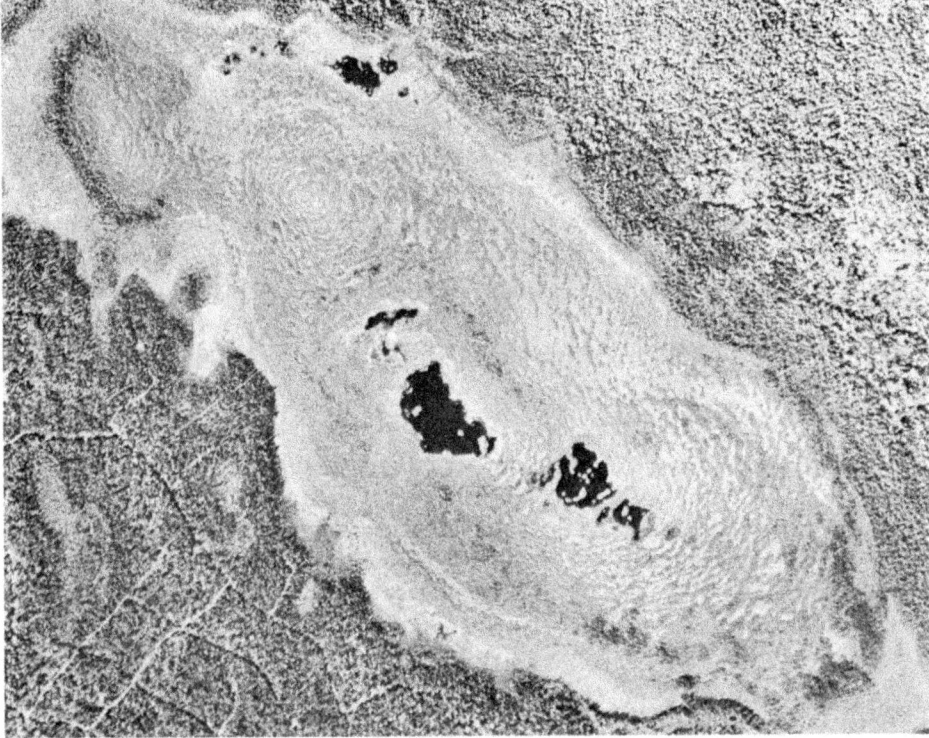

FIGURE 7   A typical raised bog in New Brunswick. Note the typical reticuloid pattern of concentric ridges. Scale: 1:15,840

different intensities in different parts of the spectrum. For example, water on the surface of muskeg, sometimes undetectable on panchromatic photographs, will register solid black in the infrared portion of the spectrum because of its absorption of the infrared radiation (Figure 8). These variations can be further enhanced by using multispectral photography with suitable filter combinations, a system which has been brought to public attention by the publication of photographs produced by the Earth Resources Technology Satellite (ERTS-1) now orbiting the earth.

Multispectral research is at the moment oriented towards finding the ideal film-filter combinations for detecting different terrain types and their states. Only a small number of works devoted to the use of infrared in combination with other film types for peatland studies have been published (Korpijaakko and Radforth, 1972; Tarnocai, 1972).

Methods Based on Electronic Recording Methods

Advanced as photographic methods may appear, in comparison with a ground survey, they have two major drawbacks: (1) they can be obtained only under good photographic conditions, i.e. they have to be taken during the daytime and when no mist or clouds are present; (2) they show nothing directly of the subsurface characteristics of the terrain.

In an effort to overcome the first problem, systems by which cloud cover can be penetrated and which are not dependent on reflected sunlight have been developed. Among these are the various types of radar (e.g. PPI and SLAR). Such systems are based on the fact that all bodies with a temperature above absolute zero emit radiation, and that this radiation can be detected by the sensor. It has been proved that this technique not only can be used in all weather 24 hours a day, but also is useful in organic terrain inventory. The specific heat of organic terrain is higher than that of most mineral terrain types and so muskeg, if covered at night by radar methods, will show up brighter than the surrounding mineral terrain.

The skin penetration of the radar into the ground is but a few centimetres at most, and in an effort to get further penetration research is now being oriented towards the use of airborne magnetometers, side-scanning sonar, and other sensors which show some degree of ground penetration. Comparatively little work in the muskeg inventory field has been done in this area, but significant steps should be made in the future.

COMPUTERIZED METHODS

The use of remote sensing methods makes it possible to accumulate great quantities of data in a very short time. However, it is impossible to process these data manually fast enough for the system to be fully efficient, so that resort must be had to computerized and automated methods. Such methods can be divided conveniently into three categories: (1) computerized and automatic methods of detection, (2) computerized methods of analysis, (3) computerized methods of display.

Computerized Methods of Detection

The large numbers of aerial photographs available have made the task of qualified photo-interpreters, which are few, difficult. Often much time is wasted in sifting through photographs which do not cover any of the features under study. In an effort to overcome this problem, some research has been oriented towards the automatic detection of muskeg from aerial photographs (Woolnough, 1970). This technique utilizes the special characteristics of the pattern displayed by muskeg on aerial photographs such as the unique pattern of a raised bog (Figure 7). Such patterns register uniquely if scanned by a microdensitometer. By programming pattern recognition techniques (which themselves may vary in the degree of simplicity) on the computer, it is possible to achieve automatic recognition of muskeg from an aerial photograph. Eventually, these techniques may be refined so that a complete detection system for muskeg, and even its classification, will be possible with partially, or even fully automatic methods, depending on the required degree of precision (Woolnough et al., 1972).

Computerized Methods of Analysis

The amount of data obtainable from a single aerial photograph, although large in itself, is small compared with that obtained from a series of multispectral photographs of a given area. The examination and comparison of different levels of the spectrum can only be a viable proposition if computer techniques are used. Such techniques as level slicing and mixing of data from different spectral bands are a few of the methods of computerized analysis.

On a completely different scale, the amount of data now being collected by investigators

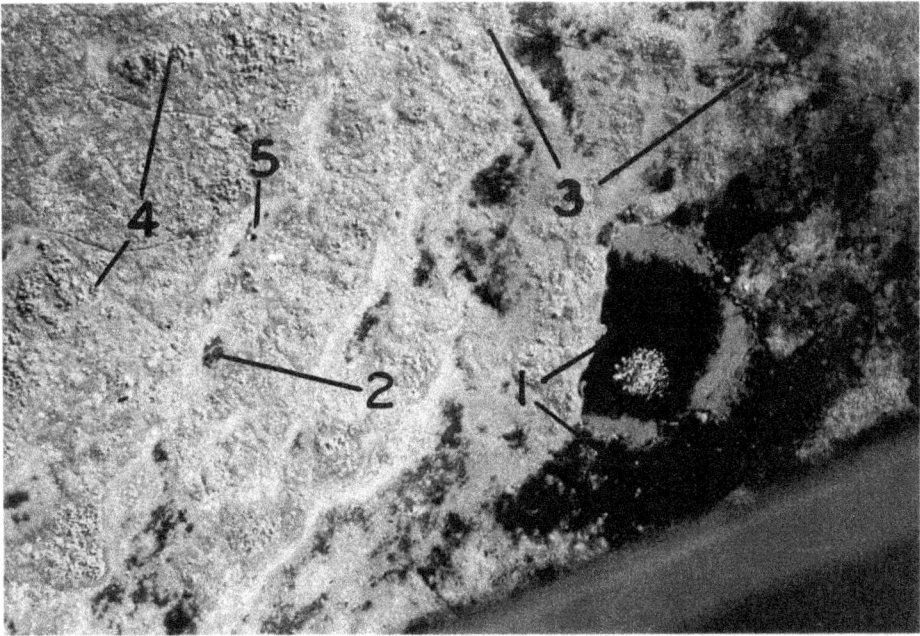

FIGURE 8 Colour infrared photograph of a bog in New Brunswick. Note the almost total absorption of infrared radiation by the water (1). In area 2, the pond is filled with floating vegetation. Number 3 points to a vehicle track, which is clearly visible because of its high water content. Yellowish to orange areas are covered mainly by Sphagna. Number 5 points to a very small (2x6ft) area of algal growth on dug-out peat (greenish tinge). Number 4 points to dead *Picea mariana* (bluish), the general surrounding area being covered by healthy shrubby vegetation (bright red). (E.O. Korpijaakko and J.R. Radforth, 1972).

throughout Canada can be kept in a manageable form only in a computerized system which stores the data in the computerized file in a readily retrievable form.

Computerized Methods of Display

Although a set of statistics on muskeg inventory, as discussed previously, is a desirable product, it is very often advantageous to have some kind of visual display. The traditional method has been to draw the locations of the peatland areas on maps of appropriate scale. Although such products are of considerable value in detailed planning, their procurement may take too long to give a quick preliminary appraisal of the situation, and they are not in a form suitable for speedy computerized analyses.

It is possible to obtain line and pollen diagrams, for example, by computer using standard computerized routines and a plotting system interfaced with the computer. Most computer systems have such an arrangement, usually with a drum or flatbed plotter. Here again, however, the speed is still slow and often the accuracy is too great for a quick visual check. To overcome this, two different methods of computerized output have been generally used. The maps required can be displayed on an oscilloscope unit, and, if a hardcopy is required, this can be obtained by either taking a polaroid photograph of the oscilloscope screen or by linking the oscilloscope to a copying machine to obtain paper copies.

Of somewhat greater range and ability is a system that uses the computer's line printer to produce presentable maps. One such system is SYMAP. Figure 9 is an example of a preliminary survey of peatlands in New Brunswick carried out by using this system. This system makes it possible to manipulate the data before computer handling and thus allows the operator to achieve variations in the end product to obtain results that suit his specific purposes. The tools required to produce the map shown in Figure 9 were 1:15,840 aerial photographs of New Brunswick, 1:250,000 topographic maps for determining the coordinates of sample sites in the UTM (Universal Transverse Mercator Grid), a dot planimeter for measuring the relative surface areas, a mirror stereoscope, and finally an IBM System/360 Computer for procuring the map by the SYMAP program. This map gives a quick and quite accurate display of the percentages of peatland in New Brunswick and is so far the only one of its kind in Canada. It is published here with permission of the Department of Natural Resources, Province of New Brunswick. The method and a few more maps will be published at a later date. This system is also now being extended to cover other areas of Canada to give a comprehensive preliminary survey of peatlands.

SYMAP is adaptable to a great variety of land inventory data. Data can also be displayed in a block diagram form to show subsurface features in a quasi-three-dimensional view. Methods to produce graphic displays like these and others are still under development, and automation of the whole system from data collection to production of the final map by providing a linkage between such instruments as the microdensitometer and the computer is possible. This will further speed up the processing of the enormous volumes of data that can be collected by modern methods.

SUMMARY AND CONCLUSIONS

Peatlands, as a natural resource with a high potential in Canada, cover large areas of land beyond the last frontier between intensive civilization and the relatively untouched wilderness. There is a need for a thorough inventory of this resource to ensure that these lands are properly utilized with no unnecessary disruption of natural processes, as has too often happened in densely populated areas situated farther south. Man should learn to live in harmony with all aspects of the natural environment and to conform to nature rather than to conquer it. Advanced technology has enabled man to conquer nature easily and has given him the power even to destroy it. At the same time, however, highly advanced technology has released man from the need for 'das Kampf ums Dasein,' enabling him to direct his attention to matters other than mere survival. It also should have released him from the burden of the old instinct to conquer. The technological advances which should help him in this respect include, among others, the new remote sensing techniques which enable him to observe large areas of the earth and with sensors that go beyond his own sensing capability.

In addition to this, techniques of dealing with data from the electronic sensors and highly developed 'electronic brains' — computers — enable him to perform calculations, store information, retrieve it, process it and manipulate it at rates and volumes hitherto unknown, helping him to survey and inventory his environment more accurately than ever before. However, even with the remote sensing techniques that have been adapted to peatland uses he still has to do much footwork and deal with final details of terrain with methods only usable by him in the field. This brings us back to the earth and to the methods described earlier in this paper. These methods encompass only the most important basic features to be surveyed. The reference to various classification systems reveals that there is still far from a common language for those dealing with peatlands. Consequently, methods and lists of objects to be surveyed can be, and should be, amended if specific needs so dictate. However, the data collected as described can be applied to a wide array of quite specific user needs.

Finally, to mention a few examples cited, to include data on peat type and its degree of humification may tell quite a lot about the suitability of the area for agriculture, forestry or peat mining. When supplemented with a nutrient analysis, the answer will become quite definite and meaningful for foresters and agronomists. Peat depth and surface contour maps will provide information for various purposes and peat mining procedures. When supplemented with knowledge of peat type, and the location of stumps and other woody erratics, planning for mining and drainage operations can be finalized.

Knowledge of the distribution of peat types, their depths, water regime and cover type, when supplemented with engineering data such as shear strength and permeability, will help directly in construction planning. Thus the generalized survey procedures described above may be applied in a great number of fields. It is hoped that this contribution will assist investigators in deciding on the methods which should be used for their surveys and inventories to serve their specific needs best.

REFERENCES

v. Bülov, K. 1925. Moorkunde (Sammlung Göschen, Berlin).
Coupal, B. 1972. Use of peat moss in controlled combustion technique. EPS4-EE-72-1, Environmental Emergency Branch, Environment Canada.
Day, J.H. 1968. The classification of organic soils in Canada. Proc. Third Internat. Peat Congr., Quebec.
Department of Agriculture and Fisheries for Scotland. 1968. Scottish Peat Surveys, vol. 4.

MUSKEG COVERAGE AS A PERCENTAGE
OF TOTAL LAND AREA IN NEW BRUNSWICK

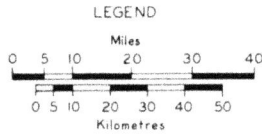

LEGEND

Miles
0 5 10   20   30   40

0 5 10   20   30   40   50
Kilometres

0 - 20%

20 - 40%

40 - 60%

60 - 80%

80 - 100%

(Maximum included in the higher level only)

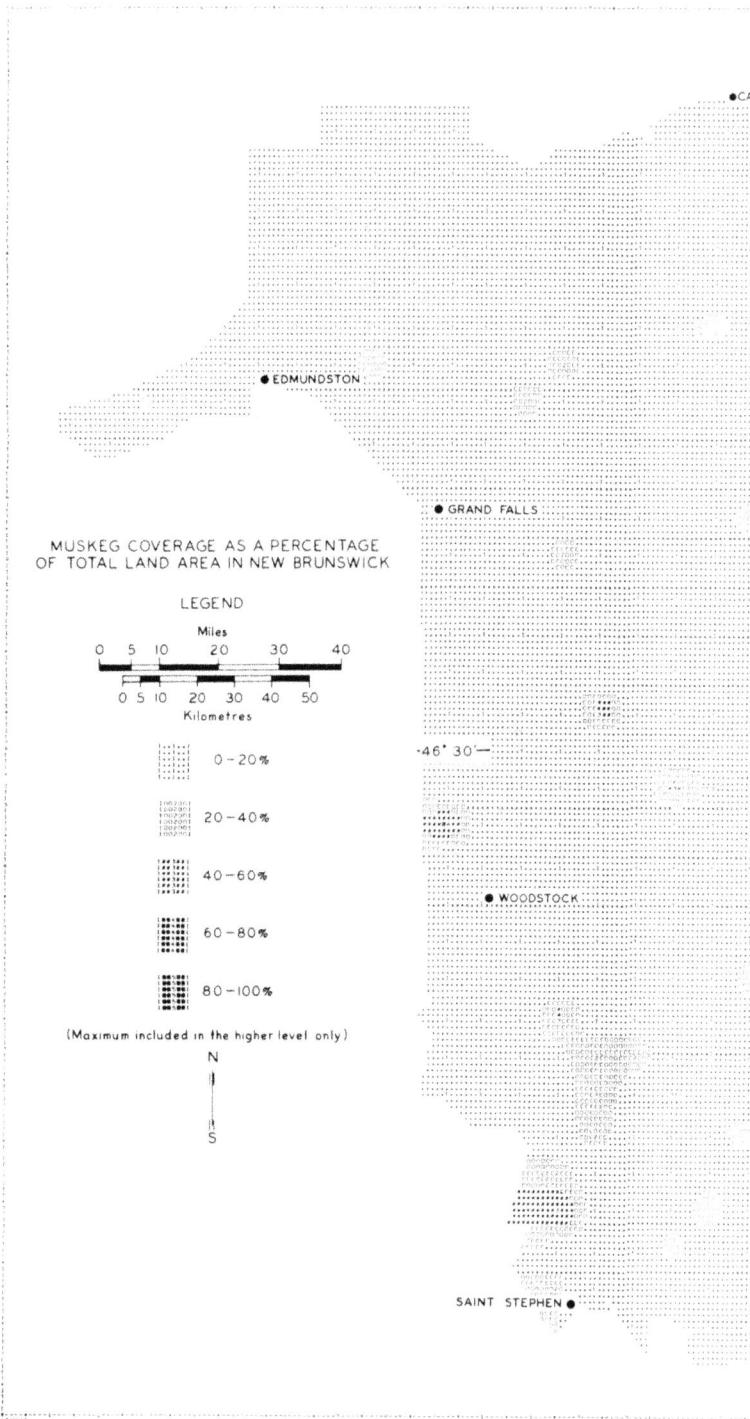

N

S

EDMUNDSTON

GRAND FALLS

-46° 30'—

WOODSTOCK

SAINT STEPHEN

CHALEUR BAY

● SHIPPEGAN

● BATHURST

MIRAMICHI BAY

POINT ESCUMINAC

● NEWCASTLE

NORTHUMBERLAND STRAIT

—46° 30'—

● SHEDIAC

● MONCTON

GRAND
LAKE

BAY OF FUNDY

● SAINT JOHN

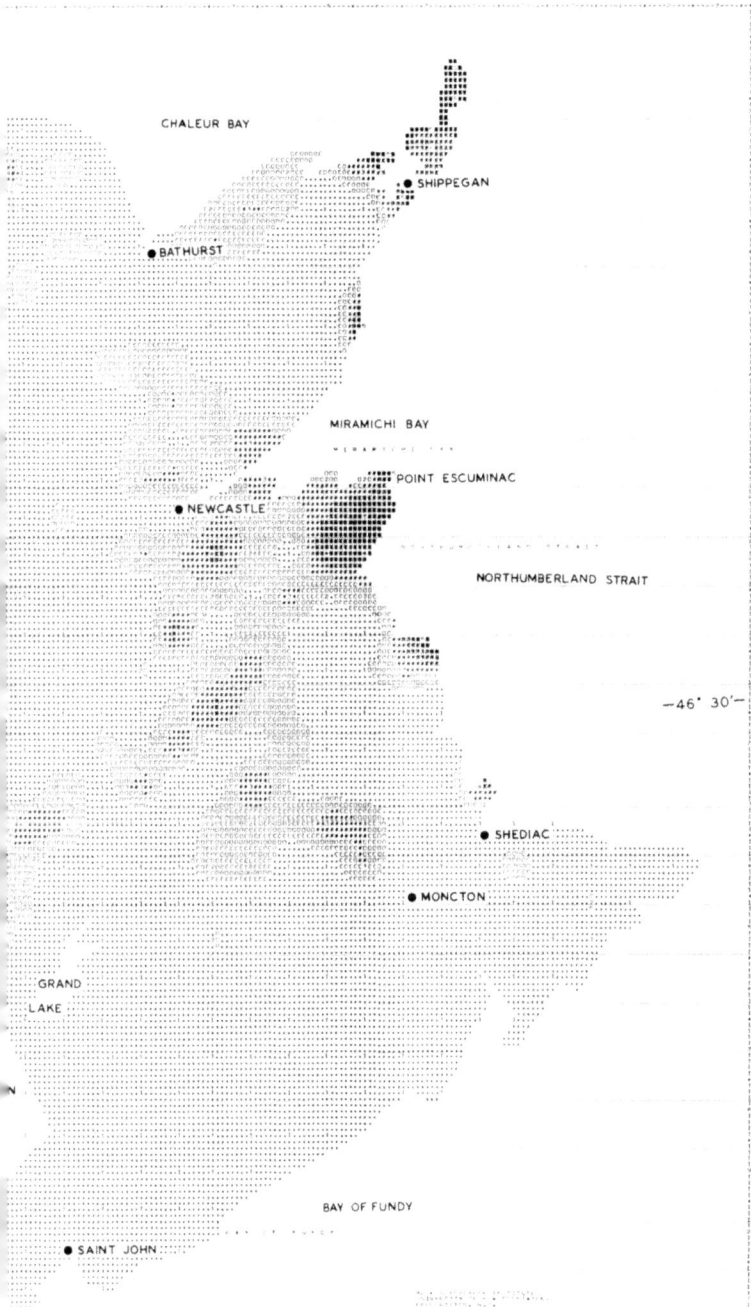

ERKKI KORPIJAAKKO
MUSKEG RESEARCH INSTITUTE
UNIVERSITY OF NEW BRUNSWICK
1973

Heikurainen, L. 1960. Metsäojitus ja sen perusteet (WSOY, Helsinki).

— 1968. Peatlands of Newfoundland and possibilities of utilizing them in forestry. Forest Res. Lab., St Johns, Newfoundland, Information Rept. N-X-16.

Kivinen, E. 1948. Soutiede (WSOY, Helsinki).

Korpijaakko, Erkki, and Radforth, J.R. 1972. Multispectral photography in the prediction of peatland conditions. Proc. Fourth Internat. Peat Congr., Helsinki.

Korpijaakko, Maija-Leena, and Radforth, N.W. 1972. On postglacial development of muskeg in the Province of New Brunswick. Proc. Fourth Internat. Peat Congr., Helsinki.

Korpijaakko, Martti, and Radforth, N.W. 1972. Studies on the hydraulic conductivity of peat. Proc. Fourth Internat. Peat Congr., Helsinki.

Pollett, F.C. 1972. Classification of peatlands in Newfoundland. Proc. Fourth Internat. Peat Congr., Helsinki.

Radforth, J.R. 1969. Preliminary engineering investigation. In Muskeg Engineering Handbook, ed. I.C. MacFarlane (Univ. Toronto Press).

Radforth, N.W. 1952. Suggested classification of muskeg for the engineer. Eng. J. 35/11.

— 1955a. Range of structural variation in organic terrain. Trans. Roy. Soc. Can., 3rd ser., sect. V, vol. XLIX.

— 1955b. Organic terrain organization from the air (altitudes less than 1,000 feet): Handbook No. 1. Defence Res. Board, Dept. Nat. Defence, DR 95.

— 1956a. The application of aerial survey over organic terrain. Proc. Eastern Muskeg Res. Meeting, NRC, ACSSM Tech. Memo. 42.

— 1956b. Muskeg access with special reference to the petroleum industry. Trans. Can. Inst. Min. Metall. Petrol. and Nat. Gas Div. 59.

— 1958. Organic terrain organization from the air (altitudes from 1,000 to 5,000 feet): Handbook No. 2. Defence Res. Board, Dept. Nat. Defence, DR 124.

— 1962. Organic terrain and geomorphology. Can. Geog. 6/3-4.

Ruuhijärvi, R. 1960. Über die regionale Einteilung der nordfinnischen Moore. Acta Bot. Soc. 'Vanamo' 31/1.

Sarasto, J. 1961. Kokeita turpeen läpäisevyydestä. Suo no. 12.

Tarnocai, C. 1972. The use of remote sensing techniques to study peatland and vegetation types, organic soils and permafrost in the boreal region of Manitoba. Proc. First Can. Symp. Remote Sensing, Ottawa.

Woolnough, D. 1970. Preliminary Investigation into Automatic Recognition of Muskeg from Aerial Photographs (M.Sc. thesis, Dept. Surveying Engineering, Univ. New Brunswick, Fredericton).

Woolnough, D., Korpijaakko, E., and Radforth, N.W. 1972. Automatic universal resource mapping as applied to muskeg. Proc. Fourth Internat. Peat Congr., Helsinki.

4
Physical and Chemical Properties of Peat

M.E. WALMSLEY

The term 'muskeg' has been and is still used in a general sense referring to a particular landscape feature or terrain type characterized by a relatively thick layer of organic materials, the origin of the organic components being dominantly the particular plant species growing in the area. Radforth (1952), however, suggested that muskeg is organic terrain comprising living cover and peat (however shallow or deep). Organic terrain or muskeg has been subdivided into environmental divisions through the use of words such as swamp, bog, fen, heath, marsh, mire, highmoor, and lowmoor. In discussions of muskeg ecology some of these terms are required to distinguish real and important differences among highly complex ecological situations. In the context of the boreal and subarctic regions throughout the northern hemisphere, bog and fen (Sjörs, 1948, 1950) are considered the major types of organic terrain or muskeg. The terms used in this chapter are similar to those described by Tarnocai (1970).

To the pedologist peat is the parent material formed by the accumulation of organic remains. Organic soils are the result of pedological processes, such as leaching and decomposition, acting upon the peat material and resulting in the formation of a particular soil profile. In particular classifications such as the Canadian System of Soil Classification (1970), the organic matter content and the thickness of the deposit necessary to qualify an organic layer as an organic soil are fully described. Classification systems such as this are cognizant of the physical and chemical characteristics of organic soils in relation to environmental factors. A particular depth is required for a soil to be termed organic, the depth depending on whether the soil is drained or not, since considerable shrinkage will take place as soon as drainage is initiated.

In general, the same edaphological principles hold for both organic and mineral soils. For example, the decomposition of organic soil material may eventually result in the formation of humus, humus being the homogeneous colloidal mass produced as the final product of soil organism and microbial activity. Processes such as lignoprotein unions and polyuronide formation take place in both organic and mineral soils during humus formation.

An idealized volume representation of organic soils as presented in Figure 1 illustrates the four major components of organic soils: organic matter, mineral material, water, and gas. A

MINERAL 1%

ORGANIC 7%

GAS   6%

WATER 86%

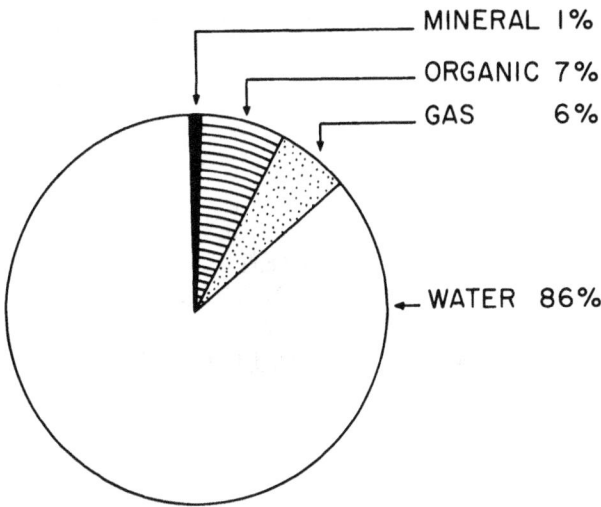

FIGURE 1   Volumetric composition of an idealized organic soil.

total pore space of approximately 92 per cent (gas and water) illustrates the need for compaction of peat rather than cultivation for good structural management. Cultivation may destroy the granular structure of the material whereas compaction will allow the roots to come into closer contact with the soil. The proportion of air and water is subject to great fluctuations under natural conditions, depending on the climate among other factors.

The physics of such an association of components defines the amount of mixing that will take place and in turn either encourages or discourages simple and complex chemical reactions to occur between the groups. An intimate mixture of the four components, as usually found in a good mineral topsoil, is generally required for the optimum growth of most plants. Only through a thorough knowledge of the physical and chemical characteristics of organic materials will an understanding of the implications of the development of muskeg for the environment be understood.

Two particular subjects that are not touched on in this chapter and which may be considered important physical properties of muskeg are the thermal characteristics and the stress deformation characteristics. Both of these subjects were covered thoroughly by MacFarlane (1969) and discussion here would not add much to that article. Although there is a section on the unique temperature characteristics of muskeg, the volumetric specific heat, the volumetric latent heat of fusion, the thermal conductivity, and the thermal diffusivity receive little attention. The stress deformation characteristics that are considered important include shear strength, tensile strength, strain characteristics, bearing capacity, consolidation, and settlement.

In general terms, this paper is centred around the following topics: (1) the fractionation of peats by chemical techniques, (2) the chemical aspects and considerations of organic materials, (3) selected physical aspects of organic materials in relation to environmental effects.

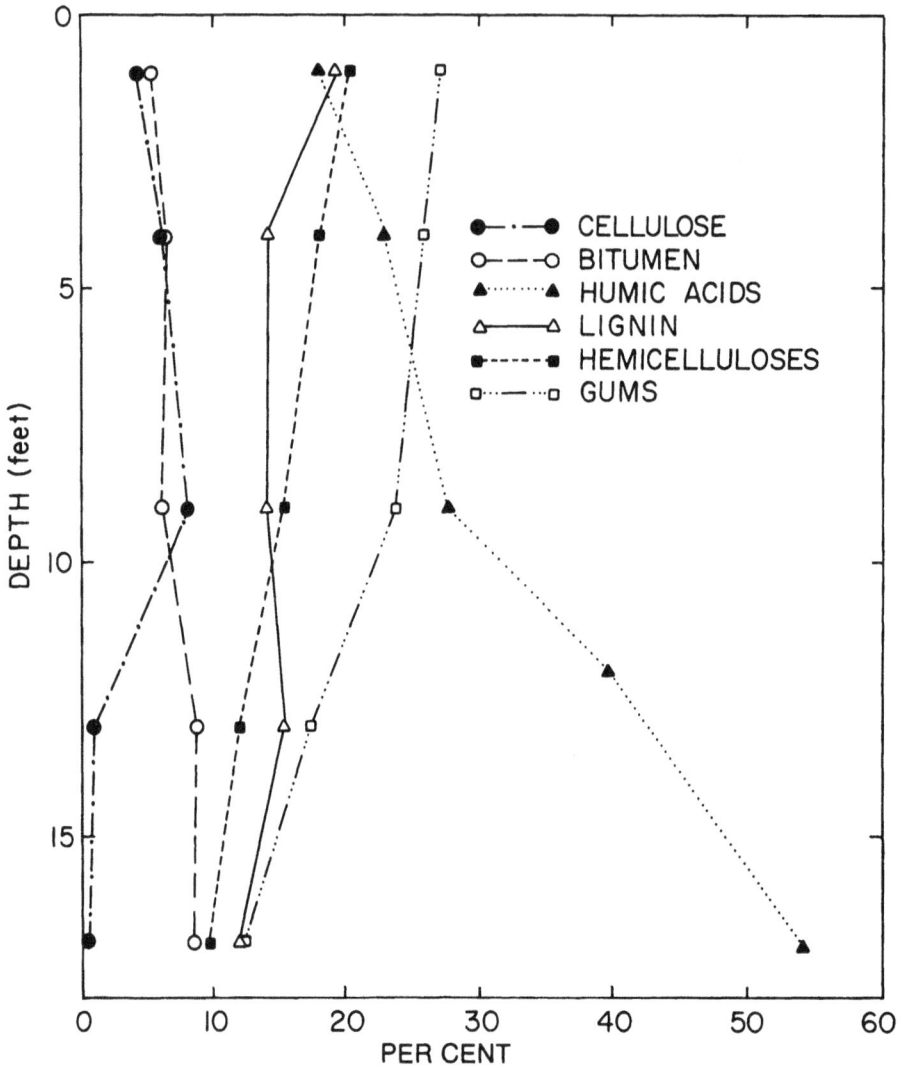

FIGURE 2   Concentration of organic constituents in an organic soil with depth (after Smith et al., 1958).

CHEMICAL PROPERTIES OF PEAT

Organic Constituents

Raw material for the formation of peat consists of cellulose, lignin, and cork-like tissues which are the main constituents of plants. Other substances include resins, waxes, proteins, dyes, and so on. Vasil'ev and D'Yakova (1958) reported the chemical components of one peat sample as follows: water-soluble material, 5 per cent; humic acids, 41.4 per cent;

1. Sugars, starches and simple proteins

2. Crude proteins

3. Hemicelluloses

4. Celluloses

5. Lignin, fats, waxes, etc.

Rapidly
decomposed

slowly
decomposed

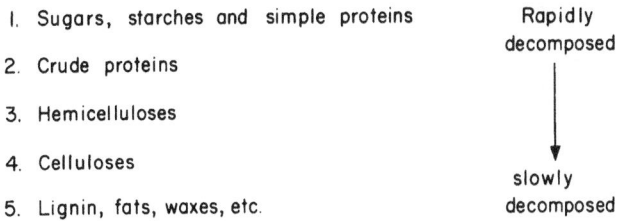

FIGURE 3   The resistance of various organic compounds to decomposition (after Buckman and Brady, 1969).

cellulose, 5.6 per cent; hemicellulose, 22.5 per cent; lignin, 10.5 per cent; bitumen, 12.9 per cent; and other substances, 2.1 per cent. Kaganovich and Rakuskii (1958) reported that a typical peat bitumen contained 52.5 per cent wax, 20.4 per cent paraffins, and 24.6 per cent resins. The proximate chemical analyses of various samples from peat bogs in eastern Canada illustrate the distribution of various organic fractions with depth (Smith et al., 1958). Figure 2 illustrates the results as reported. Assuming that decomposition increases with depth, it becomes evident that humic substances are formed by the decomposition of the carbohydrates present, especially gums, hemicelluloses, and possibly lignin.

Mineralization or decomposition of organic materials tends to be an ongoing process producing either new compounds or the formation of carbon dioxide, water, and energy. The organic matter content of a soil will either increase or decrease until a steady state is reached (Broadbent, 1953; Jenny, 1941). Figure 3 shows the relative resistance of various organic compounds to oxidation according to Buckman and Brady (1969). As would be expected, the more resistant groups accumulate with time. Buckman and Brady (1969) group the newly synthesized complexes together and refer to them as humus. It is the highly colloidal nature of this material that is significant in determining the chemical and physical properties of organic soils.

Although little is known about the detailed composition of peat waxes, even less is known about the substances constituting peat resins. It is believed that paraffin predominates in the wax fraction but peat also may contain aromatic hydrocarbons (Volarovich and Gusev, 1956). Substances such as hydrocarbons, hydroxy compounds, and small amounts of acids are known to occur in peat resins but little has been done toward their elucidation. Humic acids, which constitute the bulk of organic material in peat, are considered to be dark amorphous substances, readily soluble in dilute aqueous alkaline solutions. They are acid by nature and are formed as the primary products of the decomposition of plant material. Their molecular weights are known to range from 675 to 9,000. Nitrogen and sulphur form part of the humic acid structure as well as carbon, hydrogen, and oxygen (Wollrab and Streible, 1969). Although attempts have been made to elucidate the structure of humic acids, few concrete results have been obtained. Water-soluble material in peat includes aromatic hydroxy or methoxy compounds, ketones, and acids. In some cases diamino and monamino acids have been found in peat extracts (Shacklock and Drakeley, 1927).

Various schemes have been used for the fractionation of peat into distinct chemical groupings. Waksman and Stevens (1928a, 1929, 1930) used a combined scheme of solvent extraction and hydrolytic techniques (see Figure 4). The major drawback to this scheme appears to be the fact that the 'lignin' fraction is still partially soluble in alkaline reagents, and

PEAT (air dry)

Ethyl Ether

soluble

Lipids

95% Ethyl Alcohol

soluble

Lipids, tannins,·etc.

Cold and Hot Water

soluble

Free organics, starch, protein, tannins, pectins

2% HCl, reflux

soluble

Hemicelluloses

80% H_2SO_4 , cold
6% H_2SO_4 , reflux

soluble

Cellulose   Lignin (insoluble)

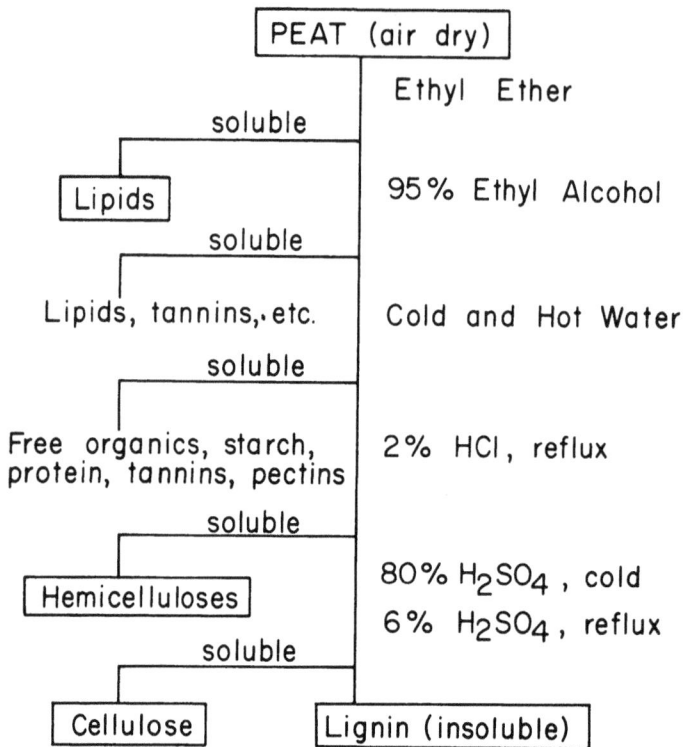

FIGURE 4   Fractionation scheme for organic soils (after Waksman and Stevens, 1928).

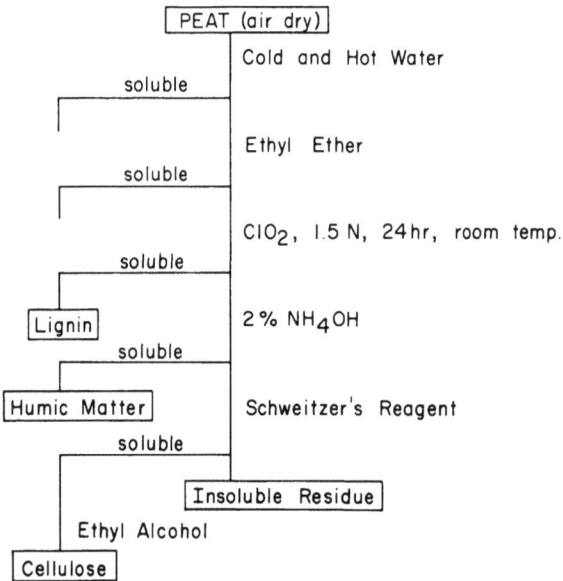

PEAT (air dry)

Cold and Hot Water

soluble

Ethyl Ether

soluble

ClO_2, 1.5 N, 24hr, room temp.

soluble

Lignin   2% NH_4OH

soluble

Humic Matter   Schweitzer's Reagent

soluble

Insoluble Residue

Ethyl Alcohol

Cellulose

FIGURE 5   Fractionation scheme for organic soils (after Thiessen and Johnson, 1929).

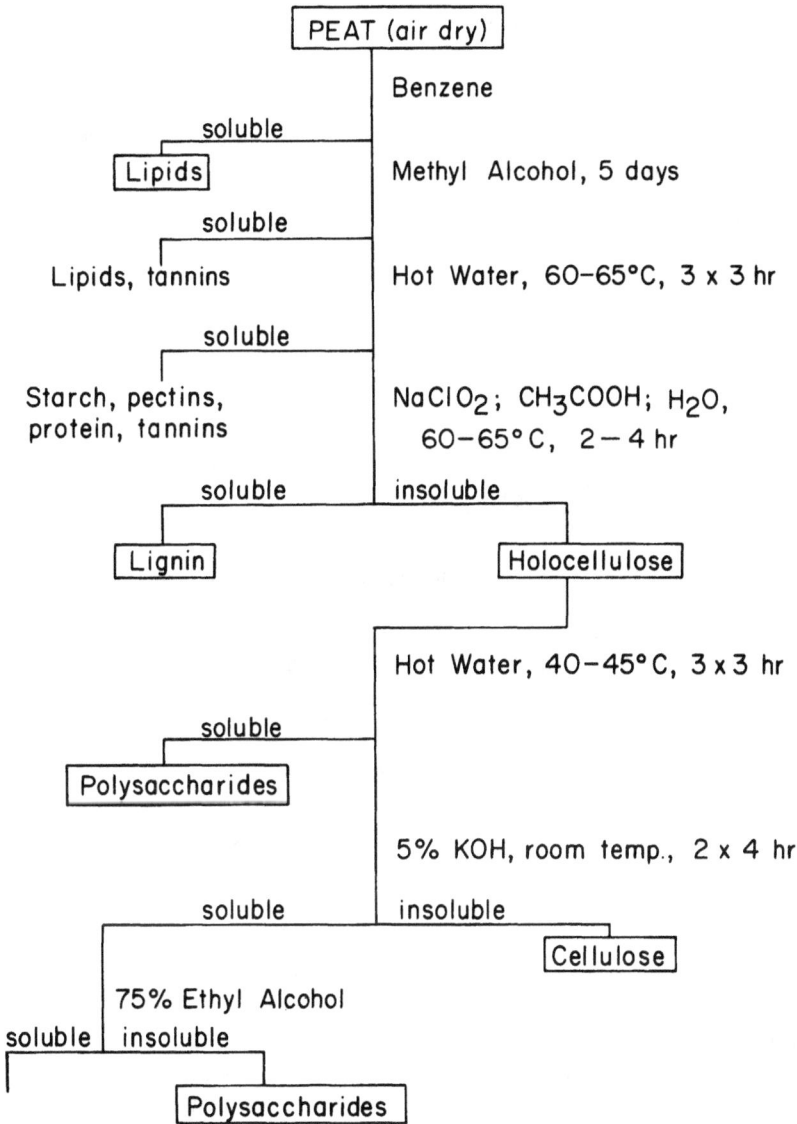

FIGURE 6   Fractionation scheme for organic soils (after Theander, 1954).

no attempt has been made to distinguish this 'humic acid' material from the lignin. About the same time Thiessen and Johnson (1929) proposed a scheme (Figure 5) which fully recognized the importance of humic acids. They were able to solubilize lignin through prolonged oxidation with chlorine dioxide and obtained a certain cellulose product. The major criticism of the scheme appears to be that the sequence of extraction leads to some ill-defined fractions. Theander (1954) applied a scheme very similar to that of Thiessen and Johnson except that he gave no consideration to the possible presence of humic acids (see Figure 6). In 1961, Parsons and Tinsley adopted a completely different approach to fractionation. Their

PEAT (air dry)

98 % HCOOH; 0.2 N Li Br
reflux   2 x 1/2 hr

soluble | insoluble

diisopropyl ether   Cellulose, lignaceous matter,
                          high mol. wt. humic acids

soluble | insoluble

Polysaccharides, protein, low
mol. wt. humic acids, nucleoprotein

0.5M Li Cl / Li₂CO₃ (acqueous)

soluble | insoluble

Polysaccharides, humates,   Humic acids, nucleoprotein,
      protein   protein

20% (W/V)   Cetavalon

soluble | insoluble

Polysaccharides, humates   some humic acids, proteins, etc.

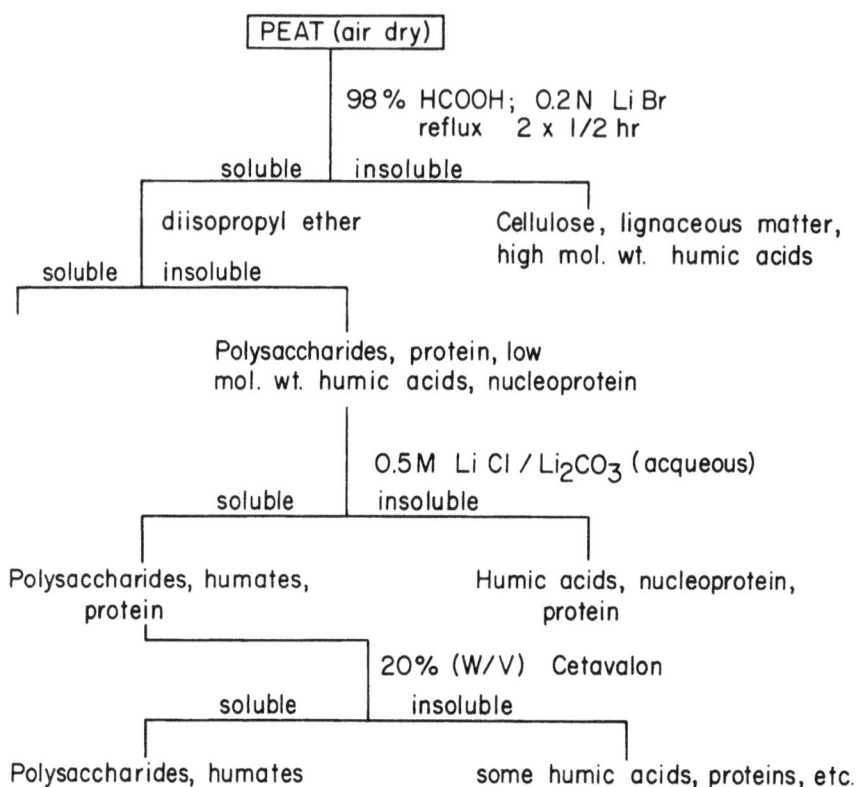

FIGURE 7   Fractionation scheme for organic soils (after Parsons and Tinsley, 1961).

scheme, shown in Figure 7, involved an initial extraction with a mixture of boiling anhydrous formic acid and lithium bromide. This procedure, although possibly good for polysaccharide analysis, is believed to produce ill-defined extracts since the action of boiling anhydrous formic acid on organic matter is not fully understood. One of the most promising schemes is that proposed by Passer et al. (1963), which seems to include the best features of the previous work. The main objection to the scheme (Figure 8) seems to be in its application to sediments of low organic matter content or where the polysaccharide content is low. Schlungbaum's (1968) fractionation scheme (Figure 9) combines the hydrolytic procedures of Waksman and Stevens (1928a, 1929) with many solvent-extraction steps. The result of this is a good scheme for quantitative characterization but one that is undesirable for preparing fractions for further chemical study. More recently, Lucas (1970) proposed a scheme for peat fractionation. He recognized that the more recent fractionation schemes for peat made no attempt to isolate all fractions from the same original sample. Fractions isolated from different samples would not seem to be significant for further chemical analysis. The scheme is given in Figure 10.

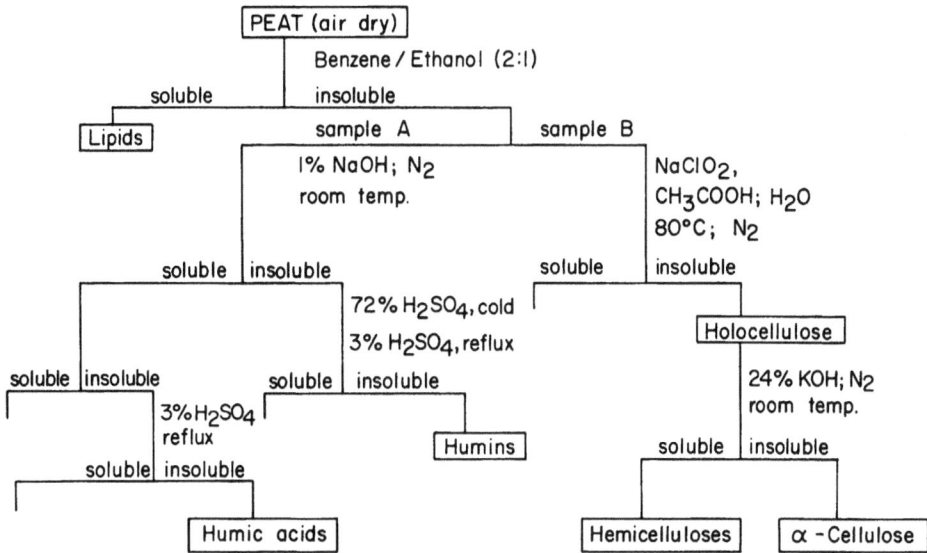

FIGURE 8   Fractionation scheme for organic soils (after Passer et al., 1963).

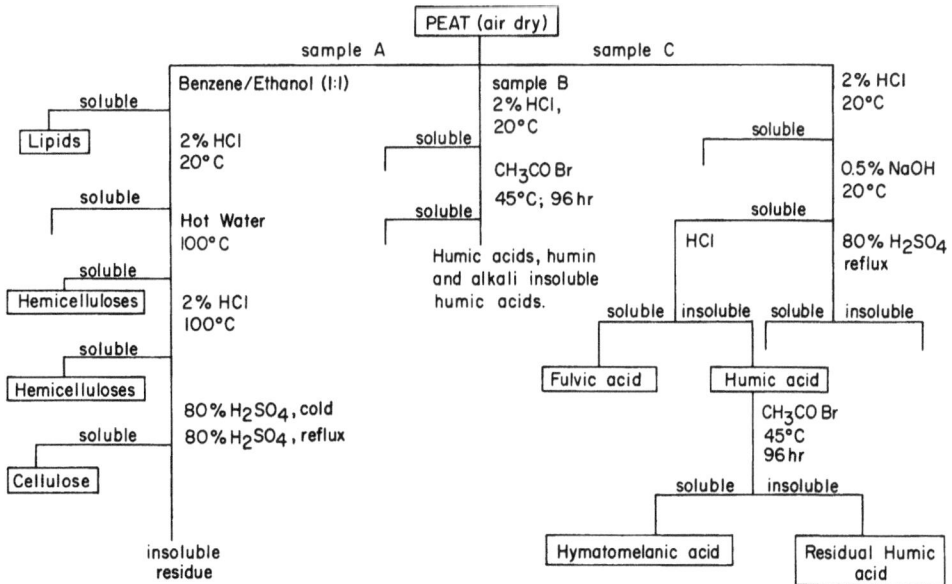

FIGURE 9   Fractionation scheme for organic soils (after Schlungbaum, 1968).

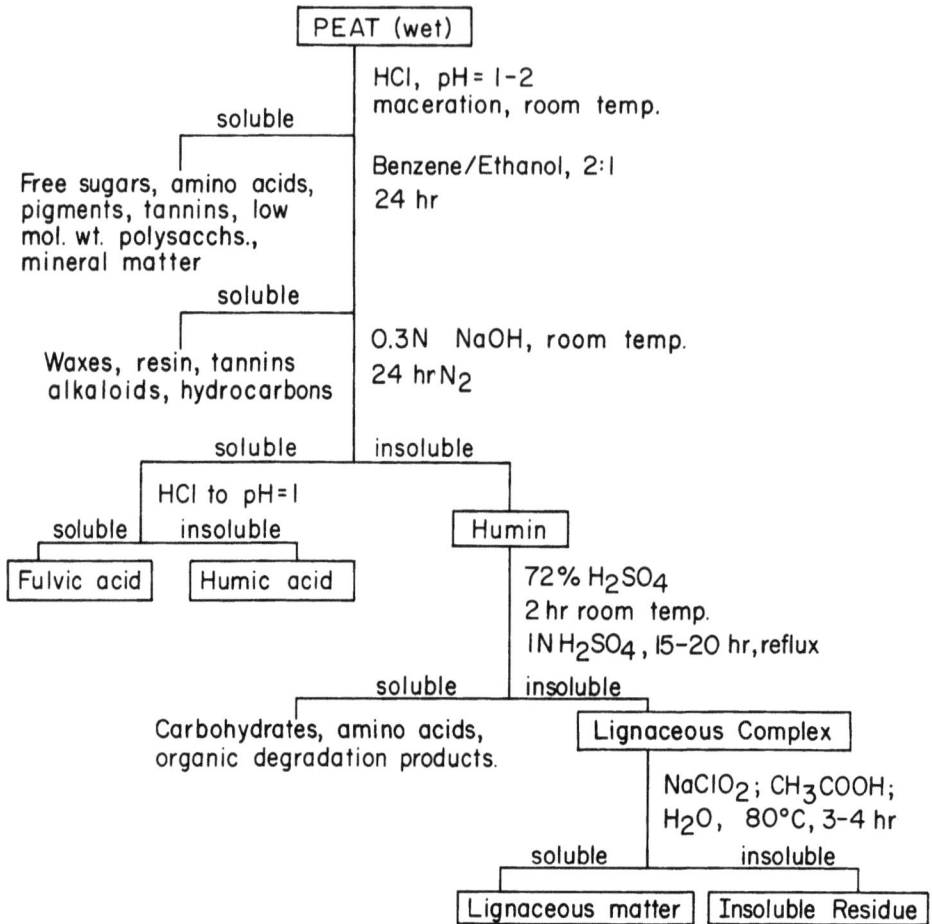

FIGURE 10   Fractionation scheme for organic soils (after Lucas, 1970).

Ash Content

The mineral constituents of peat are derived dominantly from the original peat-producing plants as well as from extraneous sources such as water flowing from mineral soils into bogs and atmospheric dust in the form of wind-borne minerals deposited by rainfall. This fraction of peat is necessarily incombustible and ash-forming. Values as low as 1 per cent, based on an oven-dry weight, have been measured for peat that is mainly free of extraneous mineral matter, with every gradation from this minimum to the point where the soil is no longer considered to be predominantly peat. Appendix 2 shows that values range anywhere from near 1 per cent in fibric undecomposed material to over 40 per cent in highly decomposed peat under cultivation.

In most peat types, there is a positive correlation between the ash content and the degree of decomposition. This is mainly due to the accumulation of mineral elements as a result of mineralization during decomposition. Yefimov (1961) showed that in acid peats there is generally little enrichment of the peat with ash elements because they are absorbed by plants or may possibly become mobile and be removed from the profile.

The usual procedure for determining the ash content is to fire an oven-dried sample in a muffle furnace at 700°c for three hours and express the loss on ignition as a per cent by weight of absolute dry peat. Several authors have indicated that loss in weight may occur at certain temperatures because of the disintegration of specific substances. Schnitzer and Hoffman (1966) explained that a 540°c loss in weight is due to the disintegration of calcium-organic complexes, whereas at 700°c the loss of weight is attributed to the disintegration of carbonates.

In engineering practices, the ash content has commonly been used to determine the organic content of peat. Because of the laborious nature of procedures such as the Walkely-Black method (Black et al., 1965), the organic matter content (ignition-loss ratio) is equated to the difference between 100 per cent and the ash content. Errors of as much as 5 to 15 per cent can result using this method, since more than organic carbon can be combusted during the heating process.

Organic Carbon and the Carbon:Nitrogen Ratio

Table 1 illustrates that the organic carbon content (total soil basis) is similar for different peat types. Values range normally from 27 to 53. The usual laboratory procedure is the dry combustion method, where carbon dioxide is absorbed in an excess of sodium hydroxide, which is precipitated as barium carbonate and then back-titrated with hydrochloric acid. A much simpler procedure, and one that is being used more often, is to use a Leco Induction Furnace (Leco Corporation, 1954) which also combusts the dry sample, but allows for a direct readout of the percentage of carbon.

Since nitrogen levels are different for different types of peat, it is expected that the carbon:nitrogen ratios would also be characteristic of the peat type. As Appendix 2 illustrates, the carbon:nitrogen ratios of the moss peats are normally greater than 20:1 whereas those of the sedge peats are less than 20:1. Because of the effect of pH on nitrogen release and availability, carbon:nitrogen ratios of 60:1 are common for very acid organic soils.

Total and Available Nutrients

Nutrient contents of peat soils vary with the particular type of peat. On a total soil basis, the total nitrogen values of moss peats range from 0.5 to 2.0 per cent, whereas the nitrogen values of sedge peats range from 2.5 to 3.9 per cent (Davis and Lucas, 1959). The total phosphorus content ranges from 3 per cent in the bog type to 12 per cent in the fen, and the total potassium content is generally 5 per cent in the bog and near 10 per cent in the fen.

Organic soils contain very little nitrogen in inorganic forms, although some nitrate may be present in well-drained soils. The higher degree of decomposition in sedge-grass peats led Waksman and Stevens (1928b) to postulate that it is the decomposition of the plant proteins and the synthesis of microbial cell substances that leads to an accumulation of organic nitrogenous complexes. In contrast to this, the low level of microbial activity and the extremely low content of nitrogen in *Sphagnum* (Gorham, 1953) lead to a low nitrogen content in *Sphagnum* bogs. In many instances, *Sphagnum* peat has less nitrogen than the original plants.

TABLE 1
Organic carbon content as related to peat type (after Padbury, 1970)

Peat type	Organic carbon, % (total soil basis)
Moss-dominated: *Sphagnum-Hypnum*	27.0
Moss-dominated: *Sphagnum*	47.2
Sedge-dominated: *Carex*	49.8
Sedge-dominated: *Carex-Sphagnum*	52.5

The two main factors which influence nitrogen release and availability are the total nitrogen content of the soil and the microbial activity. Both of these factors are in turn influenced by the pH. Organic soils of negligible ash content, with pH values below 4, often have a nitrogen content of less than 1 per cent, but soils with a pH above 5 usually contain over 2 per cent nitrogen. If the pH falls below 5, there is a marked reduction in the nitrate nitrogen released by microbial activity. Turk (1939) illustrated that the nitrifying capacity can be increased four-fold by liming an organic soil with an initial pH below 4. Lucas and Davis (1961) indicated that the favourable pH range is 1 to 1.5 pH units lower than that usually observed for mineral soils. The Kjeldahl method (Jackson, 1958) is normally used to determine the total nitrogen content and the steam distillation method (Black et al. 1965) for available nitrogen.

Unlike nitrogen, phosphorus in peat soils is present mainly in the form of mineral compounds (Yefimov and Donskikh, 1969a,b). The main forms of mineral phosphorus are aluminum, iron, and calcium phosphates. Most of the forms of phosphorus in peat soils are considered immobile, with increasing decomposition causing some of the immobile forms to become soluble. Total and available phosphorus are normally determined by the chlorostanous reduced molybdophosphoric blue colour method (Jackson, 1958).

Organic soils tend to be low in potassium (Davis and Lucas, 1959). This is due mainly to the fact that organic soils contain little mineral matter that is capable of releasing potassium over time or fixing applied potassium (Shickluna, Lucas, and Davis, 1972).

Joffe and Levine (1947) and Jones (1947) have shown that humus colloids are essentially unable to fix potassium, and organic matter cannot fix potassium in a non-exchangeable form. As a result of this, potassium is lost from organic soils as a result of leaching in areas of high rainfall or in fields subject to flooding. Such a situation necessitates the carrying out of many soil tests throughout the year to determine concentration levels, in order to ensure that the land use is the best possible in terms of biological production. The total potassium content is normally determined by dry ashing at 500°C and dissolving the ash in 6N hydrochloric acid, and the available form is extracted with 1N ammonium acetate. Atomic absorption spectrophotometry is commonly used to determine the concentration levels.

Table 2 shows the results of analysing several peat types as published by Pollett (1972). Like the total nitrogen, the total calcium, manganese, and iron content increased from bog to fen, which is no doubt due to the influx of groundwater from mineral soils. This trend is reversed for available phosphorus and manganese, with higher values being recorded from bog soils. Lucas and Davis (1961) believe this may be due to the fixation of phosphorus by iron and aluminum in fen soils, and the increased solubility of manganese at the low pH of the bogs.

TABLE 2
Total (T) and available (A) nutrients of major peat types (after Pollett, 1972) (values expressed in mg/g except available Fe and Zn, which are given in ppm)

Association	N T	N A	P T	P A	K T	K A	Ca T	Ca A	Mg T	Mg A	Mn T	Mn A	Fe T	Fe A	Zn T	Zn A
Blanket bog	6.7	0.036	0.34	0.030	0.58	0.43	1.24	0.54	1.17	0.83	0.060	0.043	0.74	4	0.031	5
Raised bog	8.4	0.034	0.41	0.019	0.37	0.26	3.09	1.74	1.87	1.14	0.045	0.049	0.92		0.030	5
Treed fen	20.9	0.057	0.92	0.012	0.41	0.59	16.13	7.91	1.31	0.51	0.295	0.021	10.07	6	0.044	3
Limestone barren fen	16.0	0.057	0.54	0.012	0.92	0.33	19.40	9.40	2.80	1.43	0.122	0.010	5.02		0.078	2
Heath fen	20.4	0.063	1.03	0.037	1.01	0.67	30.63	11.05	3.21	1.38	0.240	0.013	7.24	5	0.073	3

In frozen organic soils, it has been shown by Tarnocai (1972) that nutrients like calcium are freed from the ice during the process of ice formation and occupy the exchangeable sites on the organic soils, resulting in higher nutrient concentrations and pH in frozen soil.

Cation Exchange Capacity

Although the absorption capacity of peat soils is known to be comparatively high, there is little known about the exchange capacities of various organic or humus types. A range of 131 to 200 me per 100 g is recorded for moss-dominated peats and a range of 100 to 192 for sedge-dominated peats using a barium chloride procedure reported by Padbury (1970). Owing to the importance of the adsorptive complex to the physical and chemical properties of organic soils, it is important to know the factors which determine the magnitude of the exchange capacity of peat. In some cases, such as sedge-grass peats, where there are relatively high concentrations of adsorbed cations, it is possible to have chemical and physical properties which do not correspond to the botanical characteristics of the peat-forming material.

Puustjärvi (1956) believes that the cation exchange capacity of organic material is a result of the substitution of the dissociable hydrogen ions in certain organic groups by other ions. According to Broadbent and Bradford (1952) and also to Schnitzer and Desjardines (1965), the main sources of exchangeable hydrogen in soil organic matter are the carboxyl (COOH) and hydroxyl (OH) groups associated with humic acids and hemicellulose. However, McGeorge (1931) found no relationship between either the nitrogen content or the carbon:nitrogen ratio and the exchange capacity, although he did find a linear relation between the lignin content and the exchange capacity of the soil. Because of this, he concluded that the base exchange compounds of the organic matter are non-nitrogenous, and that lignin was the most likely compound for base exchange.

The fact that many investigators have different opinions about the phenomenon of cation exchange stems from the complexity of the subject. It is the heterogeneity of the material, as expressed by Waksman (1938) when he states that each peat type represents a specific natural humus formation, that causes this complexity. Owing to the differences in the chemical composition of the plants that constitute the origin, each of the several distinct types of peat is characteristic in chemical and physical composition. For example, com-

TABLE 3
Mean cation exchange capacities of 17 organic soils as measured by different procedures (after Mac-Lean et al., 1964)

Procedure	Consecutive replacing solutions	Measurement	Mean CEC (me/100 g)
1	1N NH₄OAc, NaCl	NH₄ by distillation	163
2	0.5N HCl, 1N NH₄OAc	H by titration to pH 7	186
3	0.5N HCl, 1N NH₄OAc, NaCl	NH₄ by distillation	188
4	0.5N HCl, 1N Ba(OAc)₂	H by titration to pH 7	200
5	0.5N HCl, 0.5N Ba(OAc)₂	H by titration to pH 7	190
6	0.5N HCl, 0.5N KOAc	H by titration to pH 7	182
7	0.5N HCl, 0.5N KOAc, 0.5N HCl	K by flame photometry	177

plexes such as hemiculluloses, cellulose, lignins, proteins, and waxes are different in *Sphagnum* plants than in sedges, reeds, trees, and algae.

Puustjärvi (1956) reported that the magnitude of the cation exchange capacity in peats is a function of the following factors:

1. The moisture content of the bog type; the exchange capacity increases with increasing dryness of the bog type; this is attributed to both the possible transformation of the colloids into an irreversible form and the oxidation of some inorganic groups.

2. The type and species of plant composing the peats; as a general rule, sedges tend to have a lower exchange capacity than *Sphagnum* mosses. *Sphagnum cuspidatum* has a lower exchange than *Sphagnum papillosum*, which Puustjärvi (1956) feels may be due in part to the fact that *S. papillosum* prefers the drier sites.

3. Base content; using the amount of exchangeable calcium as a measure of the base content, there is a significant positive correlation with exchange capacity in sedge-dominated peats. No such correlation exists with *Sphagnum*-dominated peats, but it is felt that this may be due to the fact that the amount of exchangeable calcium is small in relation to the exchange capacity.

In relation to the clay mineral fraction, the stoichiometry of cation exchange in the organic fraction is somewhat arbitrary. It is a known fact that the exchange capacity of a material under one set of conditions is not necessarily identical with the measured value under other conditions. Differences such as the cations present, the pH, the degree of dissociation of the reactive groups, the mechanisms of cation retention, and the stability of the complexes formed indicate that only values obtained from the same procedure can be compared. The concept of cation exchange has no unambiguous meaning (Puustjärvi, 1956). MacLean et al. (1964) also illustrated that it is necessary to adhere to a particular procedure if valid comparisons are to be made.

Determination of the cation exchange of organic soils is performed by different methods. The standard procedure is the ammonium acetate method, which involves the replacement of the exchangeable cations by the ammonium ion. A second method as described by Puustjärvi (1956) and by MacLean et al. (1964) involves producing a hydrogen-saturated peat sample in order to remove the basic cations. The exchangeable hydrogen is replaced by barium acetate and the solution titrated to pH 7 with sodium hydroxide. Since some exchangeable ions are more easily replaced than others, the incompleteness of replacement of some cations will depend on the particular salt used. Table 3 illustrates the results found by MacLean et al. (1964). As can be seen, the different procedures gave exchange capacity values of different magnitude. MacLean et al. determined that the 1N barium acetate procedure was convenient and possibly the most suitable of those employed, since it gave the greatest replacement of hydrogen.

Base Saturation (Base Status)

The base saturation is considered to be the percentage fraction of basic cations of the total cation exchange capacity. Among all factors important to agriculture and forestry, the base content of the soil may be regarded as having the greatest significance. The normal procedure is to consider the total amount of exchangeable cations as the difference between the cation exchange capacity and the exchangeable hydrogen ions. This value is used as the numerator in a ratio with the total cation exchange capacity. In fertility studies it is known that only the easily soluble base fraction is important and not total quantities. Methods used to determine this are: (1) determination of exchangeable calcium, (2) determination of dialysable cations, and (3) determination of the alkalinity of the ash.

Puustjärvi (1957) determined that the dialysable cations, the ash alkalinity, and the exchangeable calcium show close correlation in sedge-dominated peats, but that in *Sphagnum*-dominated peats the ash alkalinity method proved to be the most reliable for the determination of bases.

Acidity

Gorham (1966) believes that acidity in peats arises primarily from two processes: (1) the production of sulphuric acid from the oxidation of organic sulphur compounds in the peat and from atmospheric pollution, and (2) the exchange of metal cations brought down in the rain for the hydrogen adsorbed on the organic colloid produced by bog plants and peats. The hydrogen ions are produced metabolically either by the living bog plants or in the course of their decomposition. Correlations, such as that reported by Puustjärvi (1957) between pH and exchangeable hydrogen, indicate that exchangeable hydrogen ions are the main cause of acidity (Figure 11). The fact that this correlation is better for sedge-dominated peats indicates that other factors such as water-soluble acids and different degrees of dissociation of the exchangeable hydrogen ions are important in moss-dominated peats.

The types of vegetation, the rate of decomposition, and the products of decomposition are greatly influenced by and in turn influence the chemical composition of peat. Heinselmann (1963) states that in a *Sphagnum* bog acidity can be attributed to the *Sphagnum* itself, being accomplished by some ion-exchange mechanism. As a result, acidity often has been used as an index in estimating the nutrient status of the soil. Lucas and Davis (1961) studied the relationship between the pH values of organic soils and the availabilities of 12 plant nutrients. Their results indicated that for wood-sedge organic soils the ideal pH falls in the range of 5.5 to 5.8 and for *Sphagnum* peats it is about 5.0. This range is 1.0 to 1.5 pH units lower than that generally considered to be most desirable for mineral soils.

Waksman and Stevens (1929) illustrated that acidity has an overriding influence on the chemical composition of peat because of its effect on vegetation and the rates and products of

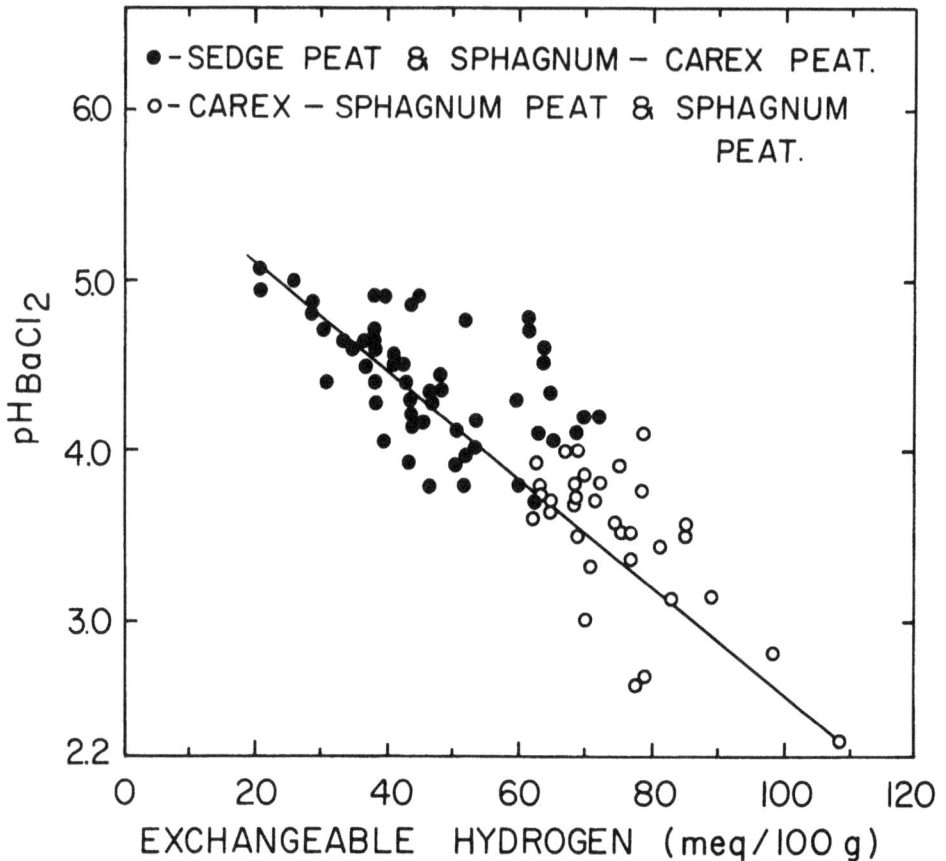

FIGURE 11 The relationship betweeen pH in barium chloride and exchangeable hydrogen (after Puust-järvi, 1957).

decomposition. For example, highly acidic sites indicate slow decomposition rates, low levels of microorganism activities, and the presence of certain celluloses and hemicelluloses. Less acidic sites, characterized by sedges and grasses, normally show a higher rate of decomposition of cellulose and hemicellulose with an accumulation of lignin, proteins, and minerals.

The hydrogen ion activity or pH value is normally used to measure acidity. Values range from 3.0 to 8.0 in most organic soils. Low values of pH indicate the presence of sulphides and possibly iron and aluminum sulphate, whereas high values are indicative of adsorbed calcium or sodium. Many small battery-operated pH meters are commercially available for pH measurement in the field. Alternatively, soil and water pH can be measured simply in the field with specially prepared indicator paper. Accurate and more sophisticated electronic equipment is available for laboratory measurements of pH.

The actual measurement of pH must be considered in the light of such parameters as the water content of the sample, the suspension ratios (solid:liquid), the type of salt suspension, and the equilibrium time. Also, many authors (Gorham, 1958, 1961; Isotalo, 1951) have

TABLE 4
Water-soluble constituents in muskeg (ppm)

Reference and discussion	Ca	Mg	Na	K	NO₃	O₂
Gorham (1956a): Great Britain						
Willow and *Carex* fen	13.2		5.3	0.17		
Sphagnum bog	1.6		4.9	0.67		
Gorham (1966)						
Falkland Islands	2.1	5.4	38.6	1.6		
West Ireland	0.8	1.8	12.8	0.5		
North Scotland	0.5	1.1	13.9	0.6		
North England	1.0	1.1	5.3	0.5		
North England	0.9	0.6	3.4	0.3		
West Scotland	0.3	0.3	2.2	0.1		
Poland	0.5	0.3	0.2	0.2		
Tarnocai (1973): Mackenzie River, NWT						
Spring fen	54.91	11.80	29.89			
Patterned fen	18.64	22.25	7.59			
Bog plateau	2.00	0.73				
Peat polygon	1.40	0.12				
Ice wedge	1.00					
Walmsley and Lavkulich (1973): NWT						
Fen	14.00		47.7		6.5	1.8
Transitional fen	100.00		14.5		6.3	2.0
Transitional bog	99.0		9.7		2.7	5.1
Bog	7.2		0.6		3.4	6.1

indicated that pH values obtained from a water extract are higher than pH values obtained by direct insertion of the glass electrode into the peat sample. This is attributed to the fact that in the fresh peat sample the electrode measures hydrogen ions adsorbed on the organic colloids as well as free acid in solution. In all cases, the suspension effect is known to produce measurement problems. The water content of the sample was shown by Duch (1963) to have a great effect on the pH value. Air drying of samples can cause differences of as much as half a pH unit. Suspension ratios normally employed range from 1:4 to 1:8 soil:water. Puustjärvi (1957) illustrated that a better correlation exists between exchangeable hydrogen and exchangeable calcium, ash alkalinity, and pH determined in a salt suspension than in a water suspension. Values for pH using barium chloride were 0.05 to 0.1 unit lower than when potassium chloride was used. Equilibrium times normally used range from 20 to 40 minutes with perhaps 30 minutes being optimum.

One of many land-use applications of pH stems from the corrosive action that bog waters, which are often low in amounts of dissolved salts, can have on concrete and other engineering structures. Mainland (1960) states that since most peats can be essentially free of dissolved oxygen, when a metal structure such as a bridge footing or a pipeline crosses from oxygen-rich mineral soil to the peat, a galvanic cell can be set up, with the area of lowest oxygen concentration being the anode. Such an electrochemical cell can produce very high corrosion rates.

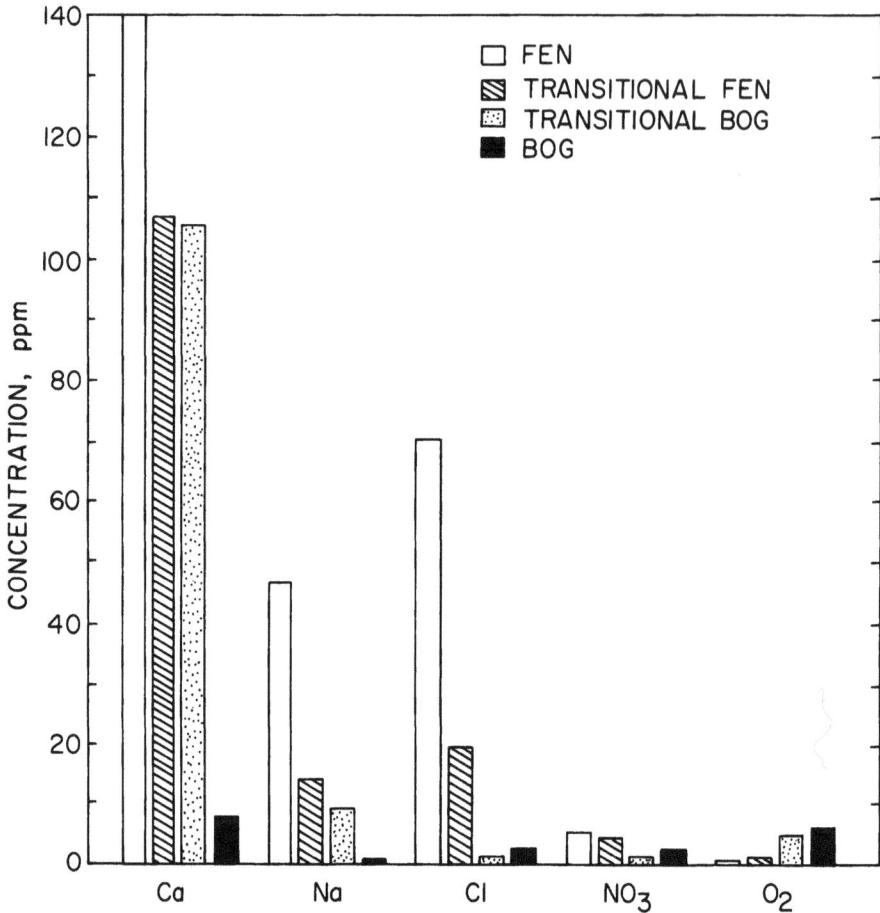

FIGURE 12 Dissolved material as an indicator of terrain type.

Water Chemistry

In northern regions, peat deposits often originate with the filling in of a lake and the subsequent swamping of large areas of mineral soil (Sjörs, 1961). Areas which retain some degree of silting or drainage from mineral soils are called minerotrophic organic deposits or fen. Those areas which no longer receive any inflow of water from mineral soils are designated ombrotrophic organic deposits or bog (Sjörs, 1961). The difference in the supply of groundwater has a marked effect on the vegetation of these areas. Bogs are usually vegetated by *Sphagnum* mosses, cottongrasses, deersedge, and heaths, whereas fens are dominated by reeds, sedges, grasses, and hypnoid mosses. This difference in vegetative composition can strongly influence soil properties. For example, the mineral ash content averages about 5.9 per cent for fen plants and only 2.5 per cent for bog plants (Gorham, 1966).

Several authors have studied the ionic composition of muskeg waters and its relationship to vegetation and peat characteristics (Gorham, 1955, 1956a, 1956b, 1960, 1966; Gorham and

Pearsall, 1956; Tolpa and Gorham, 1961; Sjörs, 1950; Walmsley and Lavkulich, 1973; Tarnocai, 1973). Results have indicated that ombrotrophic bogs receive essentially all their mineral supply from atmospheric precipitation and are necessarily low in mineral nutrients. Table 4 indicates the concentration range of some ionic species and illustrates the low levels in bog organic landforms. The pH of the water is also lower in the bog, ranging from 3.8 to 4.4, while in the fen values range from 4.8 to 6.9. Recent studies (Walmsley and Lavkulich, 1973) have also illustrated that the dissolved oxygen concentration is much lower in the bog than in the fen while the redox potential tends to be substantially higher. Figure 12 illustrates, in graphic form, the concentration levels of some ions in fen areas, transitional sites, and bog areas. The data were obtained from organic landforms in the Fort Simpson area, Northwest Territories. A general trend is present, with increasing concentrations from bog to fen.

Such chemical relationships are undoubtedly very important when one is considering the use of muskeg. For example, the fact that the bog water has an extremely low pH and low oxygen concentration makes this landform a potential site of high corrosion for engineering structures. Also, any aforestation methods must necessarily take water characteristics into account, since only very low productivity could be expected in the bog environment.

PHYSICAL PROPERTIES OF PEAT

Fibre Content

Many authors have noted the relationship between the content of particles or fibres larger than a particular size and the stage of decomposition. Farnham and Finney (1965) arbitrarily selected 0.1 mm as their criterion in determining the decomposition of organic deposits, calling everything with more than two-thirds of the total mass of fibres larger than 0.1 mm fibric, from two-thirds to one-third hemic, and less than one-third sapric. This inverse relationship between fibre content and degree of decomposition is often used to group organic materials into various decomposition stages for the purpose of classification. In the System of Soil Classification for Canada (1970), a value of 0.15 mm is arbitrarily chosen. Organic material is considered fibric if 67 per cent of the fibre content is greater than 0.15 mm, mesic if 33 to 67 per cent, and humic if less than 33 per cent. Frazier and Lee (1971) indicated that the fibre content is the most useful morphological criterion in classifying organic soils.

Fibre contents are normally measured by a wet sieving process. A known weight of undisturbed peat is soaked in a dispersing agent (e.g. 1 per cent sodium hexametaphosphate) for approximately 15 hours. The material is then washed through sieves by the application of a gentle flow of water, ensuring that it is neither (physically) broken down nor forced through the sieve. In some instances, it is necessary subsequently to wash the material with a 2 per cent hydrochloric acid solution, to dissolve any carbonates. The fibrous material collected on the sieves is then oven-dried, weighed, and expressed as a percentage of the oven-dried weight of the original sample. In certain instances, a hand-rubbed fibre content is used to group organic soils. Its application is with material that may be highly decomposed but which contains fibres that retain their original structure. These remnant fibres break down quite easily when rubbed, allowing for a more accurate estimation of the degree of decomposition. Boelter (1972) found that for Minnesota peats the rubbed fibre content was, on the average, 8.7 per cent lower than the unrubbed fibre content. For the purpose of placing a soil into a particular decomposition class, Padbury (1970) found no difference between the rubbed and unrubbed fibre content determinations.

Errors in fibre content determination are usually the result of mineral contamination. Since the specific gravity of mineral particles is essentially twice that of the organic material, the magnitude of the error can be quite large.

Bulk Density

Bulk density or volume weight is defined as the weight of a given volume of soil. It is used for converting water and nutrient contents measured on a weight per cent basis to a weight per unit volume basis. Normally bulk density is reported on a dry-volume basis, expressed as the mass per unit dry bulk volume of soil, dried to constant weight at 105°c. Lately, it has been recognized that soil volume changes with the water content, so that the bulk density must be calculated on the basis of the wet bulk volume.

The bulk density depends upon both the organic and water content of the peat. Excessive mineral contamination is reflected in somewhat higher bulk density values. High water contents result in quite low bulk densities when they are expressed on a wet-volume basis. Appendix 1 gives the values of the bulk density, which range from 0.02 g/cc to 0.34 g/cc on a dry basis and from 0.4 g/cc to 1.2 g/cc on a wet-volume basis.

Owing to its nature, bulk density is related to the degree of compaction and decomposition. Elzen (1961) determined that there is a drastic change in the pore size distribution as the bulk density increases even though the total pore volume may change very little. Sturges (1968) found that samples of low bulk density contain many large pores which release water easily whereas samples with a high bulk density have smaller pores and retain more water at higher suctions. As decomposition progresses, the larger pores normally disappear (because of the breakdown of the fibre), being replaced by many smaller pores. Because of this, bulk density has often been used as a measure of decomposition. Also, it gives the engineer a measure of the trafficability of the particular material.

Bulk density is usually measured by retaining a core sample of known volume of material and oven-drying it to constant weight at 105°c. Irwin (1966) measured the bulk density for drained agricultural organic soils using the formula:

$$D_B = \frac{W}{V(1 + w)}$$

where W is the weight of wet soil in g, V is the bulk volume of wet soil in cm³, and w is the water content in per cent by weight. He determined that the method based on wet volume was satisfactory for the range of water contents found in the field.

Specific Gravity

Specific gravity is defined as the ratio of the mass of a body to the mass of an equal volume of water, usually at 4°c or some other specified temperature. Since the value depends on the organic and inorganic content of the material, a high specific gravity for an organic soil would indicate mineral contamination. Appendix 1 indicates that the specific gravity of peat ranges from 1.1 to 2.7. It is generally believed that a specific gravity greater than 2.0 indicates considerable mineral contamination.

The procedures used for the determination of specific gravity must all take into account the occurrence of large amounts of entrapped air. Irwin (1966) simply boiled the soil:water mixture in vacuo whereas MacFarlane (1969) felt that pulverizing the oven-dried peat with a mortar and pestle was required. The standard procedure involves the use of a water pycnometer using a 'Pril' solution to reduce the surface tension forces. In cases where liquids other than water are used, such as acetone or kerosene, the specific gravity may be calculated from the equation

$$G = \frac{\text{weight of dry material}}{\text{weight of liquid displaced}} \times \text{specific gravity of liquid.}$$

In some instances it is more meaningful to express the specific gravity on an ash-free basis. Segeberg (1955) developed an equation relating the specific gravity of ash-free peat to the specific gravity of natural peat:

$$G_f = G \left(0.00836 \, X - 0.012 + \frac{X}{100} \right)$$

where G_f is the specific gravity of ash-free peat, G is the specific gravity of natural peat, and X is the mineral content in per cent.

Miyakawa (1959) developed a relationship between the specific gravity and ash content, assuming a mixture of two materials. The equation is

$$G = \frac{G_s G_p}{(G_s - G_p) n + G_p}$$

where G is the average specific gravity of the peat soil, G_s is the average specific gravity of the mineral solids, G_p is the average specific gravity of the organic solids, and n is the ash content. Cook (1956) used values of 2.70 and 1.50 respectively for G_s and G_p, assuming the mineral matter to be composed mainly of clay minerals and the organic material to be primarily lignin and cellulose. Irwin (1966) determined that values up to five per cent higher than the true specific gravity are calculated by this formula.

Void Ratio and Total Porosity

The void ratio is defined as the ratio of the volume of spaces to the volume of soil solids, and the total porosity or total pore space is normally expressed as a percentage of the total volume. The total porosity is usually considered to be equal to the total volume of water contained at saturation. This includes both the space between particles and the pores within the organic particles which contain water. An indication of the compressibility of a material is given by the void ratio. The higher the void ratio, the greater the potential compressibility. Void ratios as high as 25 have been reported for fibrous peats (Hanrahan, 1954), and values as low as 2 have been reported for amorphous peats. Reported values for porosity range from 80.7 to 95.2 per cent with an average of about 92 per cent.

As a general rule, all peat types, regardless of plant source or degree of decomposition, contain more than 80 per cent water by volume at saturation (Boelter, 1966). Although this is indicative of a high total porosity, there is a difference in the nature of the porosity for different peat types. For example, whereas undecomposed moss peats contain more than 95 per cent water by volume at saturation and release 50 to 80 per cent of this water to drainage, herbaceous peats yield only 10 to 15 per cent to drainage, even though they may contain between 80 and 90 per cent water at saturation. In undecomposed moss peat, the water is held in large, easily drained pores, but in herbaceous peat, water is retained in many small pores that are not easily drained.

These properties of organic soils have direct application in determining use constraints. In the context of land productivity, Kravchenko (1963) reported that the capillary rise in cultivated peat bog soils did not exceed 60 cm, leading him to the conclusion that groundwater should be allowed to drop more than 80 to 100 cm below the surface. The same problem may occur in undrained bogs with shallow-rooted tree seedlings, if the water table drops below 20 to 30 cm in undecomposed moss peats. Topography is often the major cause of this problem.

Total porosity is usually calculated by the formula

$$n = \frac{(1-D_B) \times 100}{G}$$

where n is the percentage porosity, D_B the saturated bulk density, and G the specific gravity. Using the value of n obtained, the void ratio can be calculated from the equation

$$e = \frac{n}{100-n}$$

where e is the void ratio and n the percentage porosity.

Water Retention Properties

Feustel (1938) described the use of peat materials as a supplement to improve the physical properties of minerals soils used in greenhouses. Investigators have always recognized the high water retention of peat materials (Feustel and Byers, 1930; and MacFarlane, 1959). The degree of decomposition determines, to a large extent, the moisture-retaining characteristics of organic soils. The water content decreases with an increase in the degree of decomposition (Amaryan et al., 1966; Feustal and Byers, 1936; and Yamazaki et al., 1957). Dyal (1960) showed that although the amount of water retained at low suctions decreased as the decomposition increased, at higher suctions more water is retained by the highly decomposed materials.

The water content of peat is often expressed as the weight of water per unit weight of oven-dry soil. Appendix 1 indicates that values over 3,000 per cent have been reported. Boelter and Blake (1964) indicated that, because of the variable bulk densities of peat and the volume reduction caused by drying, water contents must be expressed on a wet-volume basis. Furthermore, the actual state of wetness or dryness should be stated; for example, water content at 1/3 bar tension. When water content is expressed on a volume basis there is a relationship to decomposition at all suctions except at or near saturation (Padbury, 1970).

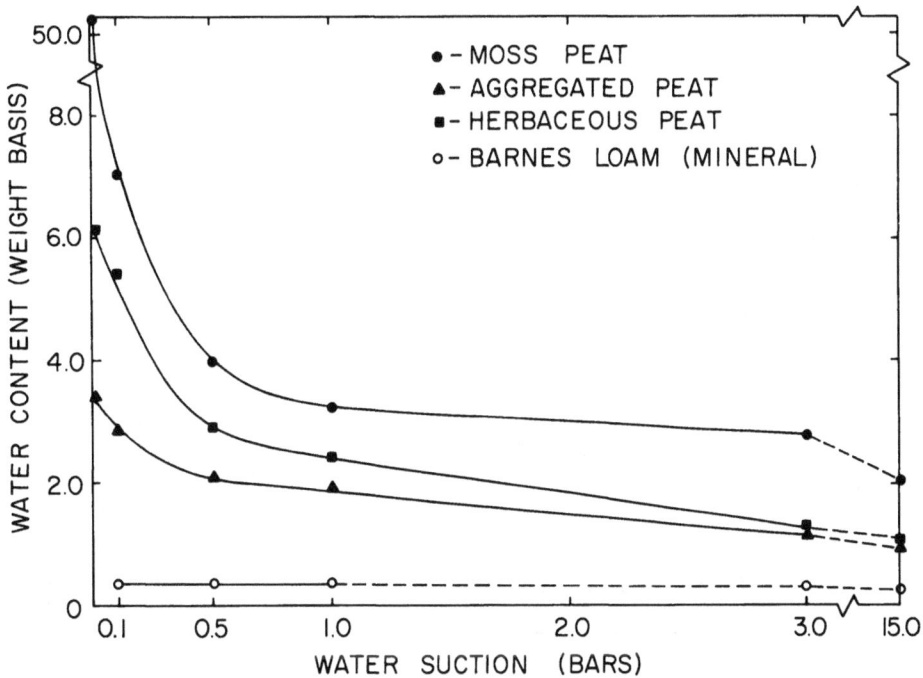

FIGURE 13   Water retention curves of three peats and a mineral soil (after Boelter and Blake, 1964).

Boelter and Blake (1964) presented water retention curves which illustrate the results of expressing the water content on a weight basis (Figure 13). On a weight basis, the water contents were much higher for organic soils than for mineral soils. Using the same water contents, but expressed on a volume basis (Figure 14), the positions of the curves are shifted. It is now seen that mineral soil has a volumetric water content similar to that of herbaceous peat at higher suctions. Figure 14 illustrates that the pore size distribution is perhaps the most significant parameter when comparing water retention curves of various peat types at water contents less than saturation. Since the undecomposed peats contain many easily drained pores they will lose most of their saturated water content at very low suctions. The more decomposed peats retain more water at higher suctions as a result of the presence of finer pores. Information such as this is of prime importance from a hydrologic point of view. For example, lowering of the water table would drain very little water from a decomposed peat but a great deal of water from an undecomposed peat. This has direct application for the use of a bog as a storage reservoir.

Water contents at suctions ranging from 0.0 to 15 bars are determined in the laboratory by means of soil water extraction techniques. Up to 0.1 bar suction, water contents are normally determined using a pressure cell apparatus described by Reginato and Van Bavel (1962). For suctions of 0.2 to 15.0 bars, pressure membrane extractors are used. Great care must be exercised when measuring the water retention of undecomposed peats at low suctions. Because of the large quantity of water released, a small change in suction will result in a significant change in water content (Boelter, 1965).

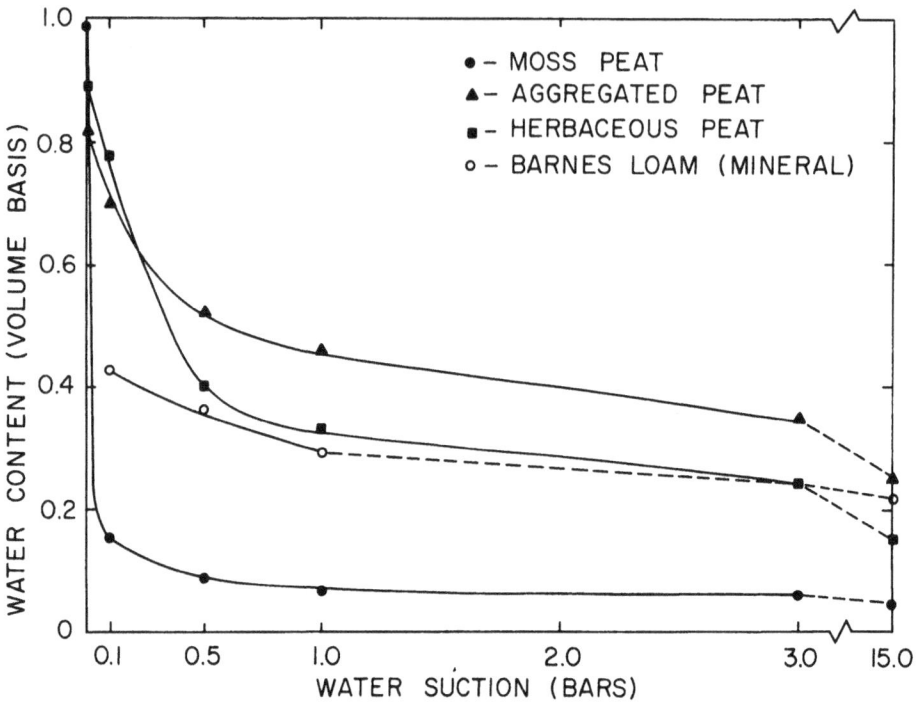

FIGURE 14   Water retention curves of three peats and a mineral soil (after Boelter and Blake, 1964).

In perennially frozen soils, Tarnocai (1973) showed that the water content of the active layer is very different from that of the frozen layers. Figure 15 illustrates the slow increase in water content in the active layer until it reaches the permafrost table. The ice content below the permafrost table shows no real increase until just above the mineral contact, where it reaches a maximum and then quickly decreases. An interesting feature of this information is that these soils never reach saturation, which has an important implication in all fields dealing with biological productivity.

Hydraulic Conductivity

The hydraulic conductivity or permeability to water of a material is a measure of the rate of water movement through the material. The relationship between the specific discharge rate and the driving force was initially proposed by Darcy (1856). He discovered that the flux density and the hydraulic gradient were linearly related by a factor which is called the hydraulic conductivity, the formula being

$$q = K \frac{\Delta H}{L}$$

where q is the flux density (volume of water flowing through a cross-sectional area in a unit of

WATER & ICE CONTENT, % (VOLUME BASIS)

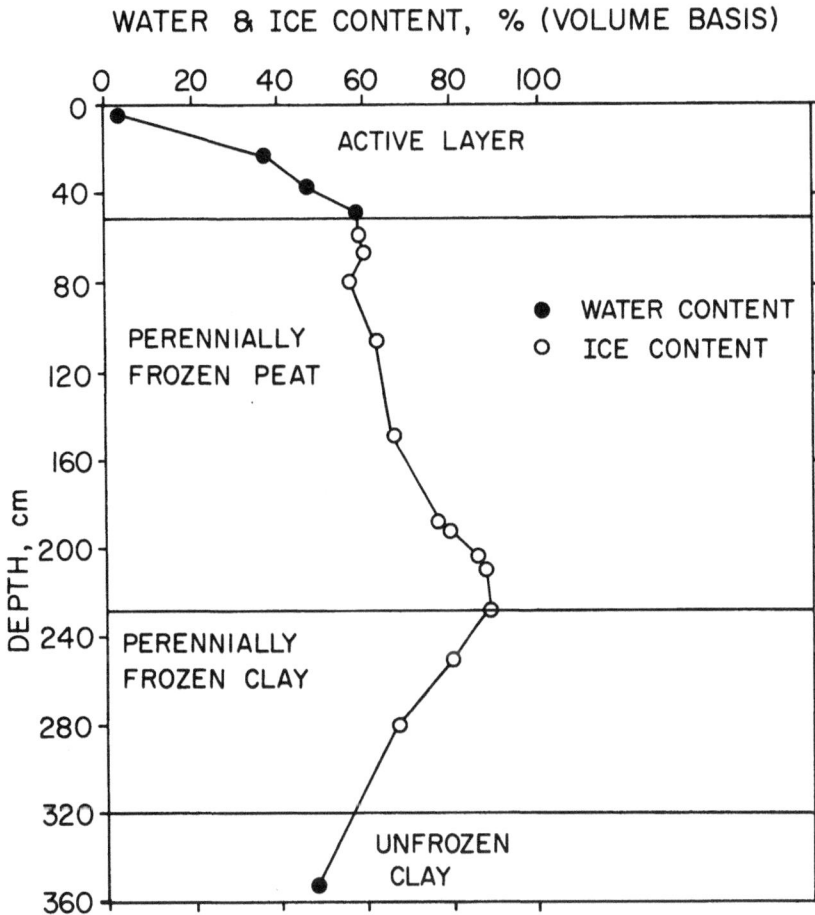

FIGURE 15   Water and ice content with depth of a perennially frozen soil (after Tarnocai, 1973).

time), K the hydraulic conductivity, ΔH the difference between the two heads, and L the length of column containing material ($\Delta H/L$ is the hydraulic gradient).

Hydraulic conductivity is considered an important physical property of organic soils since it will determine the runoff characteristics of the material. For some time it has been known that more decomposed peats show restricted water movement than undecomposed fibrous peats (Hanrahan, 1954). More recently it has been noted that the difference in saturated hydraulic conductivity between samples is due primarily to differences in the pore size distribution (Boelter, 1965). As decomposition progresses, there is a small decrease in total porosity and a considerable decrease in the effective diameter of the pores (Elzen, 1961). It is the size and continuity of the pores as affected by the arrangement of the particles that causes a wide range of hydraulic conductivities in peats. Boelter (1965) reported values as low as 7.5 x 10^{-6} cm/sec for dense, decomposed and herbaceous peat, which is lower than for some

clays or glacial tills. Undecomposed moss peats, which have many large pores, have hydraulic conductivities of more than $3,810 \times 10^{-5}$ cm/sec. Appendix 1 shows the wide range
of permeability values for a variety of peat types. Because of the development of a horizontal
laminar structure in some peat types, many authors have concluded that the horizontal and
vertical hydraulic conductivities differ. Colley (1950) and Miyakawa (1960) reported that the
horizontal hydraulic conductivity was greater than the vertical hydraulic conductivity.
Boelter (1965) determined that there was no significant difference between them in any of the
peat types he examined. It seems reasonable to assume that significant differences could
result, depending on the type of peat and the method of analysis.

Saturated hydraulic conductivity is normally measured in the field by the use of a piezometer when horizontal hydraulic conductivity values are desired. Vertical hydraulic conductivity values are normally measured by the tube method. In both methods, a tube or conduit (of
larger diameter for the tube method) is sunk to the peat horizon in question, and the peat is
removed from the tube with an auger. In the peizometer method a cavity is augered below the
end of the tube so that the cavity is in the middle of the horizon in question. The rate of rise of
water, after flushing several times, is taken as a measure of the hydraulic conductivity by use
of the formula

$$K = \frac{2.30\pi R^2}{A(t_2-t_1)} \log_{10}(h_1/h_2)$$

where K is the hydraulic conductivity in cm/sec, R is the radius of the tube or peizometer in
cm, A is a geometric function in cm (Luthin and Kirkham, 1949; Frevert and Kirkham, 1948),
h_1 and h_2 are the distances from the water table to the water level in tubes at times t_1 and t_2,
respectively, and $t_2 - t_1$ is the time interval in seconds over which the rise in water level
was measured.

Kirkham (1949) described a method for determining the hydraulic conductivity from a
drain tile discharge. He used the formula

$$K = \frac{0.366Q}{d \log_{10} 8d} \text{ ft/day}$$

where Q is the drained tile discharge and d is the head causing flow. Irwin (1966) applied this
method to several organic soils and found values ranging from 7.1×10^{-3} to 8.6×10^{-4}
cm/sec. He reported that although the values were affected by the head of water over the
drain the results were comparable to those obtained using the auger-hole method (Measland,
1958).

Laboratory measurements usually employ a core sample of the peat, mounted on a porous
surface. A constant head of water is maintained above the surface of the sample and water
discharge rates are measured. Hydraulic conductivity is then calculated from the formula

$$K = \frac{Q\Delta L}{At\Delta H}$$

where K is the hydraulic conductivity in cm/sec, Q the volume in cc passing through the core
in time t, t the length of time of measurements in seconds, A the cross-sectional area of the

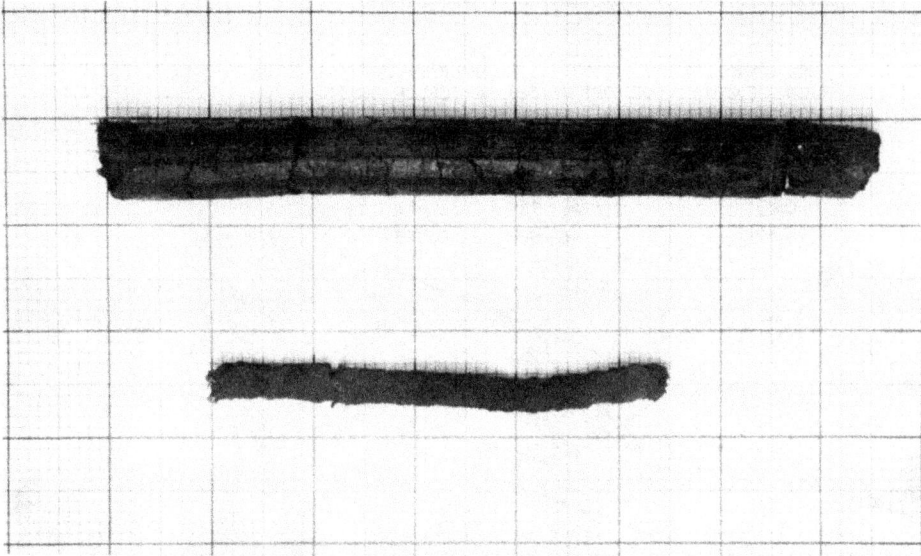

FIGURE 16   Shrinkage of an aquatic peat subsoil (after Maas, 1972).

core in cm², ΔL the length of the core in cm, and ΔH the length of the core plus the head of water in cm. Boelter (1965) assumed that field measurements gave better estimates of the actual hydraulic conductivities of peat soils because of the drying effect on the sample when it is brought into the laboratory.

Hanrahan (1954) determined that the permeability is affected by the magnitude and duration of loading. The permeability of all peat types decreased with time. He showed that a peat under a load of 8 psi for 7 months had its permeability reduced 50,000 times. Any engineering use of muskeg must take this into consideration.

Shrinkage

Volume reduction upon drying and volume recovery upon rewetting are characteristic of peat material. Shrinkage of up to 87 per cent of the original volume has been observed. Man's use of peat for cultivation, or simply excavation, must take this shrinkage into account in the development of good land use practices.

It is the type of organic material and its characteristic bulk density that determine the shrinkage and volume recovery. Maas (1972) reported that the lower the bulk density and the more gelatinous the peat particles, the greater the shrinkage on drying. Figure 16 illustrates that an aquatic peat subsoil which contains a large amount of gelatinous material will shrink drastically upon drying. In this example an 87 per cent volume reduction was reported.

Maas (1972) also showed the effect of bulk density on shrinkage and volume recovery with increasing depths for a peat under pasture. Figure 17 shows that the surface peat (bulk density = 0.42 g/cc) shrunk to 75 per cent of its original volume on drying at 30°c but recovered its entire volume upon rewetting. In contrast, the peat at 45 to 60 cm depth (bulk density = 0.17) shrunk to approximately 24 per cent of its original volume and recovered only

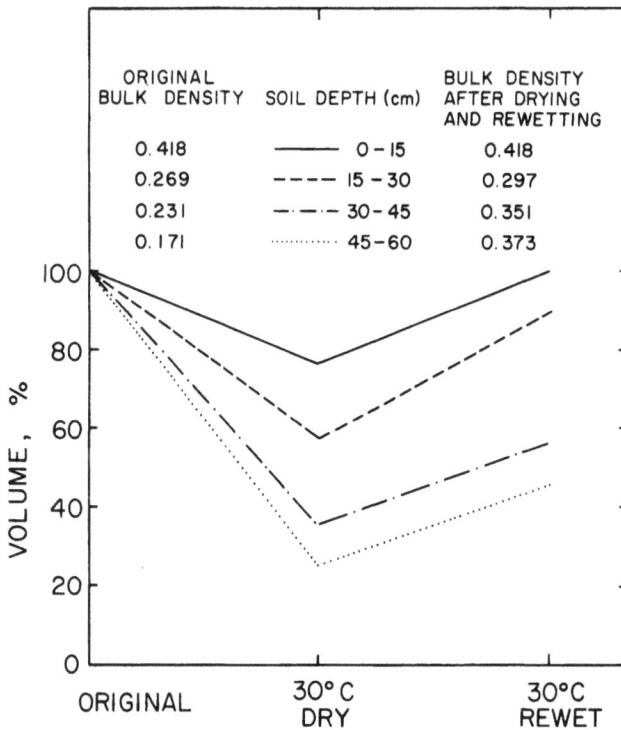

FIGURE 17　The effect of bulk density on shrinkage and volume recovery on rewetting (after Maas, 1972).

45 per cent of its original volume upon rewetting. Feustel and Byers (1930) reported that, upon drying, *Sphagnum* peat will take up (on the average) 55 per cent of the amount of water it is capable of holding, whereas a heath peat will take up only 33 per cent of what it is capable of holding. Such data indicate the hazards of draining organic soils to excessive depths.

Temperature

Soil temperature is considered an important environmental variable as it affects not only the physical conditions of materials but also the chemical and biological properties. Soil temperature influences the metabolic activity of tree roots and soil organisms and the availability of moisture (Fraser, 1957). Kramer (1949) indicated that soil temperature has a large effect on water absorption. He stated that low soil temperatures cause: (a) a change in the viscosity of cell protoplasm, (b) decreased permeability of cells, (c) decreased rates of water movement to roots, (d) retarded root growth, and (e) decreased vapour pressure. Thermal properties of

TABLE 5
Some thermal properties of organic soils at different water contents (after Van Wijk, 1963)

Water content (cm³ water/cm³ soil)	Conductivity (10^{-3} cal/cm sec °C)	Heat capacity (cal/cm³ °C)
0	0.14	0.4
0.4	0.70	0.9
0.8	1.2	1.0

peat, such as the low coefficient of heat conduction, and topographic situations that affect air drainage cause differences in the soil and air temperatures for peatlands and mineral soils. For example, it has long been recognized that if the organic matter above perennially frozen ground is disturbed even without actual removal thawing will result (Brown, 1963).

Approximate values for some thermal properties of peat soils are presented in Table 5. It can be seen that heat capacity is strongly correlated with water content, as is conductivity. The specific value of both the heat capacity and the conductivity is also related to the porosity of the material. Some of these properties were reported by Wilson (1939) in his study of the temperature profile of a peat deposit over sand in Wisconsin. He discovered that although the temperature at the surface of the bog closely paralleled the prevailing temperature of the air, the soil temperature dropped sharply at progressively lower levels. This continued until near the bottom of the deposit, where the temperature was several degrees warmer than at levels slightly nearer the surface. The reason for this phenomenon is the greater conductance of heat in sand than in peat, as well as the fact that fibrous peat is less of a conductor than finely divided limnic peat (the limnic peat was in direct contact with the sandy material, with the more fibrous type above). The consequences of such physical parameters are far reaching not only in the obvious case of construction on frozen organic soils, but also in the cultivation or disturbance of peat. Since disturbance normally increases the decomposition rate, the resultant warmer soil temperatures may have beneficial consequences in terms of plant growth but may result in the subsidence of frozen peats.

CORRELATION OF PHYSICAL AND CHEMICAL PROPERTIES OF PEAT

As has been discussed in previous sections, certain chemical and physical properties of peat bear specific relationships to one another. The impetus to discover these relationships lies in their use for the prediction of other variables which are more difficult to determine. In no manner do these relationships remove the need for detailed investigations, but they do provide the investigator with an indication of the type of material that he is dealing with, the kind of problems that might occur, and perhaps a first approximation of a method for solving the problems.

Three of the most convenient parameters used to measure properties of organic soils are the ash content, water content, and fibre content. This is not to negate the importance of other properties of peat. It is the botanical origin, degree of decomposition, and amount of mineral contamination that gauge, to a large extent, the range of physical and chemical properties of muskeg. As a result, a knowledge of these three properties and their relation-

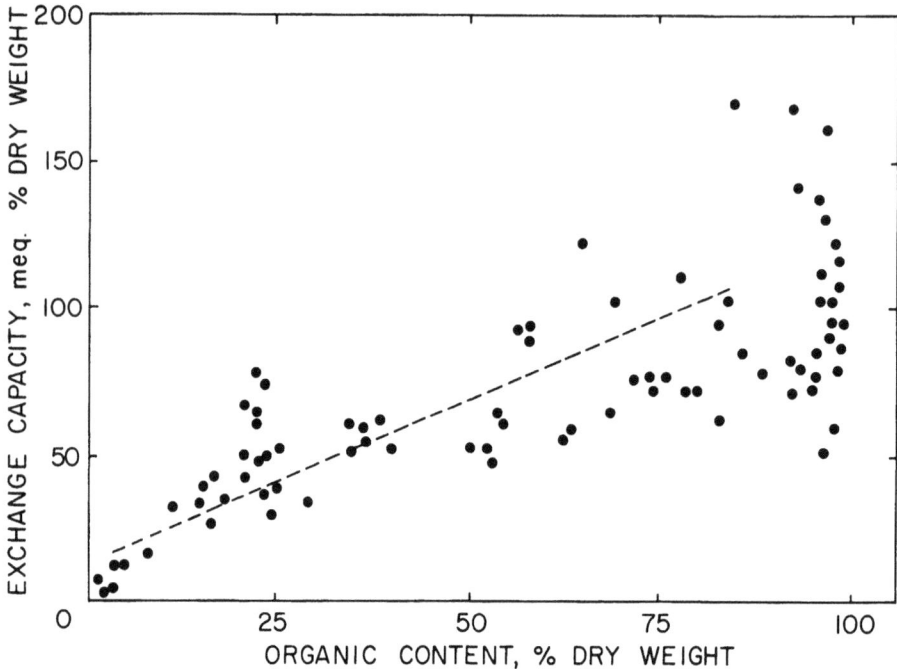

FIGURE 18   The relationship between exchange capacity and organic content (after Gorham, 1953).

ships with other parameters will provide a good indication of the use characteristics of a particular peat. The following section is divided into three parts: ash content, water content, and fibre content. Relationships are discussed under each of these headings.

Ash Content

(a) *Exchange capacity.* Figure 18 shows that the capacity of the soil to absorb cations is closely dependent on the humus content. Values range from a few me/100 g in those soils of lowest organic matter content to between 52 and 63 me/100 g in highly organic peats. Extremely high values of the exchange capacity are due to the highly colloidal or gelatinous nature of some peat types. In some cases, where the peat material is open to an influx of mineral-rich waters, there may be free bases in solution resulting in a high estimate of cation exchange capacity.

(b) *Specific gravity.* Although there is considerable scatter to the values, Figure 19 indicates a trend toward increasing specific gravity with decreasing organic matter content (increasing ash content). This would be expected since the ash elements of peat represent its heaviest fraction. Above a point of approximately 75 per cent organic content, however, the relationship tends to break down. The indication is that near this point the specific gravity is no longer wholly controlled by the mineral fraction.

(c) *Bulk density.* Like the specific gravity, the bulk density is determined to a large extent

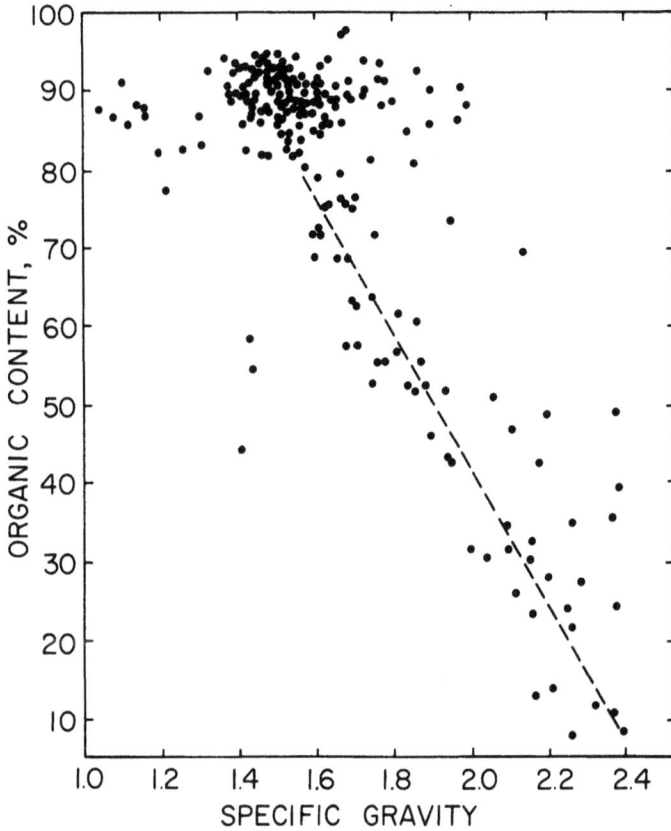

FIGURE 19   The relationship between specific gravity and organic content (after MacFarlane, 1969).

by the amount of mineral contamination. As Figure 20 illustrates, there is a general trend indicating an increase in bulk density with a corresponding increase in ash content. It would be expected that there would be little evidence for a relation between these two properties at high ash contents.

(d) *Fibre content.* As has been discussed previously, a positive correlation is expected between the ash content and the degree of decomposition. Figure 21 indicates that the ash content does increase with decreasing fibre content (increasing degree of humification). The scatter of the points indicates, however, that factors other than the degree of decomposition influence the ash content.

(e) *Acidity.* Figure 22 illustrates that there is a trend toward more acidic conditions as the organic content increases. As was indicated previously for specific gravity, there is a breakdown in this relationship at approximately 80 per cent organic matter content. It seems reasonable to assume that in both cases the mineral content has a great influence up to some upper limit, where its low concentration causes it no longer to be felt.

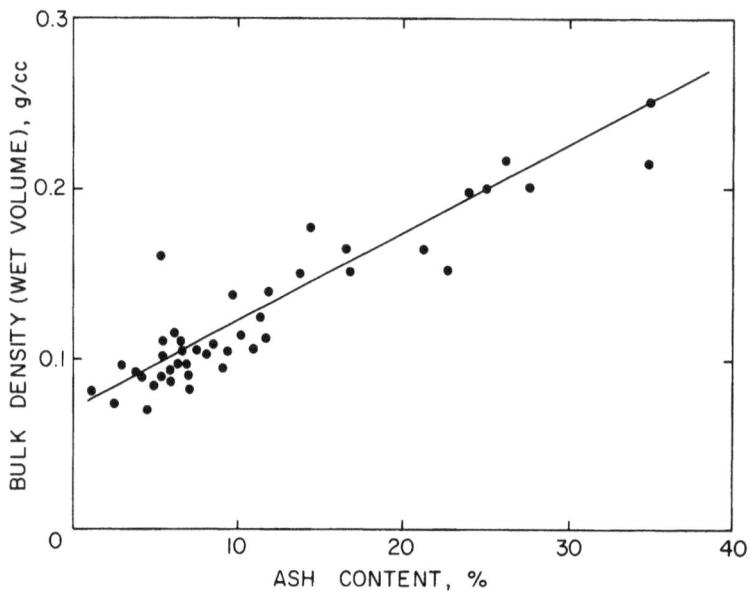

FIGURE 20   The relationship between bulk density and ash content (after Irwin, 1966).

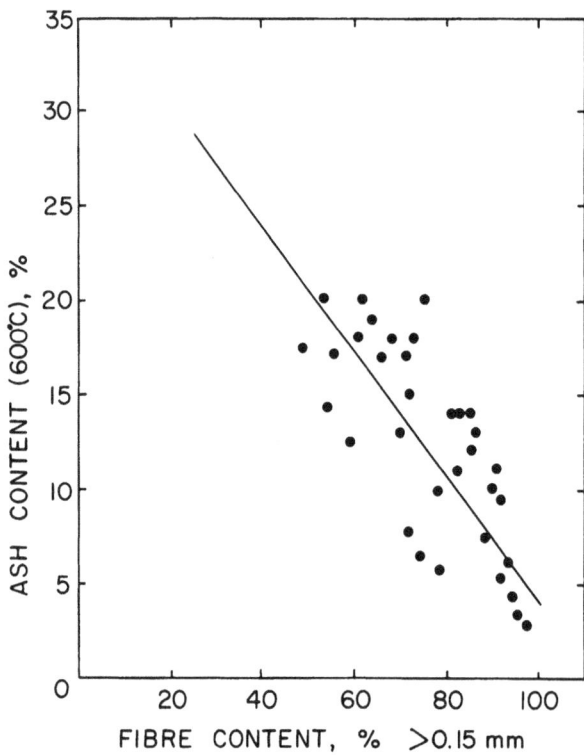

FIGURE 21   The relationship between ash content and fibre content for moss-dominated peats (after Padbury, 1970).

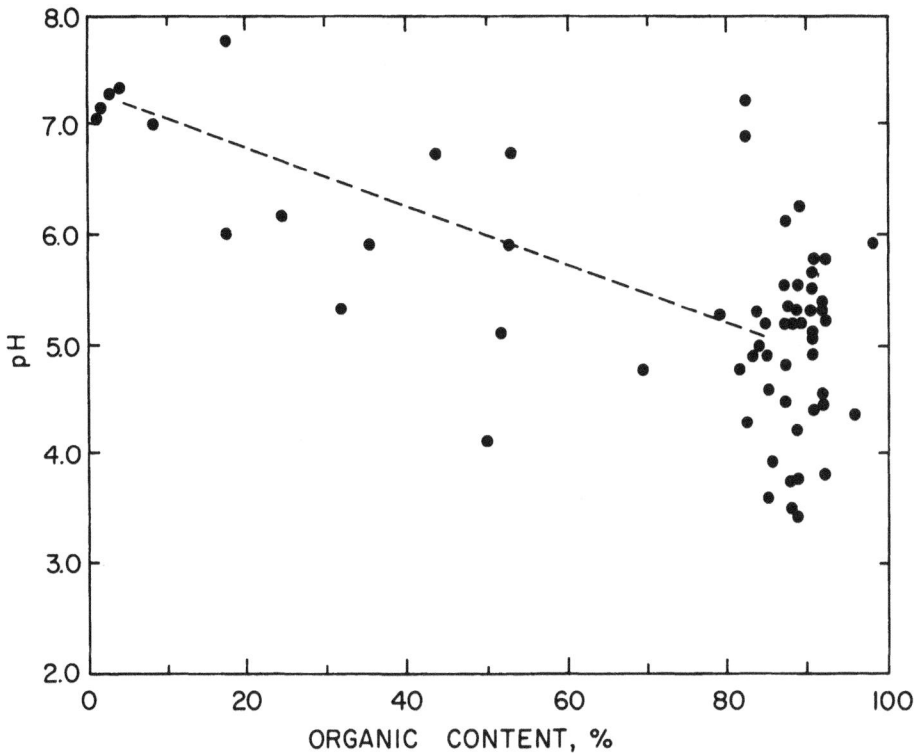

FIGURE 22   The relationship between organic content and pH (after MacFarlane, 1969).

Water Content

(a) *Bulk density*. Figure 23 shows that there is a curvilinear relationship between the water content and the bulk density at various suctions. Kuntze (1965) indicated that the water retention characteristics of peat are dependent on the degree of decomposition so it is not surprising to find a relationship between the water content and the bulk density.

(b) *Unrubbed fibre content*. As was stated previously for bulk density, it has been shown that the water retention characteristics are dependent on the degree of decomposition. Hence, one would expect a relationship between the water content and the unrubbed fibre content. Figure 24 illustrates this relationship for water contents at different values of suction. As discussed by Boelter (1972), the two relationships of bulk density and unrubbed fibre content to water content indicate that the most significant differences in physical properties occur with the least decomposed materials. Figures 23 and 24 show that the steepest part of most of the curves occurs at bulk densities less than 0.1 g/cc and fibre contents greater than 67 per cent. If a soil structure begins to develop, the indication is that increased decomposition may ultimately improve the physical properties of the organic soil.

(c) *Specific gravity*. Normally the water content decreases as the organic content decreases or the mineral content increases. As a result, it is expected that the specific gravity will increase as the water content decreases. Such a trend is indicated in Figure 25 up to a certain water content. Beyond this point the mineral content is so low that the material can be considered to be essentially pure peat, with a specific gravity of approximately 1.6.

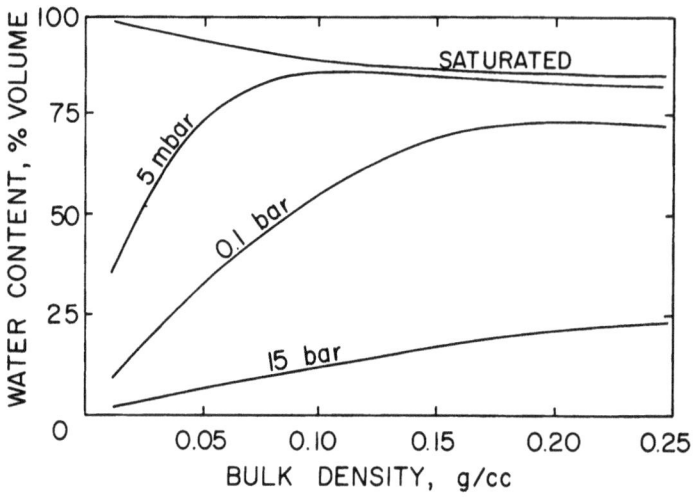

FIGURE 23   The relationship between the water content at saturation for 5 mbar, 0.1 bar, and 15 bar suctions and the bulk density (after Boelter, 1966).

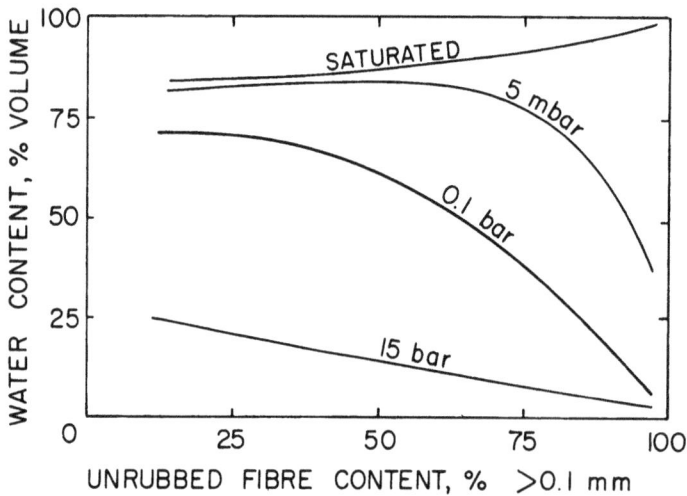

FIGURE 24   The relationship between water contents at saturation for 5 mbar, 0.1 bar, and 15 bar suctions and the unrubbed fibre content (after Boelter, 1966).

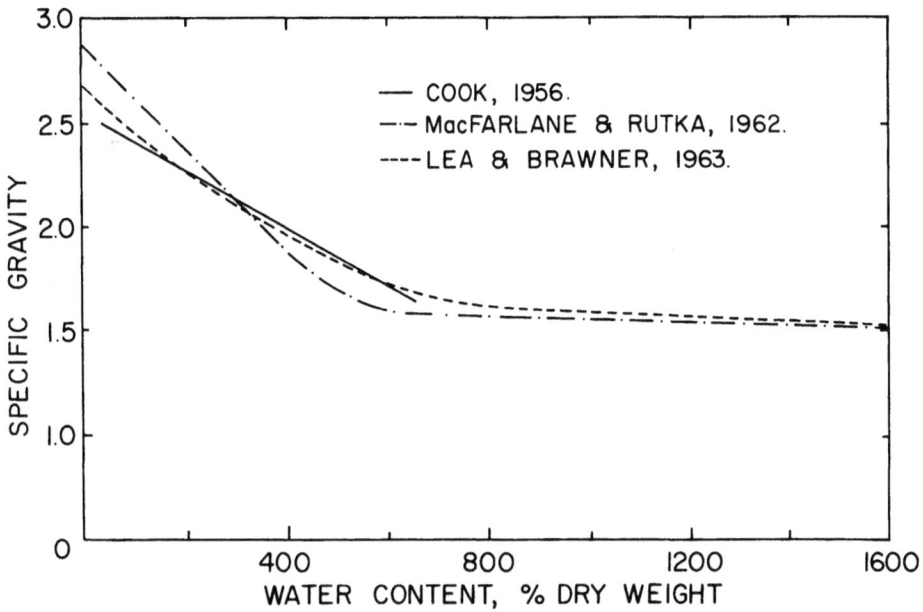

FIGURE 25   The relationship between specific gravity and water content.

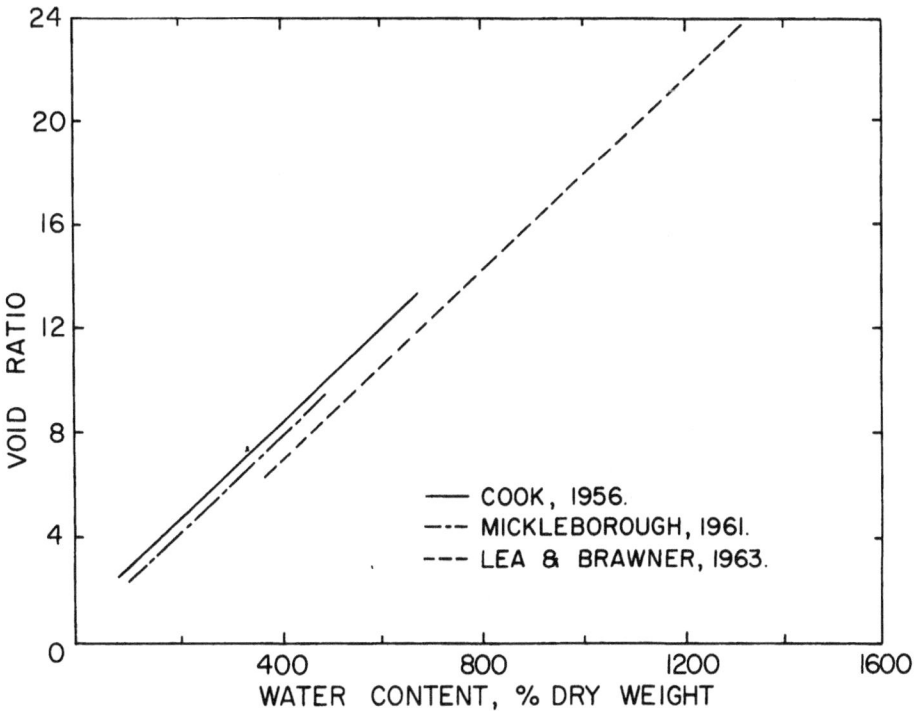

FIGURE 26   The relationship between void ratio and water content (% dry weight).

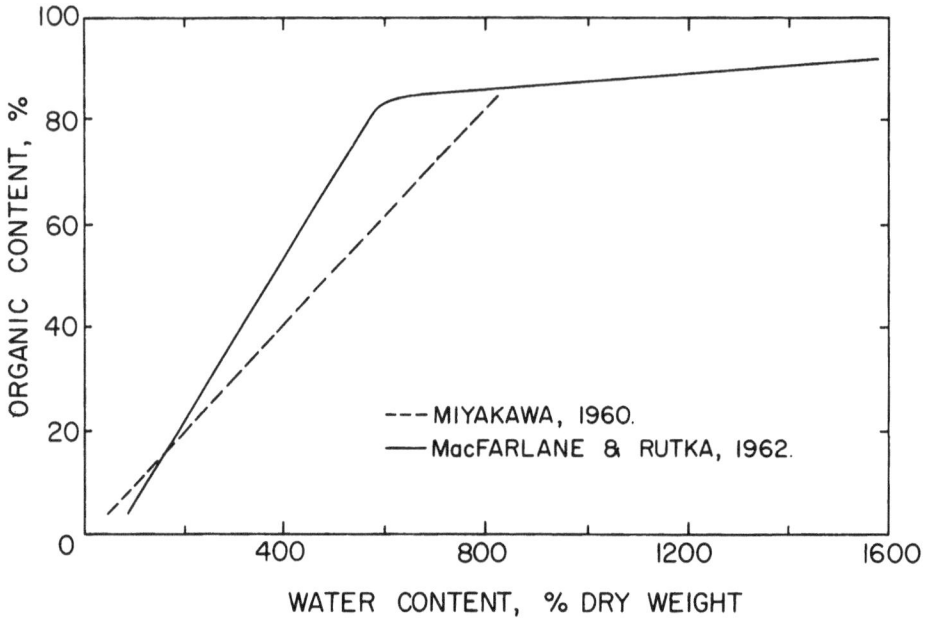

FIGURE 27   The relationship between organic content and water content.

(d) *Void ratio*. As stated by MacFarlane (1969), a linear relationship is expected between the void ratio and the water content since it is a simple mathematical relationship involving the specific gravity and gas content. Figure 26 illustrates this relationship.

(e) *Organic matter content*. Figure 27 indicates that at around 90 per cent organic matter content there appears to be no relation between the water content and organic matter content, whereas below approximately 90 per cent ash content there is a linear relationship with water content increasing as organic matter content increases. Again, this is the result primarily of mineral contamination.

Fibre Content

(a) *Bulk density*. Figure 28 shows that an approximately linear relationship exists between the bulk density and fibre content. Since the fibre content is a measure of the degree of decomposition, it is expected that a degradation of plant and moss fibres would cause an increase in the bulk density. The System of Soil Classification for Canada (1970) suggests that fibric layers usually exhibit bulk densities of less than 0.1 g/cc, whereas mesic layers usually exhibit bulk densities between 0.1 and 0.2 g/cc. Such a classification system recognizes the relation between bulk density and degree of decomposition.

(b) *Specific gravity*. Specific gravity has a positive curvilinear relationship with ash content, as was shown previously, but a negative curvilinear relationship with fibre content. Figure 29 indicates that as the fibre content decreases (decomposition increases) the specific gravity increases, since it has been shown that the ash content increases with increased decomposition. As was stated previously, ash elements constitute the heaviest fraction of peat.

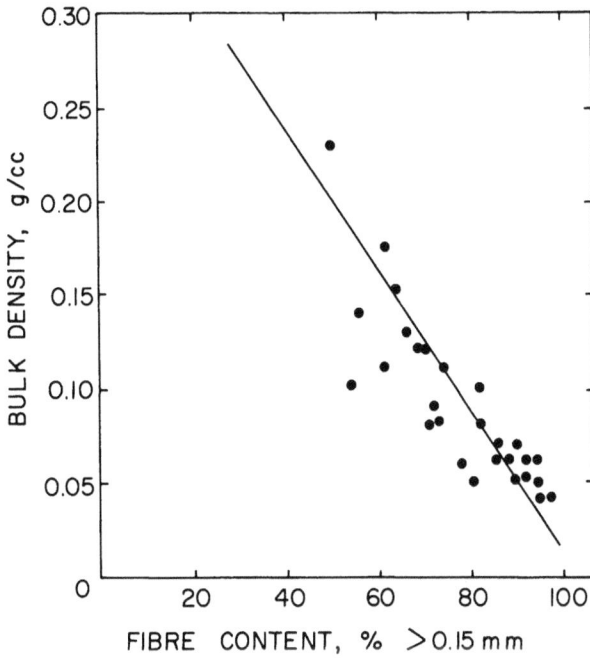

FIGURE 28   The relationship between the bulk density and fibre content of moss-dominated peats (after Padbury, 1970).

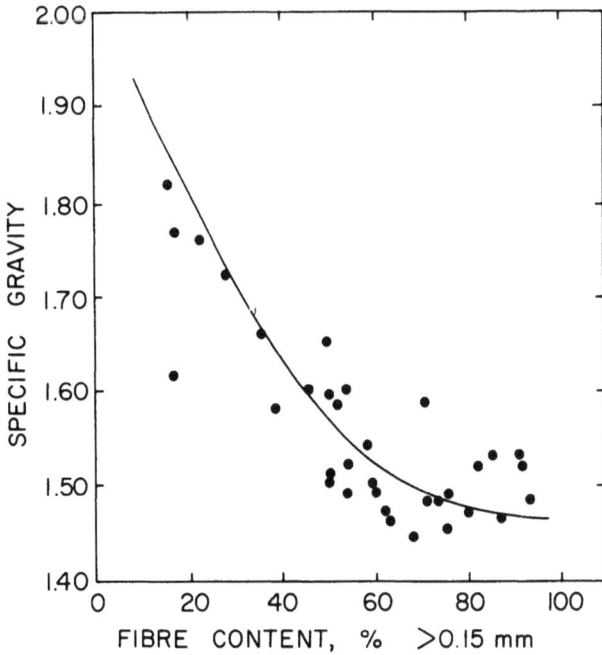

FIGURE 29   The relationship between the specific gravity and fibre content (after Padbury, 1970).

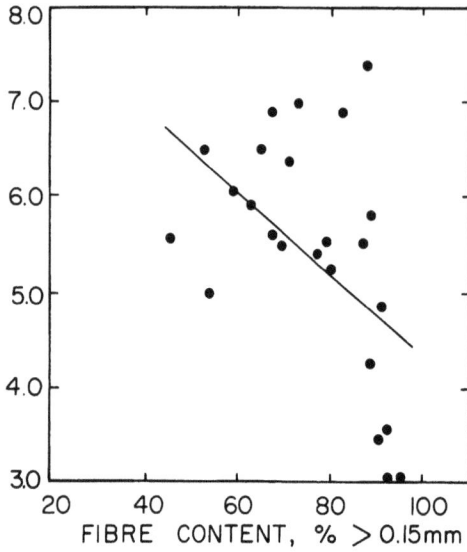

FIGURE 30   The relationship between pH and fibre content (after Padbury, 1970).

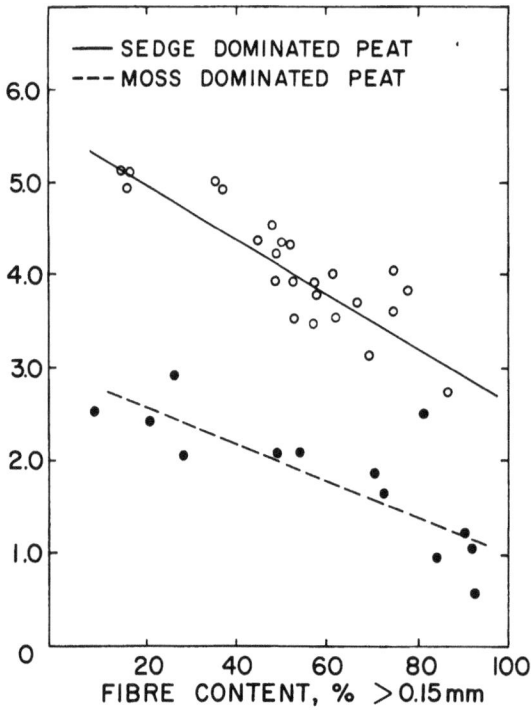

FIGURE 31   The relationship between total nitrogen content (ash-free) and fibre content (after Padbury, 1970).

(c) *Acidity*. Because of its control over the type of microorganisms which occur and their level of activity, pH plays an important role in the rate of decomposition. Highly acidic types, such as *Sphagnum* peats, are known not to contain certain microorganisms found in less acidic peats (Waksman and Stevens, 1928b). As a result, there should be a relation between the degree of decomposition as measured by fibre content and the pH. Figure 30 shows the relation for moss-dominated peats. Although there is much scatter to the points, there is a general increase in pH with a decrease in fibre content. Sedge-dominated peats do not exhibit this relationship, since the pH is sufficiently high to exclude it as a limiting factor for the growth of microorganisms.

(d) *Total nitrogen content*. When expressed on an ash-free basis, the total nitrogen content increases with an increase in the degree of decomposition. This relationship is not apparent unless the nitrogen content is expressed on an ash-free basis because of the variability among samples due to mineral contamination. Figure 31 illustrates that the total nitrogen content (ash-free basis) increases with increased degree of decomposition, as measured by the fibre content, for both major peat types.

CONCLUSION

In recent years many researchers have recognized the fact that as well as being affected by its surrounding environment, muskeg has the ability to create its own environment. As a consequence of this, muskeg, as a terrain type, expresses itself in a different manner than mineral soils. To enable an understanding of how muskeg relates to the environment as a whole, it becomes imperative that the physical and chemical properties of the material be measured consistently and understood. Due to its complex nature, an ecosystematic approach to environmental land use decision-making in muskeg areas is essential.

As an example, hydrologists have long known that by controlling the water content of muskeg by artificial drainage the rate of decomposition could be controlled. Once the water table is lowered, the balance between accumulation and deposition is so altered that decomposition is favoured (Waksman and Purvis, 1932; Isotalo, 1951). It is important to reduce the rate of decomposition as much as possible, past a certain stage, thereby prolonging the useful agricultural life of the peat (Harris et al., 1962; Neller, 1944). If decomposition is allowed to continue, organic material can essentially disappear. Hydrologists have also recognized the significant role organic soils play in determining the amount and distribution of water resources. In some areas, organic soils cover large portions of the headwater catchment of streams and rivers. Depending on the type of organic material, a slow release of water may result over the summer or all of the water may be released at freshet, as in mineral soils.

As illustrated by physical and chemical information, highly decomposed organic material is able to restrict the flow of groundwater, causing a different set of environmental conditions to prevail. *Sphagnum* mosses, under specific climatic conditions which are common in bog formations, cause a drastic reduction in pH and possess the ability, once established, to out-compete almost any other plant species. This is partially explained by their extremely low nutrient requirement, especially for nitrogen. The thermal characteristics of mosses enable them to insulate subsurface soils, thereby retaining cold temperatures throughout the year.

Climatic conditions are known to have a great influence on the formation and continuation of peatland conditions. Most muskeg in Canada is located in northern areas, mainly north of latitude 53°N. Recent exploration in the area between latitude 60°N and 64°N, in the plains area east of the Mackenzie Mountain front, has indicated that although peatland and therefore permafrost is encroaching upon post-glacial silty point bar material of the Liard River, it is degrading (thermokarsting) in a few areas, mainly in the Camsell Bend area of the Mackenzie River (Rutter, 1973). The indication that peatland is expanding on at least the glaciated surface of this area with the filling in of pond and fen areas is a concrete example of how muskeg is able to create its own environment. The impression left is that, at least in this area, muskeg or peatland is not in equilibrium with the living and non-living environment.

ACKNOWLEDGMENTS

Sincere appreciation is extended to the many colleagues who contributed information for the preparation of this paper, in particular: Dr N.W. Rutter, Geological Survey of Canada; Dr D.H. Boelter, USDA Forest Service; and Mr C. Tarnocai, Canada Department of Agriculture. The able technical assistance of Ms. B. Loughran in drafting the many figures is also appreciated.

REFERENCES

Alberta Soil Survey. 1967. Organic soils tour. Soils Division, Research Council of Alberta.

Amaryan, L.S. Bazin, Y.E.T., and Lishtvan, I.I. 1966. Swelling mechanisms of weakly decomposed peat. Pochvovedenia, 4: 391-5.

Black, C.A., Evans, D.D., White, J.L., Ensminger, L.E., and Clark, F.E. 1965. Methods of soil analysis. Agron. Ser. No. 9, Am. Soc. Agron.

Boelter, D.H. 1965. Hydraulic conductivity of peats. Soil Sci. 100: 227-31.

— 1966. Important physical properties of peat materials. Proc. Third Internat. Peat Congr. pp. 150-4.

— 1972. The hydrologic characteristics of undrained organic soils in the lake states. Proc. Organic Soils Symp., Soil Sci. Soc. Am., Florida.

Boelter, D.H., and Blake, G.R. 1964. Importance of volumetric expression of water contents in organic soils. Soil Sci. Soc. Am. Proc. 28:176-8.

Broadbent, F.E. 1953. The soil organic fraction. Adv. Agron. 5: 153-83.

Broadbent, F.E., and Bradford, G.R. 1952. Cation-exchange groupings in the soil organic fraction. Soil Sci. 74: 447-57.

Brochu, P.A., and Paré, J.J. 1964. Construction de routes sur tourbières dans la Province de Québec. Proc. Ninth Muskeg Res. Conf. NRC, ACSSM Tech. Memo 81: 74-108.

Brown, R.J.E. 1963. Influence of vegetation on permafrost. Proc. Permafrost Intern. Conf., Lafayette, Indiana. Natl. Acad. Sci. NRC Publ. 1287, Washington, DC, pp. 20-25.

Buckman, H.O., and Brady, N.C. 1969. The Nature and Properties of Soils (Macmillan Co., New York).

Canada Dept. of Agriculture, 1970. The System of Soil Classification for Canada (Queen's Printer, Ottawa).

Casagrande, L. 1966. Construction of embankments across peaty soils. J. Boston Soc. Civil Engrs. 53(3): 272-317.

Clayton, J.S., Padbury, G., St. Arnaud, R.J., and Janzen, W.K. 1967. Organic Soils Tour (Saskatchewan Inst. Pedology).

Colley, B.E. 1950. Construction of highways over peat and muck areas. Am. Highways 29(1): 3-6.

Cook, P.M. 1956. Consolidation characteristics of organic soils. Proc. Ninth Can. Soil Mech. Conf. NRC, ACSSM Tech. Memo. 41: 82-7.

Darcy, H. 1856. Les fontaines publique de la ville de Dijon (Dalmont, Paris).

Davis, J.F., and Lucas, R.E. 1959. Organic soils, their formation, distribution, utilization and management. Michigan State Univ. Agric. Exptl. Sta. Bull. 425, East Lansing, Michigan.

Duch, J. 1963. A tentative method of determining the pH in organic soils. Reczn. Glebozn. 13: 501-512. Abstr. Soils and Fert. 27 (1964), Bibliog. No. 3345.

Dyal, R.S. 1960. Physical and chemical properties of some peats used as soil amendments. Soil Sci. Soc. Am. Proc. 24: 268-271.

Elzen, J.A. 1961. Effect of agricultural utilization of fen bogs on the properties of bog soils under the conditions of the Estonian S.S.R. Pochvovedenie, 8: 840-5.

Everett, K.R. 1971. Composition and genesis of the organic soils of Amchitka Island, Aleutian Islands, Alaska. Arctic and Alpine Res. 3(1): 1-16.

Farnham, R.S. 1957. The peat soils of Minnesota. Minnesota Farm and Home Sci. 14(2): 12-14.

Farnham, R.S., and Finney, H.R. 1965. Classification and properties of organic soils. Adv. Agron. 17: 115-62.

Feustel, I.C., 1938. The nature and use of organic amendments. US Dept. Agric. Yearbook 1938: 462-8.

Feustel, I.C., and Byers, H.G. 1930. The physical and chemical characteristics of certain American peat profiles, U.S. Dept. of Agr. Tech. Bull. 214.

— 1936. The comparative moisture-absorbing and moisture-retaining capacities of peat and soil mixtures. Tech. Bull. USDA (US Govt. Printing Office, Washington, DC).

Fraser, D.A. 1957. Annual and seasonal march of soil temperatures on several sites under a hardwood stand. Canada Dept. Northern Affairs and Natl. Resources, Forestry Branch, Res. Div., Tech. Note 56.

Frazier, B.E., and Lee, G.B. 1971. Characteristics and classification of three Wisconsin histosols. Soil Sci. Soc. Am. Proc. 35: 776-80.

Frevert, R.K., and Kirkham, D. 1948. A method for measuring the permeability of a soil below a water table. Highway Res. Bd. Proc. 28: 433-42.

Goodman, L.J., and Lee, C.N. 1962. Laboratory and field data on engineering characteristics of some peat soils. Proc. Eighth Muskeg Res. Conf., NRC, ACSSM Tech. Memo. 74: 107-29.

Gore, A.J.P., and Allen, S.E. 1956. Measurement of exchangeable and total cation content for H, Na, K, Mg, Ca and Fe in high level blanket peat. Oikos 7: 48-55.

Gorham, E. 1953. Chemical studies on the soils and vegetation of water-logged habitats in the English lake district. J. Ecol. 41: 345-60.

— 1955. On some factors affecting the chemical composition of Swedish fresh waters. Geochem. et Cosmochem. 7: 129-50.

— 1956a. The ionic composition of some bog and fen waters in the English Lake district. J. Ecol. 44(10): 142-52.

— 1956b. On the chemical composition of some waters from the Moor House Nature Reserve. J. Ecol. 44(2): 375-832.

— 1958. Free acid in British soils. Nature, Lond. 181-6.

— 1960. The chemical composition of some bog water from the Falkan Islands. J. Ecol. 48(1): 175-81.

— 1961. Water, ash, nitrogen and acidity of some bog peats and other organic soils. J. Ecol. 49(1): 103-6.

— 1966. Some chemical aspects of wetland ecology. Proc. 12th Muskeg Research Conf. Tech. Memo. 90: 20-38.

Gorham, E., and Pearsall, W.H. 1956. Acidity, specific conductivity and calcium content of some bog and fen waters in northern Britain. J. Ecol. 44(1): 129-41.

Hanrahan, E.T. 1952. The mechanical properties of peat with special reference to road construction. Bull. Inst. Civil Engrs. Ireland 78(5): 179-215.

— 1954. An investigation of some physical properties of peat. Geotechnique 4(3): 108-23.

Hardy, R.M., and Thomson, S. 1956. Measurement of the shearing strength of muskeg. Proc. Eastern Muskeg Res. Meeting, NRC, ACSSM Tech. Memo. 42: 16-24.

Harris, C.I., Erickson, H.I., Ellis, N.K., and Larson, J.E. 1962. Water level control in organic soils related to subsidence rates, crop yields and response to nitrogen. Soil Sci. 94: 158-61.

Heinselmann, M.L. 1963. Forest sites, bog processes, and peatland types in the glacial lake Agassiz region, Minnesota. Ecol. Monogr. 33: 327-74.

Hillis, C.F., and Brawner, C.O. 1961. The compressibility of peat with reference to major highway construction in British Columbia. Proc. Seventh Muskeg Res. Conf., NRC, ACSSM Tech. Memo. 71: 204-77.

Hortie, H.J., Smith, R.E., and Russel, A. 1967. Organic Soils in Manitoba (Manitoba Soil Survey Unit, Univ. Manitoba).

Irwin, R.W. 1966. Soil water characteristics of some Ontario peats. Proc. Third Internat. Peat Congr., pp. 219-23.

Isotalo, I. 1951. Studies on the ecology and physiology of cellulose decomposing bacteria in raised bogs. Acta. Agralia. Fenn. 74: 106. Abstr. Comm. Bur. Soils. Bibliog. No. 1759.

Jackson, M. 1958. Soil Chemical Analysis (Prentice-Hall, Englewood Cliffs, NJ).

Jenny, H. 1941. Factors of Soil Formation (McGraw-Hill, New York).

Joffe, J.S., and Levine, A.K. 1947. Fixation of potassium in relation to exchange capacities of soils: III. Factors contributing to the fixation process. Soil Sci. 63: 241-7.

Jones, U.S. 1947. Availability of humate potassium. Soil Sci. Soc. Am. Proc. 12: 373-8.

Kaganovich, F.L., and Rakuskii, V.E. 1958. Selective extraction of peat bitumens at low temp. Vesti. Akad. Navuk Belarusk. SSR, Ser. Fiz. Tekhn. Navuk.: 117-22; Abstr. Chem. 57, 3721c.

Kirkham, D. 1949. Flow of ponded water into drain tubes in soil over-lying an impervious layer. Trans. Am. Geophys. Union 30: 369-85.

Kramer, P.J. 1949. Plant and Soil Water Relationships (McGraw-Hill, New York).

Kravchenko, V.P. 1963. Evaporation from the open surface of reclaimed peat soils. Visn. sil's kogospodar. Nauk. 6: 83-90. Ukranian Russ. summary. Abs. 175, Soils and Fert. 28: 34.

Kuntze, H. 1965. Physikalische Untersuchungsmethoden für Moor und Ammoorboden (English Summary). Landwirtschaft. Forsch. 18: 178-91.

Lavkulich, L.M. 1971. Soils-vegetation-landforms of the Fort Simpson area, N.W.T. Rept. 1971. Land Use Res. to Dept. Indian Aff. N. Develop. — ALUR Programme.

Lea, N.D., and Brawner, C.O. 1963. Highway design and construction over peat deposits in lower British Columbia. Highway Res. Board, Res. Rec. 7: 1-33 (Washington, DC).

Lewis, W.A. 1956. The settlement of the approach embankments to a new road bridge at Lockford, West Suffolk. Geotechnique 6(3): 106-44.

Lo, M.B., and Wilson, N.E. 1965. Migration of pore water during consolidation of peat. Proc. Tenth Muskeg Res. Conf., NRC, ACSSM Tech. Memo. 85: 131-42.

Lucas, A.J. 1970. Geochemistry of carbohydrates in some organic sediments of the Florida everglades (unpublished Ph.D. Thesis, Penn. State Univ.).

Lucas, R.E., and Davis, J.F. 1961. Relationship between pH values of organic soils and availabilities of 12 plant nutrients. Soil Sci. 92: 177-82.

Luthin, J.N., and Kirkham, D. 1949. Piezometer method for measuring permeability of soil in situ below a water table. Soil Sci. 68: 348-9.

Maas, E.F. 1972. The organic soils of Vancouver Island. CDA Res. Station, Sidney, B.C., Cont. no. 231.

MacFarlane, I.C. 1959. A review of the engineering characteristics of peat. Soil Mechanics and Foundations Dir., Proc. Am. Soc. Civil Engrs. 85 (SM1): 21-35.

— (ed.) 1969. Muskeg Engineering Handbook (Muskeg Subcommittee of NRC, Assoc. Comm. on Geotech. Res., Univ. Toronto Press).

MacFarlane, I.C., and Rutka, A. 1962. An evaluation of pavement performance over muskeg in northern Ontario. Highway Res. Board, Res. Bull. 316: 32-43 (Washington, DC).

MacLean, A.J., Halstead, R.L., Mack, A.R., and Jasmin, J.J. 1964. Comparison of procedures for estimating exchange properties and availability of phosphorus and potassium in some eastern Canadian organic soils. Can. J. Soil Sci. 44: 66-75.

McGeorge, W.T. 1931. Organic compounds associated with base exchange reactions in soils. Ariz. Agric. Exptl. Sta. Tech. Bull. 31.

Mainland, G. 1960. Corrosion in muskeg. Proc. Eighth Muskeg Res. Conf., pp. 100-6.

Measland, M. 1958. Auger hole method of measuring the hydraulic conductivity of soil and its application to tile drainage problems. Water Conservation and Irrigation Comm. NSW, Australia, Bull. 2.

Mickleborough, B.W. 1961. Embankment construction in muskeg at Prince Albert. Proc. Seventh Muskeg Res. Conf. NRC, ACSSM Tech. Memo 71: 164-85.

Miyakawa, I. 1959. Soil engineering research on peaty alluvia: reports 1-3. Civil Eng. Res. Inst. Hokkaido Development Bur., Bull. 20, Sapporo. English transl. by K. Shimizu, NRC Transl.

— 1960. Some aspects of road construction in peaty or marshy areas in Hokkaido, with particular reference to filling methods. Civil Eng. Res. Inst., Hokkaido Development Bur., Sapporo.

Neller, J.R. 1944. Oxidation loss of low moor peat in fields with different water tables. Soil Sci. 58: 195-204.

Padbury, G.A. 1970. Properties of organic soils in relation to classification (unpublished M.SC. thesis, Dept. Soil Sci., Univ. Saskatchewan).

Parsons, J.W., and Tinsley, J. 1961. Chemical studies of polysaccharide materials in soils and composts based on extraction with anhydrous formic acid. Soil Sci. 92: 46-53.

Passer, M., Bratt, G.T., Elberling, J.A., Piret, E.L., Hartman, L., and Madden, A.L. 1963. Trans. Second Internat. Peat Congress, Leningrad.

Phalen, T.E. 1961. Recent investigation concerning a high void ratio soil. Discussion in Proc. Fifth Internat. Conf. on Soil Mech. and Foundation Eng., 3: 379-80.

Pollett, F.C. 1972. Nutrient contents of peat soils in Newfoundland. Fourth Internat. Peat Congress, Helsinki.

Puustjärvi, V. 1956. On the cation exchange capacity of peats and on factors of influence upon its formation. Acta Agric. Scand. 6(4): 410-49.

— 1957. On the base status of peat soils. Acta Agric. Scand. 7(2): 190-223.

Radforth, N.W. 1952. Suggested classification of muskeg for the engineer. Eng. Jour. Nov.

Reginato, R.J., and Van Bavel, C.H.M. 1962. Pressure cell for soil cores. Soil Sci. Soc. Am. Proc. 26: 1-3.

Ripley, C.F., and Leonoff, C.E. 1961. Embankment settlement behaviour on deep peat. Proc. Seventh Muskeg Res. Conf., NRC, ACSSM Tech. Memo. 71: 185-204.

Rutter, N.W. 1973. Personal communication.

Schlungbaum, G. 1968. Bergbautechnik. 18: 203.

Schnitzer, M., and Desjardines, J.G. 1965. Carboxyl and phenolic hydroxyl groups in some organic soils and their relation to the degree of humification. Can. J. Soil Sci. 45: 257-64.

Schnitzer, M., and Hoffman, I. 1966. A thermogravimetric approach to the classification of organic soils. Soil Sci. Soc. Am. Proc. 30: 63-6.

Segeberg, H. 1955. The specific weight of peat from low moor. Z. Pflerhanahr Dung. 71: 133-41. Abstr. Soils and Fert. 19 (1956), Bibliog. No. 773.

Shacklock, C.W., and Drakeley, T.J. 1927. A preliminary investigation of the nitrogenous matter in coal. J. Soc. Chem. Ind. 46: 478-81.

Shea, P.H. 1955. Unusual foundation conditions in the Everglades. Trans. Am. Soc. Civil Engrs. 120: 92-102.

Shickluna, J.C., Lucas, R.E., and Davis, J.R. 1972. The movement of potassium in organic soils. Fourth Internat. Peat Congress, Helsinki.

Sjörs, H. 1948. Mire vegetation in Bergslagen, Sweden. Acta Phytogeographica Seucica 21 (Swedish, Engl. summ.)

— 1950. On the relation between vegetation and electrolytes in north Swedish mire waters. Oikes 2(2): 241-58.

— 1961. Surface patterns in boreal peatland. Endeavour 20: 217-24.

Smith, A.H.V. 1950. A survey of some British peats and their strength characteristics. Army Operational Res. Group, Rept. 32-49, London.

Smith, D.B., Bryson, C., Thompson, E.M., and Young, E.G. 1958. Chemical composition of the peat bogs of the Maritime Provinces. Can. J. Soil Sci. 38: 120-7.

Sneddon, J.I., and Luttmerding, H.A. 1967. Organic Soils Tour (Research Branch, Canada Dept. Agriculture, Vancouver).

Sturges, D.L. 1968. Hydrologic properties of peat from a Wyoming mountain bog. Soil Sci. 107: 262-4.

Tarnocai, C. 1970. Classification of Peat Landforms in Manitoba (CDA Res. Station, Pedology Unit, Winnipeg).

— 1973. Soils of the Mackenzie River area. Canada Dept. Agriculture (unpublished).

Tessier, G. 1966. Deux exemple-types de construction de routes sur muskegs au Québec. Proc. Eleventh Muskeg Res. Conf., NRC, ACGR Tech. Memo 87: 92-141.

Theander, O. 1954. Acta. Chem. Scand. 8: 989.

Thiesen, R., and Johnson, R.C. 1929. Ind. Eng. Chem. Anal. Ed. 1: 216.

Thompson, J.B., and Palmer, L.A. 1951. Report of consolidation tests with peat. Symp. Consolidation Testing of Soils, ASTM Spec. Tech. Publ. 126: 4-8.

Tolpa, S., and Gorham, E. 1961. The ionic composition of waters from three Polish bogs. J. Ecol. 49(1): 127-33.

Turk, L.M. 1939. Effect of certain mineral elements on some microbiological activities in muck soil. Soil Sci. 47: 425-45.

Tveiten, A.A. 1956. Applicability of peat as an impervious material for earth dams (Norwegian with English summary). Norwegian Geotech. Inst. Publ. 14.

Van Wijk, W.R. 1963. Physics of Plant Environment (Wiley, New York).

Vasil'ev, S.F., and D'Yakova, M.K. 1958. Thermal solvent extraction as a method of processing peat into gas, chemical products and motor fuels. Novye Metody. Rats. Ispol'z Mestnykch Topliv, Tr. Soveshch. Riaa, pp. 37-46; Abstr. Chem. 56, 2666d.

Volarovich, M.P., and Gusev, K.F. 1956. X-ray studies of the structure of peat bitumens and waxes. Kolloidn. Zh. 18: 643-6; Chem. Abstr. 51, 6980a.

Waksman, S.A. 1938. Humus (Williams and Wilkins Co., Baltimore).

— 1942. The peats of New Jersey and their utilization. N.J. Dept. Conservation and Development, Bull. 55, Geol. Series, Trenton, NJ.

Waksman, S.A., and Purvis, E.R. 1932. The microbial population of peat. Soil Sci. 34(2): 95-113.

Waksman, S.A., and Stevens, K.R. 1928a. Contribution to the chemical composition of peat. I. Chemical nature of organic complexes in peat and methods of analysis. Soil Sci. 28(2): 113-37.

— 1928b. Contribution to the chemical composition of peat. II. Chemical composition of various peat profiles. Soil Sci. 26(4): 239-51.

— 1929. Soil Sci. 26: 113.

— 1930. Ind. Eng. Chem. Anal. Ed. 2: 167.

Walmsley, M.E., and Lavkulich, L.M. 1973. In situ measurement of dissolved materials as an indicator of organic terrain type. Can. J. Soil Sci. (in press).

Ward, W.H. 1948. A slip in a flood defence bank constructed on a peat bog. Proc. Second Internat. Conf. on Soil Mech. and Foundation Eng. 2:19-23.

Wilson, L.R. 1939. A temperature study of a Wisconsin peat bog. Ecol. 20: 423-33.

Wollrab, V., and Streible, M. 1969. Earth waxes, peat, montan wax and other organic brown coal constituents. In Organic Geochemistry, ed. G. Eglinton and M.T.J. Murphy (Springer-Verlag, Berlin), pp. 576-98.

Yamazaki, F., Soma, K., and Furaya, C. 1957. On the pF curve of a peat soil. J. Agr. Eng. Soc. Japan 25: 214-17. Abstr. Soils and Fert. 22 (1959), Bibliog. No. 2163.

Yefimov, V.N. 1961. Forms of accumulation and migration of substances in bog soils. Pochvovedenie, 6: 643-9.

Yefimov, V.N., and Donskikh, I.N. 1969a. The mobility of nitrogen, phosphorus and potassium in peat soils. Soviet Soil Sci. 2: 178-87.

— 1969b. Mobility of nitrogen, phosphorus and potassium in peat soils. Soviet Soil Sci. 178-87. Translated from Agrokhimiya 3 (1969): 44.42.

APPENDIX I
Summary of physical properties of peat

Reference	Bulk density (g/cc)	Hydraulic conductivity (cm/sec x 10⁻⁴)	Void ratio	Water content (wt. %)	Specific gravity	Fibre content (%)
Alberta Soil Survey (1967): Alberta						74.7
Terric mesic humisol Of₁						80.3
Of₂						41.9
Oh₁						44.1
Oh₂						
Boelter (1965: Minnesota	0.040	381.0				
Undecomposed *Sphagnum* mosses	0.137	49.60				
Moderately decomposed woody peat	0.156	0.07				
Moderately decomposed herbaceous peat	0.261	0.04				
Well-decomposed peat						
Boelter (1966): Minnesota	0.02			2.000		
Sphagnum mosses	0.24			167		
Aggregated peat	0.16			250		
Herbaceous peat						
Boelter (1972): Minnesota						
Macell Bog	0.07	500				70-75
Fibric *Sphagnum* moss peat (surface)	0.17	0.22				35-45
Humic peat (subsurface)						
Floodwood Bog	0.07	1,600				85
Fibric *Sphagnum* moss peat (surface)	0.08	720				67
Fibric peat (subsurface)						
Brochu and Paré (1964): Quebec	0.147-0.194		7-11	300-650	1.3-1.75	
Fibrous peat		0.3	7-1	250-800		
Casagrande (1966): Massachusetts						
Clayton et al. (1967): Saskatchewan						
Terric mesic fibrisol Of₁	0.06					74.5
Of₂	0.12					70.0
Of₃	0.09					89.5
Om	0.14					56.4
Colley (1950): Florida	0.20		4.6-10.3	485-910		
Cook (1956): Vancouver			2.8-13.1	120-800	1.85-2.45	
Everett (1971): Alaska						
Fluventic borofibrist	0.09			790-2,640		
Fluventic borohemist	0.16			320-1,250		
Lithic folist	0.26			160-932		
Lithic folist	0.36			104-400		
Feustel and Byers (1930): U.S.A.						
Sphagnum peat	0.486*			3,235	1.588	
Sphagnum peat	0.645*			2,640	1.501	
Heath peat	0.851*			798	1.491	
White cedar forest peat	1.158*			643	1.557	
Saw grass peat	0.985*			1,255	1.560	
Saw grass peat	1.028*			1,485	1.490	
Goodman and Lee (1962): New York	0.24-0.41		5.2-5.5	280-320	1.51-1.62	
	0.24-0.26		4.8-5.5	240-575	1.53	
Hanrahan (1952): Ireland	0.961-1.01*				1.2-1.7	
Hanrahan (1954): Ireland	0.064-0.13	0.4				
Hardy and Thomson (1956): Alberta						
Fibrous peat				470-760	1.4	
Hillis and Brawner (1961): B.C.			2-15			
Hortie et al. (1967): Manitoba			5-20			
Unic mesisol Of						
Om₁	0.07					76
Om₂	0.11					61
Om₃	0.13					54
Irwin (1966): Ontario	0.15					45
Holland Marsh 0-15 cm	0.172	6.1	10.0		1.733	
15-30 cm	0.099	5.0	15.1		1.578	
30-45 cm	0.096	1.1	15.6		1.573	
45-60 cm	0.096	4.3	15.3		1.567	
60-75 cm	0.096	1.5	15.8		1.591	
75-90 cm	0.094	1.9	16.3		1.597	

Reference	Bulk density (g/cc)	Hydraulic conductivity (cm/sec x 10^{-4})	Void ratio	Water content (wt. %)	Specific gravity	Fibre content (%)
Lea and Brawner (1963): Vancouver	0.06	0.01-1.0		500-1,500	1.5-1.6	
Lewis (1956)	1.05*			520		
Lo and Wilson (1965): Ontario				600-700	1.95-2.05	
Maas (1972): Vancouver Island						
Cultivated marsh peat, top 6 inches	0.45					
Virgin marsh peat, top 6 inches	0.21					
Cultivated bog peat, top 6 inches	0.25					
Virgin bog peat over marsh peat, top 6 inches	0.07					
MacFarlane and Rutka (1962): Ontario				105-2,780		
Mickleborough (1961): Saskatchewan						
Cat. 9 (Radforth)	0.04-0.34		3.2-9.9	145-480	1.6-1.8	
Miyakawa (1960): Japan	0.10-0.27		6.5-11.4	155-810	1.33-1.91	
Parallel to horizontal layer		20-130				
Perpendicular to horizontal layer		0.49-0.57				
Padbury (1970): Saskatchewan						
Moss-dominated peats	0.04			1,690	1.48	93.5
	0.23			460	1.65	50.3
Sedge-dominated peats	0.05			1,430	1.46	86.9
	0.26			360	1.60	46.1
Phalen (1961)	0.11	0.1-0.001	4.2	205	1.9	
Ripley and Leonoff (1961)						
Cat. 9 (Radforth)				100-2,100		
Shea (1955): Florida	0.08-0.16			500-1,000		
Smith (1960)						
Highly humified peat				150-535		
No recognizable plant remains				145-625		
Plant remains still recognizable				495-1,340		
Fresh peat				610-1,715		
Sneddon and Luttmerding (1967): B.C.						
Unic mesisol Of	0.19					61.53
Om$_1$	0.12					59.13
Om$_2$	0.11					42.47
Oh$_1$	0.15					27.69
Oh$_2$	0.21					
Tessier (1966): Quebec	0.09-0.15		5-16	200-890		1.9-2.7
Thompson and Palmer (1951)	0.15-0.19		7-11	300-650		1.9-2.68
Degree of sat. = 82.5-95.6	0.23-0.33		5.1-7.1	240-340		1.79-2.03
Tveiten (1956)						
Sphagnum peat		1-9				
Mixed medium peat		0.01-0.05				
Grassy medium peat		0.01-0.03				
Dark grassy peat		0.01-0.02				
Ward (1948): Wales			6-17	800-1,000	1.2-1.5	

*Assumed to be calculated on a dry-volume basis.

APPENDIX 2
Summary of chemical properties of peat

Reference	pH	Ash content (%)	Base saturation (%)	C:N	Total N (%)	Exchangeable cations (me/100 g) K	Mg	Ca	CEC (me/100 g)
Alberta Soil Survey (1967): Alberta									
Terric mesic humisol Of₁	5.0*	16.9	66	31	1.3				
Of₂	5.6*	9.8	85	25	1.1				
Oh₁	5.6*	15.1	88	34	1.2				
Oh₂	5.8*	19.3	87	33	1.1				
Clayton et al. (1967): Saskatchewan									
Terric mesic fibrisol Of₁	7.3	19.8							157.4
Of₂	5.8	13.3							183.6
Of₃	5.7	10.1							178.4
Om	5.2	17.6							188.7
Everett (1971): Alaska									
Borofibrist 3-8 cm	5.0					2.0	24	17	
8-13 cm	4.7					2.0	18	13	
13-21 cm	4.8					0.5	9	7	
21-54 cm	4.7					0.3	9	9	
54-58 cm	4.8					0.2	8	10	
58-60 cm	4.9					-	1	1	
60-70 cm	5.0					0.1	7	9	
70-88 cm	5.1					0.1	7	9	
Farnham (1967): Minnesota									
Amorphous peat	6.2	66.6							
Aggregate peat	7.5	55.9							
Moss peat	3.8	4.5							
Herbaceous peat	5.4	15.8							
Aquatic peat	7.5	43.9							
Woody peat	6.0	12.5							
Goodman and Lee (1962): New York	4.7-8.6	24.8-59.8							
Gore and Allen (1956): Great Britain	6.1-7.1	25-58							
Gorham (1961): Great Britain						10.1	4.8	4.3	
Ombrogenous *Sphagnum* box peats	3.8	3.2		1.3					
Lacustrine bog peats	5.1	11.6		1.9					
Reedswamp and *Carex* fen peat	5.7	20.0		2.2					
Oakwood mor humus layer	3.2	17.0		2.4					
Gorham (1966)									
Falkland Islands	4.1**								
West Ireland	4.4**								
North Scotland	4.5**								
North England	3.9**								
West Scotland	4.5**								
Poland	3.9**								
Hortie et al. (1967): Manitoba									
Unic mesisol Of	6.3*	10.8		26	1.8	0.7	31.5	124.2	179.5
Om₁	6.1*	12.3		20	2.4	2.3	34.3	130.4	213.8
Om₂	5.9*	13.1		20	2.5	0.4	29.7	127.3	219.6
Om₃	5.8*	16.2		17	3.0	0.4	26.0	100.3	164.7
Lavkulich (1971): Fort Simpson Area, NWT									
Cyric fibrisol Of₁	4.0		27.0	49	0.9	1.5	5.4	25.0	118.9
Of₂	4.1		25.7	32	1.4	0.5	5.6	28.4	158.9
Cyric mesisol Of	6.2		76.3	15	2.6	-	24.5	75.0	134.9
Om₁	6.5		72.9	20	1.9	0.5	34.0	81.2	166.3
Om₂	5.5		70.9	31	1.5	2.6	19.2	42.5	93.3
Maas (1972): Vancouver Island									
Cultivated marsh peat, top 6 inches	5.4	41.0			1.8	2.4	15.1	101.3	143.0
Virgin marsh peat, top 6 inches	4.6	11.0			2.9	0.6	16.5	70.9	146.6
Cultivated bog peat, top 6 inches	3.3	6.0			2.4	1.6	6.4	22.6	143.0
Virgin bog peat over marsh peat, top 6 inches	3.5	3.0			1.4	1.3	15.0	14.4	146.6

	pH	Ash content (%)	Base saturation (%)	C:N	Total N (%)	Exchangeable cations (me/100 g)			CEC (me/100g)
						K	Mg	Ca	
Puustjärvi (1956): Scandinavia									
Birch fen					3.6			61	135
Carex-Sphagnum box								32	85
Water-logged *Sphagnum* bog								7	106
Puustjärvi (1957): Scandinavia									
Sedge peat	5.4		37.2				2.8	13.9	73
Eutrophic *Sphagnum-Carex* peat	5.5		80.4				16.1	80.0	130
Forest sedge peat	4.7		37.7				2.3	15.4	90
Sphagnum-Carex peat	4.5		36.9						101
Forest *Sphagnum* peat	4.2		33.2						94
Sphagnum peat	3.2		13.6						125
Smith (1950)									
Sphagnum moss		2.0							
Cotton grass peat		0.6							
Reed peat		14.5							
Birchwood peat		2.5							
Heath humus		10.0							
Sneddon and Luttmerding (1967): B.C.									
Unic mesisol Of	3.5	1.5	25.0		1.4	1.2	13.2	22.1	152.3
Om₁	3.6	1.9	19.4		0.7	1.2	15.4	12.7	159.3
Om₂	5.0	4.8	33.4		1.3	1.2	35.4	13.4	154.3
Oh₁	5.3	40.6	57.2		1.1	1.2	24.8	19.9	82.3
Oh₂			77.1		1.0	1.2	33.1	16.1	67.3
Tarnocai (1973): Mackenzie River area, NWT									
Typic mesisol (Om)	5.7	14.1		26	2.3	0.1	9.8	74.7	109
Cryic mesisol (Of)	2.7	1.3		90	0.7	0.9	5.0	14.1	103
Cryic fibrisol (Of)	2.5	2.6		38	1.6	0.9	3.0	8.1	99
Spring fen	7.3**								
Patterned fen	7.8**								
Bog plateau	4.7**								
Peat polygon	3.9**								
Ice wedge	5.9**								
Tessier (1966): Quebec	6.0-7.0	7-13							
Waksman (1942)	5.0-7.0	15-50							
Moss peat		2.2-2.5							
Decomposed heath peat		6.9							
Sedge and wood peat		2.2-37.5							
Walmsley and Lavkulich (1973): NWT									
Fen	6.5**								
Transitional fen	4.8**								
Transitional bog	4.4**								
Bog	3.8**								
Yefimov and Donskikh (1969: USSR									
Upland peat	2 6*	1.3	31.4	67	0.6				
Transitional *Sphagnum* peat	3.6*	5.3	53.7	26	1.9				
Transitional woody sedge peat	4.4*	7.6	71.2	19	2.4				
Low and woody sedge peat	5.7*	13.8	95.2	27	1.9				

* pH in salt solution.
** pH of groundwater.

5
Muskeg Hydrology

N.W. RADFORTH

Muskeg hydrology was never more significant than it is now when an understanding of the northern environment is so vital for proper resource development. Sound knowledge is required for the consideration of problems in many areas: in the preparation of national land inventory; mining; creation of national parks; planning for agriculture and forestry; road building and maintenance and railway construction; designing foundations, airstrips, and pipelines; planning for continental water reserves; assessing land use; and engaging in off-road transportation. Such matters as understanding the biotic-abiotic interaction in muskeg and studying muskeg properties (for example its thermal and permeability behaviour) are basic to a study of any of these other problems in the north.

The classification of muskeg forms a part of the basic material, but, whatever the method or system used for this, it will be incomplete if it does not account for its water relations. Indeed to attempt the development of northern North America beyond present levels without considering in ordered fashion the kinds of hydrological behaviour that occur would be to invite a national disaster.

The importance of the water factor can be emphasized by the following example. One manufacturer of primary peat in Canada produces half a million tons of product per year. For every hundred tons put back on the land for whatever reason, approximately 1,000 tons of water can be locked into the soil for immediate use. The company, therefore, creates annually the possibility of holding, for use on otherwise deprived soils, up to one billion, one hundred and ninety-eight million gallons of water. Of course, at the mining site, this potential has been destroyed by drainage canals and general disturbance of the surface. Equivalent units of loss are likely to be experienced in connection with the James Bay Development Project for one category of peat structure alone, not to mention the large volumes of muskeg that will be inundated through headpond development.

In large pipeline construction, it is claimed that environmental reporting should account for the disturbance effect for at least 75 terrain factors that commonly recur in muskeg. In almost all cases, when the environmental disturbance is assessed, it turns out that water in either the solid or liquid phase is involved.

When muskeg is flooded, usually some of the organic overburden, but not all, floats. If it is not flooded, but its water table is changed substantially by artificial means, changes in the living cover arise and these are accompanied by gross structural modification in the peat. Where this occurs on a large scale, involving one to numerous natural muskegs, reservoir drainage relations, including the capacity of the muskeg to hold or deliver water, will be altered for very large areas. Also, the quality of the drainage water will be affected. The degree of alteration will depend upon the gross peat type or types affected.

A review of this reasoning points to urgent need for a hydrological classification of muskeg.

THE APPROACH FOR CLASSIFICATION

Often definitions lack qualifications, the definition of muskeg being a case in point since it makes no reference to water: '*muskeg:* The term designating organic terrain, the physical condition of which is governed by the structure of peat it contains, and its related mineral sublayer, considered in relation to topographic features and the surface vegetation with which the peat co-exists.' This definition does, however, establish the point that peat is an essential material component of muskeg and, of course, there can be no peat unless water is present, so the presence of water is implicit in the definition.

Water is the primary fossilizing agent through which peat is preserved. In this sense it is misleading to suggest that peat is decayed vegetation. The water reduces aerobic activity wherever peat is either forming or maintaining itself. Changes in constitution within the peaty matrix are therefore either chemical or mechanical in nature. Breakdown or decomposition here does not infer dissociation by bacterial-fungal action as it might for the organic components of a mineral soil complex or for forest 'duff' or 'mull,' where aerobic activity is high and complex organic constituents are recycled into ultimate chemical products which recharge the mineral budget of the soil bed.

The question has often arisen as to whether 'wetlands' is a synonym for muskeg. If the terrain condition in question implies the presence of peat, then, by definition, the wetland is muskeg. However, a great variety of wetland conditions exist in which peat is absent. Dead organic matter, if present, undergoes at some time either decay or displacement, be it on a continuous gradual basis or intensively and periodically. Seasonal lowering of the water table often exposes plant remains that are otherwise submerged. At such times, normal processes of decay ensue until the reservoir is recharged. In some instances where the organic deposit is submerged, wave action or, on sea coasts, the tidal effect induces aeration. Such conditions in the English-speaking world are referred to as swamps or marshes, although usage would favour the word 'marsh' for a condition in which the dead vegetation, which is usually constantly under replacement, is frequently either just above or just below the groundwater level depending upon erosion, rainfall, or runoff conditions.

Despite the need for precision in classification, the expressions muskeg, swamp, marsh, and wetlands continue to be used interchangeably by some investigators and reporters (Pollett, 1972). This causes no serious misunderstanding provided the pertinent terrain conditions are specified, but where development of the land is anticipated and where land inventory is involved (of hydrological conditions in particular), greater precision is required. Engineering practice is dependent on knowledge of whether conditions of preservation exist to produce and maintain peat deposition. If they are absent sensu stricto, land is something other than muskeg, according to the accepted definition.

In Europe, far more emphasis has been placed on designating, describing, and accounting for hydrological phenomena in peatlands than in other countries. This situation is undoubtedly due to the more intensive use of the land in Europe, where the demand for horticultural and forest products and the exploitation of peat resources forced attention on muskeg problems at a relatively early date.

The early records refer primarily to the accumulation processes involved in peat deposition — 'terrestrialization' as it is expressed by Bellamy (1968). For example, Weber (1908) uses the German expressions *Niedemoore, Ubergangmoore,* and *Flochmoore,* expressions which have since been applied elsewhere as 'lowmoor,' 'transitionmoor,' and 'highmoor.' The hydrological implication is that a natural catchment basin collects the water from a remote source, either from surface flow, directly from precipitation, or from an underground supply. Whatever the source of water, its behaviour within the catchment, as terrestrialization ensues, will depend upon the evolving hydrological development. This suggests that 'categories' of hydrological expression will ensue, which brings us back to the hydrological classification of muskeg.

An examination of the need for classification reveals two major problem areas: one concerns the characterization of the major hydrological system, the muskeg, in its entirety, and the other involves hydrological determinations at specified sites within the system. For the latter, the requirements are often so specific as to be satisfied simply by the measurement of the water content.

In Europe, for local sampling within the system, use has been made of the von Post approach (in Kivinen, 1948). The qualitative assessment is expressed by the factor 'B' in the von Post formula for describing the peat, thus:

B_1 = air dry
B_2 = slightly dried out
B_3 = average wet
B_4 = very wet
B_5 = mainly water

Later, in rendering the quantitative expression, the water content is given as a percentage of either the dry or wet weight. These values, however, do not clearly indicate the bulk amounts of water, and the tendency is now to measure as a percentage of the total wet volume. This is the approach now popular in North American technology.

In the northern part of the continent, the presence of ice-mass configurations within the organic terrain is now well known, but, in fact, they occur in the south too (beyond the 55th parallel), 1,500 miles from icecap country. When this cryogenic phenomenon persists in muskeg, as it does in the north, it makes marked differences in the topographic and hydric characteristics of the organic lands and their biotic components. In the south, the seasonal ice has its characteristic effect too, but it differs from that due to permafrost.

The nature of the cryogenic features and their distribution present special aspects of hydrology which are not yet fully understood. Enough is known about them, however, to enable classification to be initiated. The variability arising from cryogenic factors applies to probably more than 40 per cent of Canada's land area. The aspects of national and provincial development which were mentioned at the beginning of this chapter are all dependent on a wise assessment of the cryogenic aspects of muskeg environment. Therefore, however little is known, it must be revealed and categorized for convenient reference. Where knowledge is imperfect, the problem should at least be identified.

CLASSIFICATION METHODOLOGY

The inevitability of circumstances requiring a number of classification methods to account for natural phenomena can readily be countered for muskeg if natural relationships characterizing the phenomenon are used to coordinate the otherwise distinct methods. Examination of Figure 1 (Radforth, 1972) suggests a taxonomic rationale for dealing with terrain states and the materials involved. The natural feature in the system is expressed by the ultimate palynological typification of microfossil conspecti.

In the present context, emphasis is placed on hydroform and cryoform (Figures 4 and 5). Both these designations relate to pattern: clearly, the first for land form as related to the effect of the water factor on biotic development, and the second for land form as related to the effect of subsurface ice on biotic development. Understanding of these patterns is always attainable by a consideration of the biotic history and reference to the contemporary vegetation.

If, in classification, abiotic taxa are to be emphasized, the expressions will suggest behavioural attributes (cf. Figure 1, no. 10, physiogenic, and no. 11, chemogenic) as well as pattern. These taxa are the ones often emphasized in European approaches whereas, until recently, the water factor received only indirect reference. Bellamy (1972) was somewhat less direct in putting forth his 'Climatic Templates' when explaining his use of the expressions Primary, Secondary, and Tertiary Mires. He modifies and augments Bulow's (1929) methods of referring to organic terrain systems as distributed in Europe. So far little emphasis has been placed on the reservoir from a strictly hydrological point of view. Essentially, Bellamy's concept of templates is oriented toward explanations of peat formation rather than reservoir genesis and behaviour of water. He concludes that 'The Primary, Secondary and Tertiary peats are in essence the "fen", "sedge" and "*Sphagnum*" peats of commerce.'

Both Bellamy and the writer, together on field investigations in North America in 1971 and 1972, noted afresh the significance of airform patterns (Radforth, 1956a) as real and recurring phenomena. Such patterns served as convenient expressions in locating and accounting for catchments, muskeg-reservoir development, drainage gradient, and directional change in gradient. It was agreed in these studies that classification should accommodate ontogenetic differences in given hydrological systems. The basic claim is that 'any area in the world in which water collects on its way down from the catchment to the sea constitutes a "template" for the development of a peat producing ecosystem' (Radforth and Bellamy, 1973). Thus patterns in muskeg, on whatever scale, express the hydrological status of the terrain, and an entire conspectus of recurring patterns, continent-wide, is the interpretive key to the understanding of continental water.

On the first muskeg map of Canada (Radforth, 1969a) the main feature is a central belt in which muskeg is usually unconfined. If it is interrupted, the breaks are major slopes. Both to the south and north of the central belt the muskeg becomes progressively more confined, closely adhering to drainage features. The terms 'unconfined' and 'confined' are already known (Radforth, 1969a), but Bellamy et al. (1973) have introduced another expression, 'thermal blanket mire (muskeg),' to describe the belt of the essentially unconfined state.

The distributional relationship between confined and unconfined muskeg along the north-south axis is shown in Figure 2, which also shows their relation to the distribution of airform patterns. These conditions, namely confined and unconfined, have been noted as descriptive

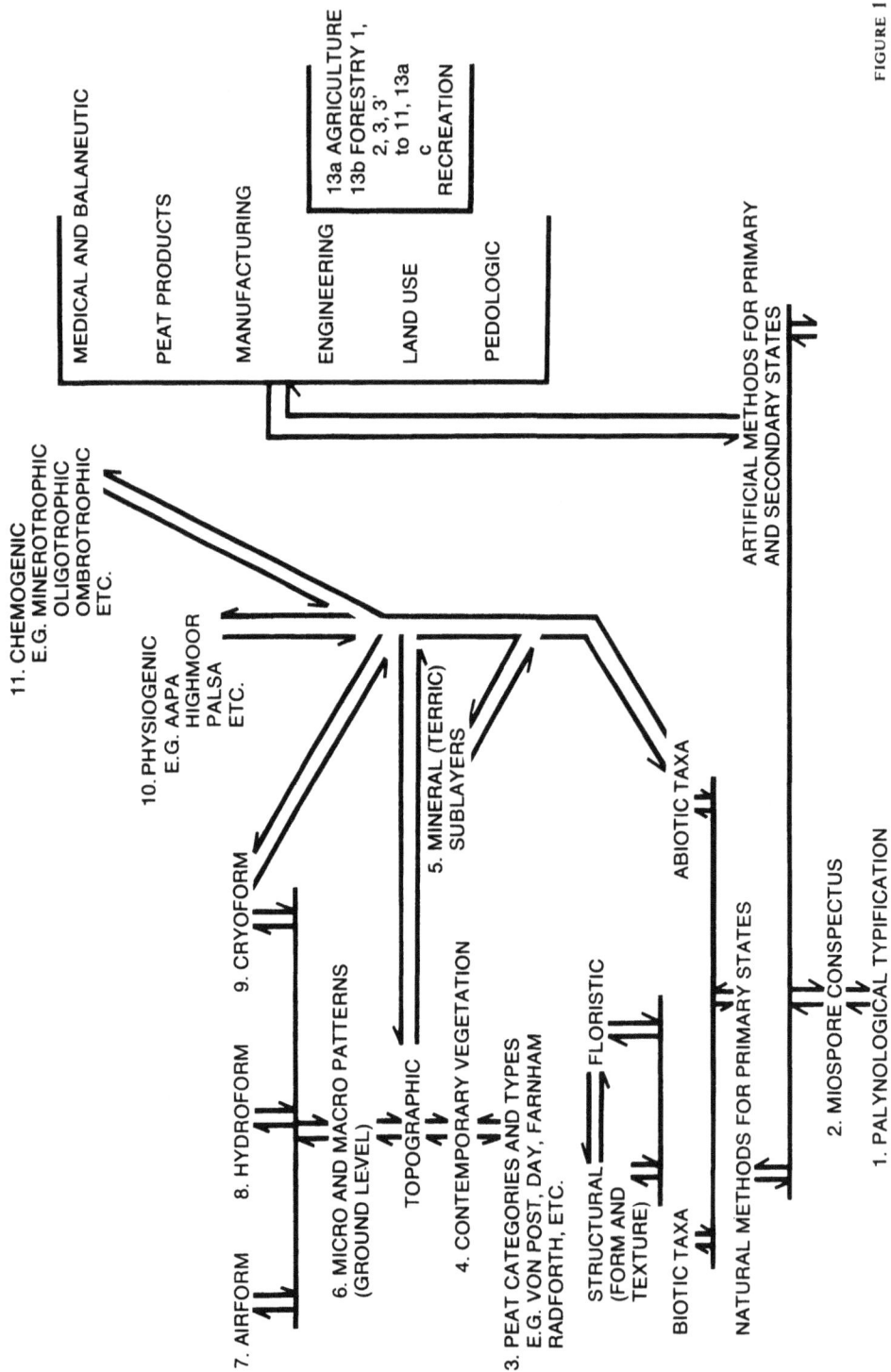

FIGURE I

"ANY TERRAIN IN WHICH WATER COLLECTS BETWEEN THE CATCHMENT AND THE SEA CONSTITUTES A TEMPLATE FOR PEAT FORMATION. THAT IS FOR THE DEVELOPMENT OF THE TERRAIN CONDITION TERMED MUSKEG."

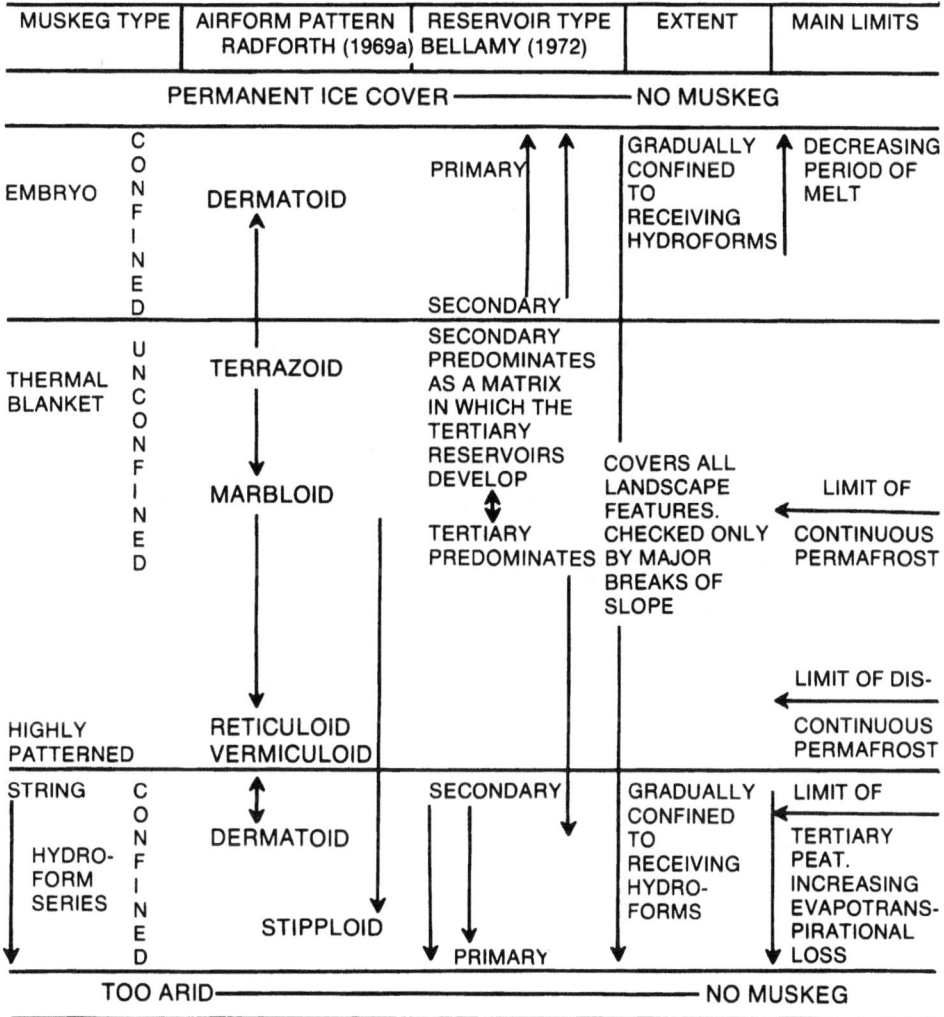

FIGURE 2   Zonal relationships of muskeg, macroclimate, and hydrology, basically along a north-south axis.

elements in the definition of airform patterns given in Table 1.

For convenience, the type photographs depicting the high-altitude airform patterns (scale 1:60,000) were first published by the writer in 1956. An account of low-altitude airform patterns (for example Vermiculoid, as in Figure 2) appears in the *Muskeg Engineering Handbook* (Radforth, 1969a), where it is also indicated how interpolation between the high-altitude patterns and ground-level vegetal conditions is achieved.

In Figure 2, other conditions bearing on patterns are indicated. In the first column, the confined state is considered as 'embryonic' in that climatic conditions and the collection of

TABLE 1

Definition of airform patterns

Dermatoid	Macro-topographically, relatively unchanging expanses differing in shape from about 0.25 mile to several miles at the narrowest dimension; infrequently interrupted by outcrops of mineral terrain; bearing microridges and clumps of hummocks, the latter where the gradient is zero; usually nearly flattened in general appearance but sometimes also on major slopes with little or only a gentle change in macro-topography. Micro-topography when caused by ridging forms a micro-net unless a major gradient is imposed (slope), and then micro-topography is oriented to form parallel ridging crossing the major gradient and often curving. Depressions may occur either on more or less level expanses or on slopes.
Marbloid	Lobed plateaus, domed when small (5 to 15 feet across), flattened with irregular micro-contour when large; plateaus usually marked by ice-wedges (polygons), and elevated from 1 to 10 feet above associated narrow tortuous flats with irregular and alternating, rarely opposite, lobes each different in size and collectively forming a complex; irregular micro-contour of mounds usually contiguous and differing in amplitude of from 0.5 to 3 feet.
Reticuloid	Networks, variable in area, each constituting wide ranges of more or less characteristic size order; ridge size difference usually a feature of length rather than width; ridges with more or less parallel sloping sides with micro-topography often formed by mounds which are sometimes contiguous; ridges separated by attenuated flats often bearing either single or clumps of hummocks.
Stipploid	Systems of gentle slopes (less than 5 per cent gradients and multidirectional); micro-topography of coalescing mounds forming secondary prominences 5 to 20 feet wide; equidimensional (nearly) and peaked to height of 2 to 4 feet at a tree or shrub base; sometimes (when prominences low) random, shallow, 1 foot deep (about) depressions occur at average intervals of about 25 feet (at centre); terrain roughness decreasing with diminishing density of stippling. Systems (as above) sometimes occur on major slopes, i.e. on mountains and foothills, Alaska, British Columbia.
Terrazoid	Angular peat plateaus, several to numerous, situated in flats and usually widely separated by distances exceeding the width of the more or less equidimensional plateaus. Plateaus often with gradients locally changing and abrupt due to mounding in micro-contour. Flats with shallow gradients occasionally reduced to zero, sometimes with single or grouped hummocks or with mounds (1 to 2 feet high) often contiguous forming tortuous ridges 1 to 5 feet in width, 15 to 30 feet in length and often joining.

water from the catchments play their part in terms of short season. This, of course, limits muskeg formation, so that usually only 'juvenile' stages are expressed and persist in the Dermatoid airform patterns. Farther south, the thermal blanket which largely succeeds the embryonic condition is represented by several airform patterns. Still farther south, beyond the predominating zone of Reticuloid and its constituent Vermiculoid, low-altitude patterns (Radforth, 1958), single examples of hydric pattern (hydroform) become obvious to the observer because the muskeg, in its response to the local hydrological conditions, is confined to local drainage axes.

The ontogenetic aspects of reservoir-muskeg development are explained best in relation to Bellamy's allusion to Primary, Secondary, and Tertiary Muskeg (Bellamy, 1972). Primary

Peats and associated Primary Muskeg occur when the surface retention of the reservoir is reduced as peat displaces water at the surface. Secondary Peat and Secondary Muskeg are established when the surface of the peat invades the mineral terrain beyond the normal delineating boundary of the land receptacle. The peat thus resists outward flow, and increases the reservoir effect and therefore the surface retention within the basin or depression area. Tertiary Peats and Tertiary Muskeg are those in which surface retention is in evidence at a level higher than that of the normal groundwater characteristic for the area.

It is considered that for Primary Muskeg the water balance is governed by the expression:

$$\text{inflow} = \text{outflow} + \text{retention}.$$

For some Secondary and all Tertiary examples of muskeg:

$$\text{inflow} + \text{precipitation} = \text{outflow} + \text{evaporation} + \text{retention}.$$

For the latter case, the muskeg will neither be generated nor persist where the macroclimate is unfavourable.

Thus the discussion so far, concerning the first three columns of Figure 2, suggests that pattern in muskeg is a reflection of the hydrology of the land. The tangible expression of this relationship is the new term 'Hydroform.'

There is, however, need to rationalize the use of this word (Hydroform) for regional circumstances in northern North America because of the wide influence of low temperatures coupled with a vast continental effect. Without this, the categorization of Hydroform might be either unrealistic or disproportionate in accounting for hydric behaviour peculiar in the main to northern Canada and the northern United States. In the fifth column of Figure 2, where the extent of muskeg is conveyed, reference is made to 'Receiving Hydroforms' predominating beyond permanent ice cover. Usually these embryonic primary muskegs are either in favourable microclimatic locations (for example, Lake Hazen area, Ellesmere Island) or, in the sense of macroclimate, in coastal situations. In either case, the linear distribution of the muskeg is easy to see. A Rheophilous condition (Kulczynski, 1949; Bellamy, 1968) exists: the muskeg develops in water moving along the linear gradient. For parts of each day during the growing season water movement may be quite rapid. As these peat deposits increase, there is a change in the local water quality as indicated by the nutrient supply. The biotic development is unusually slow and is sometimes transient as the axial water supply shifts in the course of change in catchment development. In this context (of rheophily) terms such as Niedermoore, Lowmoor, Minerotrophic, and Oligotrophic are of limited use for direct hydrological inference, however appropriate they may be for designating a muskeg condition. Associated with the concept of hydroform, they are thus less attractive than Rheophilous.

Southwards, where as in Figure 2 (fifth column) unconfined multipatterned thermal blanket muskeg is indicated as predominant, the hydrological conditions are quite different than for embryonic muskeg. The effective hydrological system utilizes meltwater from the active 'layer' and the current precipitation. This is not true on the east and west continental coasts where an unconfined condition also often exists. Rainfall here may exceed 40 inches per year in contrast to the 15 to 20 inches in the Central Subarctic. The hydroform in these contrasting situations can be expected to differ as does the airform pattern. In both cases the condition, in classical terms, is directly or indirectly ombrophilous. This term literally

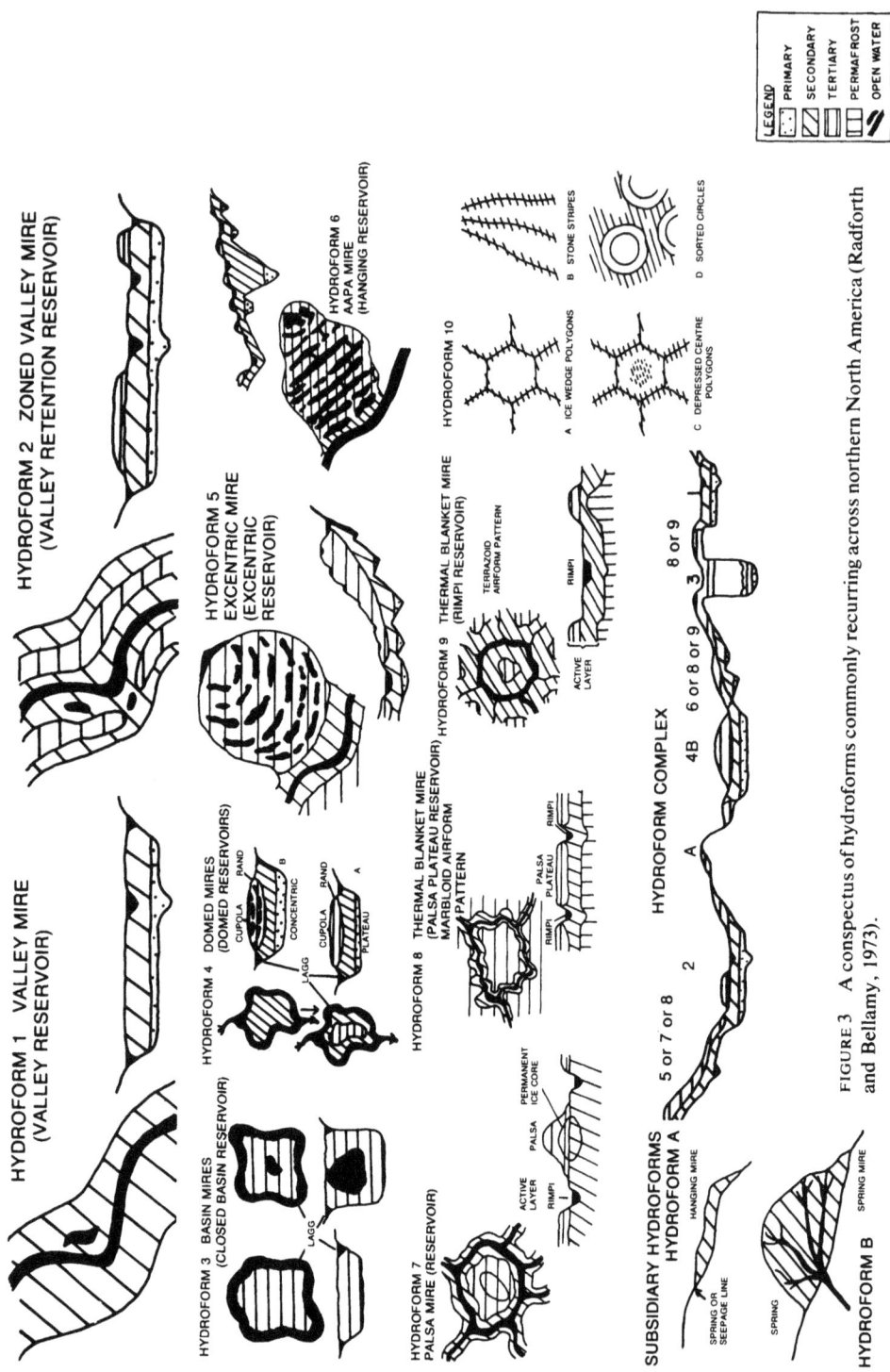

FIGURE 3   A conspectus of hydroforms commonly recurring across northern North America (Radforth and Bellamy, 1973).

translated means 'rain loving,' which, in the hydrological sense, still leaves begging a good part of the question and leaves unexplained the multi-patterned expression characterizing the thermal blanket to which the hydric factor makes a substantial contribution.

Southwards, as indicated in Figure 2, column five, the hydrological effect, though still involving the volume of the active layer and period of melt as for the thermal blanket, is directly limited by an adverse precipitation/evaporation balance. Beyond the permafrost, as is stated by Radforth and Bellamy (1973), the 'existence of a particular type of muskeg in any landscape feature must mirror closely the balance between supply, drainage and evapo-transpirational loss.'

In defining hydroforms, since they are proposed as functions of peat/landscape/reservoir interaction, it is important to recognize that the value of the surface retention within the landscape receptacle is modified to produce the new landscape feature.

Commonly recurring hydroforms are shown diagrammatically in Figure 3 as adapted from Radforth and Bellamy (1973). To facilitate comparison with other terminology not directly hydric in context but associated with peat development, the hydroforms are cross-referenced with the relevant expressions in Table 2.

An appreciation of the cryophilic influence is gained from an examination of Figure 2, sixth column, which ranges from permanent ice cover in the north to the arid limitation in the south. Hydrological assessments requiring evaluation of available free water, etc. for various hydroforms and studies of their water quality are yet to be made. In this connection the biotic factor (fossil and living) in the different landscapes will impose variability. Within the thermal blanket zone where permafrost is universal it was noted that by mid-August 1970 the active layer contained free water as shown by values in Table 3, column B. Maximum depths varied with the vegetal cover formula as shown (cf. the Appendix).

To recognize hydroform types a knowledge of the associated vegetal structure is helpful. To express this relationship, Table 3 is useful as it portrays architecturally the vegetal cover zones conforming to characteristics of the hydroforms. The conditions shown relate primarily to confined muskeg of the Canadian south. The hydroform types found there are grouped at the lower right of Figure 4.

In the condition of unconfined muskeg where, in the central part of the Subarctic, Reticuloid and later (farther north) mature Marbloid (cf. Korpijaakko and Radforth, 1968) become successively predominant, hydroforms 5 to 8 associate in various combinations depending largely upon other abiotic phenomena.

It is thus between the two major arcs in Figure 4 that these combinations arise. The configuration for hydroform 11, Figure 3 (Hydroform Complex), therefore varies constitutionally, generally towards the more northern latitudes.

Northerliness, however, is not a consistently influential factor in the progression of hydroform complex combinations. For one thing, basic catchment and reservoir form, distribution, and soil composition differ from southeast to northwest. Also, features of macroclimate change relative to this directional phenomenon. This effect is, in part, expressed spatially in Figure 4, where the oblique axis of the diagram bifurcates, leading to embryo hydroforms in the Arctic and High Arctic and true blanket muskeg (in contrast to thermal blanket muskeg) on the coasts where the cryogenic factor is not limiting the hydroforms.

The generalization of the precipitation, edaphic, and thermal influence would appear to have a 'synthetic' nature which is difficult to portray. Figure 5 is an attempt to show this. Airform, rather than hydroform, has been used to reflect the proportionate effect on the

TABLE 2
Tabulation of thirteen hydroforms arranged to show relationship with descriptive expressions defining muskeg formation

Hydroform numbers	Ontogenetic types	Associated physiographic designations
1	Secondary	Rheophilous, mesotrophic, flood-plain muskeg, valley fen Uber-schwemmungsmoor, Niederungsmoor valley reservoir
2	Tertiary	Rheophilous, oligotrophic, ombrotrophic, valley bog zoned valley muskeg, transition muskeg, valley retention reservoir, Flachmoor, Niederungsmoor
3	Tertiary	Ombrophilous, ombrogenous, oligotrophic, basin muskeg, bog, floating bog, raised bog, Schwingmoore — in part Waldhachmoor
4	Tertiary	Ombrophilous, rheogenous, oligotrophic, domed muskeg, bog, domed reservoir, Hochmoor
5	Tertiary	Ombrophilous, rheogenous, oligotrophic, excentric muskeg, excentric reservoir, transition muskeg, Kermi muskeg
6	Tertiary	Ombrophilous, rheogenous, oligotrophic, transition muskeg, hanging reservoirs, string bog, Strangmoor, aapamoor
7	Tertiary	Rheophilous-ombrophilous, cryogenic, mesotrophic, sometimes minerotrophic or eutrophic, palsa bog, palsa muskeg, Hugelmoor
8	Tertiary (sometimes no secondary condition in profile)	Ombrotrophic, oligotrophic-mesotrophic, cryogenic, transitional but often not. thermal blanket muskeg, arctic muskeg (moor)
9	Secondary	Rheophilous (cryogenic when tertiary), mesotrophic to minerotrophic (eutrophic), thermal blanket muskeg, rimpi reservoir, arctic muskeg (moor)
10	Primary	Rheophilous, cryogenic in part, minerotrophic (eutrophic) embryo muskeg
10A	Secondary	Rheophilous (subterranean), sometimes ombrophilous in part, minerotrophic (eutrophic) to mesotrophic, sometimes cryogenic, hanging type of Hangmoor
10B	Secondary	Rheophilous (subterranean), sometimes ombrophilous in part, minerotrophic (eutrophic) transition muskeg, type of Hangmoor, Quellmoor
11	Hydroform Complex Tertiary	Rheophilous-ombrophilous, cryogenic-ombrophilous, bog (usually in Europe), blanket muskeg, thermal blanket muskeg (cf. hydroforms 2, 3, 4, 5, 7, 8, 10A, 10B)

TABLE 3
Relationship between cover formulae (cf. Appendix) and depth to permafrost in inches

Cover formulae	Column A (max depth to permafrost for thermal blanket mire)	Column B (mean depth to permafrost, Fort Churchill area, August 1970)
BFI	48	37 ± 11.1
DEI	36	NR*
FI	36	NR
EI	30	32 ± 6.1
FEH	24	NR
BEI	20	18 ± 4.3
AEI	16	NR
BEH	16	14 ± 3.6
EH	15	11 ± 1.4
BHE	14	NR
AEH	12	11 ± 2.6
AHE	12	NR
HE	12	9 ± 1.9

*NR means "no record."

pattern of muskeg. The hydroform, because of the composite nature of the hydroform complex, would make the diagram too complicated and difficult to comprehend.

Also, not enough is known concerning the distributional detail of hydroforms and the relative degree to which subtypes of the hydroform complex abound.

The future of hydrological research calls for close attention to quantitative expression for without considering the concepts initially and treating the taxonomic descriptions carefully, as can be done by using the concept of hydroform, unrelated empirical forms of artificial expression will naturally arise.

An important objective of hydrological research on muskeg is to gain basic understanding for undertaking a national inventory of the hydrological potential effect in terms of the major areas of the national landscape. Without this little wisdom will be displayed in dispensing continental water.

There is also another major basic prerequisite. Too little is known about surface retention and hydraulic conductivity to consider differences in gross structure (cf. Radforth, 1969a, 17 peat categories) within the hydroform configurations. In attempts to initiate a remedy for this deficiency, Korpijaakko (Korpijaakko and Radforth, 1972) has developed new approaches for investigating water relations of peats which express their different mechanical constitution. He points out the need for 'exact recognition' of peat types. MacFarlane (in MacFarlane and Radforth, 1968), in his descriptions of 'axons,' first drew attention to the need to emphasize the physical features of the effective constituents of peat. Korpijaakko prefers the botanical reference as specified by von Post. Burwash (1972), in accounting for thermal properties of peat types, was able to differentiate them on non-botanical terms. Strictly, botanical definition is incomplete if the expression of physical behaviour is desired in terms of ultimate particulate arrangements.

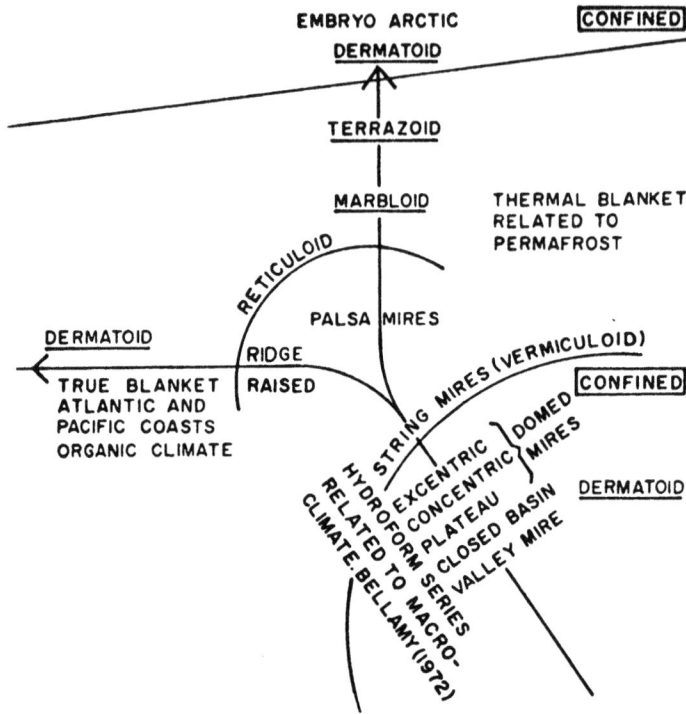

FIGURE 4   Airform pattern relationships in confined and unconfined muskeg.

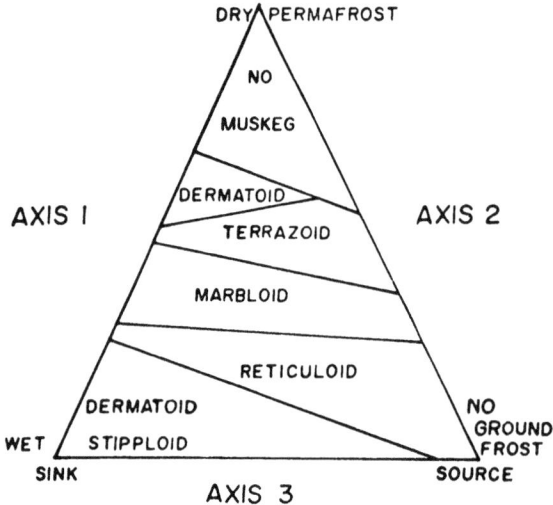

AXIS I   Precipitation Macroclimate linked Edaphic
AXIS 2   Thermal Macroclimate linked Edaphic
AXIS 3   Geomorphological

FIGURE 5   Diagram indicating relationships between abiotic (template) factors and airform pattern.

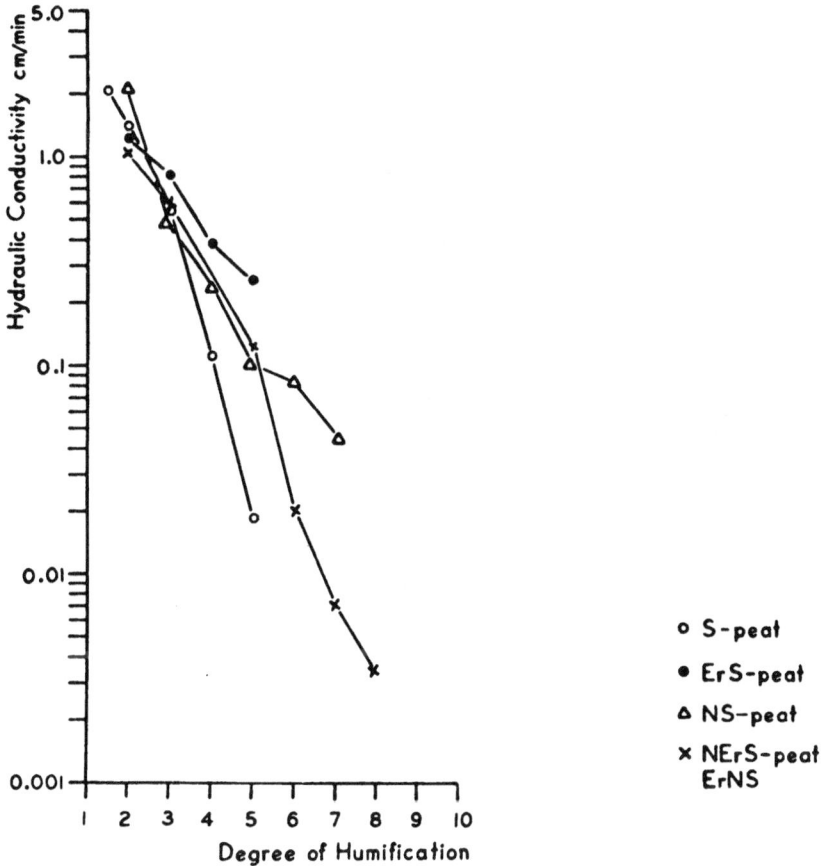

FIGURE 6   Hydraulic conductivity of different kinds of *Sphagnum* peat versus von Post's degree of humification (S = *Sphagnum*, Er = *Eriophorum*).

Korpijaakko's results substantiate this view in that when physiochemical changes occur in given peat types (botanically designated) the hydraulic conductivity changes (Figure 6). The change known as humification induces conductivity changes characteristic of the botanical type. Also, for a given peat type differing in the degree of humification, the water content, as a percentage of both the dry weight and the volume, varies depending upon the dry density (cf. Figure 7). Finally, Korpijaakko showed that the vertical hydraulic conductivity for a given peat type also varies with the dry density. When the latter is compared with the degree of humification (Figure 8), the hydraulic conductivity (vertical) appears as shown by the curve marked by the computed values.

The conductivity behaviour for different degrees of humification and for variable depth with no change in peat types (S = *Sphagnum*; von Post, 1922, 1924) is indicated in Figure 9. It would be challenging and instructive if, for inventory purposes, one could characterize hydroforms in this way. With this information, controls for translocation of water for muskeg in the pancontinental sense would be understood. Only then could our water resources be wisely manipulated with reasonable assurance that our water and land resources were being properly managed.

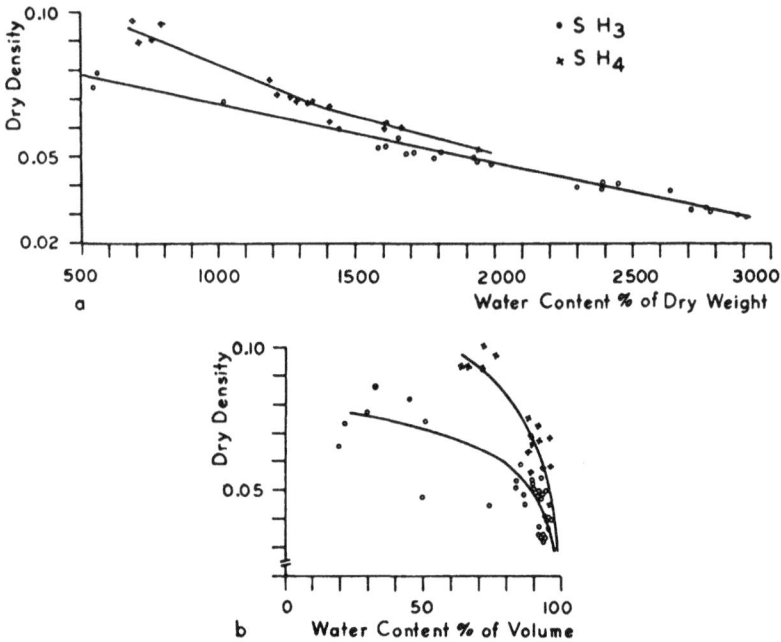

FIGURE 7   Dry density of *Sphagnum* peat (g/cc) for two degrees of humification vs. water content as a percentage of (a) the dry weight and (b) the volume at the time of sampling (S = *Sphagnum*, H = degree of humification).

APPENDIX

Radforth Muskeg Cover Classification System

An important feature of muskeg is that it is characterized by a finite number of vegetation structural types. These types consist of nine basic cover classes, each designated by one of the first nine letters of the alphabet, which are listed below in combinations called cover formulae.

Cover class	Description
A	Trees over 15 feet high
B	Trees up to 15 feet high
C	Non-woody, grass-like 2-5 feet high
D	Woody, tall shrubs or dwarf trees 2-5 feet high
E	Woody shrubs up to 2 feet high
F	Sedges and grasses up to 2 feet high
G	Non-woody broad-leaf plants up to 2 feet high
H	Leathery to crisp mats of lichen up to 4 inches high
I	Soft mats of moss up to 4 inches high

The cover classification does not depend on the naming of plant species, and since these classes can be used to describe any muskeg, the system has worldwide application.

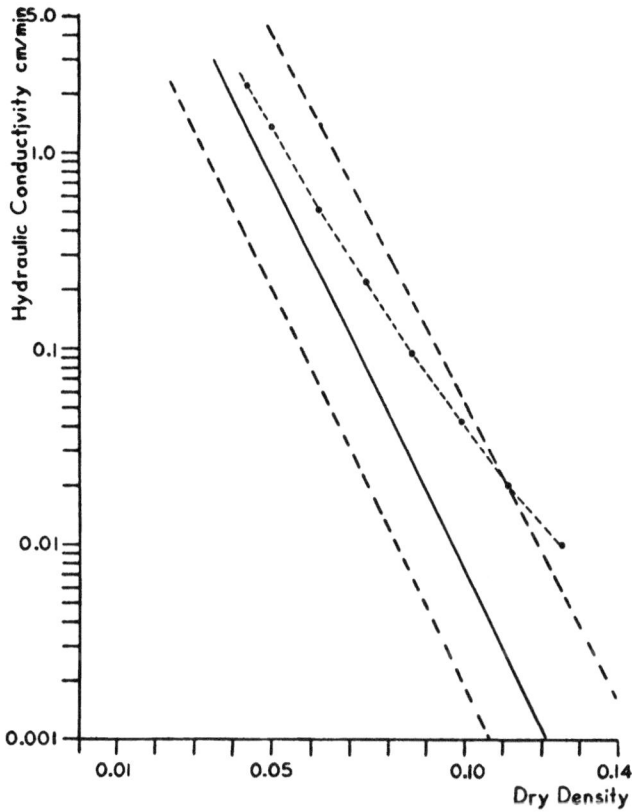

FIGURE 8   Hydraulic conductivity of *Sphagnum* peat in vertical direction vs. dry density (a/cc).

On any given muskeg area, certain of these cover classes are found in combination, and the cover formula used to describe that muskeg area is derived by listing in descending order of prominence the cover classes which can be observed to be growing there. To be included in the cover formula, any cover class must represent at least 25 per cent of the total cover, based on a visual estimate. This estimate may seem to be a rather subjective type of evaluation, but in practice there is rarely any doubt that a given cover class represents more or less than 25 per cent of the total cover.

As an example of the application of this classification system, a muskeg area covered by tall spruce trees over 15 feet high, woody shrubs 2-5 feet high, and woody shrubs up to 2 feet high would be referred to as ADE. A muskeg area covered by grasses and sedges up to 2 feet high and a carpet of moss would be called FI, and so on.

It might appear that all the combinations of the nine cover classes in groups of two or three would lead to an unwieldy number of possible cover formulae. In reality many possible combinations, such as GAD and BGC, do not occur. The most common include: ADE, ADF, ADI, AEH, AEI, AFI, BDE, BDF, BEF, BEH, BEI, BFI, C, CI, DEF, DFI, DI, EFI, EH, EI, FI, FEI, FIE, G, HE.

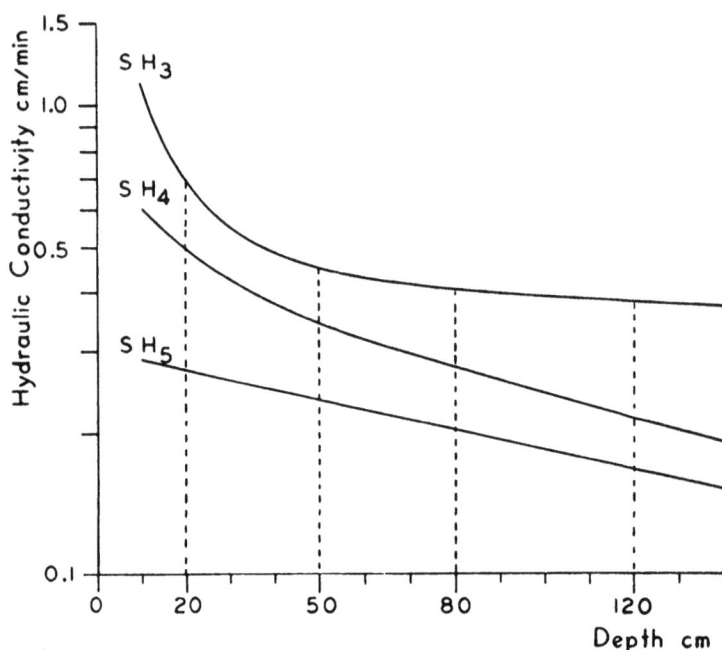

FIGURE 9   Hydraulic conductivity of *Sphagnum* peat for three degrees of humification vs. depth.

REFERENCES

Bellamy, D.J. 1968. An ecological approach to the classification of European mires. Proc. 3rd Internat. Peat Congr., Quebec, pp. 74-9.

Bellamy, D.J., Marshall, C., Waughman, G.J., and Radforth, N.W. 1973. Pattern, productivity and performance in peat producing ecosystems (in press).

— 1972. Template of peat formation. Proc. 4th Internat. Peat Congr., Helsinki, pp. 7-16.

Bulow, K. 1929. Allgemeine Moorgeologie. Handbuch der Moorkunde (Bdl., Berlin).

Burwash, A.L. 1972. Thermal conductivity of peat. Proc. 4th Internat. Peat Congr., 2: 243-54.

Kivinen, Erkki. 1948. Suotiede, Porvoo (Helsinki).

Korpijaakko, E.O., and Radforth, N.W. 1968. Development of certain patterned ground in muskeg. Proc. 3rd Internat. Peat Congr., pp. 69-73.

Korpijaakko, M., and Radforth, N.W. 1972. Studies on the hydraulic conductivity of peat. Proc. 4th Internat. Peat Congr., pp. 323-34.

Korpijaakko, M.L., Korpijaakko, M., and Radforth, N.W. 1972. Ice-biotic relationship in frozen peat. Proc. 14th Muskeg Research Conf. 1971, NRC, ACGR Tech. Memo, pp. 111-22.

Kulczynski, S. 1949. Peat bogs of Polesie. Mem. Acad. Sci. Cracovie, Ser. B, pp. 1-356.

MacFarlane, I.C., and Radforth, N.W. 1968. Structure as a basis of peat classification. Proc. 3rd Internat. Peat Congr., pp.91-7.

Pollett, F.C. 1972. Classification of peatlands in Newfoundland. Proc. 4th Internat. Peat Congr., Helsinki, pp. 101-10.

Radforth, N.W. 1956a. The application of aerial survey over organic terrain. Proc. Eastern Muskeg Research Committee, NRC, ACSSM Tech. Memo. 42: 25-30.

— 1956b. The application of aerial survey over organic terrain. Roads and Engineering Construction, August, pp. 1-3.

— 1958. Organic terrain organization from the air (altitudes 1,000-5,000 feet). Handbook No. 2, Defence Res. Board, Dept. National Defence, DR 124, Ottawa.

— 1969a. Classification of muskeg. In Muskeg Engineering Handbook, ed. I.C. MacFarlane (Univ. Toronto Press), chap. 2.

— 1969b. Airphoto interpretation. In Muskeg Engineering Handbook, ed. I.C. MacFarlane (Univ. Toronto Press), chap. 3.

— 1972. Relation between artificial and natural classification of organic terrain. Proc. 4th Internat. Peat Congr. 1: 389-400.

— 1973. Relationship between artificial and natural classification of organic terrain. Proc. 4th Internat. Peat Congr. Helsinki, pp. 389-400.

Radforth, N.W., and Bellamy, D.J. 1973. A pattern of muskeg — a key to continental waters. Can. J. Earth Sci. (in press).

von Post, L. 1922. Sveriges Geologiska Undersoknings torvinventering och nagra av dess hittils vunna resultat. Sartryck ur Sv. Mosskulturforeningens Tidskr. 1 (Jonkoping).

— 1924. Das genetische System der organogenen Bildungen Schwedens. Com. Intern. Pedol. IV Comm. 22: 287-304.

Weber, C.A. 1908. Uberrezente Eisenocker und ihre Microorganismengemeinshaften Englers. Bot. Jahrb. 90 (Leipzig).

6
Muskeg and Permafrost

R.J.E. BROWN

Vast areas of Canada are covered by muskeg, the estimated coverage being about 800,000 square miles extending from the southernmost parts of the country to the Arctic. In recent years this unique terrain, variously referred to as bogland, muskeg, organic terrain, and peatland, has been subjected to increasing scientific and engineering study as northern development progresses. In 1969 the *Muskeg Engineering Handbook* was published to bring together this knowledge and experience (MacFarlane, 1969). Much of the muskeg territory of Canada is underlain by permafrost. The interaction of muskeg and permafrost results in the development and occurrence of special surface features and ground thermal patterns. Variations in these features and patterns occur across the discontinuous and continuous permafrost zones and in the different physiographic regions of northern Canada. Remote sensing techniques such as aerial photography and infrared thermal imagery are being employed with increasing effectiveness to improve the ability to predict permafrost occurrence in muskeg and interpret surface features and ground thermal patterns.

RELATIONSHIP BETWEEN DISTRIBUTION OF MUSKEG AND DISTRIBUTION OF PERMAFROST IN CANADA

No detailed mapping of the distribution of muskeg in northern Canada has been undertaken but a small-scale map of the whole country was published in 1960, under the title 'Areas of Organic Terrain (Muskeg) in which Engineering Problems Occur' (Radforth, 1961); see Figure 1. According to Radforth this map represents areas in Canada where engineering problems associated with peat will occur. The frequency of occurrence of muskeg is divided into three categories: high, medium, and low. Radforth's muskeg classification types which are most predominant in various areas are shown.

Permafrost, or perennially frozen ground, extends over the northern half of Canada comprising an area of about two million square miles (Brown, 1967). The southern limit of the permafrost region and the discontinuous permafrost zone — a zone where permafrost does not occur everywhere beneath the land surface — lie in the belt of greatest muskeg concentration (Figure 1). Farther north in the continuous permafrost zone — a zone where permafrost occurs everywhere beneath the land surface — muskeg is limited in extent because of the more severe arctic conditions. Within these zones, the character of the permafrost and

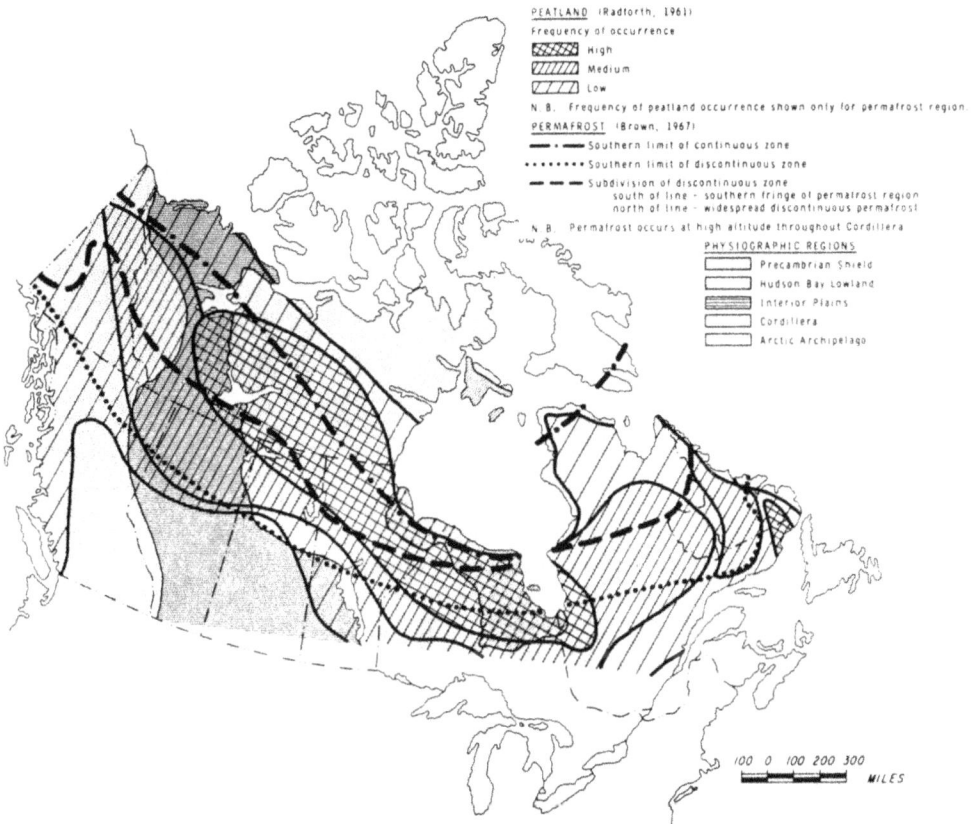

FIGURE 1   Distribution of muskeg and permafrost in Canada.

extent of muskeg varies from one physiographic region to another.

Three categories of muskeg are described in the *Muskeg Engineering Handbook* relative to the distribution of permafrost. The first, 'seasonally frozen,' is found south of the permafrost boundary. The peat here ranges in thickness from 1 foot to more than 20 feet. As the name implies, the surface-frozen layer is entirely seasonal. The depth of frost penetration or thickness of the frozen layer is extremely variable, depending upon climatic factors such as air temperature, solar radiation, and snow cover.

The second category is muskeg in the discontinuous permafrost zone. Here the observed peat thickness varies from about 1 foot to 10 to 15 feet in rock basins and palsas. In the southern fringe of this zone permafrost occurs in scattered islands a few square feet to several acres in size and is confined mainly to peatland. Northward it becomes increasingly widespread and occurs in other terrain types also. Where permafrost exists in muskeg the depth to the permafrost table is usually less than 5 feet. The active layer usually extends to the permafrost table, but a variation of this category may occur where an unfrozen layer exists through the winter between the seasonally frozen peat and the permafrost.

Muskeg in the continuous permafrost zone comprises the third category. Observations have shown that these peat deposits are generally relatively shallow, their thickness seldom exceeding a few feet. Permafrost occurs everywhere beneath the ground surface, including

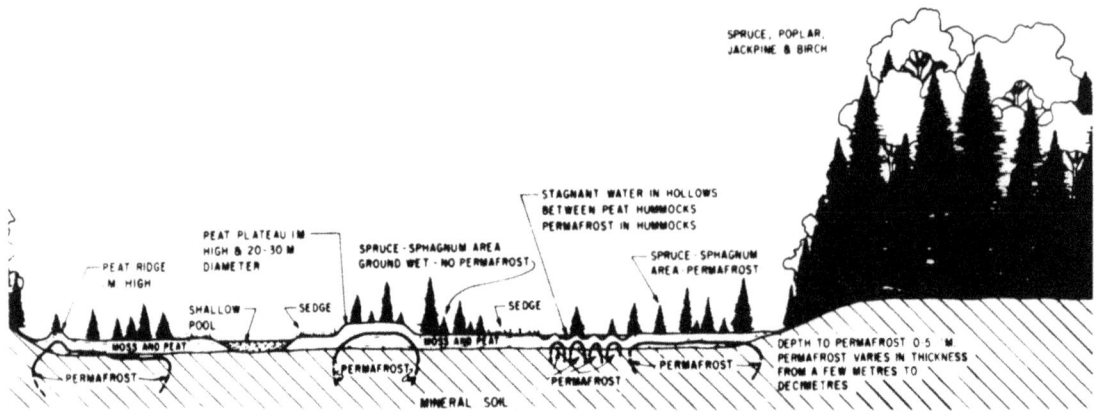

FIGURE 2   Profile through typical muskeg in southern fringe of discontinuous zone showing interaction of permafrost and terrain factors.

all muskeg. The active layer generally varies in thickness from about 1 to 3 feet in the continuous zone depending on local climatic and terrain conditions and it almost invariably extends to the permafrost table. The active layer is thinnest in muskeg, an annual depth of thaw of only 6 inches being observed in the Arctic Archipelago.

THERMAL ASPECTS

Development and Occurrence of Permafrost in Muskeg

The development of permafrost islands in muskeg in the southern part of the discontinuous zone appears to be related to changes in the thermal conductivity of the peat through the year (Tyrtikov, 1959; Brown, 1966). During the summer the surface layers of peat, except for those in proximity to frost, become relatively dry through evaporation. The thermal conductivity of the peat is low and warming of the underlying soil is impeded. The lower peat layers therefore gradually thaw downward and become wet as the ice layers in the seasonally frozen layer melt. In the autumn there tends to be more moisture in the surface layers of the peat because of a decreased evaporation rate. When it freezes the thermal conductivity of the peat is increased considerably. Thus the peat offers less resistance to the cooling of the underlying soil in winter than to the warming of it in summer. As a result, the mean annual ground temperature under peat will be lower than under adjacent areas without peat. When conditions under the peat are such that the ground temperature remains below 32°F throughout the year, permafrost results and is maintained as long as the thermal conditions leading to this lower temperature persist.

Such a close relationship exists among the environmental factors in peatland that it is difficult to single out the significant effect of each one on the thermal regime. In the southern fringe of the discontinuous zone, variations in permafrost occurrence are indicated by its patchy distribution (Figure 2). The importance of drainage (hydrology) is shown by the absence of permafrost in areas where the water table is at or near the ground surface. Surface water, even between individual small hummocks, inhibits the existence of permafrost between the hummocks but not beneath them. Permafrost occurs also in other peat features such as plateaus and palsas. These features constitute a microrelief several feet high in the

FIGURE 3 Muskeg near Dawson, YT, in northern part of discontinuous permafrost zone supporting scattered spruce up to 30 feet high. Permafrost table is at a depth of 1 foot 9 inches in 1 foot high scattered peat, plateaus 2 feet 7 inches in wet depressions and 3 feet 6 inches beneath pools of water 1 foot 3 inches deep.

muskeg. These elevated features are relatively well drained compared to the surrounding wet terrain. Variations in snow cover are presumed to be a significant factor in the patchy distribution of permafrost. The summits of palsas and other peat relief features are frequently more exposed to wind, resulting in less snow accumulation and greater frost penetration than in the surrounding low-lying terrain.

Northward, in the discontinuous zone and in the continuous zone, permafrost becomes widespread and occurs even in the wet and poorly drained portions of muskeg (Figure 3). Here the climate is sufficiently severe to cause the formation and persistence of permafrost despite the thermal effects of the water. The thickness of the active layer is influenced by the water, however, in that the depth to the permafrost table beneath shallow pools is usually greater than in neighbouring microrelief features.

Freezing and Thawing Regimes

Annual freezing and thawing of muskeg is different from that of mineral soil terrain owing to vegetation, microrelief, high water table, and thermal properties. In thermal calculations it is particularly difficult to deal with heat conduction wherever there is porous *Sphagnum* at the surface because an appreciable percentage of the heat transfer is by non-conductive processes. The underlying peat exerts an influence out of proportion to the amounts of it that may be present in the soil profile because of its unusual thermal properties. Highly variable moisture contents in surface peat layers at a single site lead to great variations in the rate and depth of freezing and thawing. Field studies have shown that snow cover is also extremely important in determining seasonal heat flow and depth of freezing.

Field studies at Thompson, Manitoba (Brown and Williams, 1972) indicate that heat flow meters give useful values of surface heat flow during winter for use in assessing freezing patterns in muskeg. Summer heat flow determinations are handicapped by moisture movements in the thawed layer. Other surface energy exchange measurements have also provided useful data. Problems occur in all calculations of the freezing and thawing regimes because the quantities of heat involved and the differences from one site to another are usually very small and therefore difficult to measure accurately. The fragile nature of muskeg also makes this type of terrain very susceptible to disturbance caused by frequent field measurements.

SURFACE FEATURES ASSOCIATED WITH MUSKEG AND PERMAFROST

The most prominent surface features associated with muskeg and permafrost are peat plateaus and palsas. The origin, development, and nature of these features have been discussed by numerous authors in North America and Europe (Brown, 1968; Forsgren, 1964, 1965; Hustich, 1957; Lundqvist and Mattson, 1965; J. Lundqvist, 1969; S. Lundqvist, 1962; P'yavchenko, 1955; Railton and Sparling, 1973; Ruuhijärvi, 1960; Salmi, 1968; Sjörs, 1959a, b, 1961; Svensson, 1961-62; Tarnocai, 1970; Thie, 1971; Tyrtikov, 1966; Zoltai, 1972; Zoltai and Tarnocai, 1971). They are widespread in the discontinuous permafrost zone, especially the southern portion where the permafrost is patchy. They are found across Canada in this subzone, having been reported in all physiographic regions from Labrador (Brown, in press) to the Western Cordillera (Hughes et al., 1972).

They are particularly prevalent in the Hudson Bay Lowland (Brown, 1973; Zoltai, 1973) and have been mapped from aerial photographs at a scale of 10 miles to 1 inch (Bates and Simkin, 1966). All stages of development can be observed. It appears that peat plateaus and palsas are morphological variations of the same process, i.e., the same mechanism is responsible for the formation of both these features, which pass through a life cycle of growth and degradation. Svensson (1961-62) and Tyrtikov (1966) both appear to support this contention.

Initially these features appear as low mounds or upwarpings of peat protruding above water level in the middle of shallow ponds less than about 3 feet in depth (Figure 4). The mechanism of their formation and control of their distribution is uncertain, but it is suggested that the ponds freeze to the bottom in winter and the underlying saturated peat is domed up at random locations by intensive frost action and ice lens growth. The elevation of the peat plateau or palsa surface above the surrounding muskeg is caused by the layers of segregated ice in the mineral soil immediately beneath the peat layer.[1] When elevated above the pond level, the dry layer of exposed peat insulates the underlying frozen mass from summer thawing, thus marking the initiation of a perennially frozen or permafrost condition. The elevation of the peat surface above the general level of the surrounding flat level surface devoid of relief exposes it to winter winds which reduce or remove the insulating snow cover. Winter frost penetration is therefore greater than in the surrounding low, flat areas, which contributes to further permafrost accumulation.

1 Detailed studies of palsas in northern Manitoba, including a cross section through the perennially frozen core 15 feet thick, showed that the surface peat layer is the same thickness as in the surrounding terrain.

FIGURE 4   Small, youthful palsas containing permafrost in wet muskeg with no permafrost, located at southern limit of discontinuous zone near south end of James Bay.

As the peat continues to accumulate year after year,[2] accompanied by an increase in permafrost thickness each winter, the mounds grow and coalesce to form ridges and plateaus. In the youthful stage there is little or no living vegetation on the peat surface. When it matures *Sphagnum* and other mosses and lichens become established along with Labrador tea and spruce (Figure 5). Old age and degradation begin when the insulating ground cover ruptures as a result of biological oxidation and general deterioration, and thawing penetrates into the underlying perennially frozen core. The surface of the palsa or peat plateau becomes very uneven because of differential thawing of the underlying ground ice and large blocks of thawed peat break off the margins. It is not certain whether rejuvenation can occur.

The critical factors in palsa and peat plateau formation are possibly climate, snow cover, and water supply. Little work has been done on the climatic requirements for the formation of these features but some observations are available. It has been recorded in Sweden that palsas occur where the air temperature remains below 32°F during more than 200 days per year. They also appear to be present only where the precipitation during the period from November to April is less than 12 inches (Lundqvist, 1962). Hamelin and Cailleux (1969) studied several groups of palsas at Great Whale River, PQ, near the east shore of Hudson Bay, observing their relationship to the local mean annual air temperature of 23°F to 25°F. In Manitoba and Saskatchewan Zoltai (1971) identified palsas and peat plateaus, indicating

---

2 Rates of peat accumulation in the permafrost region have been estimated from radiocarbon dating by the Geological Survey of Canada. In the Hudson Bay Lowland rates have been measured varying from 1 foot per 300 to 850 years, and in the Yukon Territory from 1 foot per 450 to 1,450 years.

FIGURE 5   Mature peat plateaus formed from coalesced palsas with permafrost on aerial photograph in Figure 7. Note dense coverage of burned spruce trees, dense Labrador tea, and hummocky lichen and *Sphagnum*-covered peat surface. No permafrost exists in sedge-covered depression in foreground.

the presense of permafrost now or in the recent past, in the field and on aerial photographs. The southern limit of their occurrence coincides with the 32°F mean annual air isotherm; this is south of the 30°F mean annual air isotherm designated as the southern limit of permafrost in Canada but coincident with the observed southern limit of permafrost in Alaska.

Snow cover is considered one of the critical terrain factors in palsa development once formation of the feature has begun. As mentioned above by Lundqvist, the amount of snowfall in the region through the winter may not exceed a certain quantity. In addition, snow accumulation on palsa summits is significantly less than on the surrounding terrain, because of their exposed position to wind and this results in greater frost penetration.

Drainage conditions and water supply are very important factors in the development of palsas and peat plateaus. The ponds in which they begin to grow should be sufficiently shallow to freeze to the bottom so that a frozen zone may develop below. The process of growth is not clear but the gradual updoming of the peat proceeds possibly because of its high capillarity, which draws considerable quantities of water to the freezing front from the surrounding wet areas. These conditions occur widely in the Hudson Bay Lowland, where palsas and coalescing palsas forming peat plateaus grow to heights of 12 to 15 feet. The lower peat plateaus which occur extensively in the Hudson Bay Lowland and in muskeg in the other physiographic regions may be limited in height for several reasons. There may be less

FIGURE 6   Treeless palsas at Cartwright in Labrador having a peat layer 2 to 4 feet thick overlying sandy soil. The depth to the permafrost table is 1 foot 6 inches.

water available for ice accumulation. Also dense tree growth tends to reduce exposure to wind and to encourage snow accumulation, which reduces frost penetration.

A detailed ecological study of a palsa bog was carried out in the Hudson Bay Lowland south of Hudson Bay near Winisk, Ontario (Railton and Sparling, 1973). The palsa bog was found to be ombrotrophic, the frozen cores of the palsas causing damming and perched water levels. Most of the better developed palsas were found in the drier regions of the bog. Evapotranspiration was the main source of heat loss, while exposure (reduced snow cover) and the low thermal conductivity of the peat were the main factors in palsa development. Palsa collapse was attributed to deterioration of the vegetation cover by rain and wind erosion causing melting of the ground ice.

One regional variation of palsas which has received little attention is that mature palsas in the Hudson Bay Lowland and elsewhere in Canada west of Hudson Bay generally support a dense forest growth (Figure 5), in contrast to mature palsas east of Hudson Bay and in northern Scandinavia, which are generally devoid of tree growth (Figure 6). This difference may be related to the more continental climatic conditions prevailing in Canada west of Hudson Bay. The vegetation of many mature palsas in the Hudson Bay Lowland has been burned over by forest fires, but this appears to have little influence on the permafrost because only the top 1 or 2 inches of the peat is affected.

REMOTE SENSING OF PERMAFROST IN MUSKEG

Airphoto Interpretation

The recognition on aerial photographs of permafrost occurrences in muskeg is possible where distinctive patterns exist. The largest and most distinctive permafrost features in the southern fringe of the discontinuous zone are peat plateaus and palsas. The most intensive work in recent years has been carried out in northern Manitoba, where Tarnocai (1970) and

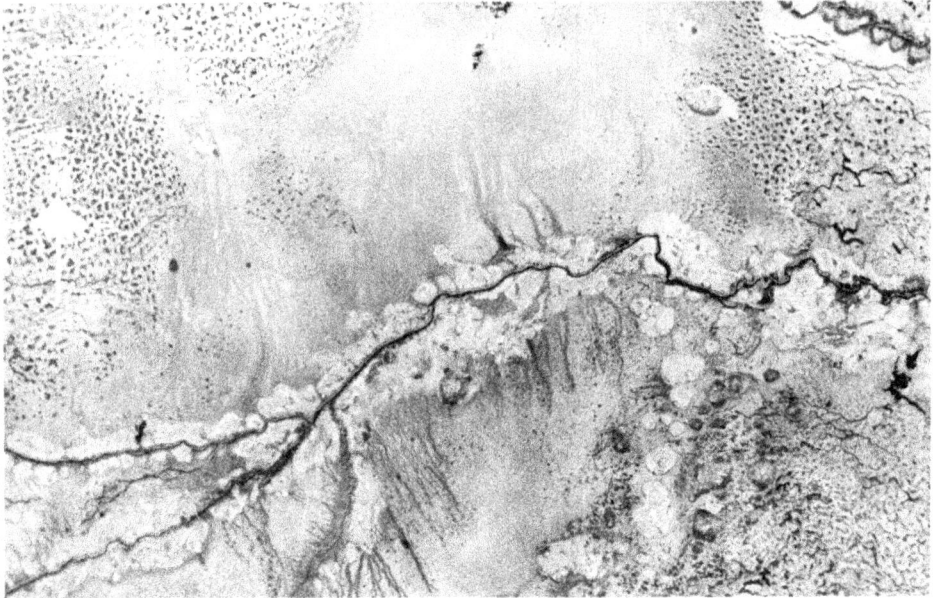

FIGURE 7   Section of Royal Canadian Air Force aerial photograph A14961-128 of terrain in Figure 5. Location in Hudson Bay Lowland 30 miles west of James Bay in southern fringe of discontinuous permafrost zone.

Thie (1971) have classified peatland forms in the permafrost region and described their airphoto characteristics.

A typical aerial photograph from the southern fringe of the discontinuous permafrost zone, taken in northern Ontario, 30 miles west of James Bay, is shown in Figure 7 (see also Figures 4 and 5). The entire area is peatland, but three main patterns are evident:

1. Medium grey with smooth texture covering the central and southwest portions of the photograph on both sides of the stream which flows from west to east. This is a low, wet, poorly drained flat sedge area. The black pepper-like flecks are small pools of water less than 50 feet in diameter. No permafrost occurs in this pattern.

2. Fine network of closely spaced dark grey to black flecks in a light grey mesh-like matrix covering the northwest and eastern sections of the photograph. This is a low, wet, poorly drained and flat area consisting of shallow pools up to 100 feet in diameter separated by low narrow sedge, and moss-covered peat ridges about 2 feet high. The dark grey circular areas in the southeast corner are spruce (*Picea* sp.) islands 300 feet or more in size. No permafrost occurs in this pattern.

3. Light grey circular and irregularly shaped areas with white patches adjacent to the stream and bordering pattern 2 in the southeast portion of the photograph. These areas are large, mature palsas 10 to 15 feet high and high peat plateaus and coalesced palsas. The light grey tone is caused by the dense cover of Labrador tea (*Ledum groenlandicum*) growing on the lichen cover of the palsas. The white patches are lichens (*Cladonia* sp.). Permafrost occurs in these features. The peat is about 5 feet thick overlying grey silty clay with sand and

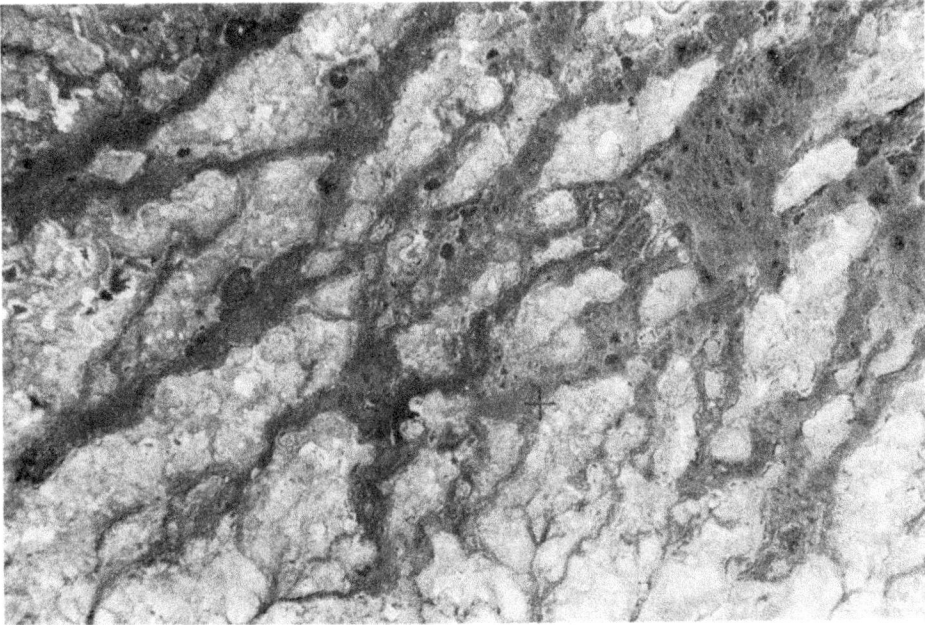

FIGURE 8   Section of Royal Canadian Air Force aerial photograph A14188-119 of terrain in Figures 9 and 10. Location near Nelson River, northern Manitoba, in northern portion of discontinuous permafrost zone.

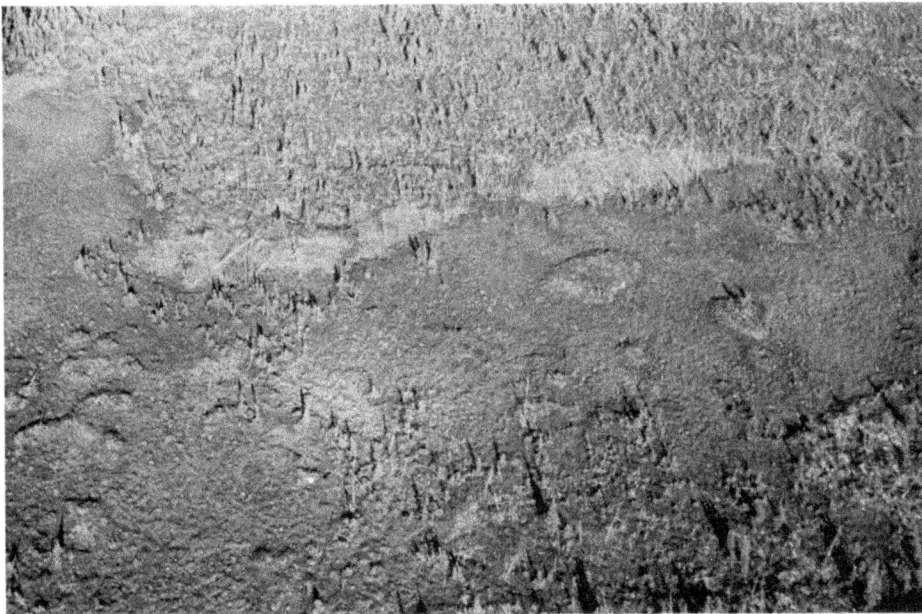

FIGURE 9   Aerial view from altitude of 500 feet of terrain on aerial photograph in Figure 10 showing forested peat plateau with permafrost and low, wet treeless areas without permafrost.

FIGURE 10   Ground view of terrain on aerial photograph in Figure 8 showing forested peat plateaus in background with permafrost and low, wet treeless area in foreground with no permafrost.

small stones. The permafrost table is about 28 inches below the ground surface and the permafrost is probably about 20 to 30 feet thick.

In the northern part of the discontinuous zone, surface features associated with permafrost are more widespread. In some localities, the features with permafrost are quite distinct from those with no permafrost. Such a situation exists in the northern portion of the discontinuous zone in northern Manitoba on the Nelson River about 100 miles west of Hudson Bay. The peatland in this region is a mosaic of peat plateaus interspersed with wet depressions (Figures 8, 9, 10). The aerial photograph in Figure 8 shows clearly the two main patterns:

1. Dark grey with small black circular and elongated areas. These are low, flat, wet, grass- and sedge-covered depressions with shallow pools of water. Tree growth is virtually non-existent. On the oblique aerial view (Figure 9) this pattern comprises most of the dark grey smoother textured lower half of the photograph. In the ground view (Figure 10), the low wet area in the foreground is typical of this terrain type. Probings and drilling indicate that no permafrost occurs under these wet depressions and pools.

2. Light grey with small white irregular-shaped areas. These are peat plateaus rising 4 to 5 feet above the level of pattern 1 covered with dense spruce forest and ground cover of hummocky thick *Sphagnum* and other mosses, lichens, and Labrador tea. On the oblique aerial view (Figure 9) this pattern comprises the coarse-textured upper third of the photograph. In the ground view (Figure 10) the forested peat plateau in the background is typical of

this terrain type. Probings and drilling indicate that permafrost exists in the peat plateaus to a depth of about 80 feet. The active layer varies from about 1½ to 2 feet in depth.

The identification of features in muskeg containing permafrost is not always as straightforward as in the two examples cited above. Most palsas are relatively small features and cover very small areas on all but aerial photographs flown at low levels. Peat plateaus are more extensive and thus more easily discernible. Where peat microrelief features are absent, the existence of permafrost can only be inferred in relation to visible terrain factors such as drainage. These terrain relationships have to be considered in the broad framework of climate and location of the area in the permafrost region. For example, it is very unlikely that permafrost will be found in wet muskeg in the southern fringe of the discontinuous zone. Its occurrence is quite probable in similar terrain conditions in the northern porition of the discontinuous zone; and in the continuous permafrost zone, it will invariably exist.

Other Remote Sensing Techniques

During the past few years, considerable experimentation has been underway in the use of remote sensing techniques other than the usual black and white aerial photography to detect and delineate areas of permafrost in the discontinuous zone and determine the characteristics of perennially frozen ground and related features throughout the permafrost region. In July 1969 the National Research Council of Canada made an airborne survey over the discontinuous zone in northern Manitoba using infrared (false) colour photography and infrared thermal imagery (4-7 u and 11-14 u). The area covered included extensive tracts of muskeg with peat plateaus and palsas. Both techniques detect thermal patterns on the ground surface, but they cannot penetrate into the ground to the permafrost table. Variations in vegetation are accentuated by differences in thermal emission. Thus peat plateaus and palsas showed up very clearly where their vegetative cover differed from the surrounding terrain. Areas of permafrost with no surface expression in the vegetation or other terrain feature did not show up on the imagery or by the use of other techniques. The same results were reported by Thie (1972), who examined the same infrared thermal imagery.

Tarnocai (1972) also analysed multispectral imagery in northern Manitoba to determine the usefulness of remote sensing techniques in studying muskeg and permafrost. Dependable differences were found in the multispectral response patterns obtained from infrared thermal imagery, infrared colour, infrared black and white, as well as the usual black and white, and colour aerial photographs. These differences made possible the separation and mapping of the peat landforms, vegetation, organic soils, and permafrost.

The most recent development in remote sensing of the terrain is the Earth Resource Satellite Programme (ERTS). The first satellite (ERTS-1), launched in July 1972, has a design life-span of 1 to 3 years and a capability of multispectral scanning. The second (ERTS-B), to be launched in 1976, has an additional infrared thermal imagery capability. Vegetation, geomorphology, and permafrost mapping of an area in the Hudson Bay Lowland near Winisk, Ontario, were attempted (Şuffling, 1973; Palabekiroglu, 1973). Vegetation patterns and geomorphological features could be mapped at a small scale with considerable accuracy. Very small features, such as peat plateaus and palsas, were difficult to detect but may become more discernible with improvement in the imagery. Possible solifluction features and thermafrost lakes were detected although ground truth was not available for verification.

VARIATIONS IN PERMAFROST AND MUSKEG RELATIONSHIPS
BY PHYSIOGRAPHIC REGIONS

Mention has been made of the high concentration of muskeg occurrence in the Hudson Bay Lowland compared with the other much larger physiographic units in the permafrost region of Canada. Virtually all of the Hudson Bay Lowland is composed of unconfined muskeg, except the banks of the major rivers. The amount of permafrost varies from a very patchy occurrence in the south to continuous in the north. Peat plateaus and palsas in all stages of development abound in this region. In the discontinuous zone, no permafrost is found in the wet fens, which comprise most of the muskeg. In the narrow strip of continuous permafrost on the Hudson Bay Coast, permafrost occurs everywhere beneath the ground surface.

In the Precambrian Shield, muskeg is much less extensive, consisting mostly of confined bogs in rock basins. An exception to this pattern is the existence of extensive muskeg east and north of Lake Winnipeg in the glacial Lake Agassiz plain. String bogs are widespread throughout the Shield; their occurrence is attributed to the underlying acidic rocks. The existence of permafrost in these features in the discontinuous zone, as in the patterned fen in the Hudson Bay Lowland, is inhibited because they are too wet, but permafrost occurs in them in the continuous zone. Permafrost occurrence varies from south to north in the same manner as in the Hudson Bay Lowland. Palsas are much less numerous but peat plateaus are widespread. In the southern fringe of the discontinuous zone, permafrost islands in muskeg are less numerous in Quebec than west of Hudson Bay possibly because of the higher snowfall.

Extensive unconfined muskeg occurs in some areas of the Interior Plains. In northern Alberta and northeastern British Columbia, for example, tracts of peat terrain extend over several tens of square miles. Very few palsas have been noted in these areas but peat plateaus with permafrost are widespread. Although this muskeg resembles the Hudson Bay Lowland more than the Precambrian Shield, the permafrost features are more similar to those in the latter region.

Peat terrain in the Cordillera is mostly confined to rock basins. The mountainous relief causes variations in permafrost occurrence with elevation in addition to the usual latitudinal changes. The occurrence of permafrost in the peat areas depends on the elevation as well as the latitude. As in the two previous regions, palsas are infrequent but peat plateaus are common. In the northern part of the discontinuous zone, permafrost is found even in wet swampy areas.

The Arctic Archipelago actually comprises three regions, the Arctic Lowlands and Plateaus, Innuitian Region, and the northern part of the Precambrian Shield, but in this paper it is considered as one region. It lies entirely within the continuous permafrost zone except possibly the southeast tip of Baffin Island. Fairly extensive muskeg occurs in the interiors of some of the islands and string bogs have been noted in some areas. Permafrost is found everywhere beneath the ground surface in these areas, regardless of drainage conditions.

Measurements of peat thickness and depth of active layer have been carried out in the southern fringe of the permafrost region in all physiographic regions and in the Arctic Archipelago. The thickest peat occurs in the Precambrian Shield and Hudson Bay Lowland, to maximum depths exceeding 20 feet. The generally greater thickness of peat in the Precambrian Shield may be related to the existence of many rock basins in which peat can accumulate to considerable depths. In the Hudson Bay Lowland, the maritime climate with cool summers combined with poor drainage over vast expanses contributes to considerable peat accumulation. The thinnest peat layers occur in the Arctic Islands although local accumulations of 6 feet have been encountered.

Although peat thicknesses vary considerably from one physiographic region to another, the thickness of the active layer, or depth to the permafrost table, is fairly uniform. In the discontinuous zone, the thinnest active layer occurs in the Hudson Bay Lowland, and average thicknesses are similar in the other three regions. The average percentage deviation is similar in all four regions and much lower than the deviations observed for the peat thickness. Active layers are generally thinner in the Hudson Bay Lowlands, probably because of the cooler summers and accompanying lower thawing index values compared to the other three more continentally situated regions. In the Arctic Archipelago, the active layer is generally about one-half as thick, because of the much cooler summers, being roughly 6 to 12 inches. The insulating properties of the peat are manifested by the fact that the permafrost table occurs usually in the peat layer and not beneath it.

CONCLUSION

In northern Canada, muskeg and permafrost are very closely related, especially in the southern fringe of the permafrost region. The distribution of permafrost and its thermal regime in this fringe are governed to a considerable extent by the thermal properties of the peat. Mapping the distribution of permafrost in muskeg is facilitated to some extent by the occurrence of it in such distinctive microrelief features as peat plateaus and palsas. It is frequently difficult to identify these features on aerial photographs because they are usually small and varied in shape, and many variations occur in photographic tone and texture. Once recognized, they are fairly reliable indicators of the existence of permafrost. Developments in other remote sensing techniques appear to hold promise for the rapid detection and mapping of muskeg and related permafrost features. Some differences in the occurrence of permafrost in muskeg appear to exist among the various physiographic regions but more observations are required before completely reliable correlations can be made.

REFERENCES

Bates, D.N., and Simkin, D. 1966. Vegetation patterns of the Hudson Bay Lowlands. Map published by Dept. Lands and Forests, Prov. of Ontario, 10 miles to 1-inch scale.

Brown, R.J.E. 1966. The influence of vegetation on permafrost. Proc. Internat. Conf. on Permafrost, us Nat. Acad. Sci., Nat. Res. Council, Washington, DC, Pub. 1287, pp. 20-5.

— 1967. Permafrost map of Canada. Div. Building Res., Nat. Res. Council of Canada, NRC 9769, and Geol. Survey Canada, Map 1246.

— 1968. Permafrost investigations in northern Ontario and northeastern Manitoba. Nat. Res. Council of Canada, Div. Building Res. Tech. Paper 291, NRC 10465.

— 1973. Permafrost — distribution and relation to environmental factors in the Hudson Bay Lowland. Proc. Symp. Physical Environment of the Hudson Bay Lowland, Univ. Guelph, March 30-31.

— 1975. Permafrost Investigations in Quebec and Newfoundland (Labrador). Nat. Res. Council of Canada, Div. Building Res. Tech. Paper (in press).

Brown, R.J.E., and Williams, G.P., 1972. The freezing of peatland. Nat. Res. Council of

Canada, Div. Building Res. Tech. Paper 381, NRC 12881.

Forsgren, B. 1964. Notes on some methods tried in the study of palsas. Geog. Ann. 46 (3): 343-4.

— 1965. Tritium determination in the study of palsa formation. Geog. Ann. 48A (2): 102-10.

Hamelin, L.E., and Cailleux, A. 1969. Les palses dans le bassin de la Grande Rivière de la Baleine. Rev. Géog. Montr. 23 (3): 329-37.

Hughes, O.L., Rampton, V.N., and Rutter, N.W. 1972. Quaternary geology and geomorphology, southern and central Yukon (northern Canada). Field Excursion All, 24th Session, Internat. Geol. Congr. Guidebook.

Hustich, I. 1957. On the phytogeography of the subarctic Hudson Bay Lowland. Acta Geog. Fenn. 16 (1): 1-48.

Korpijaakko, E.O., and Radforth, N.W. 1966. Development of certain patterned ground in muskeg as interpreted from aerial photographs. Proc. Third Internat. Peat Congr., Quebec.

Lundqvist, J. 1969. Earth and ice mounds—a terminological discussion. In The Periglacial Environment, ed. T.L. Péwé, pp. 203-16 (McGill-Queen's University Press, Montreal).

Lundqvist, S., and Mattsson, J.O. 1965. Studies on the thermal structure of a palsa. Lund Studies in Geography. Ser. A, Physical Geography, No. 34 (The Royal Univ. of Lund, Sweden, Dept. Geography), pp. 38-49.

MacFarlane, I.C. (ed.). 1969. Muskeg Engineering Handbook (Univ. Toronto Press).

Palabekiroglu, S. 1973. ERTS-1, Imagery for geomorphological studies. Remote Sensing in Ontario, CCRS Remote Sensing Office (Guelph), Newsletter No. 4, May 31.

P'yavchencko, N.I. 1955. Bugristye Torfyaniki (Hummocky peat bogs) (Akad. Nauk SSSR, Institut Lesa [Academy of Sciences of the USSR, Forestry Institute], Moscow). (In Russian)

Radforth, N.W. 1952. Suggested classification of muskeg for the engineer. J. Eng. Inst. Canada, p.9.

— 1961. Distribution of organic terrain in northern Canada. Proc. Seventh Muskeg Res. Conf., Nat. Res. Council of Canada, Assoc. Comm. on Soil and Snow Mech., Tech. Memo. 71:8-11.

Railton, J.B., and Sparling, J.H. 1973. Preliminary studies on the ecology of palsa mounds in northern Ontario. Can. J. Bot. 51(5): 1037-44.

Ruuhijärvi, R. 1960. Uber die regionale Einteilung der Nordfinnischen Moore (Regional distribution of north Finnish bogs). Ann. Bot. Soc. Zool. Bot. Fenn. 'Vanamo' 31 (1).

Salmi, M. 1968. Development of palsas in Finnish Lapland. Proc. Third Internat. Peat Congr., Quebec, Aug. 18-23, pp. 182-9.

Sjörs, H. 1959a. Bogs and fens in the Hudson Bay Lowlands. Arctic 12 (1): 3-19.

— 1959b. Forest and peatlands at Hawley Lake, Northern Ontario. Nat. Mus. Canada, Contrib. Botany, Bull. 171, pp. 1-31.

— 1961. Surface patterns in boreal peatlands. Endeavour xx (80): 217-24.

Suffling, P. 1973. Vegetation mapping in the Hudson Bay Lowlands using ERTS-1 imagery. Remote Sensing in Ontario, CCRS Remote Sensing Office (Guelph), Newsletter No. 4, May 31.

Svensson, H. 1961-62. Nägra Iattagelser från Palsområden Flygibildanalys Fältstudier: Nordnorska Frostmarksomraden (Observations on palsas — photographic interpretation and field studies in northern norwegian frost ground areas). Norsk Geog. Tidsskr. XVIII (5-6): 212-27. (In Norwegian)

— 1964. Structural observations in the minerogenic core of a palsa. Särtryck Från Lunds Univ. Geog. Inst. 17: 138-42.

— 1971. Observations on Icelandic polygon surfaces and palsa areas. Photo interpretation and field studies. Lund Studies in Geog., ser. A, Phys. Geog. 51: 115-45.

Tarnocai, C. 1970. Classification of Peat Landforms in Manitoba (Canada Dept. Agriculture, Res. Station, Pedology Unit, Winnipeg).

— 1972. The use of remote sensing techniques to study peatland and vegetation types, organic soils and permafrost in the boreal region of Manitoba. Proc. First Canadian Symp. on Remote Sensing, Feb., pp. 323-35.

Thie, J. 1971. Air photo analysis and description of surficial deposits of an area north of Lake Winnipeg, with special reference to the occurrence and melting of permafrost (Dept. Soil Science, Univ. Manitoba).

— 1972. Application of remote sensing techniques for description and mapping of forest ecosystems. Proc. First Canadian Symp. on Remote Sensing, Feb., pp. 149-69.

Tyrtikov, A.P. 1959. Perennially frozen ground and vegetation. In Principles of Geocryology, ed. P.F. Shvetsov (Academy of Sciences of the USSR), vol. I, pp. 299-421. Nat. Res. Council of Canada Tech. Transl. 1163 (1964).

— 1966. Formirovaniye i Razvitiye Krupnobugristykh Torfyanikov v Severnoy Tayge Zapadnoy Sibire (Formation and development of large hummocky peat bogs in the northern taiga of Western Siberia). Merzlotnyye Issledovaniya (Permafrost Investigations), VI: 144-54. (In Russian)

Zoltai, S.C. 1971. Southern limit of permafrost features in peat landforms, Manitoba and Saskatchewan. Geol. Assoc. Canada, Special Paper 9: 305-10.

— 1972. Palsas and peat plateaus in central Manitoba and Saskatchewan. Can. J. Forest Res. 2 (3): 291-302.

— 1973. Vegetation, surficial deposits and permafrost relationships in the Hudson Bay Lowlands. Proc. Symp. on the Physical Environment of the Hudson Bay Lowland, Univ. Guelph, March 30-31.

Zoltai, S.C., and Tarnocai, C. 1971. Properties of a wooded palsa in northern Manitoba. J. Arctic and Alpine Res. 3 (2): 115-29.

PART II

Utilization of Muskeg

7
Forests, Muskeg, and Organic Terrain in Canada

P.J. RENNIE

Both nationally and internationally the Canadian forest is large and important. It occupies 48 per cent of the land area of Canada or $1,115 \times 10^6$ acres (450×10^6 ha). Some 55 per cent of this forest, or 614×10^6 acres (248×10^6 ha), is productive and contains 749×10^9 cubic feet (21.1×10^9 m³) of merchantable timber. In 1969, 2.25×10^6 acres (0.91×10^6 ha) of forest were harvested, yielding 4.3×10^9 cubic feet (122×10^6 m³) of timber. This harvest amounted to 5.7 per cent of the world's total and provided the raw material for forest-product exports worth three billion dollars, or 18 per cent of all Canadian exports. In addition, both productive and non-commercial forests have incalculable value for aesthetic and recreational purposes, for provision of shelter and habitat for wildlife, and because of their ability to stabilize soil and water supplies.

No very precise figure has been available for muskeg, peatland, or organic terrain in Canada, but MacFarlane's recent estimate (MacFarlane, 1969) puts the total area of undifferentiated peatland at a conservative 320×10^6 acres (130×10^6 ha), or nearly 14 per cent of the total land area.

It is the purpose of this paper to explore the relationship that exists between these areal estimates: the $1,115 \times 10^6$ acres of forest on the one hand and the 320×10^6 acres of peatland on the other. Are they discrete or do they overlap? If the latter, by how much and in what way? What is meant by peatland? Does it, in fact, support a significant part of the Canadian forest? And what sort of forest? Has it special characteristics and, in view of the generally increasing exploration of our northern and more remote areas, are there special features that need to be investigated or safeguarded?

Clearly, a glance at certain existing maps, especially those of Radforth (1960) and Rowe (1959), can readily provide answers to some of these questions, though not, on the whole, very precise ones. Again, numerous classical papers and publications are available on the Canadian forest (Halliday, 1937; Halliday and Brown, 1943; Wilson, 1969; Manning and Grinnell, 1971); they do not, however, take peatlands as a particular point of orientation. This in no way signifies an omission, but merely — as will become apparent in the present paper — that our knowledge of the Canadian land-base is insufficient or too vague to permit a definite treatment from this viewpoint.

The present paper, therefore, is no more than an interim measure and an approximation. It commences with a short general description of the forest resource and its setting. This is followed by a more detailed examination of those forest regions and sections which, in whole or in part, are supported by peat soils. Productivity and problems are then outlined and a summarizing discussion attempts a look into the future.

THE CANADIAN FOREST AND ENVIRONMENT

The Forest

The distribution of the 1,115 x 10⁶ acres (450 x 10⁶ ha) of essentially natural forest may be seen in Figure 1. It forms a belt 600 to 1,000 miles (950-1,600 km) wide stretching through all territories and provinces of Canada from the Yukon in the northwest to Newfoundland in the east. Except for a grassland triangle in the Prairie Provinces, it borders the southern frontier of Canada. Its northern boundary runs roughly southeast from latitude 68N on the Macken-zie River delta to latitude 60N on the west coast of Hudson Bay; on the east coast of Hudson Bay it continues from latitude 57N in a roughly northeasterly direction to Ungava Bay.

Approximately 80 per cent of the forest is coniferous and 20 per cent broadleaved. Coniferous species include the spruces (*Picea mariana* (Mill.) BSP, *P. glauca* Moench (Voss), *P. rubens* Sarg., *P. engelmanni* Parry, *P. sitchensis* (Bong.) Carr), 44 per cent; the pines (*Pinus banksiana* Lamb., *P. contorta* Dougl. var. *latifolia* Engelm., *P. resinosa* Ait.), 17 per cent; the firs (*Abies balsamea* (L.) Mill.), 15 per cent; the hemlocks (*Tsuga heterophylla* (Raf.) Sarg., *T. canadensis* (L.) Carr.), 10.4 per cent; the cedars (*Thuja plicata* Donn, *T. occidentalis* L.), 5.6 per cent; Douglas fir (*Pseudotsuga menziesii* (Mirb.) Franco), 5.5 per cent; and others, including tamarack (*Larix laricina* (Du Roi) K. Koch), 2.7 per cent. Poplars (*Populus tremuloides* Michx., *P. balsamifera* L.) account for 57 per cent of broad-leaved species; white birch (*Betula papyrifera* Marsh.), 22 per cent; yellow birch (*B. alleghaniensis* Britt.) 8 per cent; and several others, including alders, 13 per cent.

As might be expected for a forest of such large geographical distribution, there is consider-able floristic and ecological diversity. In the first place, though, there can be variations due to the different definitions used for 'forest.' For forest-production purposes, it is convenient to separate 'forest land,' amounting to 796 x 10⁶ acres (321 x 10⁶ ha), from 'wildland,' amount-ing to 1,296 x 10⁶ acres (522 x 10⁶ ha); the criterion is that 'forest land' is capable of producing trees 4 inches (10 cm) dbh (diameter at breast height) and over on 10 per cent or more of the land, whereas 'wildland' embraces 'barrens, muskeg, rock and land with scrub and/or land with forest substandard to "forest" as specified above' (Forest Economy Research Insti-tute, 1972). As the total area of the forest regions and sections of Figure 1 is 1,440 x 10⁶ acres (582 x 10⁶ hectares), it may readily be appreciated that not all of this area, nor all of the 1,115 x 10⁶ acres of 'forest' spoken of at the outset, measures up to this definition of forest land.

Accepting the more limited definition of 'forest' given above, the percentage distribution of the 796 x 10⁶ acres throughout the territories and provinces is:

Yukon	6.7%	Ontario	14.9%
Northwest Territories	16.0%	Quebec	22.0%
British Columbia	17.3%	New Brunswick	2.0%
Alberta	8.0%	Nova Scotia	1.2%
Saskatchewan	3.1%	Prince Edward Island	0.1%
Manitoba	4.9%	Newfoundland	3.8%

FIGURE 1   The major vegetation types of Canada.

LEGEND
PRINCIPAL TREE SPECIES

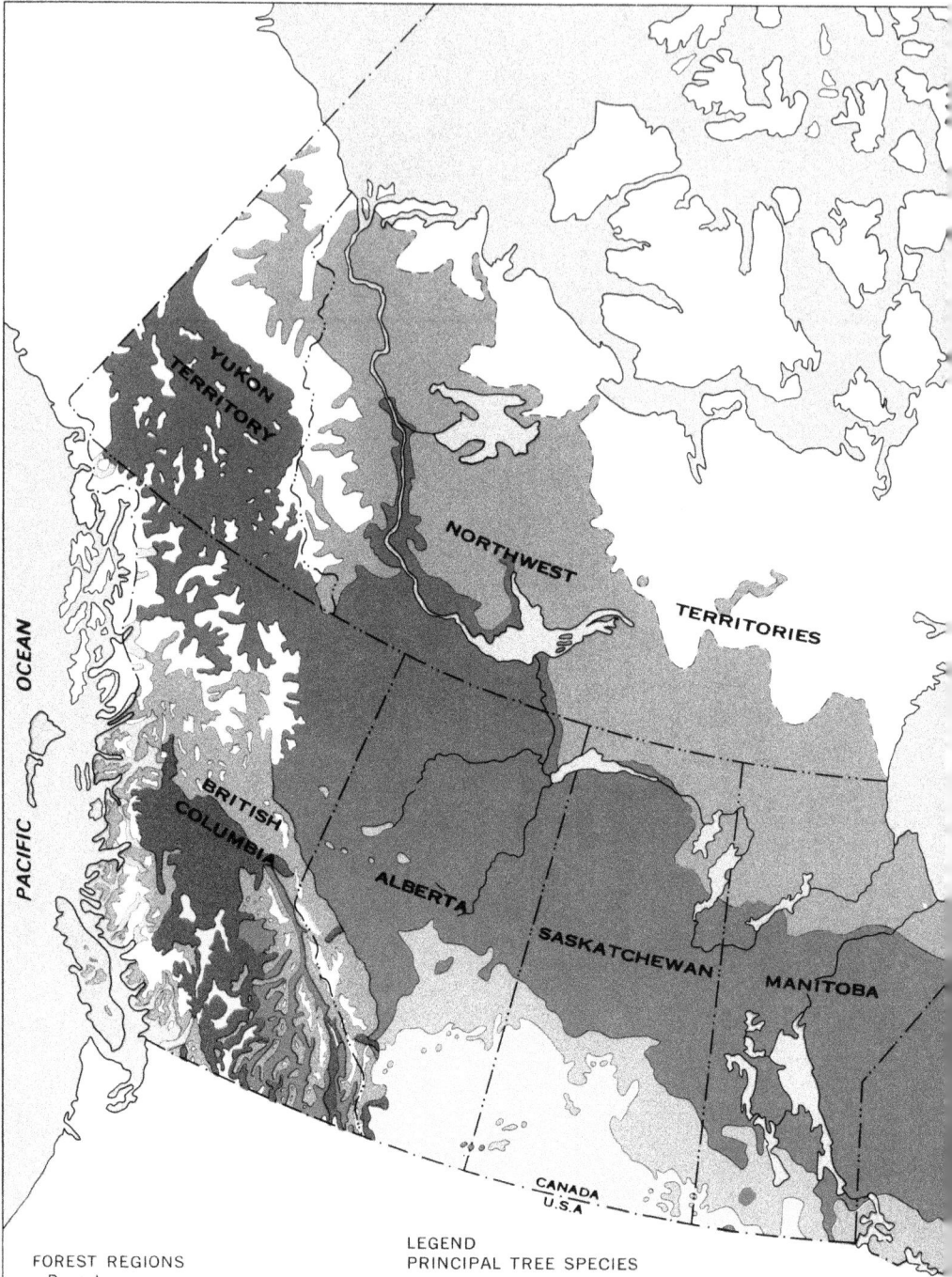

FOREST REGIONS		
Boreal		
Predominantly Forest	White Spruce, Black Spruce, Balsam Fir, Jack Pine, White Birch, Trembling Aspen	
Forest & Barren	White Spruce, Black Spruce, Tamarack	
Forest & Grass	Trembling Aspen, willow	
Subalpine	Engelmann Spruce, Alpine Fir, Lodgepole Pine	
Montane	Douglas-fir, Lodgepole & Ponderosa Pine, Tr. Aspen	
Coast	W. Red Cedar, W. Hemlock, Sitka Spruce, Douglas-fir	
Columbia	W. Red Cedar, W. Hemlock, Douglas-fir	
Deciduous	Beech, maple, Black Walnut, hickory, oak	
Great Lakes-St. Lawrence	Red Pine, E. White Pine, E. Hemlock, Yellow Birch, maple, oak	
Acadian	Red Spruce, Balsam Fir, maple, Yellow Birch	
GRASSLAND	Trembling Aspen, willow, Bur Oak	

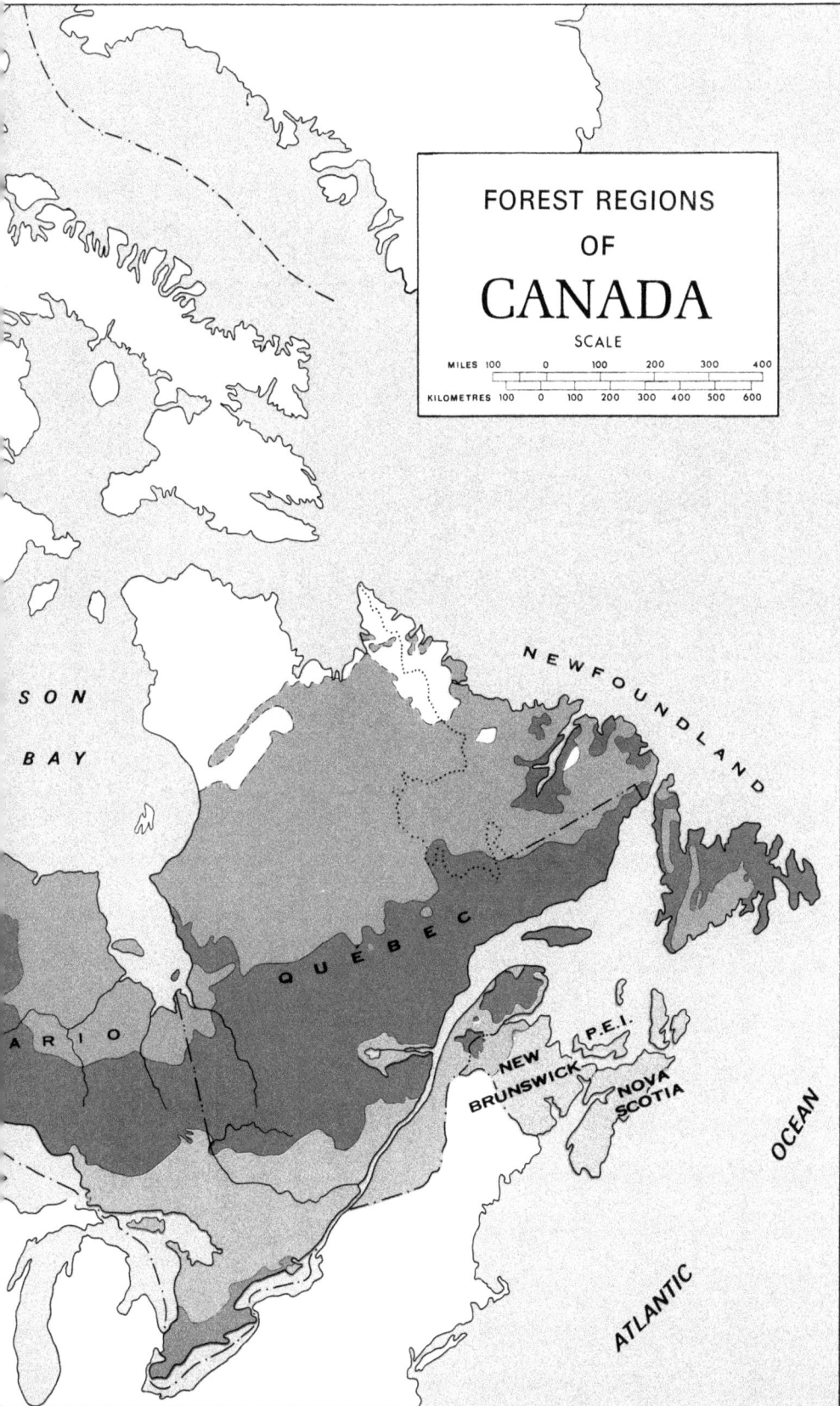

FOREST REGIONS

OF

CANADA

SCALE

MILES	100	0	100	200	300	400
KILOMETRES	100	0	100	200	300	400	500	600

NEWFOUNDLAND

SON

BAY

QUÉBEC

ARIO

NEW
BRUNSWICK

P.E.I.

NOVA
SCOTIA

OCEAN

ATLANTIC

The over-all classification of Figure 1 comprises ten major divisions. Two — the Tundra of the north and of certain alpine regions, and the Grassland of the Prairies and of interior British Columbia — do not include trees. The remaining eight all include forest growth. They are the Deciduous Region of southern Ontario (occupying 0.8 per cent of the whole), the Columbia Region of British Columbia (0.8 per cent), the Acadian Region of the Maritime Provinces (2.0 per cent), the Coast Region of British Columbia (2.2 per cent), the Montane Region of central British Columbia (2.3 per cent), the Subalpine Region of the Cordilleran (3.7 per cent), the Great Lakes — St Lawrence Region of southeastern Canada (6.4 per cent), and the transcontinental Boreal Region (81.8 per cent). The Boreal Region comprises three components: a predominantly forested central zone, a more southerly grassland-transition zone which flanks it in the Prairies, and a more northerly tundra-transition zone which runs the entire length of the central transcontinental span. The eight forest regions are further divided into 90 subregions or sections. More will be said on certain of these later.

The various forest regions and sections display particular attributes of composition and of ecological parameters which permit their delineation and characterization. Throughout many, however, may be found particular tree species of unusually wide distribution and abundance. Shown in Figures 2 to 6 are Halliday and Brown's frequency-of-distribution diagrams for the spruces, the pines, the true firs, the poplars, and two birches. These diagrams are of value in simplifying approaches to the correlation of site characteristics and forest growth, and it will be seen that they are particularly useful in the consideration of the more northerly peatland areas.

The growth of the Canadian forest displays considerable variation. As a starting point, it can be best and most simply explained by an appreciation of the annual net radiation (Hare and Ritchie, 1972). Values reproduced from J.E. Hay's unpublished 1969 thesis by Hare and Ritchie vary from 15 kilolangleys per year in the Mackenzie River delta, through 20-25 kiloangleys per year in much of the Boreal-Tundra transition, to 35-40 in the Boreal-Grassland transition, in coastal British Columbia and in the Atlantic Provinces. As net radiation is closely related to potential productivity the over-all constraints on biomass productivity within the different localities can be seen. Forest yields parallel these values approximately, but they are dependent upon a number of other factors, such as length of rotation, density of stocking, and utilization standards. For pulpwood stands with high utilization and short rotations, annual productivity figures can vary from 100 cubic feet of merchantable produce per acre per year (7 $m^3ha^{-1}yr^{-1}$) for the Acadian Region and better parts of the Boreal, to around 20-30 cubic feet per acre per year (1.4-2.1 $m^3ha^{-1}yr^{-1}$) for much of the Boreal. For longer rotations and more selective utilization, values vary from around 50 cubic feet per acre per year (3.5 $m^3ha^{-1}yr^{-1}$) in coastal British Columbia, through 20 cubic feet per acre per year (1.4 $m^3ha^{-1}yr^{-1}$) in Newfoundland, to 10 cubic feet per acre per year and less (0.7 m^3/ha per year and less) in the main Prairie Boreal Region (Bedell, 1958). The lower values for the Boreal Region are not out of line with those published for northern Scandinavia (Ilvessalo, 1949; Mikola, 1970).

Because of these variations in natural productivity, the distribution among the provinces of the 749 x 10⁹ cubic feet (221.1 x 10⁹m³) of standing merchantable timber does not exactly parallel the areal distribution tabulated earlier. Excluding the Yukon, Northwest Territories, and Labrador, the percentage distribution is:

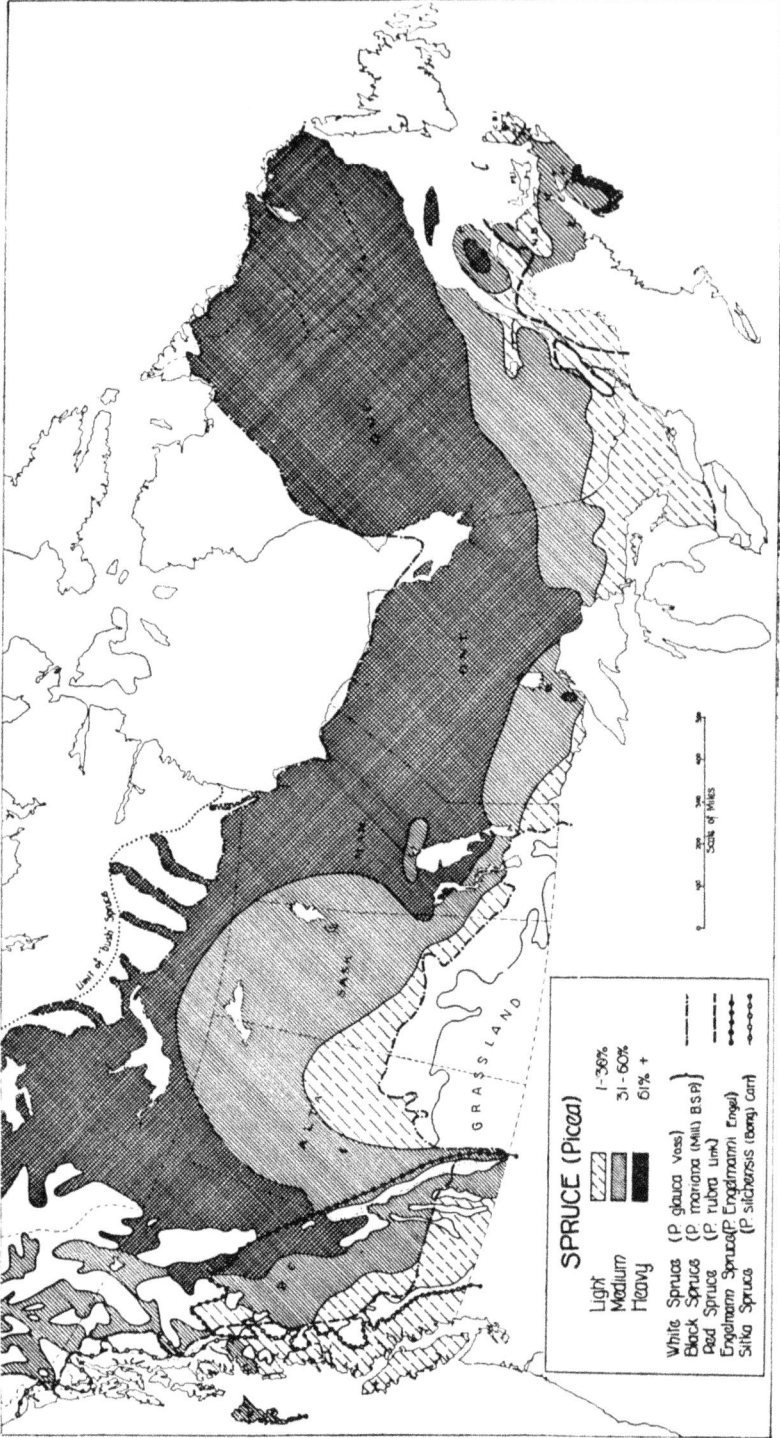

FIGURE 2   The distribution and population intensity of spruce in Canada (from Halliday and Brown, 1943).

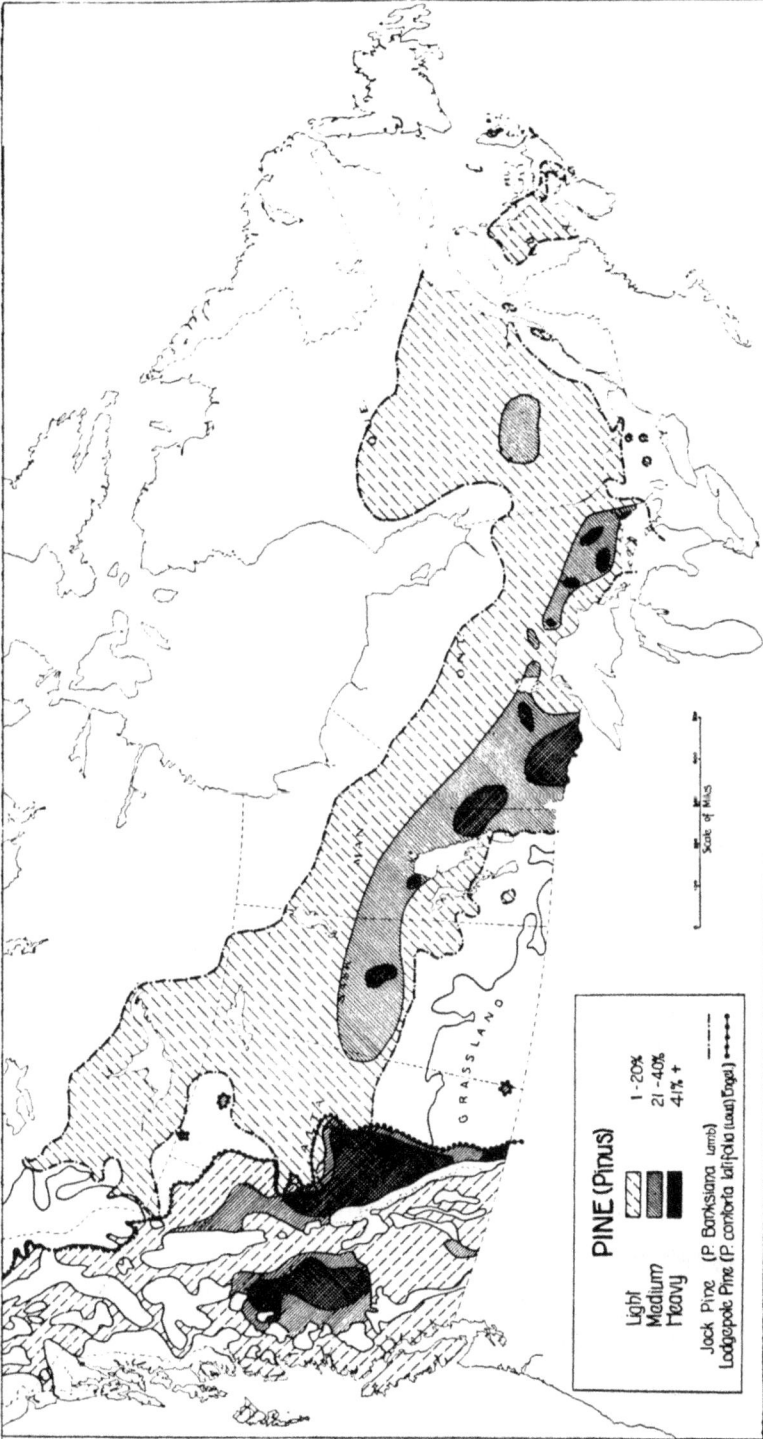

FIGURE 3 The distribution and population intensities of jack and lodgepole pines in Canada (from Halliday and Brown, 1943).

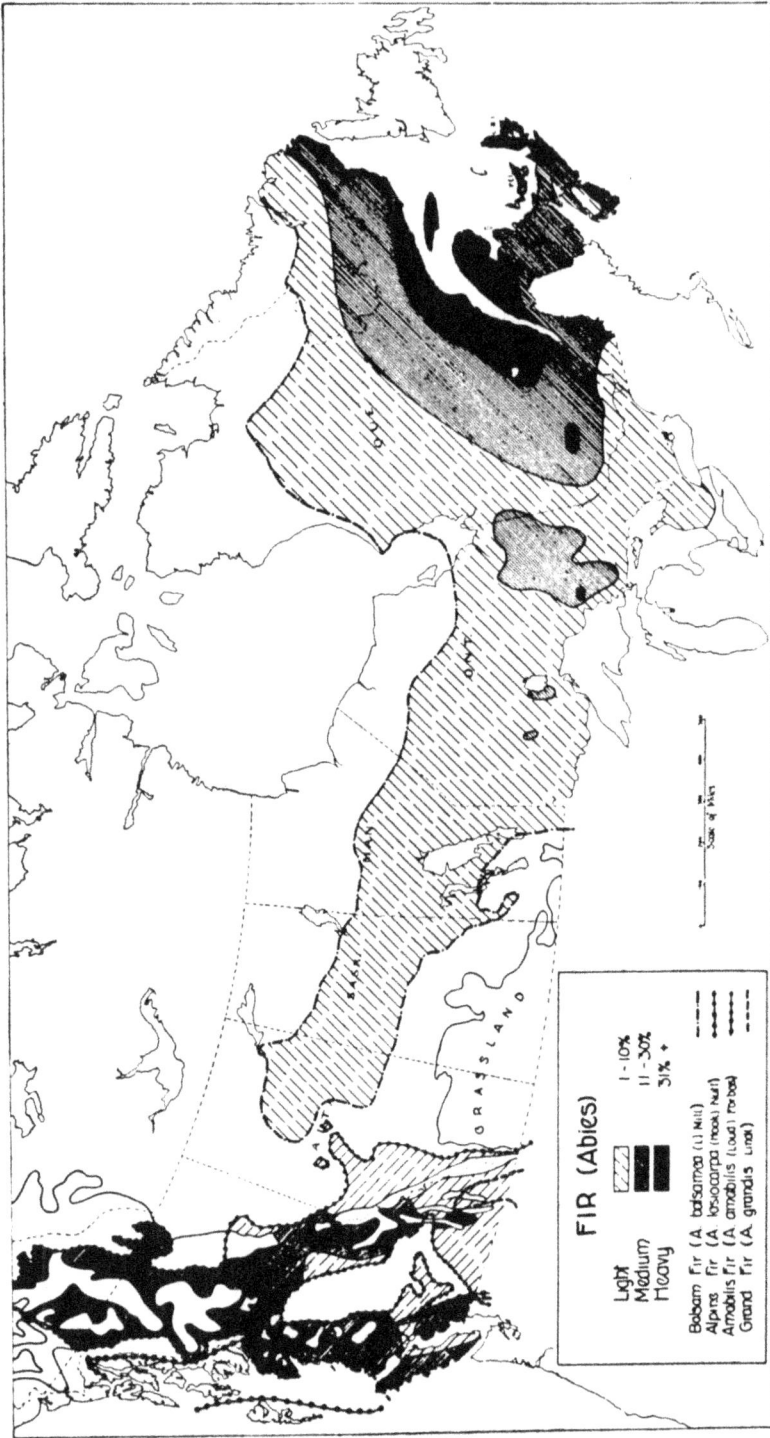

FIGURE 4   The distribution and population intensity of fir in Canada (from Halliday and Brown, 1943).

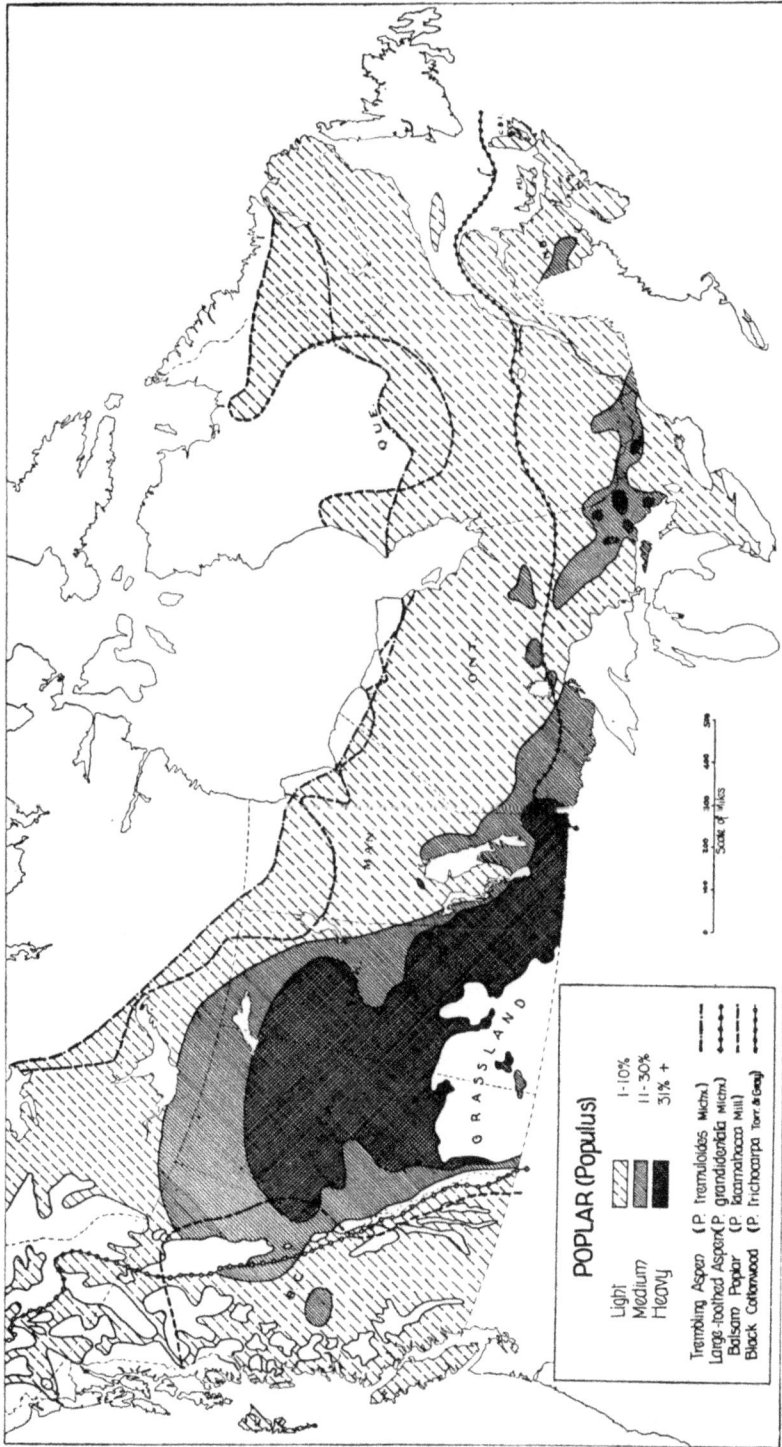

FIGURE 5   The distribution and population intensity of poplar in Canada (from Halliday and Brown, 1943).

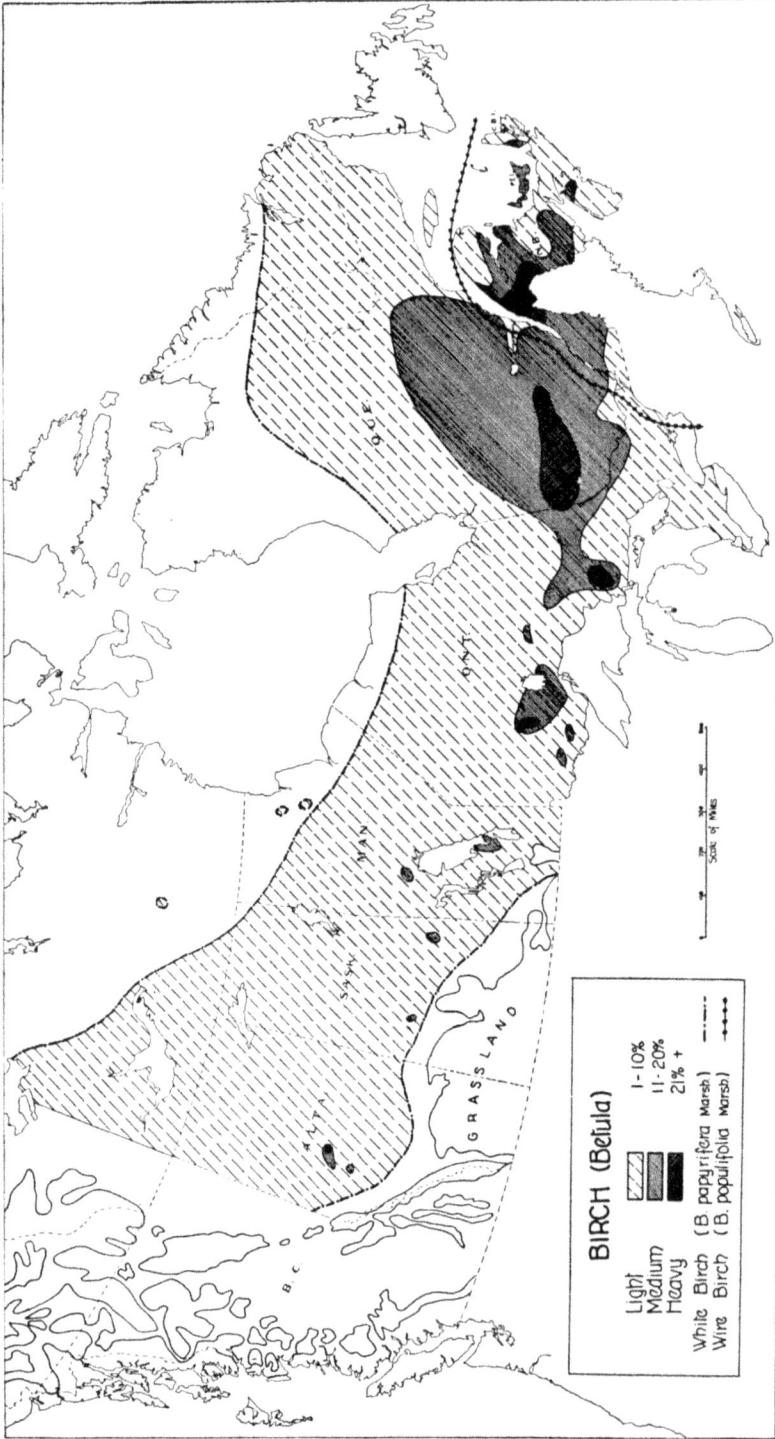

FIGURE 6   The distribution and population intensities of white and wire birches in Canada   (from Halliday and Brown, 1943).

British Columbia	42.7%	Quebec	20.7%
Alberta	9.5%	New Brunswick	2.7%
Saskatchewan	2.7%	Nova Scotia	1.4%
Manitoba	2.0%	Prince Edward Island	(0.02)%
Ontario	17.7%	Newfoundland	0.6%

For reasons of accessibility and others the distribution of the actual harvest among the provinces and territories differs again. The following percentage distribution is based on the five-year average for the period 1964-68 (Manning and Grinnell, 1971):

Yukon and		Ontario	15.5%
Northwest Territories	0.1%	Quebec	25.6%
British Columbia	41.9%	New Brunswick	5.5%
Alberta	3.3%	Nova Scotia	2.9%
Saskatchewan	1.4%	Prince Edward Island	0.2%
Manitoba	1.1%	Newfoundland	2.5%

Although not entirely up to date Wilson's 1962-63 map of the operable commercial forestry area in Canada (Wilson, 1969) reveals the approximate expanse of northern Boreal forest which has not so far been worked.

Further information on the ecology and distribution of Canadian tree species may be found in *Native Trees of Canada* (Hosie, 1969), and information on the properties and utilization of various Canadian timbers has been summarized by McElhanney (1951).

The Environment

The various areas of Canada display considerable variation in climatic parameters, but only a few statistics can be presented here. The length of the growing season — i.e. the numbers of days when the mean daily-temperature exceeds 4°F (5.6°C)—varies from 260 days in coastal British Columbia, through 190 days for the Atlantic Provinces, to around 140 days for much of the Boreal Region. The Boreal-Tundra transition corresponds to 80-100 days. For the same four locations the number of degree-days — i.e. the total number of degrees above 42°F for all days during the growing season — varies from 3,000 through 2,500 to 2,000, with a value approaching 1,000 corresponding to the Boreal-Tundra transition. The annual precipitation for the first three areas is over 80 inches (>2,000 mm), 40 inches (1,100 mm), and 10-40 inches (250-1,000 mm), it being necessary to distinguish between the drier Boreal west of Manitoba and the increasingly wetter Boreal east of the Manitoba-Ontario border. A similar split has to be made in the annual hydrologic-runoff figures, which are useful indicators of that portion of the total precipitation which is not evaporated and can thus serve as a potential agent for soil-leaching and other pedological processes (Canadian National Committee, International Hydrological Decade, 1969). The figures for the first three areas are over 70 inches (>1,800 mm), 20-40 inches (500-1,000 mm), and 5-30 inches (130-750 mm). It is again possible to see a striking difference in this hydrologic parameter for different parts of the Boreal. In Quebec, values fall from 40 to 15 inches (1,000 to 380 mm) across a northwesterly transect connecting Anticosti Island in the Gulf of St Lawrence to Belcher Islands in Hudson Bay. Across the Boreal of Ontario an analogous northwesterly transect reveals a drop from 18 inches (460 mm) at Lake Abitibi to 6 inches (150 mm) at Sandy Lake. From north of Lake Winnipeg to as far as the Mackenzie River delta there is remarkable constancy, the annual hydrologic-runoff remaining at around 5 inches (130 mm).

There are eight major soil orders represented throughout Canada: the Chernozemic,

Solonetzic, Luvisolic, Podzolic, Brunisolic, Regosolic, Gleysolic, and Organic (Canada Department of Agriculture, 1970). There is, of course, progressively further division of these orders into 22 great groups, 138 subgroups, 800 to 1,000 families, around 3,000 series, and around 4,000 types. The Podzolic Order, for example, comprises three great groups, the Humic Podzol, Ferro-Humic Podzol, and Humo-Ferric Podzol, each of which includes a number of subgroups, the last, for example, having eight, two of which are the Orthic Humo-Ferric Podzol Subgroup and the Gleysol Humo-Ferric Podzol Subgroup.

As might be expected, the Organic Order embraces numerous organic soils, which normally must have a surface layer of organic material not less than 24 inches (60 cm) thick for fibric sphagnum moss or 16 inches (40 cm) thick for organic debris otherwise derived. It must be noted, however, that in the national Canadian system organic or mucky phases are classified within certain other orders. These are the Gleysolic Order, covering all three Great Groups and their 18 Subgroups; the Gleyed Gray Wooded (or Gleyed Gray Luvisol) Subgroup of the Gray-Wooded (or Gray Luvisolic) Great Group of the Luvisolic Order; and two Subgroups of the Podzolic Order: the Placic Ferro-Humic Podzol of the Ferro-Humic Podzol Great Group and the Gleyed Humo-Ferric Podzol of the Humo-Ferric Podzol Great Group. In all these cases the surface organic matter may attain thicknesses not exceeding 16 or 24 inches (40 or 60 cm) depending, as with the Organic Order itself, upon the nature of the plant material from which the organic layer is derived. As other authorities have taken muskeg and peat to be organic terrain where the thickness of the organic surface layer exceeds a few inches (MacFarlane, 1969) or 30 to 40 cm (Ketcheson and Jeglum, 1972), it is obvious that the classification of land and the areal estimates of organic terrain in Canada will vary considerably according to the criteria selected.

Canadian forest is supported predominantly by soils of the Podzolic Order, with those of the Luvisolic, Brunisolic, Gleysolic, and Organic orders playing relatively smaller roles. Nevertheless, in absolute terms soils of the latter four orders support considerable areas of forest, and over vast areas any one soil order or subgroup could be the prime focus of attention. Except for parts of the Yukon and Northwest Territories, no forest areas of Canada were unglaciated: soils have been derived from medium to coarse morainic materials, often with varying additions of finer materials stemming from glacial-lake or marine inundations (Geological Survey of Canada, 1968). Glacial-lake inundation occurred extensively from the Rockies eastwards to Lake Nipigon, Ontario, in a large area common to Ontario and Quebec centred on Lake Abitibi, and in southern Ontario, but elsewhere it was largely absent. Marine inundation mainly affected the Hudson Bay Lowlands, parts of the Ottawa and St Lawrence river valleys, the Lac St Jean basin, and coastal New Brunswick.

Bedrocks vary according to the five main physiographic regions: the Cordilleran of the west; the Interior Plains stretching from the Rockies to Lake Winnipeg; the Shield, covering all areas east of Lake Winnipeg and north of the Great Lakes and the St Lawrence River (except for the Hudson Bay Lowlands and southern Ontario); and the Appalachian, covering all areas east of the St Lawrence estuary (Geological Survey of Canada, 1969, 1970).

The coastal Cordilleran bedrocks are mainly uplifted sedimentaries with considerable igneous intrusions. The Interior Plains are horizontally bedded limestones, shales, and sandstones. The Shield is composed of granites, quartzites, and other ancient hard siliceous rocks. The Appalachians, although formed by folding and faulting processes analogous to those of the coastal Cordilleran, differ from the latter by comprising essentially Mesozoic rather than Palaeozoic strata. The bedrocks of the Appalachians vary from the granites and quartzites of Nova Scotia and central Newfoundland, through the limestones and shales of

New Brunswick, the Gaspé, and western Newfoundland, to the red sandstone of Prince Edward Island.

The best general account of the soils of Canada is represented by the 14 special regional and other articles making up a special issue of the *Agricultural Institute Review* (vol. 15, no. 2, 1960). To characterize Canadian forest soils in a few words is not easy. Many are of coarse to medium texture with appreciable admixture of stones on tills or glacio-fluvial outwash plains: the clay content is greater on lacustrine and marine-influenced sites. Most display a surface accumulation of organic matter with an incipient or well-developed eluvial horizon: below is a darker illuvial horizon enriched in clay or in organic matter and sesquioxides. In general, profiles are shallow without the indurated ironpans seen in other north-temperate forest soils. Soils are usually acidic, pH values ranging from 4 or sometimes less in the surface organic horizons to 5.5 or sometimes more in the subsoil. Nitrogen contents vary down the profile from 1.5 to 0.02 per cent. Except in the surface organic layers, exchange-capacities are low: throughout the profile base-saturation is low, with calcium being the dominant base in the surface organic layers but with potassium becoming relatively more important in the mineral horizons.

A few recent notable papers are Pollett (1972a) for Newfoundland peat types, Page (1971) for Newfoundland podzolic and brunisolic soils, Dumanski et al. (1970) for Alberta luvisolic soils, Scotter (1971) for northern Prairie and Northwest Territories podzolic and brunisolic soils, and Wilde et al. (1954) for northern Ontario organic, gleysolic, and podzolic soils. Atkinson's (1971) recent comprehensive bibliography of Canadian soils studies, Leggett's (1961) collection of symposium papers, and many of the detailed Soil Survey of Canada Reports are useful sources for more local data. Some information on soil may be obtained from the 1:250,000 site-capability maps of the Canada Land Inventory, which cover approximately one million square miles (2.6×10^6 km²) of the more southerly or settled parts of Canada (McCormack, 1967; Duffy, 1971). In a wider context, Canadian subarctic soils are described and reviewed in Tedrow's recent comprehensive treatment (1970) of all circumpolar subarctic soils.

FOREST-ORGANIC TERRAIN RELATIONSHIPS

Methodology

Three main maps and five supplementary ones were used to examine and attempt to analyse relationships between forest cover and organic terrain. The first of the three is Radforth's map at a scale of 1:6,000,000 (94 miles = 1 inch) showing the frequency of occurrence of areas of organic terrain (muskeg) in Canada (Radforth, 1960). This map is reproduced as Figure 7. Letters on the map denote types of vegetative cover: A signifies a forest or woody cover of spruce or tamarack more than 15 feet (4.5 m) high; B and D signify woody or shrub covers 5-15 feet (1.5-4.5 m) and 2-5 feet (0.6-1.5 m) high, respectively; and all other letters denote various non-woody covers less than 5 feet (1.5 m) high or a woody cover less than 2 feet (0.6 m) high. The second main map is that of the forest regions and sections of Canada, similar to Figure 1, but more detailed and at a scale of 1:5,000,000 (79 miles = 1 inch). The third main map is the new but as yet unpublished soil map of Canada, also at a scale of 1:5,000,000 (Clayton et al., 1973). The five supplementary maps shown as Figures 2 to 6 illustrate the frequency of occurrence of important northern tree species. They were originally available only at a scale of 1:26,700,000 (423 miles = 1 inch). All maps were photographed and then reproduced as

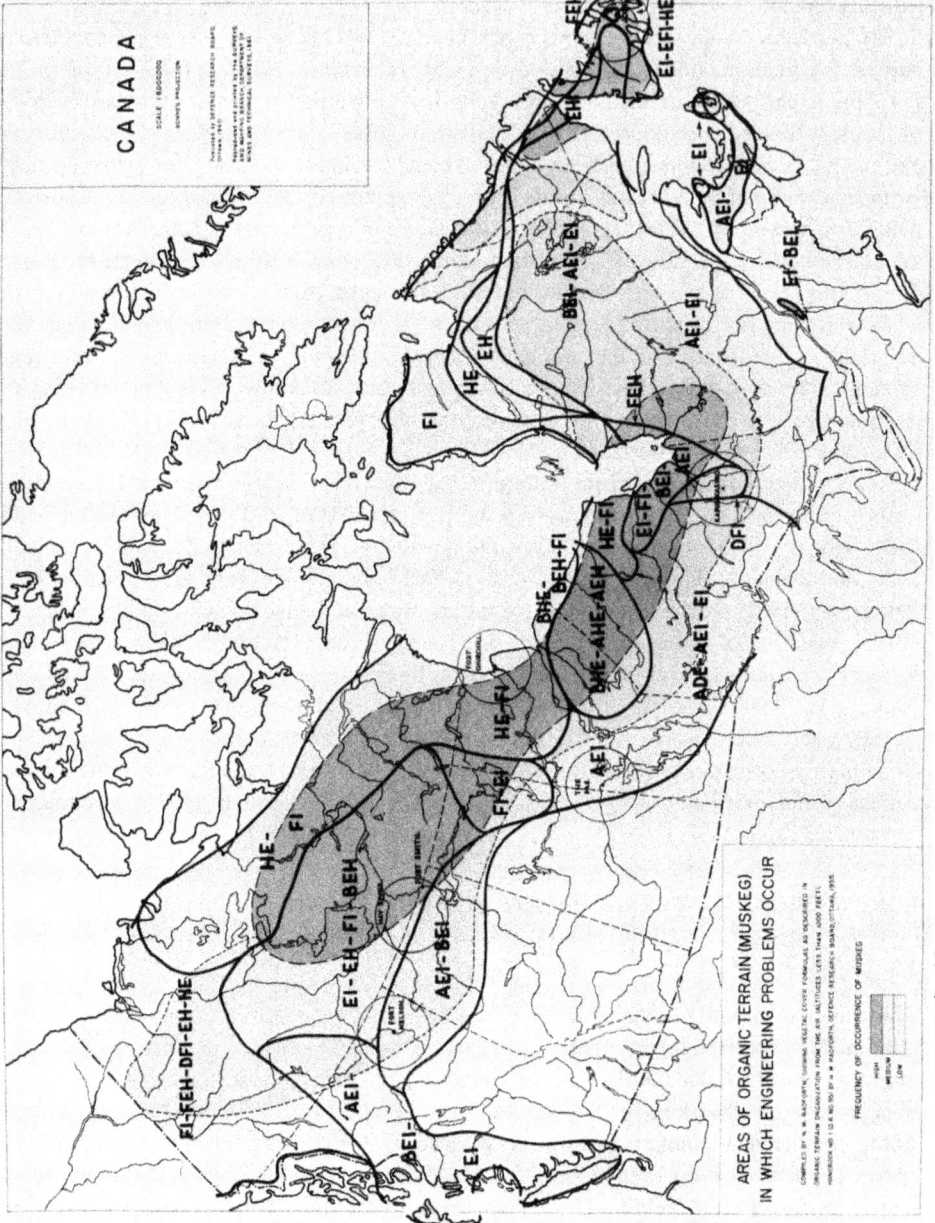

FIGURE 7 The frequency of occurrence of (organic terrain) muskeg in Canada (from Radforth, 1960).

black-and-white transparencies at a common scale of 1:5,000,000. The three main maps — of
Radforth's organic-terrain frequency, of the forest regions and sections, and of the principal
soil orders and subgroups — were then redrawn into one map, also at 1:5,000,000. This is
shown reduced as Figure 8.

In the preparation of Figure 8 comprehensiveness was the aim in the organic-terrain areas
(Radforth) and in the forest region and sections. The soils of the Prairie grassland zone, of
Ontario south of a line between Georgian Bay and Kingston, and of the arctic islands are not
shown. Also omitted are the forest regions of British Columbia south of latitude 50°N: the
patterns of distribution are so complex that analysis of the relatively small areas of organic
terrain involved would not be possible. Also omitted from Figure 8 are the organic terrain
subdivisions shown on Figure 7. On Figure 8 heavy and light discontinuous lines delineate
the forest regions and sections: the well-known reference numbers of the sections are shown
in circles (Rowe, 1972). Soil areas are delineated by continuous lines and are identified by
letters and numbers. Their interpretation is:

A Dominantly Chernozemic Soils
   A1   Brown Chernozemic
   A2   Dark Brown Chernozemic
   A3   Black Chernozemic
   A4   Dark Grey Chernozemic
B Dominantly Solonetzic Soils
   B1   Brown Solonetz
   B2   Black Solonetz
   B4   Solod
C Dominantly Luvisolic Soils
   C1   Grey Brown Luvisolic
   C2   Grey Luvisolic (Grey Wooded)
D Dominantly Podzolic Soils
   D3   Humo-Ferric Podzol
E Dominantly Brunisolic Soils
   E1   Melanic Brunisol
   E2   Butric Brunisol
   E3   Dystric Brunisol

F Dominantly Regosolic Soils
   F1   Orthic Regosol
   F2   Cumulic Regosol
   F3   Cryic Regosol
G Dominantly Gleysolic Soils
   G1   Humic Gleysol
   G2   Gleysol
   G3   Cryic Gleysol
H Dominantly Organic Soils
   H1   Fibrisols and Mesisols
   H3   Cryic Fibrisol
I   Dominantly Icefields
R   Dominantly Rockland
   R1

On the original soil map every area delineated also has an identifying number which serves as
a reference to an accompanying text. The numbers are not shown in Figure 8, but relevant
information from the text is used in the analysis below.

The procedure used in the analysis is to take in turn each forest section that falls within the
Radforth organic-terrain area and to bring together the information about forest-cover and
soils that can be obtained from Figure 8, from the other figures, and from the supplementary
texts referred to (Bedell, 1958; Clayton et al., 1973; Rowe, 1972).

The Analysis

A condensation of the above comparison for the three main zones of the Boreal Forest
Region is shown in the Appendix. The more northerly Forest and Barren Sections are listed
first, followed by those of the central predominantly Forest areas, and terminated by the two
sections of the more southerly Forest and Grassland Zone.

The various sections are as identified by Rowe (1972). A very short abstract of their
description is taken from the individual accounts and the locality is indicated. Abbreviations

used are: Y, Yukon; NWT, Northwest Territories; BC, British Columbia; A, Alberta; S, Saskatchewan; M, Manitoba; O, Ontario; Q, Quebec; NB, New Brunswick; NS, Nova Scotia; PEI, Prince Edward Island; and NL, Newfoundland and Labrador. In the abstracts tree species names are abbreviated thus: jP, jack pine; wS, white spruce; bS, black spruce; T, tamarack; bF, balsam fir; C, cedar; wB, white birch; bPop, balsam poplar; tA, trembling aspen; and lP, lodgepole pine. The total land areas of the sections are shown, although in some instances sections are combined. The forest area of a section is shown as a percentage and as an absolute number of square miles. A third entry shows the absolute area of productive forest and its percentage of the total forest area. Productivity data follow, based it is understood on 100-year rotations (Bedell, 1958). Values are for mean annual increment, in cubic feet of merchantable lumber per acre per year.

A simple numbering system has been used to indicate the location on the Radforth map (1960); '3' signifies an area of high frequency of occurrence of organic terrain, '2' the medium area, '1' the low area, and '0' areas where muskeg areas are not said to occur.

The soil number, soil subgroup, soil parent material, and vegetation data are all taken from the Soils of Canada reports (Clayton et al., 1973). In the soil number column the first number, e.g. the '80' of '80 E2.55,' signifies the approximate percentage of the soil number (E2.55) making up the forest section (B33). These percentage estimates are by eye, and soil areas making up less than five per cent of a forest section are ignored. In certain cases the configurations of the forest section and soils boundaries are too complex to estimate in this way. The 'E2' part of '80 E2.55' signifies the soil great group, here Eutric Brunisol, and the '55' is an identifying number for the particular geographic area of this soil. In the reports (Clayton et al., 1973) supplementary information is available on the soil subgroups present in this particular area of Eutric Brunisol. For example, the next column shows that over 40 per cent (symbol 1) of area '55' is made up of the soil subgroup 'Orthic Eutric Brunisol' and more than 20 per cent (symbol 2) of 'Cumulic Regosol' or Rockland. In certain cases the symbol 3 is present, such as in 5 D3.79 of Forest Section B31. This means that from 10 to 20 per cent of soil area '79' of D3, the Humo-Ferric Podzol Great Group, is made up of a Fibrisol Great Group of the Organic Order. The classification, therefore, reveals some of the variability in make-up of different areas. Parent material is briefly indicated, abbreviations being: aeol, aeolian; alluv, alluvium; calc, calcareous; dep, deposit or deposits; fluv, fluvial; glac, glacial; glac-fluv, glacial-fluvial; lac lacustune; mar, marine; mod, moderately; non-calc, non-calcareous; sl, slightly; str, strongly; and wkly, weakly. Vegetation notes are confined to forestry aspects, abbreviations being: conifer, coniferous; for, forest; non-prod, non-productive; and prod, productive.

The Results

To the writer's knowledge, Figure 8 and the Appendix represent the first occasion when moderately detailed information on the forest regions and sections of Canada and their underlying soils have been brought together on a national scale. Within certain limits it is possible to select any locality on the map and to learn immediately its particular forest section and its particular soil great group. Reference to the Appendix, to the other figures, and to the literature cited can reveal much additional information on a locality's forest composition, forest productivity, climate, and soil characteristics.

FIGURE 8

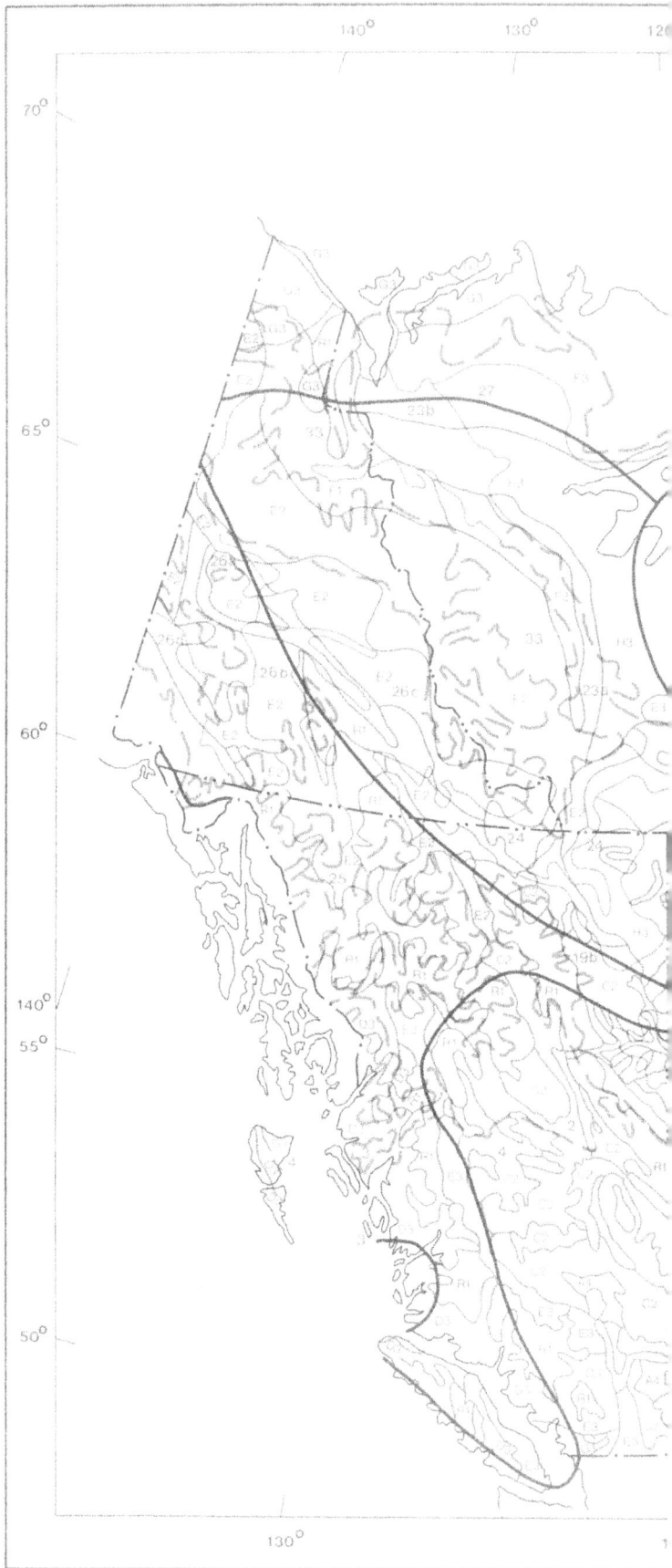

TH

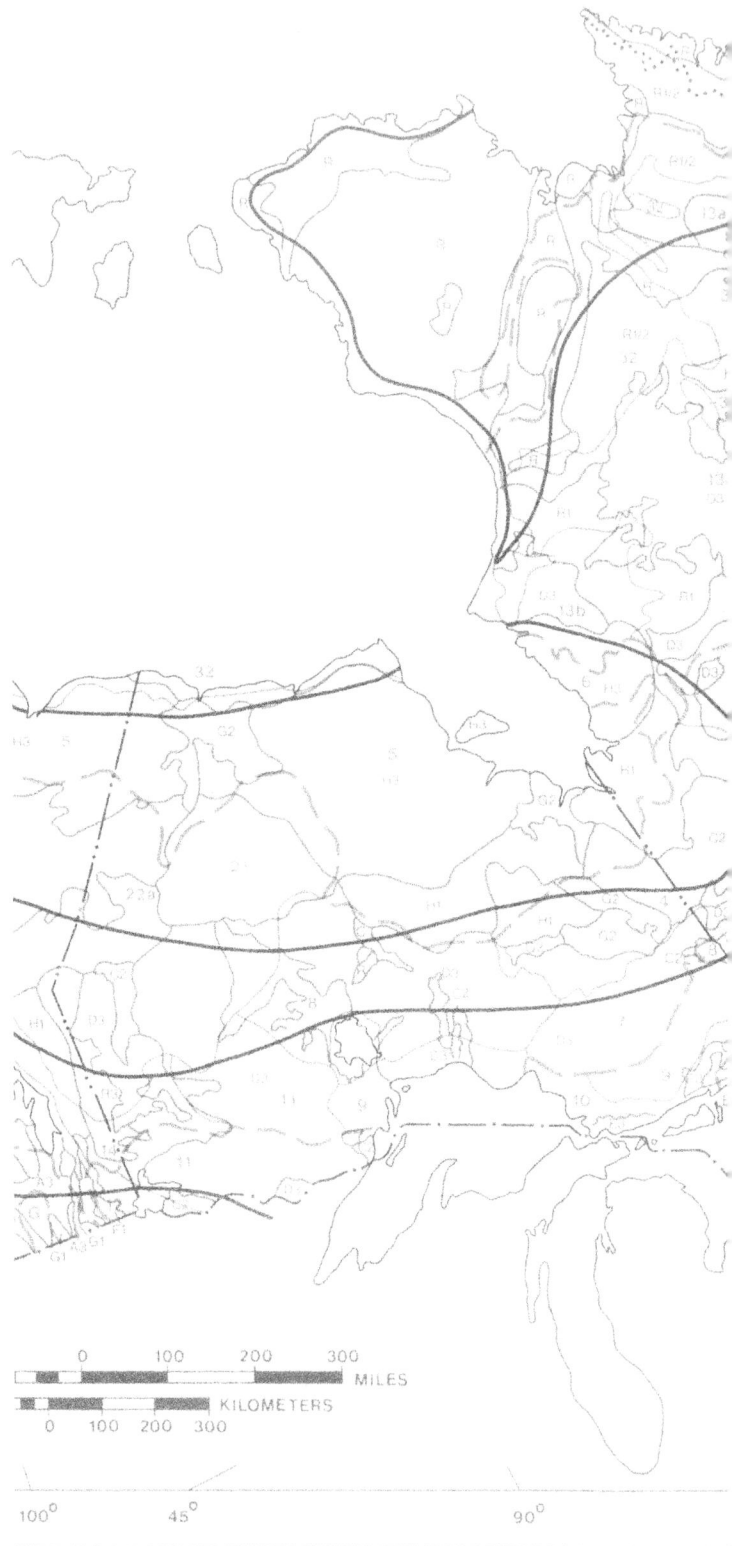

```
        0    100   200   300
                              MILES
              KILOMETERS
        0   100  200  300
```

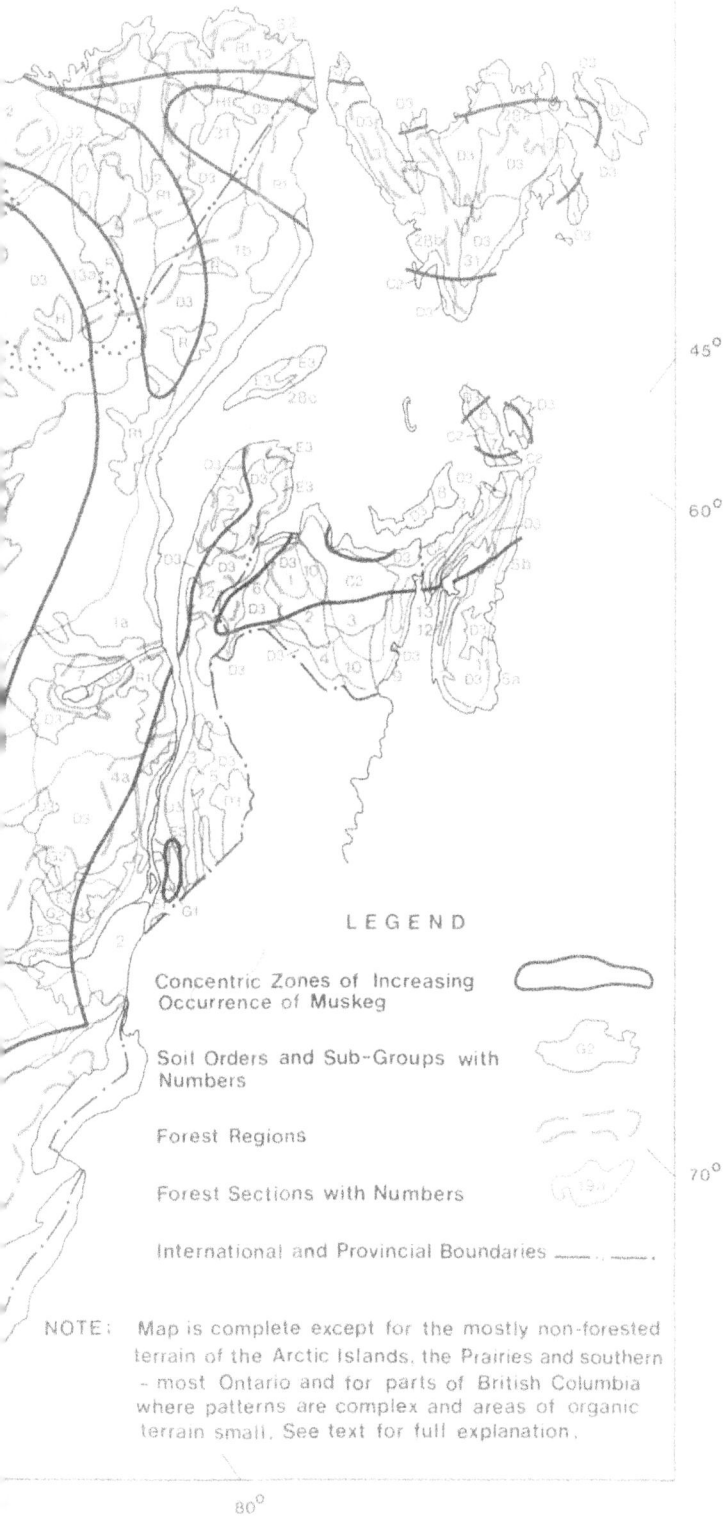

LEGEND

Concentric Zones of Increasing
Occurrence of Muskeg

Soil Orders and Sub-Groups with
Numbers

Forest Regions

Forest Sections with Numbers

International and Provincial Boundaries _____ . _____ .

NOTE: Map is complete except for the mostly non-forested
terrain of the Arctic Islands, the Prairies and southern
- most Ontario and for parts of British Columbia
where patterns are complex and areas of organic
terrain small. See text for full explanation.

Map prepared by Cartographic Section, Graphic Services
Division, 1973.

It.is possible to use this material in a variety of ways for a variety of purposes. The ecological properties of particular forest sections can be compared, the vegetation characterizing particular soil great groups may be studied, the features of certain geographical areas may be specified if a major development project is contemplated, a much more penetrating view is obtainable of geographical areas prone to a particular insect or disease attack, and the results of field research may be extrapolated to other areas with greater reliability. What could also be facilitated is the further subdivision of the forest sections for forest-management purposes, a refinement and desirability long recognized by Halliday in the rationale of his forest classification (Halliday, 1950).

It is obvious, of course, that much of the value of Figure 8 and the Appendix rests in their usefulness as a reference. Further condensation is difficult — rather in the same way that summarizing a map is difficult — and embodies the risk of omission by overgeneralization.

Nevertheless, a number of more general points may be made in the context of the aims of this paper. The first is that the frequency of occurrence of muskeg (peatland) as specified and delineated on the Radforth map (Figure 7) correlates well with the forest, other vegetation, and site descriptions presented by Rowe (1972). There is, moreover, further general agreement between the map and the descriptions relative to the location of muskeg (peatland) supporting productive forest growth and that supporting shrubs, ericaceous plants, and other non-forest vegetative cover. If the muskeg (peatland) area is viewed as a curved belt stretching from the Yukon to Newfoundland, then its southerly component supports forest growth, often productive, merchantable, and accessible (and hence of commercial significance), whereas its northerly component comprises either open or untreed muskeg (peatland) or a landscape made up largely of this with intervening patches of forest too sparse or too remote to have more than very local commercial significance. The muskeg (peatland) and Boreal Forest areas of Canada are not discrete, therefore: they overlap, though only partially, so that considerable areas of muskeg (peatland) support little or no forest whereas others support productive forest of commercial significance.

The second general point is that if the areas of predominantly 'organic soil' are inspected — that is areas defined as composed predominantly of soils of the Organic Order — it may be seen that they are not as widespread as might have been supposed. They total 348,100 square miles (900,000 km²), made up of 60 per cent of the Cryic Fibrisol Great Group and 40 per cent of the Fibrisol Great Group. As the name implies, the former great group is located in permafrost areas. Regarding the specific distribution, the areas of predominantly 'organic soil,' as defined in this way, lie mainly in that part of the Northwest Territories which is northwest of Alberta, in northern Alberta, in Manitoba east and northeast of Lake Winnipeg, in the Hudson Bay Lowlands of Ontario and Quebec, and in parts of Labrador. With the exception of northern Alberta, of small parts of Manitoba north of Lake Winnipeg, and of Ontario in the Kapuskasing area, these areas do not support forest that is productive or is currently being exploited. The principal areas of predominantly 'organic soils' in Canada, therefore, do not support the main volume of forest production.

Having said this, certain features of the generalization have to be recognized. 'Organic soils' must have the minimum thicknesses of surface organic matter described earlier and 'organic-soil areas' must contain over 40 per cent of such soils. Other soils may be present in more than a subdominant way, making up not more than 20 per cent of the area. Certain of these subdominant soils contain substantial amounts of surface organic matter and by their usually better drainage characteristics support better forest growth. In a comparable way it can be seen from the Appendix that areas delineated as predominantly Gleysolic, Podzolic, or Luvisolic may also contain up to 60 per cent of various subdominant soils, one of which could be the Organic Order, which could account for up to 20 per cent. Moreover, all the eighteen subgroups of the Gleysolic Order, two of the eighteen subgroups of the Podzolic Order, and one of the twelve subgroups of the Luvisolic Order can be present as peaty or mucky phrases, characterized by quite thick surface layers of organic matter (up to 16 or 24 inches, or 40 or 60 cm, depending upon the nature of the plant debris). As some authorities accept appreciably smaller thicknesses of surface organic matter as the criterion for peatland and organic soils, it is obvious that there is no fundamental conflict provided the different criteria are known. Thus, almost the whole of Newfoundland is predominantly Podzolic, which might be difficult to reconcile with the statement that the island contains 7,800 square miles (20,000 km^2) of Peatland (Pollett, 1972a). A glance in the Appendix at the descriptions of the Podzolic soils falling within the various Newfoundland forest sections, however, shows them to contain substantial proportions of the Fibrisol and Gleysolic subgroups.

In a similar way it may be wondered if there is a large discrepancy between the MacFarlane (1969) estimate of 500,000 square miles of muskeg (peatland) in Canada and the above Soil Survey of Canada estimate (Clayton et al., 1973) of 348,100 square miles of predominantly organic-soil areas. The writer thinks not, partly because of the latitude in the thickness criterion and partly because of the necessity to generalize over large areas. Not all the predominantly organic-soil areas comprise organic soils, hence the 348,100-square-mile figure is an overestimate; on the other hand, there are organic soils in the predominantly podzolic, gleysolic, and luvisolic areas and as the latter three are geographiclly much larger than the predominantly organic area, the gain within the three more than offsets the loss in the predominantly organic areas. Therefore, not only does the total area of soil in Canada supporting substantial thicknesses of organic matter appear to amount to somewhere in the neighbourhood of 500,000 square miles, but appreciable areas of this total either already support important forest-growth or, as in Newfoundland, are of considerable interest because of their potential for supporting forest growth.

Finally, it may be felt unsatisfactory to have such variation in the definition and understanding of peatland. In some ways this is so, but variation arises because different interests stress different characteristics and see the same material in different ways. Much effort can be wasted in adopting too inflexible an approach to identifying what peat might be. What does seem desirable, however, is that the different interests fully specify and elaborate their terminologies, so that a common position or system can be built up on the basis of shared understanding and a full recognition of the different positions that prevail.

A few remarks are now made about each of the provinces analysed.

Yukon. For Forest Sections B33, 26, and 24 in the Yukon, forest-growth covers some 20, 40, and 80 per cent, respectively, of the land area. Productivity appears minimal in B33 and moderate to low elsewhere. Brunisolic soils are well represented, followed by regosolic or rockland types. Peat or organic terrain is poorly represented. Forest production in the Yukon is relatively very small but important centres are Whitehorse, Watson Lake, and the Fort Laird areas (Karaim, 1970).

Northwest Territories. These consist for the most part of the Forest and Barren Boreal sections B33, 32, 27, and 23b, where the forest cover makes up only 20-40 per cent of the land area and where growth is minimal to low. Organic soils are strongly represented east of the Mackenzie River, supporting open shrubby vegetation and barrens or patches of mostly stunted forest. Forest Sections B18b, 23a, and 24 make up the area of Boreal forest from south and west of the Great Slave Lake to Fort Laird and down the Mackenzie valley as far as Norman Wells. Here forest covers 40-80 per cent of the land area and growth is low to fair. Some peats are present, and peaty phases of gleysolic soils, but there is appreciable brunisolic and some luvisolic representation. The very little forest production is centred on the Fort Laird area (Karaim, 1970).

Alberta. Forest Section B19a of the main Boreal forest area and Forest Section B17 of the Boreal-Grassland transition contain very few organic soils, the main representation in the two main components of the Boreal being luvisolic and chernozemic respectively. North of around Lesser Slave Lake there is much organic soil associated with Forest Sections B18a, 18b, and 23a, these areas being forested to an extent of 75 per cent of the total land-area and growth of both conifers and hardwoods being moderate. For the most part Luvisols make up the other main soil order, but in Forest Sections B22b and B23a Brunisolic and Podzolic orders are important. On the whole, therefore, organic soils form an important base to the commercial forests of northern Alberta.

Saskatchewan. Here there are large areas of all three main components of the Boreal forest: the Boreal-Tundra transition north of an arc connecting the southern shores of Lakes Athabasca and Reindeer, the main Boreal south to a northeasterly line running through Prince Albert and the Boreal-Grassland south of this to another approximately northeasterly line running through Regina and Saskatoon. Organic soils are strongly represented in Forest Section B15 centred on the La Ronge-Flin-Flon-Swan River triangle, but are also substantially present in the other main Boreal forest sections, where the main soil order is either Podzolic (B22a and 22b) or Luvisolic (B18a and 20). Productive forest-growth is restricted to the less deep organic soils.

Manitoba. North of latitude 56 Manitoba consists mostly of the two Boreal-Tundra Forest Sections B5 and 27 which are not productive but contain much muskeg (peatland) with black spruce. The main Boreal forest sections terminate some distance north of Winnipeg. They support forest-growth of moderate productivity over some 50-80 per cent of their land area. Section B21, east and north of Lake Winnipeg, is made up very largely of soils of the Organic Order, and the adjacent areas of Sections B22a, farther to the east and west, B14 to the southeast, and B15 and B18a to the west also contain substantial areas of organic soils. Whereas B15 and 18a are dominated by Luvisolic soils, B14 and B22a are essentially Podzolic. Much of the forest northeast of Lake Winnipeg is not currently exploited.

Ontario. Muskeg, peatland, or organic soils are extensively represented in the northern Boreal-Tundra transition zone made up in Ontario of almost entirely Section B5. Only 15 per cent supports forest cover and of a very low growth-rate. Within the main Boreal zone, forest of moderate to good growth covers from 80 to 90 per cent of the land area. Soil orders are largely Podzolic, but the Organic Order is represented to an extent of 25 per cent in B22a, 15 per cent in B14, and 10 per cent in B4. There are, furthermore, organic subdominants within the Podzolic and Gleysolic orders making up the rest of these forest sections. Sections B7, 8, 9, 10, and 11 contain no organic soils either as dominants or subdominants but, as was remarked earlier, there can be organic phases of the Podzolic and Gleysolics present which some authorities might classify as peatland or organic soils. This needs to be kept very much in mind when Ketcheson and Jeglum's very interesting study (1972) of peatland and black spruce in Ontario is interpreted, for these authors' data show generally much higher percentages of peatland in the various areas of Ontario. In fact, 49 per cent of the whole land area of Ontario, or 167,000 square miles (431,000 km²) is said to be peatland, 18 per cent of which is commercially productive black spruce forest and a further 60 per cent non-productive. Some idea of the importance of the 18 per cent may be gained from the further fact that this represents nearly half of all the productive black spruce in Ontario, which makes up a considerable proportion of the total forest harvest.

Quebec. In contrast to Ontario, the Boreal-Tundra transition zone, made up of Forest Sections B13a, 13b, 31, and 32, contains more Rockland and Regosolic soil orders and less muskeg, peatland, or organic soil. As with Ontario, however, forest of very low growth covers only some 20 per cent of the land area. The main Boreal zone includes two important forest sections shared with Ontario, B4 and 7, with five other sections, B1a, 1b, 2, 3, and 6, making up a very large area stretching from James Bay to the Atlantic Ocean. Only B2 and B7, centred on the Gaspé peninsula and on south-central Quebec respectively, contain no Organic orders either as dominants or subdominants but this is not to say that peaty phases of the Podzolic and Gleysolic orders making up these forest sections are absent. Sections B1a, 1b, 4, and to a small extent 3 all contain some Organic soil orders as either dominants or subdominants. Section B6 is exceptional in consisting of over 95 per cent organic soil. Sections support forest cover of moderate to very good growth over from 55 to 90 per cent of their areas. There is no doubt that, as with Ontario, organic soils either as Organic orders in their own right or as peaty phases of Gleysolic or Podzolic orders support considerable areas of important commercial forest.

Newfoundland. Newfoundland and Labrador both contain high proportions of the Boreal-Tundra zone, particularly Labrador. Both support very little forest cover — around 20 per cent of the total land area — of very poor growth. A difference is that both B13a and 31 in Labrador contain about 25 per cent of the Organic Soil Order whereas B31 on the island contains no Organic soil as a dominant. In Labrador the main Boreal zone is relatively small and mostly restricted to the Hamilton and Eagle river valleys (B12), where some good growth of black spruce occurs on Podzolic soils containing Organic subdominants. In the Boreal areas on the island forest cover makes up some 80 per cent of the land area, except in the Avalon peninsula (B30), where it is considerably less. All forest sections, B28a, 28b, 29, and 30, are made up exclusively of the Podzolic Order, except B28b where Luvisolic is also a dominant. All Podzolic dominants, however, are associated with Gleysolic, Organic, and Rockland as subdominants and all these, except the last, have Peaty phases which, in other systems, permit the soils to be viewed as peats. A recent paper (Pollett, 1972b) summarizes the work that has been done on the classification of Newfoundland peats and discusses the potential utilization of the three main types recognized: ombrotrophic, weakly minerotrophic, and euminerotrophic.

Some Utilization and Research Aspects

It will be obvious that for an area as vast as the three main zones of the Boreal forest of Canada, whose resources have been a source of wealth and industry for numerous agencies over very many years, far more than a few paragraphs would be necessary to reveal the magnitude and ramifications of this endeavour.

The distribution of the main tree species across the Boreal is shown in Figures 2 to 6 and much on the distribution of these species in relation to climate and soil is apparent from the various maps, the Apppendix, and the above provincial summaries. White and black spruces are characteristic species of the Boreal and Tamarack ranges throughout. Balsam fir and jack pine are prominent in the centre and east, and alpine fir and lodgepole pine in the west and northwest. The Boreal is primarily coniferous, but there is a general admixture of such deciduous species as white birch and poplar, especially in the central and south-central parts. In the north, as the organic surface layers of the soil thicken to imbue more the properties of muskeg, peat, and organic terrain, rather than just mineral soils with organic surface horizons, the proportion of spruce and larch increases. This in turn gives way to an open lichen-forest of very slow growth or one interspersed with open muskeg or rock barrens.

It has been shown how estimates of peatland in the Boreal might vary, depending upon the criteria used, but the forestry interest in organic terrain in the Boreal is two-fold. First, there is the interest in the area as a forested ecosystem in its fullest environmental sense. Secondly, there is in certain areas the additional interest in the forest as a renewable resource.

The first interest stems from the basic concepts of ecology and the fundamental role of the vegetative cover in contributing to the character and properties of an ecosystem. In the broadest terms the soil, the climate, and the vegetation are in dynamic equilibrium, and an imposed or man-made change made in any part of the system is reflected elsewhere. Where the system is particularly delicate and where knowledge is not extensive, as with our northern peatlands, there is considerable concern lest man-imposed actions upset these delicate systems. Degrading processes could be induced bringing incalculable damage to landscapes that are unique for their aesthetic aspects, for their provision of shelter to wildlife, and for their ability to supply stable water supplies.

It is for these reasons that foresters themselves have cautioned against the expansion of commercial forestry operations into far northern localities, without first fully exploring the risks (Sayn-Wittgenstein, 1969). It is also for these reasons that the wealth of specialist expertise that is available within the Canadian Forestry Service, for example, has already been used extensively for investigating the possible effects of major development projects upon the fragile ecosystems of the north, quite aside from its use in providing several more routine type of research services for years. The study of the 32,400 square miles (84,000 km²) of the Mackenzie valley is one example of the special investigations that have been made, and the regular monitoring for fire and insect and disease-attack is an example of the second. And the same reasons, strengthened by the present-day concern for the quality of the environment (Rennie, 1971, 1972), have led to the initiation of several large ecosystem studies in which the effects of everyday silvicultural operations such as drainage, scarification, burning, fertilization, spraying, and clear-cutting upon such environmental characteristics as water yield, water quality, and faunal populations are being tested in a comprehensive and rigorous manner.

The second main interest in the forest has been the more traditional one of production forestry, which has received very much attention. If the forest harvest outside of British Columbia is considered (where production is mostly not on organic soils) it can be seen that Ontario and Quebec alone account for over 70 per cent of the harvest, of which black spruce on organic soils makes up a very substantial proportion. Because the other principal peatland species, poplar, has been used relatively very little in Canada (Maini and Cayford, 1968), the traditional silvicultural interest in peatland might almost be said to be an interest in black spruce. Two examples of stands of this tree are shown in Figures 9 and 10.

FIGURE 9   Black spruce of good quality near Hinton, Alberta (by A.L. Potvin)

The many studies conducted on black spruce on peatland areas in Canada have been described in Vincent's well-known review (1965), from which the many aspects of black-spruce management are apparent. In a nutshell there remain three main classes of problem: adequate regeneration after harvesting, growth stimulation of existing stands, and the classification of stands. The first arises from inadequate or inappropriate regeneration after logging—often complicated by a rise in water table, fire, or invasion of weed growth. It is acknowledged that there are very large areas of black spruce peatland in Ontario and Quebec which have not regenerated well and for which there is no known economic regeneration technique. The second problem arises from the need to increase the growth of this important pulpwood tree species on sites where transportation costs are not excessive. Both drainage and fertilization offer promise, but satisfactory techniques have to be developed. And the third problem is fundamental, both for deciding on the particular ways in which individual stands should be managed and in order to extrapolate and apply the results of research. As

FIGURE 10   Black spruce of poor quality in Parc Verendry, Quebec (by A.L. Potvin)

might be expected, all problems necessitate more applied experimentation as well as support studies, the latter often for the reason that there has been little characterization of many sites in terms of soil chemistry, physics, and other specialist disciplines. The nature of the systems being managed is poorly known. Research programs by a number of authorities in these fields are in progress in both Ontario and Quebec.

Not specific to black spruce is the problem of timber extraction on forested muskeg and peatland; much thought is being given to the engineering design of appropriate vehicles in an attempt to optimize extraction and cut costs.

In Newfoundland, peatland problems are very much those of open peatland, which has a potential for forestry (Pollett, 1969). These include, among others, classification, drainage, and species trials. And reinforcing all more applied silvicultural programs centred on peatland forestry throughout Canada are innumerable more basic studies, too numerous to mention individually, and such regular programs of observation and study as those concerned with the protection of the forest against harmful insects and diseases. These are described in comprehensive annual reports (Forest Service, 1973).

DISCUSSION

Different authorities define 'forest' and 'organic terrain' in different ways, but accepting the working figures of $1,115 \times 10^6$ acres for the extent of the former in Canada and 320×10^6 acres for the latter, Canada may be seen to be both a major forest country and a major peatland one. Over 80 per cent of Canada's forest area is classified as Boreal, and this enormous zone stretches from Yukon in the west to Newfoundland in the east. Superimposed on this zone are the main peatland areas, but in such a way that considerable areas of peatland support little or no forest growth, whereas others support productive forest of commercial significance.

White and black spruces are characteristic species of the Boreal and tamarack is distributed throughout. Balsam fir and jack pine are prominent in the centre and east, and alpine fir and lodgepole pine in the west and northwest. The Boreal is primarily coniferous, but there is a general admixture of such deciduous species as white birch and poplar, especially in the central and south-central parts. In the north, the organic surface-layers of the soil thicken and imbue sites more with the properties of true muskeg, peatland, and organic terrain than those of essentially mineral soils with thick surface humus. In the north the proportion of spruce and larch increases, and the closed forest condition gives way, in turn, to an open lichen-forest of very slow growth or one of just patches scattered in vast areas of open muskeg or rock barrens.

Considerable care is necessary in the use of the terms muskeg, peatland, organic soil, and organic terrain to avoid misunderstanding. The above figure of 320 x 10^6 acres for organic terrain in Canada is in fair agreement with one of 224 x 10^6 acres for soils predominantly of the Organic Order in Canada, when certain quite reasonable causes for latitude are taken into account. Both these figures would be considerable underestimates, however, if peatland and organic terrain were taken to be soils with organic surface-layers of thicknesses appreciably less than the qualifying thresholds specified by the National Soil Survey of Canada. Thus, half of Ontario could be regarded as peatland, and as this alone amounts to 107 x 10^6 acres, some very substantial increase in the estimate of 224 x 10^6 acres for the whole of Canada would be necessary. Moreover, some considerable change would be necessary in the description of Canadian soils and sites.

As currently seen under the National Canadian System of Soil Classificaton, the Boreal Forest is supported predominantly by soils of the Podzolic Order, with those of the Luvisolic, Brunisolic, Gleysolic, and Organic orders playing relatively smaller roles. In absolute terms, of course, these 'relatively smaller roles' can be very large, for the Organic Order alone covers 348,100 square miles (224 x 10^6 acres), made up of 60 per cent of the Cryic Fibrisol Great Group and 40 per cent of the Fibrisol Great Group. Organic soils defined in this way lie mainly in that part of the Northwest Territories which is northwest of Alberta in northern Alberta, in Manitoba east and northeast of Lake Winnipeg, in the Hudson Bay Lowlands of Ontario and Quebec, and in parts of Labrador. Although all these areas support forest growth, they do not — with the exception of northern Alberta, of small parts of Manitoba north of Lake Winnipeg, and of Ontario in the Kapuskasing area — support forest that is commercially productive. The principal areas of true organic soil in Canada, therefore, may support considerable forest, but they are not the main areas of timber production.

It must be added, however, that all 18 subgroups of the Gleysolic Order, two of the 18 subgroups of the Podzolic Order, and one of the 12 subgroups of the Luvisolic Order can be present as Peaty or Mucky phases, characterized by organic surface-layers up to 16 or 24 inches (40 or 60 cm) thick depending upon plant origin, without separate identification. Moreover, the over-all classification gives only the predominant soil order for a geographical area: a particular locality may embrace truly organic soils of the system as subdominants. It is perfectly legitimate to conclude, therefore, that important areas of productive forest in Canada are located on soils with quite thick organic surface-layers. These soils cannot be classed within the Organic Order, but they may be regarded by some authorities as peats and for the purpose of forestry may exhibit characteristics making it more appropriate to regard them as peats than mineral soils.

Semantic problems of this nature are undesirable in the long term, but their existence seems explainable by the developing nature of the discipline of Canadian peat science. In this development process the risks of misunderstanding and confusion can be minimized if special care is taken to define terms and materials as unequivocally as possible. Facilitating this task are the Appendix and Chapter 2 of this volume dealing respectively with peatland classification and peatland terminology.

From a forestry viewpoint, interest in the organic soils of the Boreal is two-fold. First, there is the interest in such lands as forest ecosystems in the fullest environmental sense. In the broadest terms it is recognized that the climate, soil, and plant cover of such areas are in dynamic equilibrium: and in these northern peatlands in a state of quite delicate equilibrium. Seemingly trivial man-made changes imposed on one part of the system may well be reflected elsewhere and magnified with disastrous consequences. Foresters themselves, therefore, discourage any incautious extension of their traditional art into fragile northern peatlands. They participate, along with other specialists, in attempts to forecast the effects that the ever-increasing number of large development-projects might have on the unique environments of northern lands. And they are mounting comprehensive ecosystem studies in a number of localities where even the traditional techniques of forestry — such as drainage, partial cutting, scarification, and fertilizing — are being tested, not only for their forestry value, but for their effect upon important properties of the environment such as water yield and quality and wildlife.

The second main forestry interest is centred on organic soils as the land-base for production forestry, of which black spruce in Ontario and Quebec forms a very substantial part as a highly valued commercial tree species. Inadequate or inappropriate regeneration after harvesting remains a formidable problem over many areas, in spite of much past experimentation, and economic solutions are being actively sought by a number of authorites. Promising results are being obtained by the drainage and fertilization of black spruce sites where transportation costs are not excessive, and site classification is being developed to place management on a sounder basis.

As the pressures on the northern lands, and particularly the peatlands, steadily increase there is an increasing need to have available knowledge on the ecological characteristics of these lands. Much has to be done in the area of basic characterization and in the testing of a variety of stresses upon such ecosystems. This calls for a blend of more applied experimentation and support studies in a number of disciplines. Only in this way can the principles of sound environmental management be developed, disasters avoided, and a unique heritage and resource conserved or utilized in the optimum way.

In this paper mapping data on the forests and soils of the Boreal and other regions of Canada have been brought together for the first time. The characteristics of particular locations can be specified. The data are not detailed and much refinement can obviously be done, but it is hoped that this interim synthesis will serve as a small step towards progress. Forests and their productivity may be seen more as what they are — one ecological attribute within a landscape unit — rather than as a soulless statistic in a tallyman's table.

SUMMARY

Canada is a country of extensive forests (1,115 x 10⁶ acres, or 450 x 10⁶ ha) and extensive organic terrain (320 x 10⁶ acres, or 130 x 10⁶ ha). This paper examines these statistics and explores the relationship between the two areas. For this the Radforth map of muskeg areas, the Rowe map of the forest regions, the Halliday and Brown maps of the distribution and population intensity of important tree species and the new soil map of Canada have been compared and combined.

For the first time a new map at a scale of 1:5,000,000 (79 miles = 1 inch) has been prepared showing both the forest regions and sections of Canada and their underlying soils. The map is reproduced at a reduced scale (Figure 8). An appendix provides further summary data on each forest section of the Boreal, where most organic terrain is located. Shown for each forest section are: area, province, forest composition and coverage, forest productivity, proportion and name of each soil great group present, proportion of each soil subgroup present, and textural nature of the soil parent-material. The map and appended data have a number of important uses and have been basic to the particular analysis of this paper.

Considerable areas of Canada's organic terrain support little or no forest growth, whereas others support productive forest of commercial significance. Great care is necessary in specifying what is meant by organic terrain and its various 'synonyms,' for different authorities use the terms in different ways. Taken in its fuller sense, 'organic terrain' would add considerably to the above total of 320 x 10⁶ acres for Canada and would place very extensive areas of productive forest, particularly black spruce in Ontario and Quebec, on such soils. Using the term in the way it is specified in the National Soil Survey of Canada, which embodies thresholds for the thicknesses of the organic surface-layers of 16 and 24 inches (40 and 60 cm) depending upon origin, provides an areal estimate of organic terrain in fair agreement with the above figure of 320 x 10⁶ acres. Such areas, however, lie mainly in the Northwest Territories, in northern Alberta, in Manitoba east and northeast of Lake Winnipeg, in the Hudson Bay Lowlands of Ontario and Quebec, and in parts of Labrador. These areas support forest growth, but except for northern Alberta, for small parts of Manitoba north of Lake Winnipeg, and for the Kapuskasing area of Ontario, they are not commercially productive.

The forestry interest in organic terrain embraces its wider interpretation and is two-fold. The first stems from a concern for forest ecosystems in the fullest environmental sense. Particular interest is focused upon the fragile peatlands of the north in the face of the ever-increasing pressures of large development-projects, and much specialist expertise is being deployed to forecast the environmental impact of particular projects. This approach extends also to traditional silvicultural operations, such as partial cutting and site-preparation, whose effects are being examined not only from a production-forestry viewpoint, but also from one designed to ensure the quality of the environment. Special problems are associated with the management of black spruce on organic terrain, and research effort is being directed towards improved regeneration, growth, and classification of this highly valued resource.

For the optimum conservation and management of organic terrain there seems a need to strengthen the ecological approach and to characterize our unique northern resources in better quantitative terms. The present paper and synthesis can be a small step in this direction.

ACKNOWLEDGMENTS

The author wishes particularly to thank Dr J.H. Day and his colleagues of the Canada Department of Agriculture for making freely available as yet unpublished material. Appreciation is also extended to Mr A.L. Potvin of the Canadian Forestry Service for the use of Figures 9 and 10 from his collection.

REFERENCES

Anon. 1960. A look at Canadian soils. Agric. Inst. Rev. 15(2): 9-60.

Atkinson, M.J. 1971. A Bibliography of Canadian Soil Science (Publ. Can. Dept. Agric. 1452).

Bedell, G.H.D. 1958. Productivity of Canadian forests. Unpublished Rept. of Forestry Branch, Canada, with a supplementary table for 1964.

Canadian Department of Agriculture. 1970. The System of Soil Classification for Canada (Publ. Can. Dept. Agric.)

Canadian National Committee, International Hydrological Decade. 1969. Map of mean annual runoff. In Hydrological Atlas of Canada (Natl. Res. Council, Ottawa).

Clayton, J.S., Ehrlich, W.A., Cann, D.B., Day, J.H., and Marshall, I.B. 1973. Soils of Canada. Vol 1. Soil Report. Vol 2. Soil Inventory (with maps). Can. Dept. Agric., Ottawa (in press).

Duffy, P.J.B. 1971. Canada land inventory at midpoint. Pulp and Pap. Mag. Can. 72(4): 98, 101, 103.

Dumanski, J., et al. 1970. Geographic zonation in selected characteristics of surface mineral horizons in Alberta soils. Can. J. Soil Sci. 50: 131-9.

Forest Economy Research Institute. 1972. Canada's Forests (Dept. Environment, Ottawa).

Forest Service. 1973. Forest insect and disease survey. In Annual Report 1972 (Environment Canada; in press).

Geological Survey of Canada. 1968. Glacial Map of Canada, Map 1253A (Dept. Energy, Mines Resources, Ottawa).

— 1969. Geological Map of Canada, Map 1250A (Dept. Energy, Mines & Resources, Ottawa).

— 1970. Physiographic Regions of Canada, Map 1254A (Dept. Energy, Mines & Resources, Ottawa).

Halliday, W.E.D. 1937. A forest classification for Canada. Bull. For. Serv., Dept. Mines & Resources 89.

— 1950. Climate, soils and forests of Canada. For. Chron. 26: 287-301.

Halliday, W.E.D., and Brown, A.W.A. 1943. The distribution of some important forest trees in Canada. Ecol. 24: 353-73.

Hare, F.K., and Ritchie, J.C. 1972. The boreal bioclimates. Geog. Rev. 62: 333-65.

Hosie, R.C. 1969. Native Trees of Canada (7th ed., For Serv., Can. Dept. Fish. & For.).

Ilvessalo, Y. 1949. The forests of present-day Finland. Commun. Inst. For. Fenn. 35(6).

Karaim, B.W. 1970. A statistical review of forest resources in the Prairies region. Unpublished internal rept. Northern For. Res. Cen., For. Serv., Dept. Environment, Canada.

Ketcheson, D.E., and Jeglum, J.K. 1972. Estimates of black spruce and peatland areas in Ontario. Great Lakes For. Res. Cen., For. Serv., Dept. Environment, Canada, Inform. Rept. 0-X-172.

Legget, R.F. (ed.). 1961. Soils in Canada: Geological, Pedological and Engineering Studies (Roy. Soc. Canada Special Publ. 3; Univ. Toronto Press, Toronto).

MacFarlane, I.C. (ed.). 1969. Muskeg Engineering Handbook (Univ. Toronto Press).

Maini, J.S. and Cayford, J.H. 1968. Growth and utilization of poplars in Canada. Publ. For. Br. Dept. For. & Rur. Devel., Canada, 1205.

Manning, G.H., and Grinnell, H.R. 1971. Forest resources and utilization in Canada to the year 2000. Publ. For. Serv., Dept. Environment, Canada 1304.

McCormack, R.J. 1967. Land capability classification for forestry. Can. Land Inventory Rept. 4, Dept. For. & Rur. Devel.

McElhanney, T.A. 1951. Commercial timbers of Canada. *In* Canadian Woods, Their Properties and Uses (2nd ed., For. Br. For. Prod. Labs. Div.), 23-56.

Mikola, P. 1970. Forests and forestry in subarctic regions. *In* Ecology of the Subarctic Regions. Proc. Helsinki Symp. (1966) Ecology and Conservation (UNESCO, Paris), 1: 295-302.

Page, G. 1971. Properties of some common Newfoundland forest soils and their relation to forest growth. Can. J. For. Res. 1: 174-92.

Pollett, F.C. 1969. A program for the development of peatland forestry research in Newfoundland. Nfld. For. Res. Cen., For. Br., Can. Inform. Rept. N-X-29.

— 1972a. Nutrient contents of peat soils in Newfoundland. Proc. 4th Internatl. Peat Congr. III: 461-8.

— 1972b. Classification of peatlands in Newfoundland. Proc. 4th Intern. Peat Congr. I: 101-10.

Radforth, N.W. 1960. Areas of organic terrain (muskeg) in which engineering problems occur. *In* Organic terrain organization from the air (altitudes less than 1000 feet). Handbook no. 1, Def. Res. Bd., Dept. Natl. Def., Ottawa, DR 95.

Rennie, P.J. 1971. The role of mechanization in forest site preparation. *In* Papers Presented at the XVth Congr. Intern. Union of Forest Research Organizations, 1971, Div. 3, Publ. 1, Stockholm, pp. 63-102.

— 1972. Forest fertilization: retrospect and prospect. *In* Proc. Forest Fertilization Symp., Northeastern Forest Soils Conf., Warrensburg, NY. Publ. US For. Serv., Northeast For. Exptl. Sta., Upper Darby, Penn. Tech. Rep NE-3: 234-241.

Rowe, J.S. 1959. Forest regions of Canada. Bull. For. Br. Can. 123.

— 1972. Forest regions of Canada. Publ. For. Serv., Dept. Environment, Canada 1300.

Sayn-Wittgenstein, L. 1969. The northern forest. Can. Pulp and Paper Ind. 22(11): 77-8.

Scotter, G.W. 1971. Fire, vegetation, soil, and barren-ground caribou relations in northern Canada. *In* Proc. Fire in the Northern Environment Symp. Pac. NW For. & Range Exptl. Sta., Portland, Oregon, pp. 209-30.

Tedrow, J.C.F. 1970. Soils of the subarctic regions. *In* Ecology of the subarctic regions. Proc. Helsinki Symp. (1966), Ecology & Conservation (UNESCO, Paris), 1: 189-205.

Vincent, A.B. 1965. Black spruce; a review of its silvics, ecology and silviculture. Publ. Dept. For., Canada 1100.

Wilde, S.A., et al. 1954. The relationships of soils and forest growth in the Algoma district of Ontario, Canada. J. Soil Sci. 5: 22-38.

Wilson, D.A. 1969. Forest resources and industries of Canada. Commonw. For. Rev. 48: 127-43.

APPENDIX

Areas (square miles)			Productivity (ft³ ac⁻¹ yr⁻¹)		Rad-forth area	Soil great group	Soil subgroup proportion and identification	Parent material	Forest vegetation
Total land	Forest	Productive forest	Soft-wood	Hard-wood					
242,100	48,420 (20)*	24,210 (50)*							

Forest Section: B33 Alpine Forest Tundra (Y, NWT): open park-like stands of stunted wS, alternating with grass or barrens on N and E slopes, alone or with wS

			Soft-wood	Hard-wood	Rad-forth	Soil great group	Soil subgroup	Parent material	Forest vegetation
			2	0	2	80 E2.55	1 Orthic eutric brunisol	Stony sandy mod calc alluv & fluv dep	Non-prod conifer for
							2 Cumulic regosol Rockland Regosolic		
						10 F1.15	1 Cryic gleysol	Loamy, alluv	Non-prod conifer for
						5 G3.40	3 Orthic melanic brunisol		
						5 G3.41	1 Cryic gleysol	Loamy, till & glac-fluv	

(figures are for B33, 32, and 31 combined)

Forest Section: B32 Forest Tundra (NWT, M, O, Q, NL): tundra barrens with patches of stunted forest: wS, bS, and T are the primary species, accompanied by alder and willow shrubs
(see above)

Soft-wood	Hard-wood	Rad-forth	Soil great group	Soil subgroup	Parent material	Forest vegetation
2	0	3	60 F3.69	Cryic regosol		
			5 E3.94	1 Orthic dystric brunisol	Stony, glac till & fluv dep	Non prod conifer for
				2 Rockland		
				3 Orthic gray huvisol		
			5 E3.97	1 Cryic dystric brunisol	Stony, sandy, glac till, mar & glac-fluv dep	Barren
				2 Cryic regosol		
			20 R1.27	Rockland		
			10 R1.14	,,		

Forest Section: B31 Newfoundland-Labrador Barrens (NL,Q): moss and heath barrens with patches of open stunted bS & bF, organic soils prominent
(see above)

Soft-wood	Hard-wood	Rad-forth	Soil great group	Soil subgroup	Parent material	Forest vegetation
2	0	3	25 H1.33	1 Fibrisol	Organic dep	Non prod conifer for
				2 Rockland		
			20 D3.33	1 Orthic humo-ferric podzol	Stony, loamy glac till	,,
			20 D3.125	1 Orthic humo-ferric podzol	stony, sandy till	,,
				2 Rockland		
			15 D3.71	1 Orthic humo-ferric podzol	Stony, loamy glac till & glac-fluv dep	,,
				2 Orthic gleysol		
				3 Fibrisol		
			15 D3.63	1 Orthic humo-ferric podzol	Loamy, glac till & glac-fluv dep	Uplands prod conifer for
				2 Orthic gleysol		
				3 Fibrisol		
			5 D3.79	1 Orthic humo-ferric podzol	Stony, loamy glac till	Lowlands non-prod conifer for
				2 Orthic gleysol		Non-prod conifer for
				3 Fibrisol		

APPENDIX (continued)

Areas (square miles)			Productivity ($ft^3ac^{-1}yr^{-1}$)		Rad-forth area	Soil great group	Soil subgroup proportion and identification	Parent material	Forest vegetation
Total land	Forest	Productive forest	Soft-wood	Hard-wood					

Forest Section: B13a Northeastern Transition (Q, NL): a patchwork of lakes, rivers, swamps, bogs and muskeg, with areas of barren and forest, the latter being open park-like bS; occasional closed stands; an open woodland of bS develops on wet areas

Total land	Forest	Productive forest	Soft-wood	Hard-wood	Rad-forth area	Soil great group	Soil subgroup proportion and identification	Parent material	Forest vegetation
398,900 (20)	79,780 (35)	27,910	3	1	2	20 { H3.55	1 Cryic fibrisol / 2 Rockland	Organic dep	Non-prod conifer for
						H3.54	(as H3.55 above)	"	"
						D3.23	1 Orthic humo-ferric podzol	Sandy & loamy non-calc glac till & glac-fluv dep	"
						70 { D3.22	1 Orthic humo-ferric podzol	"	"
						10 R1.37	Rockland		

(figures are for B13a, 13b, and 27 combined)

Forest Section: B13b Fort George (Q): open and closed stands of bS with jP on morainic ridges; large areas of musket and open bog

Total land	Forest	Productive forest	Soft-wood	Hard-wood	Rad-forth area	Soil great group	Soil subgroup proportion and identification	Parent material	Forest vegetation
(see above)			3	1	3	25 H3.46	1 Cryic fibrisol / 2 Orthic humo-ferric podzol / Cryic gleysol	Organic, & sandy, loamy or clayey glac till / Lac or glac-fluv dep	Non-prod conifer for
						25 D3.22	1 Orthic humo-ferric podzol	sandy & loamy non-calc glac till & glac-fluv dep	"
						50 R1.37	Rockland		

Forest Section: B27 Northwestern Transition (M, S, NWT): areas of bog muskeg and barren with open subarctic woodland of bS with wS on less wet sites. T & wB on more northerly part

Total land	Forest	Productive forest	Soft-wood	Hard-wood	Rad-forth area	Soil great group	Soil subgroup proportion and identification	Parent material	Forest vegetation
(see above)			3	1	3	5 F3.69	Cryic regosol		
						40 H3.44	1 Cryic fibrisol / 2 Orthic gleysol / 3 Orthic eutric brunisol	Organic & clayey mod calc glac till, glac-fluv & alluv dep	Non-prod conifer for
						10 E3.94	1 Orthic dystric brunisol / 2 Rockland / 3 Orthic gray luvisol	Stony, sandy, glac till & fluv dep	"
						40 D3.139	1 Orthic humo-ferric podzol / 2 Orthic humic gleysol / Cryic fibrisol / Rockland / 3 Orthic dystric brunisol	Stony, sandy, glac till & glac-fluv dep	Non-prod conifer for & tundra
						5 C2.128	1 Orthic gray luvisol / 2 Orthic eutric brunisol / Cryic fibrisol	Loamy, glac till	Non-prod conifer for

APPENDIX (continued)

Total land	Areas (square miles)		Productivity (ft³ac⁻¹yr⁻¹)			Soil great group	Soil subgroup proportion and identification	Parent material	Forest vegetation
	Forest	Productive forest	Soft-wood	Hard-wood	Rad-forth area				

Forest Section: B5 Hudson Bay Lowlands (O, M, Q): open woodland of bS & T in the muskegs; on river leves wS, bF, bPop & wB

Total land	Forest	Productive forest	Soft-wood	Hard-wood	Rad-forth area	Soil great group	Soil subgroup proportion and identification	Parent material	Forest vegetation
130,000	19,500 (15)	7,780 (40)	2	0	3	10 H3.46	1 Cryic fibrisol 2 Orthic humo-ferric podzol Cryic gleysol	Organic & sandy, loamy or clayey glac till lac or glac-fluv dep	Non-prod conifer for
						5 H1.7	1 Fibrisol 2 Orthic humo-ferric podzol Orthic gleysol	Organic & sandy to clayey glac-fluv, glac till & lac dep	Lowland non-prod swamp conifer for
						5 C2.107	1 Orthic gray luvisol 2 Orthic gleysol Orthic humo-ferric podzol 3 Fibrisol	Sandy to clayey mod calc glac till, glac-fluv & lac dep	Prod conifer for
						5 {G2.25	1 Orthic gleysol 2 Fibrisol	Clayey to loamy mod calc glac till, lac mar & organic dep	Non-prod conifer for swamp or march
						{G2.24	(as G2.25)	"	"
						5 H1.3	1 Fibrisol 3 Orthic gray luvisol	Organic & mod calc lac & glac-fluv dep	"
						35 H3.41	1 Cryic fibrisol 3 Orthic humo-ferric podzol	Organic & sandy mod calc glac-fluv dep	Non-prod conifer for
						5 G2.35	1 Orthic gleysol 2 Cryic fibrisol Orthic gray luvisol	Clayey str calc glac till lac & organic deposits	Non-prod conifer for
						30 H3.48	1 Cryic fibrisol 2 Orthic eutric brunisol	Loamy, glac till, organic & glac-fluv dep	"

Forest Section: B1a Laurentide-Onaatchiway (Q): predominantly coniferous forests, bF on hill slopes and bS on wetter land and on thin-soiled plateaus; wB is the common hardwood associate peats characterize dwarf-shrub and moss bogs

Total land	Forest	Productive forest	Soft-wood	Hard-wood	Rad-forth area	Soil great group	Soil subgroup proportion and identification	Parent material	Forest vegetation
125,700	69,130 (55)	48,390 (70)	40	20	1	80 D3.126	1 Orthic humo-ferric podzol 2 Rockland 3 Fibrisol	Stony, sandy glac till, glac-fluv dep & peat	Prod conifer for
						10 D3.39	1 Orthic humo-ferric podzol 2 Fibrisol	Sandy glac till & glac-fluv dep	Prod conifer for
(figures include B1B)						10 R1	Rockland		

Forest Section: B1b Chibougamau-Natashquan (Q, NL): predominantly bS forest on both slopes and peaty lowland; wS & bF are scarce

Total land	Forest	Productive forest	Soft-wood	Hard-wood	Rad-forth area	Soil great group	Soil subgroup proportion and identification	Parent material	Forest vegetation
(see above)			40	20	2	5 {H1.12	1 Fibrisol 2 Orthic humo-ferric podzol 3 Orthic gleysol	Organic dep & sandy glac till	Lowland non-prod conifer for
						{H1.13	(as H1.12)	"	Upland prod conifer for "
						85 D3.126	1 Orthic humo-ferric podzol 2 Rockland 3 Fibrisol	Stony, sandy glac till Glac-fluv dep & peat	Prod conifer for
						5 D3.39	1 Orthic humo-ferric podzol 2 Fibrisol	Sandy & glac till & glac-fluv dep	"
						5 R1	Rockland		

APPENDIX (continued)

Areas (square miles)			Productivity (ft³ac⁻¹yr⁻¹)		Rad-forth area	Soil great group	Soil subgroup proportion and identification	Parent material	Forest vegetation
Total land	Forest	Productive forest	Soft-wood	Hard-wood					

$$\text{Productivity in } ft^3 ac^{-1} yr^{-1}$$

Forest Section: B3 Gouin (Q): jP & bS in open stands with bS more on the muskeg: also mixed wood of tA, bPop, wB, wS & bF on upland till

Total land	Forest	Productive forest	Soft-wood	Hard-wood	Rad-forth area	Soil great group	Soil subgroup proportion and identification	Parent material	Forest vegetation
9,100	5,000 (55)	3,500 (60)	25	10	2	50 D3.41	1 Orthic humo-ferric podzol 2 Fibrisol	Sandy, non-calc lac & glac-fluv dep	Upland prod mixed for Lowland non-prod conifer for
						40 D3.120	1 Orthic humo-ferric podzol 2 Rockland	Sandy, glac till, glac-fluv dep	Prod conifer for
						10 D3.27	1 Orthic humo-ferric podzol	Sandy, stony glac till & glac-fluv dep	Prod mixed for

Forest Section: B4 Northern Clay (Q, O): bS on gentle uplands and on lowlands mixed with fens and Sphagnum bogs. C with bS but little T

Total land	Forest	Productive forest	Soft-wood	Hard-wood	Rad-forth area	Soil great group	Soil subgroup proportion and identification	Parent material	Forest vegetation
40,100	36,090 (90)	30,560 (85)	11	6	2	10 H1.5	1 Fibrisol 2 Orthic gleysol 3 Orthic gray luvisol Orthic humo-ferric podzol	Organic & mod calc clayey Glac till & lac dep	Upland prod conifer for Lowland non-prod conifer swamp for
						G2.26	1 Orthic gleysol 2 Fibrisol 3 Orthic gray luvisol	Clayey, mod calc glac till lac and organic dep	Upland prod conifer for Lowland non-prod conifer for
						25 G2.31	1 Orthic gleysol 2 Rockland (as G2.31)	Clayey, mod calc lac dep	Prod conifer for
						G2.30			
						10 D3.149	1 Orthic humo-ferric podzol 2 Rockland 3 Orthic eutric brunisol Orthic gray luvisol	Stony, sandy glac till	Prod mixed for Non-prod swamp for
						30 C2.107	1 Orthic gray luvisol 2 Orthic gleysol Orthic humo-ferric podzol 3 Fibrisol	Sandy, to clayey, mod glac till, glac-fluv & lac dep	Prod conifer for
						20 D3.121	1 Orthic humo-ferric podzol 2 Rockland	Sandy, glac-fluv dep & glac till	Non-prod conifer for
						5 D3.41	1 Orthic humo-ferriz podzol 2 Fibrisol	Sandy, glac till & glac-fluv dep	Upland prod conifer for Lowland non-prod conifer for

Forest Section: B7 Missinaibi-Cabonga (Q, O): predominantly a mixed forest of bF, bS and wB with scattered wS and tA, bS with T on wet organic soils sometimes bS with C

Total land	Forest	Productive forest	Soft-wood	Hard-wood	Rad-forth area	Soil great group	Soil subgroup proportion and identification	Parent material	Forest vegetation
39,200	35,370 (90)	31,830 (95)	13	13	2	40 D3.149	1 Orthic humo-ferric podzol 2 Rockland 3 Orthic eutric brunisol Orthic gray luvisol	Stony, sandy, glac till	Prod mixed for and non-prod swamp for
						30 D3.27	1 Orthic humo-ferric podzol	Sandy, stony glac till and glac-fluv dep	Prod mixed for
						20 D3.126	1 Orthic humo-ferric podzol 2 Rockland 3 Fibrisol	Stony, sandy, glac till glac-fluv dep	Prod conifer for
						10 D3.150	1 Orthic humo-ferric podzol 2 Fibrisol	Stony, sandy, glac till, glac-fluv dep	Upland prod mixed Lowland non-prod conifer for

APPENDIX (continued)

Areas (square miles)			Productivity (ft³ac⁻¹yr⁻¹)		Rad-forth area	Soil great group	Soil subgroup proportion and identification	Parent material	Forest vegetation
Total land	Forest	Productive forest	Soft-wood	Hard-wood					

Forest Section: B2 Gaspé (Q): pure stands of bS and bF and mixed stands of these species with C. bS, bF, wS & wB also form a characteristic cover type: forest cover is almost continuous

9,500	5,220 (55)	4,180 (80)	33	10	0	45 D3.31	1 Orthic humo-ferric podzol	Loamy, glac till	Prod conifer for
						25 D3.29	1 Orthic humo-ferric podzol	Sandy & glac loamy till	"
						D3.73	1 Orthic dystric brunisol / 2 Orthic gray luvisol / 3 Orthic gleysol	Loamy to clayey glac till & lac dep	Not forested
						10 { Orthic humo-ferric podzol (as E3.73 above)		"	Prod mixed for
						E3.72			Lowlands prod conifer for
						10 D3.138	1 Orthic humo-ferric podzol / 2 Orthic humic gleysol / Cryic fibrisol / Rockland	Stony, sandy glac till & glac-fluv dep	Uplands non-prod conifer for
						10 D3.30	3 Orthic dystric brunisol / 1 Orthic humo-ferric podzol	Loamy glac till & glac-fluv dep	Prod mixed for

Forest Section: B12 Hamilton and Eagle Valleys (NL): conifers on poorly drained flats, dominated by sB, sometimes with bF and wB

12,000	9,000 (75)	4,500 (50)	14	2	2	40 { D3.57 / D3.?	1 Orthic humo-ferric podzol / 2 Orthic gleysol / 3 Fibrisol / Rockland / Rockland / "	Stony, sandy glac till & glac-fluv dep	Upland non prod conifer for
						60 { R10 / R11			Lowland prod conifer for

Forest Section: B29 Northern Peninsula (NL, Q): stands of bF, bS and wS with ericaceous and moss peats on poorly drained lands

22,100	17,600 (80)	14,100 (80)	20	13	3	30 D3.86	1 Orthic humo-ferric podzol / 2 Orthic gleysol / 3 Fibrisol / Rockland	Stony, loamy glac till & glac-fluv dep	Non-prod conifer for
						10 D3.64	(as D3.86 above)	"	Upland prod conifer for
						15 D3.70	1 Orthic humo-ferric podzol / 2 Orthic gleysol / 3 Fibrisol / Orthic humic podzol / Rockland	"	Lowland non-prod conifer for / Non-prod conifer for
						15 D3.59	1 Orthic humo-ferric podzol / 2 Orthic gleysol / 3 Cumulic regosol / Orthic ferric-humo podzol	Sandy, loamy glac till and glac-fluv dep	Prod conifer for
						30 D3.79	1 Orthic humo-ferric podzol / 2 Orthic gleysol / 3 Fibrisol / Rockland	Stony, loamy glac till	Non-prod conifer for

(figures include B28a, 28b and 28c)

APPENDIX (continued)

Forest Section: B28a Grand Falls (NL): most productive forest of Newfoundland: chiefly coniferous dominated by bF and bS open bS stands and barrens on peat are common

Areas (square miles)			Productivity ($ft^3ac^{-1}yr^{-1}$)		Rad-forth area	Soil great group	Soil subgroup proportion and identification	Parent material	Forest vegetation
Total land	Forest	Productive forest	Soft-wood	Hard-wood					
(see above)			20	13	3	50 D3.59	1 Orthic humo-ferric podzol 2 Orthic gleysol 3 Cumulic regosol Orthic ferric-humo podzol	Sandy & loamy glac till & glac-fluv dep	Prod conifer for
						50 D3.63	1 Orthic humo-ferric podzol 2 Orthic gleysol 3 Fibrisol Rockland Orthic ferro-humic podzol	Loamy, glac till & glac-fluv dep	Upland prod conifer for Lowland non-prod conifer for

Forest Section: B28b Corner Brook (NL): the second largest area of productive forest in Newfoundland. bF in association with bS and wS; also small stands of hardwoods in which wB is dominant and tA playing a subsidiary role

(see above)			20	13	3	85 D3.79	1 Orthic humo-ferric podzol 2 Orthic gleysol 3 Fibrisol Rockland	Stony, loamy glac till	Non-prod conifer for
						5 D3.144	1 Orthic humo-ferriz podzol 2 Orthic gleysol 3 Fibrisol Orthic ferro-humic podzol	"	Prod conifer for
						10 C2.116	1 Orthic gray luvisol 2 Orthic humo-ferric podzol 3 Fibrisol Orthic humic-podzol	Loamy & clayey glac till glac-fluv dep	Prod conifer for

Forest Section: B28c Anticosti (Q): mixed wood forest of wS, bF, bS and wB

(see above)			20	13	1	50 E3.90	1 Orthic dystric brunisol 2 Rockland	Stony, loamy wkly calc glac till	Non-prod conifer for
						50 E3.84	1 Orthic dystric brunisol 2 Fibrisol	Loamy, wkly clac glac till	Non-prod mixed for

APPENDIX (continued)

Areas (square miles)			Productivity (ft³ ac⁻¹ yr⁻¹)		Rad-forth area	Soil great group	Soil subgroup proportion and identification	Parent material	Forest vegetation
Total land	Forest	Productive forest	Soft-wood	Hard-wood					

Forest Section: B6 East James Bay (Q): wS, bS, bPop, tA, wB and bF reach merchantable size in valleys

Total land	Forest	Productive forest	Soft-wood	Hard-wood	Rad-forth area	Soil great group	Soil subgroup proportion and identification	Parent material	Forest vegetation
8,100	6,070 (75)	3,000 (50)	17	1	3	80 H3.46	1 Cryic fibrisol / 2 Orthic humo-ferric podzol / Cryic gleysol	Organic & sandy, loamy or clayey glac till, lac or glac-fluv dep	Non-prod conifer for
						15 H1.7	1 Fibrisol / 2 Orthic humo-ferric podzol / Orthic gleysol	Organic & sandy to clayey glac-fluv, glac till & glac-fluv dep	Lowland non-prod conifer for / Upland poor conifer for
						5 D3.129	1 Orthic humo-ferric podzol / 2 Rockland / 3 Fibrisol	Sandy glac till, glac-fluv dep & peat	Prod mixed for

Forest Section: B8 Central Plateau (O): jP on sandier areas; bS types well represented bS mixed with jP on better drained sites with wB and tA as associates; C & T with bS on low sites; bog and muskeg throughout; peaty phase gleysols on low areas

Total land	Forest	Productive forest	Soft-wood	Hard-wood	Rad-forth area	Soil great group	Soil subgroup proportion and identification	Parent material	Forest vegetation
34,400	32,680 (95)	29,410 (90)	13	13	2	20 D3.19 / 10 D3.148	1 Orthic humo-ferric podzol / 1 Orthic humo-ferric / 2 Rockland / 3 Orthic eutric brunisol / Orthic gray luvisol	Sandy, wkly calc lac dep / Stony, sandy glac till	Prod conifer for / "
						20 D3.147 / 40 D3.149	(as D3.147 above) / (as D3.147 above)	" / "	" / Prod mixed for & non-prod swamp for
						5 D3.17	1 Orthic humo-ferric podzol	Sandy, non-calc glac-fluv lac dep	Prod conifer for
						5 { C2.42 / C2.43 }	1 Orthic gray luvisol / 3 Orthic eutric brunisol / (as C2.42 above)	Clayey str calc lac dep / "	Prod conifer for / Prod mixed for

Forest Section: B9 Superior (O): very variable forests; on deeper soils a mixed forest of wS, bF, wB and tA; on till slopes a similar mixture but with more wB and some bS; wetter and poorly drained sites support good bS, with some C & T

Total land	Forest	Productive forest	Soft-wood	Hard-wood	Rad-forth area	Soil great group	Soil subgroup proportion and identification	Parent material	Forest vegetation
9,540	8,100 (85)	7,700 (95)	12	11	1	20 D3.149	1 Orthic humo-ferric podzol / 2 Rockland / 3 Orthic eutric brunisol / Orthic gray luvisol / (as D3.149 above)	Stony, sandy glac til	Prod mixed for / Non-prod swamp for
						70 D3.148 / 10 { C2.43 / C2.44 }	1 Orthic gray luvisol / 3 Orthic eutric brunisol / 1 Orthic gray luvisol	" / Clayey, str calc lac dep / "	Prod conifer for / Prod mixed for / Prod mixed & conifer for

APPENDIX (continued)

Areas (square miles)			Productivity (ft³ac⁻¹yr⁻¹)						

	Areas (square miles)			Productivity ($ft^3ac^{-1}yr^{-1}$)		Rad-forth area	Soil great group	Soil subgroup proportion and identification	Parent material	Forest vegetation
Total land	Forest	Productive forest		Soft-wood	Hard-wood					

Forest Section: B10 Nipigon (O): a uniform forest with bS dominant merging with jP on higher slopes and T on lower; few open bogs; organic accumulation on lower areas

1,060	900 (85)	860 (95)	12	11	1	70 D3.19	1 Orthic humo-ferric podzol	Sandy, wkly calc lac dep	Prod conifer for
						25 D3.17	"	Sandy, non-calc glac fluv lac dep	"
						5 D3.18	"	Sandy, wkly calc lac dep	"

Forest Section: B11 Upper English River (O): predominantly bS and jP with mixture of wS, bF, tA and wB; bS dominates peat-filled depressions; jP the higher drier slopes; T is not abundant

13,100	11,790 (90)	8,840 (75)	17	9	1	100 D3.148	1 Orthic humo-ferric podzol	Stony, sandy glac till	Prod mixed for
							2 Rockland		
							3 Orthic eutric brunisol		
							Orthic gray luvisol		

Forest Section: B14 Lower English River (O, M): on well-drained sites mixed stands of tA, bPop and wS are the main cover; on wetter sites bS and T predominate with jP, wB, bF on sandier areas

6,700	5,690 (85)	3,980 (70)	10	8	1	15 H1.32	1 Fibrisol	Organic & clayey, mod calc till & lac	Non-prod conifer for
							2 Rockland		
							Orthic gray luvisol		
							Rockland		
						25 R13			
						15 D3.128	1 Orthic humo-ferric podzol	Sandy, glac till, glac-fluv dep	Prod confier for
							2 Rockland		
						10 D3.148	1 Orthic humo-ferric podzol	Stony, sandy glac till	Prod conifer for
							2 Rockland		
							3 Orthic eutric brunisol		
							Orthic gray luvisol		
						35 C2.45	1 Orthic gray luvisol	Clayey, str calc lac dep	Prod mixed for
							3 Orthic gleysol		

Forest Section: B30 Avalon (NL): mainly patchy dense young conifer stands of bF or bS interspersed with extensive barrens. bS is prevalent on wetter lowlands often accompanied by T; roughly half area is moss and heath bog

8,800	3,080 (35)	2,310 (75)	20	10	3	10 D3.63	1 Orthic humo-ferric podzol	Loamy glac till	Upland prod conifer for
							2 Orthic gleysol	glac fluv dep	Lowland non-prod for
							3 Fibrisol		
							Rockland		
							Orthic ferro-humic podzol		
						40 D3.69	1 Orthic humo-ferric podzol	Loamy, sandy	Non-prod conifer for
							2 Orthic gleysol	glac till	
							3 Fibrisol		
							Orthic humic podzol		
							Rockland		
						10 D3.145	1 Orthic humo-ferric podzol	Loamy glac till	Non-prod conifer for
							2 Orthic gleysol		
						15 D3.76	1 Orthic humo-ferric podzol	Stony, loamy glac	Non-prod conifer for
							2 Orthic humo-ferric podzol	till, glac-fluv dep	
							Orthic gleysol		
							3 Fibrisol		
							Rockland		
						25 D3.77	(as D3.76 above)	Stony, loamy, sandy glac till	Non-prod conifer for

APPENDIX (continued)

Forest Section: B22a Northern Coniferous (S, M, O); bS is the predominant tree mixed T on the poorly-drained lowlands and with jP on the sandy uplands

Areas (square miles)			Productivity (ft³ac⁻¹yr⁻¹)		Rad forth area	Soil great group	Soil subgroup proportion and identification	Parent material	Forest vegetation
Total land	Forest	Productive forest	Soft-wood	Hard-wood					
162,500 (figures include B22b)	97,170 (80)	82,580 (85)	7	4	2	25 D3.85	1 Orthic humo-ferric podzol	Stony, sandy glac till & glac-fluv dep	Upland prod conifer for
							2 Fibrisol Rockland		Lowland non-prod conifer for
							3 Orthic dystric brunisol		
						5 H1.30	1 Fibrisol	Organic, clayey, str calc lac dep	″
							2 Rockland		
							Orthic gray luvisol		
						15 H1.31	(as H1.30 above)	Organic, clayey, mod calc lac dep	Non-prod conifer for
						15 C2.127	1 Orthic gray luvisol	thin loamy glac till, lac dep	″
							2 Cryic fibrisol Rockland		
						15 D3.146	1 Orthic humo-ferric podzol	Loamy glac till, sandy glac-fluv dep	Prod mixed for
							2 Orthic dystric brunisol		
						15 D3.148	1 Orthic humo-ferric podzol	Stony sandy glac till	Prod conifer for
							2 Rockland		
							3 Orthic eutric brunisol		
							Orthic gray luvisol		
						5 D3.19	1 Orthic humo-ferric podzol	Sandy, mod calc glac till, glac-fluv dep	″
						5 { H3.48	1 Cryic fibrisol	loamy, glac till,	Non-prod conifer for
							2 Orthic eutric brunisol	Organic, glac-fluv dep	
						H3.41	1 Cryic fibrisol	Organic sandy mod calc glac-fluv dep	″
							3 Orthic humo-ferric podzol		

Forest Section: B22b Athabasca South (A, S); park-like stands of jP on sandier areas; moister sandy flats support bS with T
(see B22a above)

Areas (square miles)			Productivity (ft³ac⁻¹yr⁻¹)		Rad forth area	Soil great group	Soil subgroup proportion and identification	Parent material	Forest vegetation
Total land	Forest	Productive forest	Soft-wood	Hard-wood					
			7	4	3	85 D3.43	1 Orthic humo-ferric podzol	Sandy, stony glac till, glac-fluv dep	Non-prod conifer for
							2 Fibrisol		
							3 Orthic dystric brunisol		
						5 D3.52	1 Orthic humo-ferric podzol	Sandy, glac-fluv & aeol dep	″
							2 Degraded dystric brunisol		
							3 Orthic regosol		
						5 D3.51	(as D3.52 above)	Sandy, aeol dep organic & sandy	Upland prod mixed for
							1 Fibrisol		
							2 Orthic humo-ferric podzol	glac till, fluv dep	Lowland non-prod conifer
						5 H1.8	3 Orthic gleysol		

APPENDIX (continued)

Areas (square miles)			Productivity (ft³ac⁻¹yr⁻¹)		Rad forth area	Soil great group	Soil subgroup proportion and identification	Parent material	Forest vegetation
Total land	Forest	Productive forest	Soft-wood	Hard-wood					

Forest Section: B15 Manitoba Lowlands (S, M): bS and T in patches on the poorly-drained land with intervening swamps; on the better-drained alluvial strips there are good stands of wS, tA, bPop; jP occurs on low ridges, sometimes with tA; soils are generally gleysol with peat

Total land	Forest	Productive forest	Soft-wood	Hard-wood	Rad forth area	Soil great group	Soil subgroup	Parent material	Forest vegetation
33,700	16,850 (50)	8,420 (50)	10 (50)	2	1	10 {H1.2	1 Fibrisol / 3 Orthic gray luvisol	Organic dep over coarse fluv & fine lac clay dep	Upland prod mixed / Lowland non-prod / Conifer swamp for
						H1.30	1 Fibrisol / 2 Rockland / Orthic gray luvisol	Organic & clayey, str calc lac dep	Upland prod conifer / Lowland non-prod conifer for
						5 H1.6	1 Fibrisol / 2 Orthic gleysol / Cumulic regosol	Organic & loamy alluv dep	Non-prod mixed for & non-prod conifer swamp for
						5 H1.32	1 Fibrisol / 2 Rockland / Orthic gray luvisol	Organic & clayey, mod calc glac till & lac dep	Non-prod conifer for
						30 {C2.72	1 Orthic gray luvisol / 2 Fibrisol / 3 Orthic eutric brunisol	Loamy to clayey, mod to str calc till & lac dep	Upland prod mixed for / Lowland non-prod conifer for
						C2.74	1 Orthic gray luvisol / 2 Fibrisol	"	
						30 C2.75	(as C2.74 above)		
						10 G1.15	1 Orthic humic gleysol / 2 Mesisol	Loamy, mod calc alluv & lac dep	Non-prod mixed for
						10 A4.135	Dark gray chernozemic		

Forest Section: B18a Mixedwood (A, S, M): on the well-drained uplands the characteristic association is a mixture of tA, bPop, wB, wS & bF, the last two being prominent in old stands; on drier soils jP is present; on wetter lands bS & T are dominant but the peats are not thick; the main soil is gray luvisol

Total land	Forest	Productive forest	Soft-wood	Hard-wood	Rad forth area	Soil great group	Soil subgroup	Parent material	Forest vegetation
176,400 (figures include B18b)	132,300 (75)	105,840 (80)	10	12	0	C2.39	1 Orthic gray luvisol / 3 Dark gray luvisol / Orthic humic gleysol	Loamy mod to str calc glac till	Prod mixed for
						C2.86	1 Orthic gray luvisol / 2 Fibrisol	"	Upland prod mixed for
						60 {C2.83	(as C2.86 above)		Lowland non-prod conifer for
						C2.85	1 Orthic gray luvisol / 2 Fibrisol / 3 Orthic gleysol	Loamy, mod calc glac till	"
						C2.82	1 Orthic gray luvisol / 2 Fibrisol	Loamy, wkly calc glac till	"
						C2.68	(as C2.82 above)	Loamy, sl to mod calc glac till	
						30 {H1.17	1 Fibrisol / 2 Gleysol / Orthic gray luvisol	Loamy, sl calc glac till	Not forested
						H.125	1 Fibrisol / 2 Orthic gray luvisol / 3 Orthic humic gleysol / Solod	Organic & loamy, sl to mod calc glac till	Upland prod mixed for / Lowland non-prod conifer swamp for
						10 B4.22		Organic & sandy or clayey, wkly calc glac till	"

Areas (square miles)			Productivity ($ft^3 ac^{-1} yr^{-1}$)		Rad-forth area	Soil great group	Soil subgroup proportion and identification	Parent material	Forest vegetation
Total land	Forest	Productive forest	Soft-wood	Hard-wood					

Forest Section: B18b Hay River (A., NWT, BC); a northern extension of B18a above, with less wS and tA; bS is dominant on both uplands and lowlands; some jP is present

Total land	Forest	Productive forest	Soft-wood	Hard-wood	Rad-forth area	Soil great group	Soil subgroup proportion and identification	Parent material	Forest vegetation
(see B18a above)			10	12	2	60 G2.33	1 Orthic gleysol 2 Cryic fibrisol Orthic gray luvisol 3 Orthic eutric brunisol Orthic gleysol	Clayey, sl calc lac, glac-fluv & beach dep	Upland prod conifer for Lowland non-prod swamp for
						C2.126 10 { C2.130	1 Orthic gray luvisol 2 Cryic fibrisol (as C2.126 above)	Glayey, wkly calc glac till & glac-fluv dep Loamy, wkly calc glac till & glac-fluv dep	" Non-prod conifer for
						C2.63	(as C2.126 above)	Loamy, str calc glac till & glac-fluv dep	Upland prod conifer for Lowland non-prod for Non-prod conifer for
						20 { H3.52	1 Cryic fibrisol 2 Orthic gray luvisol 3 Orthic gleysol	Organic dep, clayey or loamy wkly calc glac till	
						H1.34	1 Fibrisol 2 Orthic gleysol Eutric brunisol 3 Cryic fibrisol	Organic & sandy fluvial dep	"
						5 E2.32	1 Orthic eutric brunisol 3 Orthic gleysol Black solonetz	Clayey, mod calc stony glac till & lac dep	Prod conifer for
						5 B2.10			

Forest Section: B21 Nelson River (M): bS makes up a large part of the forest cover, but growth is generally restricted; T is present with bS in swamps; on drier areas there are good stands of wS with some bPop, wB, tA and bF

Total land	Forest	Productive forest	Soft-wood	Hard-wood	Rad-forth area	Soil great group	Soil subgroup proportion and identification	Parent material	Forest vegetation
18,000	14,400 (80)	6,480 (45)	6	2	2	15 H1.32	1 Fibrisol 2 Rockland Orthic gray luvisol (as H1.32 above)	Organic & clayey, mod calc glac till & lac dep	Non-prod conifer for
						50 H1.31	(as H1.32 above)	Organic & clayey, mod calc lac dep	"
						20 H1.30	(as H1.32 above)	Organic & clayey, str calc lac dep	Upland prod conifer for Lowland non-prod conifer for
						15 D3.35	1 Orthic humo-ferric podzol 2 Fibrisol Rockland 3 Orthic dystric brunisol	Stony, sandy glac till & glac-fluv dep	"

APPENDIX (continued)

Areas (square miles)		Productive forest	Productivity ($ft^3 \cdot ac^{-1} \cdot yr^{-1}$)		Rad-forth area	Soil great group	Soil subgroup proportion and identification	Parent material	Forest vegetation
Total land	Forest		Soft-wood	Hard-wood					

Forest Section: B20 Upper Churchill (S): extensive stands of jP on the sand plains and low ridges with bS and T on the poorly-drained intervening areas. wS and tA are present where drainage permits

Total land	Forest	Productive forest	Soft-wood	Hard-wood	Rad-forth area	Soil great group	Soil subgroup proportion and identification	Parent material	Forest vegetation
44,800	35,760 (80)	17,880 (50)	8	8	0	70 { C2.119	1 Orthic gray luvisol 2 Orthic humo-ferric podzol Fibrisol	Sandy & loamy, wkly calc glac till & glac-fluv dep	Prod conifer for
(figures include B24 and B25)						C2.123	3 Orthic dystric brunisol (as C2.119 above)	Sandy to loamy, wkly to mod calc glac till & glac-fluv dep	Prod mixed for
						10 D3.36	1 Orthic humo-ferric podzol 2 Fibrisol Orthic dystric brunisol	Sandy, glac-fluv dep & alluv	Upland prod conifer for Lowland non-prod conifer for
						20 { H1.10	1 Fibrisol 2 Orthic humo-ferric podzol 3 Orthic gray luvisol (as H1.10 above)	Organic & sandy glac till & glac-fluv dep	Upland prod conifer for Lowland non-prod conifer swamp for
						H1.11	(as H1.10 above)	Organic & sandy glac till & glac-fluv dep	"

Forest Section: B24 Upper Laird (BC, Y, NWT): characterized by 1P, bS, wS and tA; good growth of wS and bPop on alluvial flats; bS and 1P from pure and mixed stands (see B20 above)

Total land	Forest	Productive forest	Soft-wood	Hard-wood	Rad-forth area	Soil great group	Soil subgroup proportion and identification	Parent material	Forest vegetation
			8	8	2	50 { E2.57	1 Orthic eutric brunisol 2 Orthic regosol Rockland 3 Fibrisol	Sandy, glac till & alluv dep	Non-prod mixed for
						E2.55	1 Orthic eutric brunisol 2 Cumulic regosol Rockland	Stony, sandy mod calc alluv & fluv dep	Non-prod conifer for
						50 H3.52	1 Cryic fibrisol 2 Orthic gray luvisol 3 Orthic gleysol	Organic dep, clayey or loamy wkly calc glac till	"

Forest Section: B25 Stikine Plateau (BC): a scant forest cover confined to the valleys, comprising tA, wS and 1P in open form interspersed by grassy areas (see B20 above)

Total land	Forest	Productive forest	Soft-wood	Hard-wood	Rad-forth area	Soil great group	Soil subgroup proportion and identification	Parent material	Forest vegetation
			8	8	1	70 E2.56	1 Orthic eutric brunisol 2 Orthic regosol Rockland 3 Fibrisol	Sandy, mod calc glac till & alluv dep	Non-prod conifer for
						30 R1	1 Orthic eutric brunisol 2 Orthic regosol Rockland Rockland		

APPENDIX (continued)

Areas (square miles)		Productivity (ft³ac⁻¹yr⁻¹)		Rad-forth area	Soil great group	Soil subgroup proportion and identification	Parent material	Forest vegetation
Total land	Forest / Productive forest	Soft-wood	Hard-wood					

Forest Section: B23a Upper Mackenzie (NWT, A, BC): contains some of the best timber-producing land in the northwest. wS and bPop form the main cover types, with bF and wB prominent south of L. Athabasca Sands are occupied by jP and lP, and wet lands and muskeg by bS and T

Total land	Forest / Productive forest	Soft-wood	Hard-wood	Rad-forth area	Soil great group	Soil subgroup	Parent material	Forest vegetation
75,700	30,280 (40) 27,250 (90)	7	8	3	50 { E2.55	1 Orthic eutric brunisol 2 Cumulic regosol Rockland	Stony, sandy mod calc alluv & fluv dep	Non-prod conifer for
					E2.50	1 Orthic eutric brunisol 2 Orthic gleysol Cryic fibrisol	Sandy & loamy, mod calc glac till & lac dep	"
(figures include B23b)					30 G2.33	1 Orthic gleysol 2 Cryic fibrisol Orthic gray luvisol 3 Orthic eutric brunisol Orthic gleysol	Clayey, sl calc lac glac-fluv & beach dep	Upland prod conifer for Lowland non-prod swamp for
					20 H3.44	1 Cryic fibrisol 2 Orthic gleysol 3 Orthic eutric brunisol	Organic & clayey, mod calc glac till, glac-fluv & alluv dep	Non-prod conifer for

Forest Section: B23b Lower Mackenzie (NWT, Y): there is more non-forested than forested land; where the permafrost table is not high wS attains sawlog size, but on finer alluvium growth is poor, and stunted wS and bS prevail; on poorly drained land there is much stunted bS but T is absent

Total land	Forest / Productive forest	Soft-wood	Hard-wood	Rad-forth area	Soil great group	Soil subgroup	Parent material	Forest vegetation
(see B23a above)		7	8	2	30 H3.44	1 Cryic fibrisol 2 Orthic gleysol 3 Orthic eutric brunisol	Organic & clayey, mod calc glac till, glac-fluv & alluv dep	Non-prod conifer for
					60 { F1 15 F2 18	Regosolic		
					10 F3 70	Regosolic		

Forest Section: B26a Dawson (Y): wS, either pure or mixed with wB, Alaska B or tA on the valley slopes; in the valley bottoms stunted stands of bS and wS are usual

Total land	Forest / Productive forest	Soft-wood	Hard-wood	Rad-forth area	Soil great group	Soil subgroup	Parent material	Forest vegetation
97,600	39,040 (40) 35,140 (90)	9	6	1	E2.56	1 Orthic eutric brunisol 2 Orthic regosol Rockland 3 Fibrisol	Sandy, mod calc glac till & alluv dep	Non-prod conifer for
(figures include B26b, B26c and B26d)					E2.46	1 Orthic eutric brunisol 2 Orthic dystric brunisol Cumulic regosol	Loamy, glac till	Non-prod mixed for
					E2.47	(as E2.46 above)	"	Non-prod conifer for

Forest Section: B26b Central Yukon (Y): bent forest growth is on protected lowlands, falling off with altitude; wS attains sawlog size on lower slopes; in the valleys lP and wS share dominance in open or closed stands, sometimes with tA; bS occurs on organic soils

Total land	Forest / Productive forest	Soft-wood	Hard-wood	Rad-forth area	Soil great group	Soil subgroup	Parent material	Forest vegetation
(see B26a above)		9	6	1	E2.56	1 Orthic eutric brunisol 2 Orthic regosol Rockland 3 Fibrisol	Sandy, mod calc till & alluv dep	Non-prod conifer for

APPENDIX (continued)

Areas (square miles)			Productivity $(ft^3ac^{-1}yr^{-1})$		Rad-forth area	Soil great group	Soil subgroup proportion and identification	Parent material	Forest vegetation
Total land	Forest	Productive forest	Soft-wood	Hard-wood					

Forest Section: B26c Eastern Yukon (Y, BC): open barrens are common, especially on north slopes; on south and west slopes there are mixed stands of wS, tA and birches; on lower slopes wS is dominant, with 1P and tA, in the southern parts with wS, birch and bS in the northern areas; bS with T occur on bogs; organic soils are extensive

(see B26a above)			9	6	2	E2.57	1 Orthic eutric brunisol 2 Orthic regosol 3 Fibrisol Rockland	Sandy, glac till & alluv dep	Non-prod mixed for
						R1	Rockland		

Forest Section: B26d Kluane (Y): park-like forests even in the valleys; the best growth is of wS, tA and bPop; there is no 1P and bS and T are not prominent; on peat the bS and wS seem interchangeable

(see B26a above)			9	6	1	E2.44	1 Orthic eutric brunisol 2 Cumulic regosol	Loamy, mod calc glac till, fluv & alluv dep	Non-prod mixed for
						E.245	(as E2.44 above)	Clayey, mod calc lac fluv & alluv dep	Non-prod conifer for
						E2.56	1 Orthic eutric brunisol 2 Orthic regosol Rockland 3 Fibrisol	Sandy, mod calc glac till & alluv dep	"

Forest Section: B19a Lower Foothills (A, BC): 1P is distinctive, with tA and bF; wS is present in older stands, sometimes with bS

66,200	59,580 (90)	56,600 (95)	10	11	0	C2.27	1 Orthic gray luvisol 3 Fibrisol	Loamy, sl calc glac till	Upland prod mixed for
(includes also B19b and B19c)						C2.35 C2.34 C2.67	(as C2.27 above) (as C2.27 above) (as C2.27 above)	" " "	" " "

Forest Section: B19b Northern Foothills (BC): the dominant trees are wS, bS and 1P, characteristic of the high plateaus and also of lower altitudes; broadleaved trees are not abundant; bS occupies the low-lying organic soils

(see B19a above)			10	11	1	C2.58	1 Orthic gray luvisol 2 Orthic eutric brunisol 3 Fibrisol	Loamy, non-calc to wkly calc glac till & glac-fluv dep	Prod mixed for
						C2.88	1 Orthic gray luvisol 2 Fibrisol 3 Orthic gleysol	Clayey, wkly calc lac glac till & alluv dep	Upland prod conifer for Lowland non-prod conifer for
						C2.129	1 Orthic gray luvisol 2 Orthic eutric brunisol Cryic fibrisol	Loamy, wkly calc glac-till & glac-fluv dep	"
						E2.38	1 Orthic eutric brunisol 2 Orthic dystric brunisol Orthic regosol	Loamy, wkly calc glac-fluv & alluv dep	Prod conifer for
						E2.43	1 Orthic eutric brunisol 2 Cumulic regosol	Loamy, wkly calc fluv & alluv dep & glac till	Non-prod conifer for

APPENDIX (continued)

Areas (square miles)			Productivity ($ft^3 ac^{-1} yr^{-1}$)		Rad-forth area	Soil great group	Soil subgroup proportion and identification	Parent material	Forest vegetation
Total land	Forest	Productive forest	Soft-wood	Hard-wood					

Forest Section: B19c Upper Foothills (A, BC): 1P is dominant and hardwoods are scarce, tA, bPop and wB only being sparsely represented; bS is present in the north, elsewhere wS (see B19a above)

| | | | 10 | 11 | 0 | C2.61 | 1 Orthic gray luvisol | Loamy, mod to sl calc glac till | Prod conifer for |
| | | | | | | | 2 Orthic brunisol | | |

Forest Section: B16 Aspen-Oak (M, S): tA is the prevalent tree species occurring as patches ringing wet depressions or as continuous closed stands towards the northern boundary

| 19,000 | 10,450 (55) | 5,750 (55) | 7 | 3 | 1 | A3 | Black chernozemic, with transition to dark gray chernozem on well-drained and to humic gleysol on wetter sites | | |

Forest Section: B17 Aspen Grove (A, S): only tA is abundant in natural stands, with bPop on moist lowlands; wB occurs sporadically

| 56,100 | 28,050 (50) | 15,430 (55) | 5 | 5 | 0 | A3 B3 C2 | Black chernozemic is predominant, with some solonetzic and lubisolic | | |

8
Canadian Muskegs and Their Agricultural Utilization

J.M. STEWART

Very little information is available from the prescientific era on what man thought of the origins and nature of peat, although some inkling of these thoughts may be gained from the folklore of countries with extensive peatlands. To some, the peatlands were the haunt of evil spirits and to be avoided, but others used them as burial grounds and accorded them religious respect. The bleak climate and the monotony of such landscapes probably helped to create and foster these beliefs. Many peatlands acted as barriers to access, some were used as sources of fuel, and others were drained, cleared, and then used for agriculture and/or forestry.

More information is available on the economic exploitation of peatlands from this period. The scarred land surface formerly covered by peat in many areas of Europe testifies to this exploitation. The primary use of peat was as a fuel. By draining, cutting, and drying the peat during the summer, a winter's supply of fuel was obtained. Considerable attention was given also to the agricultural potential of the lands, especially their conversion into pasture or tillage after peat-cutting or drainage.

With the advent of the Industrial Revolution the search for new, cheap sources of fuel included peat as a possibility. Peat has since proved uneconomical for use as a fuel in most countries, however, although it is still exploited for fuel in those countries with extensive deposits, for example Eire and the USSR. In the past decade the world's utilization of peat has steadily increased, although its principal use is not as a fuel but as an additive to soils.

Today the peat industry is well established in many countries and employs, directly or indirectly, specialists in such diverse fields as engineering, agriculture, forestry, horticulture, geobotany, anthropology, chemistry, and medicine. This progress in the industry is a direct result of our expanding population together with our ability to exploit natural resources mechanically at an ever increasing rate. The need for more space and for more efficient production of food has led to the increasing conversion of natural peatlands to cultivation in countries where good mineral soils are scarce. Organic soils can, if managed properly, be among the most productive as witnessed by the garden crops grown in areas such as the Holland Marsh in southern Ontario. Even many of the ombotrophic peatlands in eastern Canada are used to a limited extent, for rough grazing and hay pastures. This potential as a source of forage is being studied actively in Newfoundland, where there are hopes of establishing a livestock industry based on reclaimed pastures (Rayment and Chancey, 1966).

From the Proceedings of the Third International Peat Conference (1968) it is evident that the exploration of this last land barrier in the northern hemisphere is now a fact of life. In Canada our muskegs are used principally as a substrate source for expanding horticultural industry in other countries. Still in its infancy is the reclamation of extensive peatlands for the purposes of growing crops or establishing permanent pastures.

The extent of muskeg in Canada is still not fully known, but some say there are 37,000 square miles of it (Tibbetts, 1968) while others say 500,000 square miles (Radforth, 1968). Most of these peatlands occur in remote uninhabited areas of the north; consequently most of them are considered low on the scale of priorities for cultivation, although their importance for storage of fresh water in the hydrological balance should not be underestimated.

Today active reclamation of muskeg is underway in all provinces by both government and private enterprises. An example can be seen in the Erts photograph of Figure 1, which reveals the intrusion of agricultural practices into peatlands on the outer rim of mineral soil farming in the Red River valley, Manitoba. Most provinces have published surveys of peat deposits within their boundaries and these often serve as a spring-board for exploitation, be it for moss litter, fuel, or reclamation for garden crops and pastures. Even in the Northwest Territories and Yukon revegetation and agricultural studies have indicated that a number of potential hay-producing species (e.g. red fescue, timothy, Canada blue grass, Kentucky blue grass) survive and propagate on the thin peaty soils (Younkin, 1972; Bliss and Wein, 1972). Also the work by Elliott (1970) on forage introduction into northern Alberta and southern Northwest Territories shows a large number of potentially adaptive species for this area. These species have potential for converting our northern peatlands into more productive pastures once the problem of cultivation on permafrost can be solved — if ever.

It is the object of this chapter, first, to examine the present status of agriculture in muskeg as it relates to Canada; and secondly, to survey, briefly, the experiences of other countries in the reclamation of peatlands for agriculture in terms of development and production practices.

AGRICULTURAL PRACTICES

Before we discuss the agricultural capability of peatlands it is necessary to explain some of the terms which are used throughout this chapter. For instance, what is considered as an 'agricultural practice'? The Canada Land Inventory (1970) lists the following categories of agricultural practice:

1. *Horticultural:* Land used for the intensive cultivation of vegetables and small fruits including market gardens, nurseries, flower and bulb farms and sod farms (Symbol H)

2. *Orchards and Vineyards:* Land used for the production of tree fruits, hops and grapes. (Symbol G)

3. *Cropland:* Land used for annual field crops; grain, oilseeds, sugar beets, tobacco, potatoes, field vegetables, associated fallow, and land being cleared for field crops. (Symbol A)

4. *Improved Pasture and Forage Crops:* Land used for improved pasture or for the production of hay and other cultivated fodder crops including land being cleared for these purposes. (Symbol P)

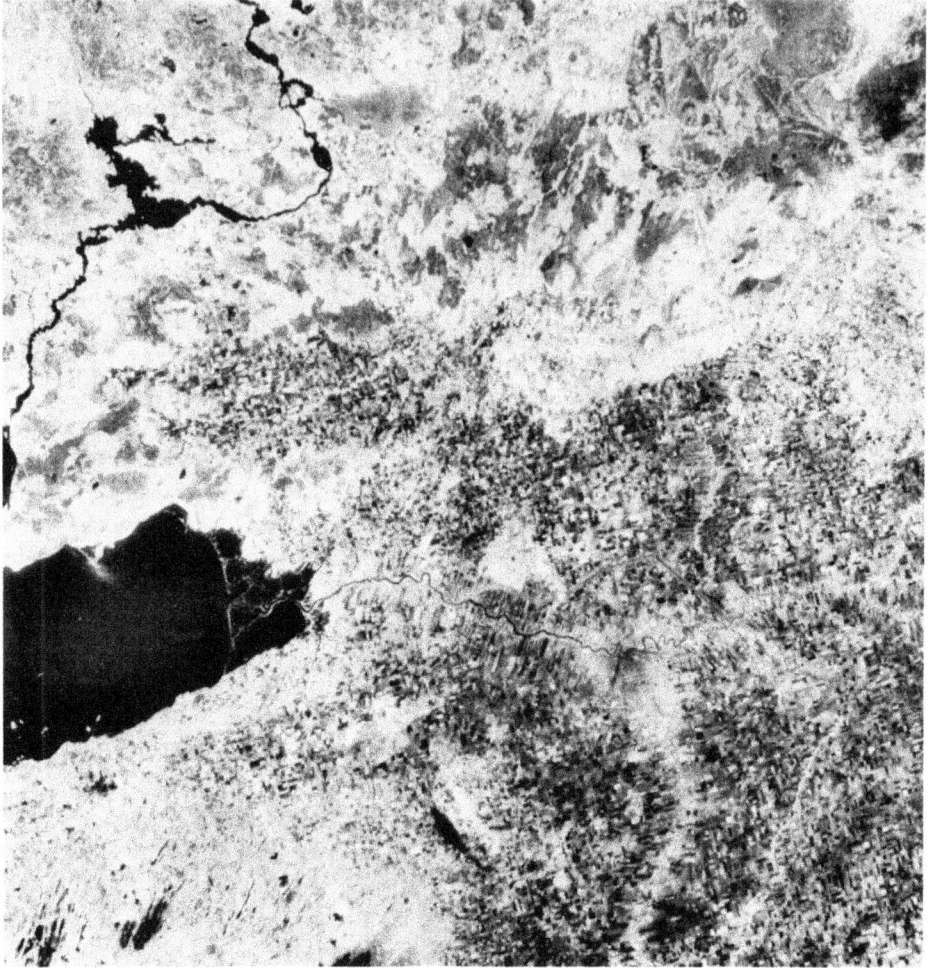

FIGURE 1   ERTS photograph of Red River Valley, Manitoba. Intrusion of agricultural practices into peatlands bordering the eastern section of the Red River Valley.

5. *Rough Grazing and Rangeland:* (a) Areas of natural grasslands, sedges, herbaceous plants and abandoned farmland whether used for grazing or not. Bushes and trees may cover up to 25 per cent of the area. If in use, intermittently-wet, hay lands (sloughs or meadow) are included. (b) Woodland grazing: If the area is actively grazed and no other use dominates, in some grassy, open woodlands, bushes and trees may somewhat exceed 25 per cent cover. (Symbol K)

Although these categories were designed for mineral soils, they can be applied to organic soils with the possible exception of category 2, 'Orchards and Vineyards.' All these agricultural activities occur to varying degrees throughout the provinces. Perhaps the only qualification need be made to the first category, Horticulture, which should include the peat moss extracted and used as a growing medium in greenhouses.

PEAT AS A SOIL

Another question often raised concerns the reference to peat as a soil. Leahey (1961) defined soil from the agricultural point of view as 'the natural medium for the growth of plants,' and from the pedological viewpoint as 'collection of *natural bodies* on the earth's surface supporting or capable of supporting plants.' These natural bodies are classified as either mineral or organic. According to the National Soil Survey Committee of Canada (Day, 1968), the highest level of abstract classification for organic soils is the *order*, i.e. 'soils saturated with water for most of the year and containing 30 percent or more of organic matter to certain depths.' The next level of abstraction is the *great group*, i.e. 'organic soils are differentiated on the basis of degree of decomposition of organic material at certain levels (surface, middle and bottom) of variable depths.' The remaining level of abstraction is the *sub-group*, which consists of up to 29 layer types within the over-all profile.

Peat, then, can be viewed as a soil although it can be defined as 'the substrate of peatlands consisting of more or less fragmented remains of vegetable matter sequentially deposited and fossilized' (MacFarlane, 1969). Peat is formed from plant remains whose decomposition is retarded by poor drainage and with the resulting anaerobic condition persisting for a long time. In most natural peatlands, plant litter formation and subsequent accumulation is greater than the rate of decomposition. Reader and Stewart (1972) found that less than 10 per cent of the annual net primary production will remain as a peat. Estimates of the rate of accumulation of peat from pollen analysis and radiocarbon dating are in the order of 0.02 inch per year of 12 inches over 500 years (Durno, 1961).

The properties of peat which make it an ideal additive to mineral soils for growing plants are: (a) it increases the water storage capacity of soils with coarse texture; (b) it provides ventilation in soils with fine textures (e.g. clay); and (c) it improves the nutrient-holding capacity of all soil types. As a substrate peat is porous, with often only 3 to 10 per cent of its volume solid. The air/water ratio per unit volume is adjustable by compression or loosening. The peat particles are surface-active and have a high cation-exchange capacity (90-150 me/100 g dry matter). Also peat, when semi-dried and baled, is light in weight and therefore easy to handle.

AGRICULTURAL USES OF PEATLANDS

In the natural state, muskegs are of little agricultural value except those that can be used for rough hay pastures. Today agricultural technology can make most natural peatlands productive by well-established drainage and cultivation techniques as witnessed by the diversity of success with peatlands throughout the world (Peat Abstracts). Examples of this widespread use of peatlands are:

1. Direct cultivation (UK, USA, USSR, Israel, Netherlands, Eire, Poland, Finland, Malaysia, Japan, Canada).
2. Forestry (UK, USA, Finland, Norway, USSR, Canada).
3. Pasturage (New Zealand, East Germany, Poland, UK, USSR, Canada).

4. Soil additives (Czechoslovakia, East Germany, Sweden, USSR, Finland, USA, New Zealand, Canada).
5. Fertilizer (Czechoslovakia, East Germany, Sweden, UK, USSR, Finland).
6. Peat composts (Czechoslovakia, Netherlands, USSR).
7. Peat litter (USA, USSR, UK, Eire, Canada).

As well as a diversity of uses, there is a diversity of crops that can be produced on cultivated peatlands. Examples include varieties of hay species, sod grass, garden crops (celery, onions, cucumbers, tomatoes, peppers, asparagus, cauliflower), strawberries, cloudberries, raspberries, hemp, bedding flowers, cut flowers, tapioca, and pineapple (Peat Abstracts). In addition experiments are being conducted in most countries to increase the productivity of the crops, for example, by using humic extracts from peat to stimulate maize growth (Ortega, 1968).

In Canada the variety of crops grown is not large since the climate, peat types, and population pressure (which differ in each of the provinces) serve to restrict the demand for the development of muskegs. In Newfoundland (Rayment and Heringa, 1972; Pollett, 1972; Rayment, 1972) the cool, moist, and windy climate combined with the low nutrient status of the peats favours the growth of forages. With the help of fertilizers, timothy, red fescue, clovers, and reed canary have been shown to be the most adaptable. Cereals have had limited success owing to their tendency to lodge. Where wind protection is present, other crops such as turnips, carrots, celery, spinach, lettuce, beets, and potatoes have been grown, but the scale of operation so far cannot be considered commercial. The use of peatlands in the Maritimes is not as intensive as in Newfoundland, but forages again dominate and some efforts have been made to cultivate cranberries and blueberries. In Quebec a few muskegs are used as pastures and some have been converted to market gardening on a limited scale. Ontario has limited pasturage on peatlands, and mostly on the peaty soils of the clay belt near Cochrane. Good examples of intensive cultivation occur in southern Ontario, e.g. the Alfred bog, Holland Marsh, Powasson bog. The best known of these is the Holland Marsh, whose 550 farmers and 20,000 acres supply the Toronto market with garden produce (Irwin, 1963). Specialized crops such as cranberries near Collingwood and wild rice near Emo in northwestern Ontario also grow on converted muskegs. Manitoba has some pasture on peatlands near The Pas, and wild rice paddies are located on the Piney and Stead bogs (Stewart, 1970). In northeast Saskatchewan there is a small pasture development and plans for wild rice paddies in the muskegs along the Carrot River. The agricultural potential of peat soils in Alberta has been reviewed by Dew (1968), who pointed out that hay crops are the most important with *Bromus inarnus* in areas of good drainage and *Phalaris arundinacreae* in areas of high moisture. Maas (1972) and Van Ryswyk (1971) have reported on the crops grown on the peaty soils of Vancouver Island and the interior of British Columbia. Blueberries and cranberries fare best of all on the acid-type peats, whereas potatoes and other vegetables and hay are most productive on sedge-type peats. Specialized crops, such as raspberries and loganberries, are grown on the sedge-type peat areas.

TYPES OF MUSKEG SUITABLE FOR AGRICULTURE

Altogether the area of organic soils in Canada capable of being cultivated is estimated at 300 x 10^6 acres (Leahey, 1961). This considerable acreage is located mostly in the north, where the climate, permafrost, and nutrient status of the peaty soils are not conducive to large-scale commercial agriculture. These remote peatlands do, however, offer a reserve for peat suitable for horticultural production and as well retain considerable volumes of fresh water. Most of the peatland that will be reclaimed in the near future is found in the southern fringe of our provinces, where closeness to markets, demand, and economic feasibility make the conversion to agricultural production highly profitable.

The status of peatlands in relation to the main soil regime and zones in Canada is presented in Figure 2, and the following is a brief description of the extent of peatlands in each soil region and zone (Underhill, 1959; Nowosad and Leahey, 1960; Leahey, 1961; Brown, 1968; Day, 1968; Radforth, 1969).

Soil Regions

1. Arctic	Peatlands with permafrost present but limited to alluvial plains, valleys, and deltas
2. Subarctic	Extensive peatlands with continuous and discontinuous permafrost
3. Canadian Shield	Peatlands with discontinuous permafrost, common
4. Cordillera	Peatlands very limited

Major Soil Zones

5. Luvisolic and Podzolic soils	5a	Grey Wooded, peatlands confined to extensive
	5b	Grey-Brown Luvisolic, peatlands confined
	5c	Eastern Podzolic, peatlands mostly confined, extensive in Newfoundland
6. Gleysolic soils	6a	St Lawrence Lowland, peatlands confined
7. Chernozemic soils	7a	Brown, peatlands non-existent
	7b	Dark Brown, peatlands very limited
	7c	Black, peatlands confined
8. Brunisolic soils	8a	Pacific Coast, peatlands confined and located mostly in alluvial flood-plains and deltas

If we combine the peat landform classification (Tarnocai, 1970) with the surface vegetation classification (Radforth, 1952; Zoltai and Tarnocai, 1969) it is possible to state the agricultural capability of muskegs south of the continuous and discontinuous permafrost region based on existing land usage of cultivated peatlands. Two terms, 'established' from existing commercial practices and 'experimental' from research station results, indicate qualitatively the potential productivity of the peatland before reclamation (Table 1). Furthermore, if we use the Radforth system (1961) based on the predominant surface vegetation classes in cover formulae, certain peat structure categories can be equated with statements of agricultural significance in terms of land preparation and cultivation practices (Table 2).

CULTIVATION PRACTICES

In the preparation of peatlands for cultivation, a variety of techniques are available, the choice of which depends on peat type, mined or cut-away status of the peatland, climate, drainage, access, nearness to markets, and the crops to be grown. The conversion of natural peatlands into productive land is both difficult and costly, which is perhaps one of the reasons why they are the last land resource to be diverted into fulfilling man's need for more food.

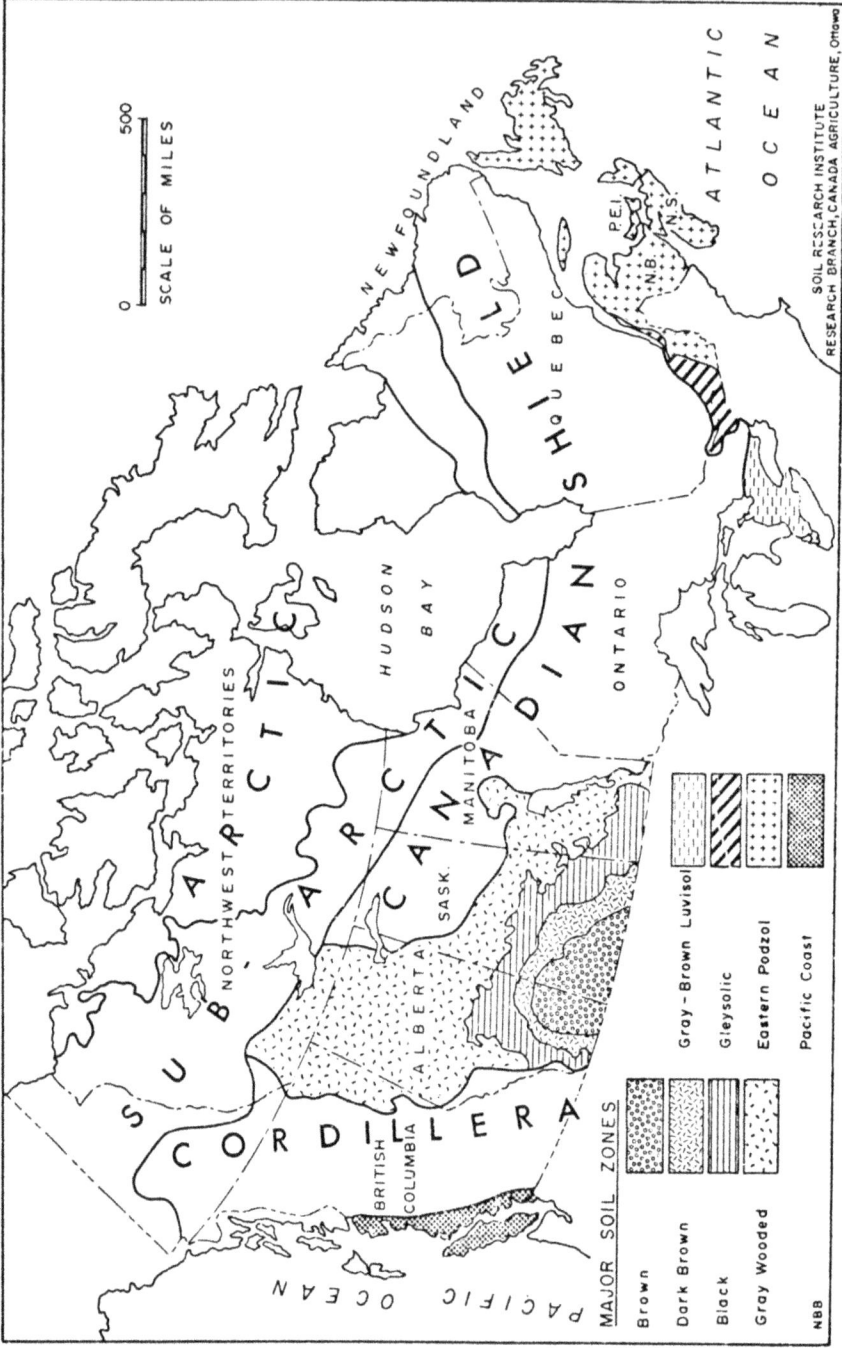

FIGURE 2   Major soil zones and regions of Canada (after Leahey, 1961). See text for brief description of extent of peatlands in each zone and region.

TABLE 1

Agricultural use of some peatland types in Canada

A. Bog

Class (Tarnocai, 1970)	Subclass (Tarnocai, 1970)	Principal cover formulae (Radforth, 1969)	Agricultural practices			
			Horticulture	Pasture forage	Garden crop	Rough grazing
1A. Ombrotrophic	Raised bog	BE1, E1, HE	Estab.	Exptl.	?	Estab.
2A. Transitional	Flat bog	AF1, AE1, BF1, BE1	Estab.	Estab.	Exptl.	Estab.
	Bowl bog	BE1	Estab.	?	?	?
	Blanket bog	AE1, BE1, E1	Exptl.	Exptl.	?	Estab.
B. Fen						
Minerotrophic	Horizontal fen	AF1, BF1, DF1, FE1, F1	Estab.	Estab.	Estab.	Estab.
	Sloping fen	DF1, FE1, F1	Estab.	Estab.	?	Estab.

TABLE 2

Peat characteristics indicated by predominant cover classes (Radforth, 1961) prior to clearance for agricultural purposes

Predominant class in formulae	Peat structure category	Agricultural significance
A & B (*Picea mariana, Larix laricina*)	17, 16, 15, 12 coarse-fibrous	Low: high cost of clearing surface vegetation
D (*Salix* spp., *Alnus* spp.)	14, 13, 11, 9 coarse-fibrous to fibrous	Low: subject to spring flooding, wood clumps interspersed between amorphous granular pockets, access difficult for tillage vehicles
	10, 8	High: peat usually uniform in structure, drainage and tillage practical
E (*Chamaedaphne calyculata, Ledum groenlandicum*)		
F (*Carex* spp., *Scirpus* spp.)	7, 6	Medium: drainage practical, low bearing capacity can cause difficulties in tillage, peat permeable, nutrient status higher than E
H (*Cladonia* spp.)	1, 3 amorphous-granular	Low: where it predominates it indicates permafrost, crops only possible where active layer allowed to deepen, surface usually uneven
I (*Sphagna, Hyphna*)	2, 4, 5 fine-fibrous to amorphous-granular	High: Excellent for horticulture, high exchange capacity, high moisture retention, high permeability, high aeration when drained, drainage and tillage practical

Muskeg and the Northern Environment in Canada

Generally speaking, to develop a peatland for agricultural purposes requires four procedures which must be carried out in sequence after a detailed survey of the area has been completed.

1. *Primary drainage*. The first prerequisite is to instal the main drains according to the contours of the peat surface. This is essential before the installation of a secondary drain network, and enables the removal of free-standing water at rates and times satisfactory to plant growth. A common method for draining bogs, at least in Europe, uses earthenware and corrugated PVC pipes in the mole draining technique with depth and spacing according to electronic gradient controls (Kuntze, 1970).
2. *Secondary drainage*. This secondary network feeding into the primary drains allows greater control of water levels, a necessity for garden crops. These drains are usually covered, thus permitting a greater surface area for crop growth.
3. *Land preparation*. Once the drainage network is installed, then land clearing, followed by land levelling, can proceed. Equipment with low bearing capacity tracks disc or plough the upper tier of peat, which ensures root penetration and plant anchorage. At this stage fertilizer may be added depending upon the requirement of the crop.
4. *Crop seeding*. For frost-sensitive species (e.g. celery), seedlings are recommended for an early crop. Where species are not frost-sensitive, then direct seeding is sufficient. It is essential that water control be possible at this stage since during the day the peat surface with its low heat capacity develops high temperature and is subject to wind erosion unless it is moist. At night the low thermal capacity allows greater heat loss, therefore increasing the chances of frost. The presence of moisture close to the surface helps to alleviate extremes in temperature.

WATER CONTROL

The preparation for market gardening on peatlands requires a much more intensive drainage pattern — shallower depths, closer lateral spacing of drainage tiles, smaller diameters of drainage tiles — than preparation for pastures. Control of water is perhaps the key factor to successful farming on organic soils. As Ayers (1963) points out, any water control in a peatland must be based on total water control of the area. The hydrological balance must be maintained in that the input of water into the system — precipitation, flowing water, or seepage — must equal the output — drainage, evaporation demand, temperature and wind control. Allowance should be made for storage whether it be channel storage, moisture in the peat, or actual groundwater storage. Of these factors only drainage is subject to human control, and the advantages of increased crop production from this control temporarily outweigh the disadvantages of accelerated decomposition.

Reclaiming exploited peatlands after the peat has been mined or cut away is simpler. Usually the primary and secondary drains are cleaned out, after which the mineral subsurface and remainder of the peat are deep-ploughed and mixed. This practice is common in Europe, where in West Germany alone over 100,000 hectares of available farming land has been created in this manner (Kuntze, 1971).

FERTILIZERS

The application of fertilizer is an accepted agricultural procedure necessitated by the need to maintain the healthy vigorous growth of the crop and to supply the nutrients essential to the health of the animals feeding on the crop. An example of the latter is the inclusion of selenium for sheep in the fertilizer applied to peat pastures in New Zealand (Van der Elst and Watkinson, 1972).

Fertilizer needs of organic soils are different from those of mineral soils. Crops grown in peat require liming followed by the application of nitrogen, phosphorus, and potassium at various ratios depending on the crop; e.g. carrots need 1:1:1.5 and kale 1:7.5:1.5 (Kaila, 1971). New slow-release fertilizers, such as Plantosan (2:1:1.5), are being used more frequently since they duplicate the slow release of elements from decomposing humic colloids (Gugenhan and Deiser, 1971). Liming has been shown to increase the exchangeable Ca^{++} and Mg^{++} and to decrease the exchangeability of K^+ (Mackai, 1971).

The role of K^+ is unique in that it moves down the peat profile rapidly in contrast to nitrogen and phosphorus which are slow. K^+ requires constant replacement, since up to 60 per cent is lost from the top 15 centimetres in one year. Heavy precipitation accentuates the leaching. The reason for this behaviour is believed to be the low chelating effect humic colloids have on K^+ (Shicklona et al., 1972).

The time of year chosen for application is important, top-dressing of fertilizer in the spring being best for plant growth (Van der Elst, 1969). The experience in Newfoundland indicates that small regular dressings of nitrogen, phosphorus and potassium are superior to one heavy application (Rayment and Heringa, 1972).

Although we tend to think of fertilizers as being beneficial to crops by stimulating their growth and eventual yield, this reaction to the fertilizer is believed to be secondary. Initially, in its natural state, peat is fibrous to amorphous granular (colloidal) in structure with the voids filled with water. The minerals in solution are assumed to be in equilibrium with the elements attached to the peat colloids. The effect of adding fertilizer is to enrich this solution, thus creating initially a bloom condition for the microflora. This results in a more rapid decomposition of particles of peat, with more nutrients being reduced to their elemental state ready for rapid root uptake.

PEST CONTROL

Weeds are an important pest problem in peatland agriculture, more so than in mineral soils. The high moisture-holding capacity of peat, its increased aeration after drainage, and its physical structure are ideal for weed growth (Cassidy, 1971). The importance of weeds in reducing crop productivity has been shown in Malaysia, where weeds reduced the yield of pineapples by as much as 40 per cent (Wee, 1972).

The standard herbicides for mineral soil cultivation do not work effectively in peats, thus diquat and paraquat have no effect on the naturally occurring species like *Carex* in cultivated peatlands (Turek and Mika, 1969). One herbicide which has been used very effectively for the pre-emergent control of weeds in peats is Lenacil (Long, 1970). When applied to sugar beet cultivation in fen peats it gave an effective 100 per cent kill when mixed in the upper 2 to 3 inches of peat. So far there is little evidence of downward leaching, and the herbicide biodegrades in the short term.

SUBSIDENCE

When peatlands are drained and brought under cultivation, rapid decomposition rates are experienced as anaerobic conditions give way to aerobic, with subsequent disappearance of the peat. The disappearance of the surface layer of peat is called 'subsidence,' and this forms one of the major handicaps to permanent agriculture on peatlands. The Fens in England (Darby, 1940) have the oldest recorded history of this phenomenon, which shows a loss of 1.5 inches each year. Irwin (1963) recorded a loss of 1.3 inches per year at the Ontario Agricultural College Muck Research Station in Holland Marsh. In other words, as Irwin so aptly states: 'Man is destroying in one year what it took nature 55 years to create.' The economic importance of subsidence was demonstrated in a recent *Financial Post* article (April 7, 1973) concerning the 1 inch per year disappearance of the Florida Everglades peats. Over 700,000 acres are affected with a cash income of over 250 million dollars per year. If those soils keep subsiding at the rate of 1 inch per year, then within the next 30 years most of the highly profitable market garden crops will no longer be grown on peat. Where will the food production from such acreages then come from?

Although subsidence is principally a phenomenon of the zone of aerobic decomposition, there is evidence that geologic subsidence of the region, compaction of mechanical equipment, shrinkage from drying, burning, and wind erosion all contribute (Broadbent, 1960). A number of methods (Smith, 1969), such as the use of plastic surface covers and straw mulches to prevent evaporation, have been tried to slow down subsidence but they have proved to be uneconomical.

CONCLUSION

In Canada, population pressures and the demand for resources will result in the reclamation, for agricultural purposes, of the muskegs surrounding our urban centres. These organic soils, especially those located in the southern fringes of Ontario and British Columbia, are among the most productive in the country, but also the most susceptible to decomposition. Less productive muskegs are capable of conversion into productive pastures, thus supporting an expanding livestock industry. Specialized crops, such as cranberries and wild rice, will be grown with increasing frequency on suitable peatlands provided the economics of production are justified.

It must be realized that the use of this resource for agricultural purposes is essentially one-way, for once peat is drained, it loses the hydrological balance which is responsible for its formation. It is a non-renewable resource and man must be made to realize that its cultivation is strictly short-term. If he wishes to continue to use this resource more research must be undertaken. Paramount to long-term utilization of our peatlands for agriculture will be research into management practices which can delay or limit rates of decomposition. Also needed, as Radforth (1968) pointed out, is research into a number of other matters: (a) the segregation of peat types for various human activities such as manufacturing, agriculture, forestry, and engineering; (b) reclamation of peatlands in accordance with specified needs, such as for garden crops, fruit, or pasture; and (c) suitability of certain peat types for growing and producing economical harvests of specific crops.

REFERENCES

Ayers, H. 1963. Engineering problems in exploitation of organic terrain for agricultural re-clamation. Proc. Ninth Muskeg Res. Conf., NRC Tech. Memo. 81, pp. 244-54.

Bliss, L., and Wein, R. 1972. Botanical studies of national and man-modified habitats in the eastern Mackenzie Delta region and the Arctic Islands, Rept. to ALUR of DIAND. See T. Babb, High Arctic Disturbances Studies, pp. 175-229.

Broadbent, F. 1960. Factors influencing the decomposition of organic soils of the California Delta. Hilgardia 29: 587-612.

Brown, R. 1968. Occurrence of permafrost in Canadian permafrost. Proc. 3rd Internat. Peat Congr., Quebec, pp. 174-81.

Canada Land Inventory. 1970. Objectives, scope and organization. Rept. no. 1, 2nd ed., DREE, Ottawa, appendix VII, p. 49.

Cassidy, J. 1971. Weed control in vegetables on peat soils. HEA Conf. Dublin.

Darby, H. 1940. The Drainage of the Fens (Cambridge Univ. Press).

Day, 1968. The classification of organic soils in Canada. Proc. 3rd Internat. Peat Congr., pp. 80-4.

Dew, D. 1968. Agricultural potential of peat soils in Alberta. Proc. 3rd Internat. Peat Congr., pp. 258-60.

Durno, S. 1961. Evidence regarding the rate of peat growth. J. Ecol. 49: 347-51.

Elliott, C. 1970. Forage introduction. Can. Agric. Res. Branch, Res. Sta. Beaverlodge, Alta., Publ. NR6 70-16.

Financial Post. April 7, 1973.

Gugenhan, E., and Deiser, E. 1971. Fertilizer experiments with Pelargoniums and Fuchsias. Hort. Abst. 41(3): 7085.

Irwin, R. 1963. Progress report on hydrological investigation of organic soils. Proc. Ninth Muskeg Res. Conf., NRC Tech. Memo. 81, pp. 226-35.

Kaila, A. 1971. Effect of application of lime and fertilizers on cultivated peat soil. J. Sci. Agric. Soc. Finl. 40: 133-41.

Kuntze, H. 1971. Progress in drainage techniques — Limits to drain efficiency. Peat Abstr. Bord Na Mona, Autumn, no. 80.

Leahey, A. 1961. The soils of Canada from a pedological viewpoint. In Soils in Canada, ed. R.E. Legget (Univ. Toronto Press; Roy. Soc. Canada Spec. Publ. 3), pp. 147-57.

Long, E. 1970. Trials of lenacil in the fens. Farmers Weekly 72 (14): 40.

Maas, E.F. 1972. The organic soils of Vancouver Island. Canada Dept. Agric. Res. Sta., Sidney, BC, no. 231.

MacFarlane, I. 1958. Guide to a field description of muskeg (based on the Radforth Clas-sification System), NRC, ACSSM Tech. Memo 44, rev. ed.

— (ed.) 1969. Muskeg Engineering Handbook (Univ. Toronto Press).

Mackai, F. 1971. Nutrient dynamics (NPK) in cultivated peat. Peat Abstr. Bord na Mona, Summer, no. 93.

Nowosad, R., and Leahey, A. 1960. Soils of the arctic and subarctic regions of Canada. Agric. Inst. Rev. 15: 48–50.

Ortega, C. 1968. Different effects on maize of humic acid extracted from peat. Isotopes and radiation in soil organic matter studies. Proc. Symp. IAEA/FAO, Vienna, pp. 541-53.

Peat Abstracts. 1970-73. Bord na Mona. Quarterly.

Pollett, F. 1972. Nutrient content of peat soils in Newfoundland. Proc. 4th Internat. Peat Congr. 3: 461-8.

Radforth, N.W. 1952. Suggested classification of muskeg for the engineer. Eng. J. 35 (11).

— 1961. Organic terrain. *In* Soils in Canada, ed. R.E. Legget (Univ. Toronto Press; Roy. Soc. Canada Spec. Publ. 3), pp. 113–39.

— 1968. New developments in peatland studies. Proc. 3rd Internat. Peat Congr., Quebec, pp. 1–3.

— 1969. Classification of muskeg. *In* Muskeg Engineering Handbook, ed. I.C. MacFarlane (Univ. Toronto Press), pp. 31-52.

Rayment, A. 1972. Rough pasture in Newfoundland. Can. Agric. 17 (2): 22-5.

Rayment, A., and Chancey, H. 1966. Peat soils in Newfoundland. Agric. Inst. Rev.

Rayment, A., and Heringa, P. 1972. The influences of initial and maintenance fertilizers on the growth and ecology of grass-clover mixtures on a Newfoundland peat soil. Proc. 4th Internat. Peat Congr., Helsinki, 3: 111-20.

Reader, R., and Stewart, J. 1972. The relationship between net primary production and accumulation for a peatland in southeastern Manitoba. Ecol. 53: 1024-37.

Shicklona, J., Lucas, R. and Davis, J. 1972. The movement of potassium in organic soils. Proc. 4th Internat. Peat Congr. 3: 131-48.

Smith, J. 1969. Soil mixing to preserve fen peats. Agric. Land 76: 612-16.

Stewart, J. 1970. Paddy production of wild rice in muskegs. Proc. 13th Muskeg Res. Conf. NRC Tech. Memo 99, pp. 91-7.

Tarnocai, C. 1970. Classification of peat land forms in Manitoba. Can. Dept. Agric. Res. Sta., Pedology Unit, Winnipeg.

Tibbetts. 1968. Canada: A brief review of peat in Canada. Proc. 3rd Internat. Peat Congr., pp. 8-10.

Turek, F., and Mika, V. 1969. Contribution to the problem of natural-grassland reclamation by means of chemical cultivation. Peat Abstr. Bord na Mona, Winter (1970), no. 68.

Underhill, F. 1959. The Canadian Northwest: Its Potentialities (Univ. Toronto Press).

Van der Elst, F. 1969. A comparison of phosphatic and sulphur containing fertilizers for pasture production on restiad peat in the Walkato District. Trans. 9th Internat. Congr. Soil Sci. 3: 407–17.

Van der Elst, F., and Watkinson, J. 1972. Selenium deficiency on high moor peat soils in New Zealand. Proc. 4th Internat. Peat Congr. 3: 149-64.

Van Ryswyk, A. L. 1971. Management and improvement of meadows on organic soils of interior BC. Publ. Branch, BC Dept. Agric., Victoria.

Wee, Y. 1972. Weed infestation of peat areas under pineapple cultivation in West Malaysia. Proc. 4th Internat. Peat Congr. 3: 165-74.

Younkin. 1972. Revegetation studies of disturbances in the Mackenzie Delta Region. In L. Bliss and R. Wein, Rept. to ALUR of DIAND, 1972, pp. 175-229.

Zoltai, S.C., and Tarnocai, C. 1969. Permafrost in peat land farms in northern Manitoba. Thirteenth Annual Manitoba Soil Sci. Meeting, Dec. 10-11, pp. 3-16.

9
Industrial Utilization of Peat Moss

M. RUEL, S. CHORNET, B. COUPAL, P. AITCIN, and M. COSSETTE

'... a great deal of new technology is not new knowledge; it is new perception. It is putting together things that no one had thought of putting together before, things that by themselves had been around a long time.' (Drucker, 1973)

Peat has always been with man. Because of its proximity it has been taken for granted for a long time, but not much could be done with it other than using it as a soil conditioner or as a heat source. To find new uses for this material, innovative research is required. Peat has inherent characteristics such as a low density, a high porosity, an ion exchange property, and thermal and acoustical insulation properties, which are worthy of a systematic evaluation with a view to solving actual problems of modern society, such as has been initiated recently in the USSR, Finland, and Canada.

Some work has been done on the extraction of various chemicals from peat, and the industrial potential of various processes has been discussed a great deal. M. Passer (1963) has made an intensive review of this subject and it will not be repeated here. We would, however, like to point out that because of the relatively low concentration of these products in peat, a systematic approach must be taken to ensure that all portions of the peat will give rise to final marketable products. Otherwise, the process cannot compete with other existing methods of obtaining the chemicals.

In this paper, we stress the areas of research which, in our opinion, show good potential for developing new uses for peat. These areas could be divided as follows: (1) treatment of polluted water, (2) obtaining of activated carbon from peat moss, (3) use of peat moss as a construction material. Each of these areas will be discussed briefly and the most recent developments presented. The advantages of, and some of the needs for, further research and development will be pointed out.

TREATMENT OF POLLUTED WATER

The physical and chemical characteristics of peat have been studied by a multitude of researchers. In 1966 MacFarlane (1969) continued the work of others by trying to establish a method for expressing the peat structure in quantitative terms. In the course of his work,

FEED

GOLF-GREEN TYPE FILTER BED
70' - 90' DIAMETER

8'-12' PEAT OR
PEAT-SAND
MIXTURE   SPRINKLER

ROTARY
ARM

PEAT
MOSS   BAFFLES

6" PEA ROCK
OR
COARSE GRAVEL

OUTLET PIPE

SUPPORT   2' FINE SAND

ADSORPTION   FILTER   SYSTEM

GOLF-GREEN TYPE FILTER SYSTEM

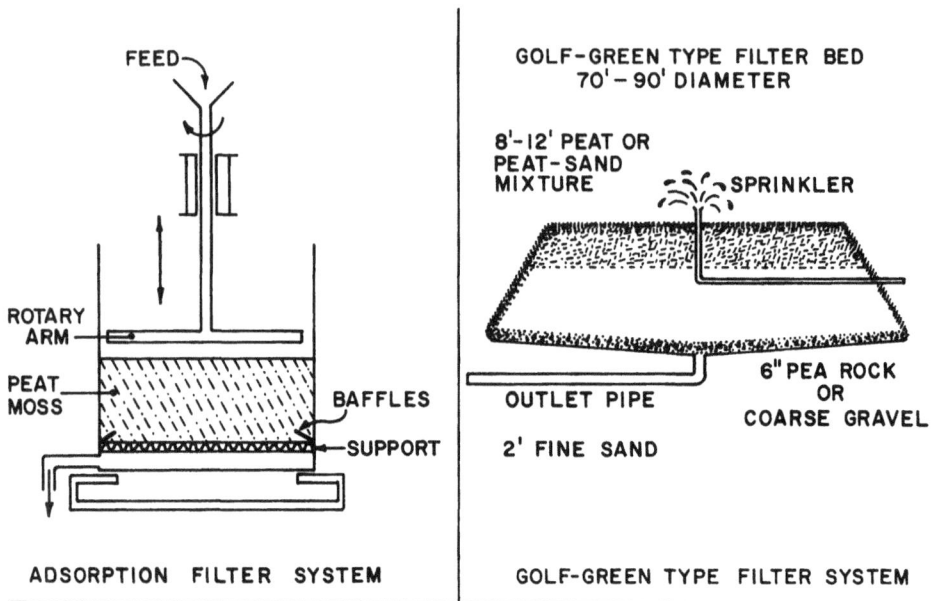

FIGURE 1   Methods of contacting waste waters and peat.

he tabulated some of the important physical characteristics of peat such as the void ratio, permeability, specific gravity, acidity, and compressibility. These properties are very important design parameters for engineers who are trying to develop new uses for peat.

From this study and others, peat was identified as having great potential as a filtering and adsorption agent. In early 1970, Ruel proposed the use of peat in the treatment of polluted water using these two principles. Various research projects were subsequently undertaken to identify the flow regime at which a filtration-adsorption bed of peat could best be used (Van et al., 1971; Dufort and Ruel, 1972a, b; Farnham and Brown, 1972). Simultaneously, the adsorption capacity of peat for various pollutants was studied and assessed in a systematic fashion, the main pollutants studied being: BOD or COD or TOD (Van et al., 1971; Dufort and Ruel, 1972; Farnham and Brown, 1972; Surakka and Kamppi, 1971; Sanyal, 1973; Mueller, 1972; Silva, 1972); detergents (Van et al., 1971; Sanyal, 1973; Silva, 1972); phosphate (Farnham and Brown, 1972; Surakka and Kamppi, 1971; Silva, 1972); bacteria (Surakka and Kamppi, 1971); solid and colloidal particles (Nguyen, 1973; Silva, 1972); iron (Ruel, 1971-72; Silva, 1972) and other heavy metals (Lalancette and Coupal, 1972; Mueller, 1972); phenol (Mueller, 1972); nitrogen (Silva, 1972); oil and fats (Mueller, 1972; Silva, 1972).

Many other studies have also been carried out in the complexing of heavy metals, nitrogen, phosphate, and other compounds with peat or components of peat, but these will not be reviewed here. Gamble (1972) reviewed the chemistry of peat and indicated the important role that the humic materials present in peat play in determining its chemical and physical properties and their potential use in the treatment of polluted water. Some of the methods used in contacting waste waters with peat are illustrated in Figures 1 and 5. Because of the low mechanical strength of peat fibres, as shown in various studies (Van et

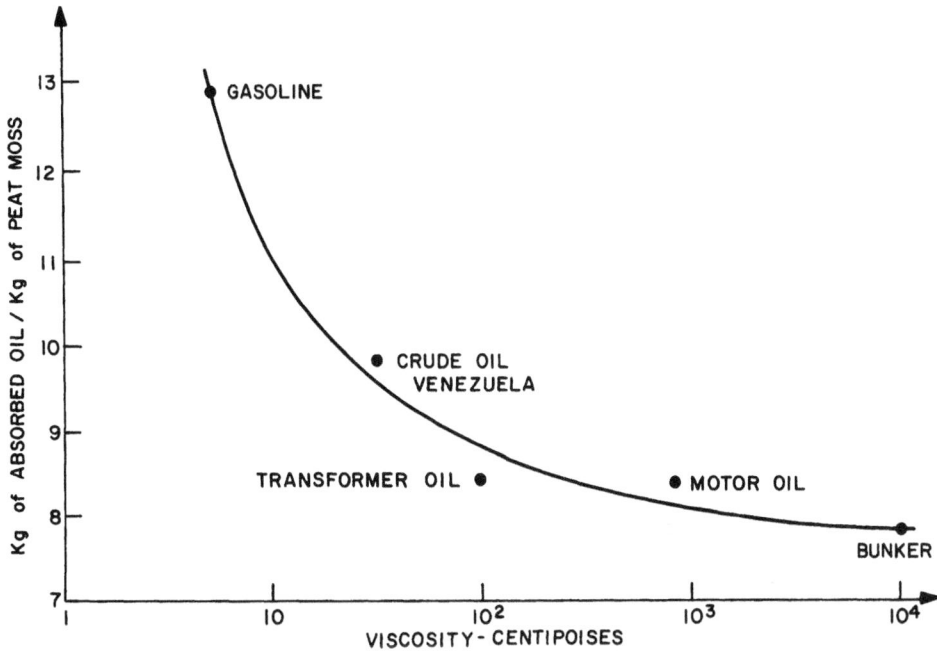

FIGURE 2   Absorbency of peat moss at 70°F as a function of the viscosity of oil.

al., 1971; Dufort and Ruel, 1972b; Farnham and Brown, 1972; Nguyen, 1973), it has been recommended that a peat bed should be gravity-fed. An adsorption-filtration column can also be used but the feed rate should be low; otherwise, the pressure drop across the bed increases too much and the bed will collapse (Sanyal, 1973; Nguyen, 1973). In all cases, the authors have stressed the importance of wetting the peat before using it as a filtration or adsorption agent.

Peat as an Absorbent and Burning Agent for Oil Spills

Oil pollution on water and land has been with us for a long time, but because of the increasing demand for oil to satisfy the huge appetites of modern technology, oil will be transported in greater amounts over the next twenty to thirty years at least. The potential for spills of oil due to accident or carelessness will be with us for the foreseeable future.

One of the methods used in the clean-up of oil spills makes use of absorbents, and a large number have been proposed at one time or another. In the evaluation of an absorbent, many factors must be considered, as illustrated in Table 1. A merit rating, based on work performed by various authors (Fuller, 1971; Ekman and Sandelin, 1971; D'Hennezel and Coupal, 1972) across the world and adapted to prevailing Canadian conditions, has been adopted to assess the value of the different absorbents, although some factors, such as the temperature effect and the type and age of the oil to be absorbed, have not been assessed systematically for all of the absorbents. The capacity of peat to absorb various types of oil is shown in Figure 2.

The criterion of absorption capacity is only one aspect to be considered in choosing an

TABLE 1
Floating absorbents: Criteria for choosing them

Materials	Availability in Canada	Preparation for use	Storage	Application	Absorption	Efficiency when wet	Leach-out by water	Harvesting from water	Oil drainage	Oil recovery	Ease of disposal	Biodegradability	Volume (ft³)[1]	Material cost ($)[1]
Straw	****	***	***	**	****	*	*	**	***	Nil	***	Yes	V.large	200
Sawdust	****	****	*	****	**	*	*	*	***	Nil	**	Yes	300	15
Siliconized sawdust	*	*	**	****	**	***	***	*	***	Nil	**	?	300	62+
Pine bark	***	***	**	***	***	***	***	*	***	Nil	*	Yes	400	250a
Peat	****	****	*	***	***	*	***	*	***	Nil	***	Yes	240	30
Chrome leather	*	****	**	***	****	****	**	**	**	**	**	No	250	15
Ekoperl	**	****	*	****	****	****	****	—	****	Nil	*	?	1,500	2,500
Vermiculite	*	****	**	****	*	Low	—	*	—	—	—	?	—	—
Natural rubber crumb	*	***	**	**	***	****	****	**	****	Nil	***	No	300	625
Natural rubber foam	*	***	*	**	****	**	*	**	*	****	***	No	3,500	190
Butyl rubber crumb	*	***	**	**	**	Low	—	—	—	—	—	No	—	—
Polystyrene pellets	*	***	*	**	*	Low	—	—	—	—	—	No	—	—
Polyurethane foam	*	*	****	**	****	****	***	**	***	****	***	No	5 (Liq)	125
Polypropylene fibre[2]	*	***	**	***	****	****	*	****	*	****	***	No	100	310

Key: the more stars, the better.

[1] For material to treat 10-ton spill.

FIGURE 3   Peat spreader (old model) Finland.

absorbent. As indicated by the merit rating in Table 1, peat presents a fairly good over-all potential as an absorbent, although certain precautions should be taken. The peat used should have a moisture content less than 35 per cent (wet basis) and preferably in the 10 to 15 per cent range. Otherwise, it will sink in a relatively short time (Ekman and Sandelin, 1971).

Peat has a tendency to break down into small particles when stored for long periods of time so that its application to an oil slick becomes difficult and messy. To obviate this difficulty, peat could be formed into granules. The agglomeration process (Ruel and Sirianni, 1972) could be a useful tool in forming granules of sufficient strength. The granulation of the peat would simplify its spreading as well as the harvesting of the oil-absorbent mixture. It is of the utmost importance, of course, that methods for applying and harvesting peat be developed. Absorbent booms and absorbent plaquettes are examples.

Peat is now being used by various agencies in Canada (Information Canada, 1971) and other countries (Ekman and Sandelin, 1971) as an oil absorbent. In view of recent developments in this field of technology, the necessity of devising methods which will simplify the application and harvesting of peat becomes apparent; otherwise this product is unlikely to remain competitive. Apparatus for spreading loose peat moss is under development in Finland (Ekman, private communication). Figure 3 illustrates a model now in use in Finland which is being modified to improve its characteristics, which will be as follows:

Spreading capacity	4,000-5,000 litres/hour
Distance of spreading	5-6 metres
Weight	70 kilograms
Capacity of peat-holding tank	75 litres
Motor	4.5 horsepower
Fuel tank capacity	3 litres
Operation time with full tank of gasoline	1.5 hours

Peat is a potential filtering agent for oily water and waste waters in vegetable oil and margarine factories. This possibility is now being studied in Canada and Finland (Mueller, 1972; Silva, 1972; Ekman, private communication).

Peat, combined with a promoter, has been used as a wicking agent to burn oil spilled on water, ice, and snow. Results obtained in Finland (Ekman and Sandelin, 1971) have been good. Studies conducted in Canada (Coupal, 1972) in an open tank showed that peat, with a promoter such as diesel oil, could burn oil on water effectively down to a thickness of around 1/16 of an inch. For thinner oil slicks, the heat loss to the water column becomes more and more important relative to the heat released by combustion. The proportion of peat to promoter required for burning Bunker c is 2 litres of diesel oil for each 4 to 5 pounds of peat. These quantities will burn effectively 9 us gallons of Bunker c. For crude oil, 5 to 6 pounds of peat are required to achieve the same efficiency.

In a more recent study (Ruel, private communication), on the St Lawrence River near Rimouski, oil was burned on water, ice, and snow using peat moss as a promoter. The efficiency of the combustion varied from 79 to 89 per cent. This technique could become an important tool in dealing with oil spills where the physical removal of the oil is not possible and where the danger of propagation of fire to vegetation and human amenities is minimal.

Removal of Heavy Metals

Heavy metals in water can create serious environmental problems, and various methods to remove them have been proposed. One of these methods uses peat moss (Lalancette and Coupal, 1972; Coupal and Lalancette, 1971, 1972) to remove heavy metallic ions in solution, such as lead, zinc, copper, mercury, cadmium, and chromium. The process can be divided into three steps: (1) contacting of waste water with peat, (2) drying of the peat by mechanical pressure, (3) burning of the peat and recuperation of the metals. These steps will now be described briefly.

Step 1: Contacting Waste Water with Peat
In this part of the process, the solution is treated with a precipitating agent such as sodium sulphide and then contacted with peat. The precipitating agent can be added to the waste

Na$_2$S or pH Control

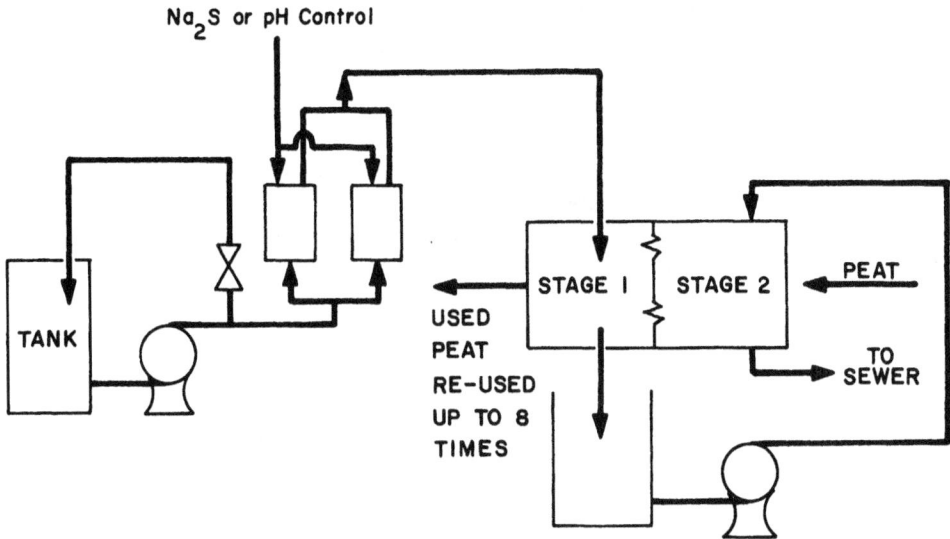

FIGURE 4   Process for the removal of heavy metals.

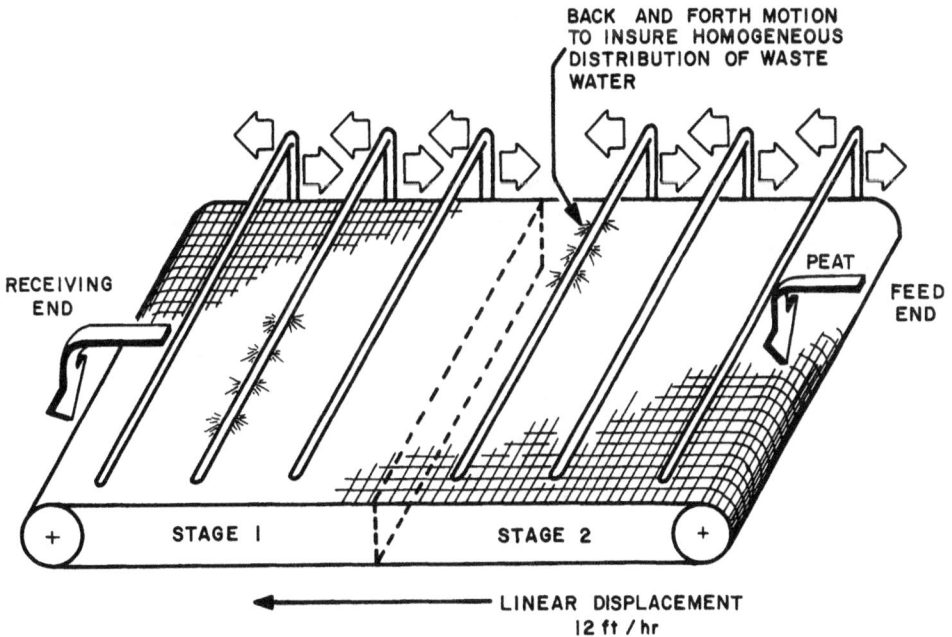

FIGURE 5   Contacting equipment.

waters before or during the actual contacting of the solution with the peat. A pilot plant, illustrated in Figure 4, has been built, and various parameters have been evaluated. The plant has a daily capacity of 17,000 US gallons and consists (see Figure 5) of a wire mesh (24 feet long by 8 feet wide) slowly moving at the rate of 12 feet per hour. A layer of peat approximately 3/4 inch thick at 90-95 per cent humidity is spread on the wire mesh, and the metal-contaminated water is then fed on to the bed and percolated through it.

After the first pass, the peat can be fed back to the distributor and reused up to eight times. Instead of using a precipitating agent such as sodium sulphide, the pH of the solution can be adjusted between 8 and 9. At this pH, metals will precipitate as hydroxides or oxides. It has been known for years that peat possesses a natural ion exchange capacity, which is due mainly to the presence of humic acid (Gamble, 1972). This property is, however, related to the pH of the solution and the nature of the metals to be removed, as illustrated in Table 2. It can be seen that the ion exchange capacity is strongly dependent on the pH and the metal involved.

TABLE 2
Natural capacity of peat (30% humidity) for ion exchange

Cu		Cd		Hg		Pb	
pH	me/100g	pH	me/100 g	pH	me/100 g	pH	me/100 g
2.0	0.11	2	7	2.0	6	2.0	16
3.0	0.21	3	18	3.0	14	3.0	36
4.0	0.23	4	19	4.2	17	4.0	38
4.75	0.24	5	19	6.0	20	5.0	40
				7.0	23		

Most hydroxides or sulphides of heavy metals form water-insoluble precipitates in water. The separation of these precipitates requires usually the addition of chelating agents followed by settling. The carboxylic and phenolic groups in peat act as a chelating agent for the complexing of the heavy metals as sulphides or hydroxides. Settling is not necessary in this process. The blanket of peat is used as a filtering and chelating agent.

The removal of cyanides is also an essential part of the process. In order to remove these toxic materials, the formation of a non-toxic ferrocyanide is required, which is achieved by adding ferrous chloride prior to the addition of the sodium sulphide. Through experiments, peat has been shown to be highly efficient in retaining ferrocyanides.

Table 3 illustrates results obtained in removing various heavy metals in a pilot plant. In every case, 1,200 US gallons of solution was treated at a rate of 12 US gallons per minute.

Step 2: Drying the Peat by Pressure
After use, the peat has the same amount of moisture as at the beginning. For easier handling (peat could be taken to a centralized incinerator) or for recuperation purposes, it is necessary to remove part of the mechanically held water from the peat. This is done by applying mechanical pressure to it. Depending on the applied pressure, the moisture content can be decreased down to 65-70 per cent. It is not economical or feasible to drive out more water by applying higher pressure, and in any case, for handling purposes, a peat with a moisture content of 80-85 per cent is not dripping. Table 4 gives the relation between the residual moisture content and the applied pressure.

TABLE 3
Removal of various metals with peat

No.	Metal	Initial concentration (ppm)	Process	Concentration after 1st stage (ppm)	Concentration after 2nd stage (ppm)
1	Hg	15	Natural	1	1
2	Cu	100	Sulphide	1	1
3	Zn	100	Sulphide	2	1
4	Hg	15	Natural	1	1
5	Fe	50	Sulphide	1	1
6	Hg	25	Sulphide	2	2
	Pb	25	''	0.5	0.5
	Cd	25	''	0.3	0.3
	Zn	25	''	2	2
7	Cu	100	Sulphide	2	2
8	Hg	55	Sulphide	1	1
9	Cr	7,000	Sulphide and	1	1
	Cu	18	ferric chloride		
10	Cu	60	Sulphide	1	1
11	Cu	50	Sulphide	1	1
12	Cu	200	Sulphide	1	1
13	Hg	100	Sulphide	1	1

Step 3: Burning of the Peat and Recuperation of the Metals

The technology related to the combustion of peat is well known and will not be discussed here. Since the ash content of peat is very small, the combustion will result in the formation and easy recuperation of metal oxides.

This ·brief description of the process illustrates the potential capacity of peat in the treatment of water containing heavy metals. The economics of the process are still being worked out. As an example, the treatment of 17,000 US gallons per day of a waste water containing 100 ppm of copper would cost $12.50 in peat and chemicals. The cost of the equipment is difficult to estimate at the present time since the design still has to be improved and discussed with potential users.

Conclusion

If the capacity of peat to filter and absorb various types of organic and inorganic matter is developed to economic feasibility, a wide potential for peat will be created. It appears that because a low flow rate per unit area must be used in contacting the waste waters with peat the most promising field of application will be in treating relatively small amounts of waste waters. The adsorption capacity of peat varies with the type of material being adsorbed but because of its relatively low cost compared to other adsorbents, it is believed that it will be able to compete successfully for particular applications.

OBTAINING ACTIVATED CARBON FROM PEAT MOSS

In a general way, active carbons refer to a variety of substances obtained from carbonaceous materials whose common characteristic is an exceptionally developed micropore struc-

TABLE 4
Humidity of peat versus applied pressure

Pressure (psi)	Humidity (%, wet basis)
0	87
2	87
5	87
10	86
20	85
50	80
100	79
125	74
150	72
175	71
200	70
500	70
1,000	67
2,000	65
3,000	61
10,000	60
20,000	62
34,000	54

ture and a large surface area. This feature makes possible their use as adsorbents (Smisek and Cerny, 1970; Hassla, 1963).

Two different methods of preparation are known. In process A the raw material is carbonized (in the absence of air) at temperatures ranging from 500°c to 800°c. This produces a charcoal which, after activation with steam or carbon dioxide at temperatures between 700°c and 1,000°c, yields the final activated carbon. This method is commonly referred to as physical activation. In process B, dehydrating agents are added to the raw materials, the mixture is heated in an inert atmosphere at 600-800°c, and the cooled solid is then leached with water and/or acids and dried. The chemicals generally used are zinc chloride, phosphoric acid, sulphuric acid, or hydrochloric acid to a lesser extent. The product of process B is sometimes further activated with steam. This method is usually referred to as chemical activation.

Raw materials for the processes mentioned above must contain a carbon-based matrix. Typically they consist of wood, coal, peat, fruit stones, and nutshells.

Canada's vast peat lands (Simard, 1972) are generally of easy access and can be harvested readily. These characteristics should encourage the production of activated charcoal in locations which are in close proximity to peat bogs, particularly when the development of the over-all active carbon industry remains, to a high degree, limited by the scarcity of other raw materials (sawdust, fruit stones, and nutshells).

Early work (Reilly and O'Donoghue, 1941) confirmed the possibility of manufacturing active carbon from peat, and recent investigations on the use of peat as starting material (Ekman, 1972; Jozef, 1972) have yielded positive results. A pilot plant for steam activation has been developed in Finland (Ekman, 1972). In Canada, some early attempts were made by Kossman (1953), but no definite conclusion on the feasibility has been communicated.

The studies reported here are aimed at describing the types of active carbon that can be produced from peat, and the method and conditions of carbonization and activation which are applicable in the laboratory.

The importance and uses of active carbon are well known in the chemical industry (Smisek and Cerny, 1970; Hassla, 1963), where the adsorption properties of the material are employed extensively. Because of the current concern over environmental factors, and the likelihood that more strict regulations on pollutant discharges or emissions will be implemented, the use of active carbon will undoubtedly expand in the near future. As the demand for active carbon increases all around the world it is more than probable that the use of peat moss as raw material will become a necessity.

Characteristics of Peat as Raw Material

Studies have been carried out at the University of Sherbrooke using sphagnum-type peat moss, which is the most abundant type in the Province of Quebec. Earlier studies (Risi et al., 1953) have reported the following average composition for this type of peat:

Fixed carbon	28.6%
Volatile matter (not including nitrogen)	67.6%
Ashes	2.8%
Nitrogen	1.0%

The moisture content varies with time from about 90 per cent for in situ material to about 30-50 per cent for the air-dried samples used as starting material for the carbonization studies.

Experimental Procedures

The methods of preparation vary according to the nature of the process. In chemical activation, known amounts of salts are added to the raw material (air-dried). The mixture is then activated in a muffler furnace with steam at temperatures ranging from 300°c to 800°c. The variables studied are: the amount and nature of the added salts, the activation temperature, and the activation time.

In physical activation, the air-dried samples are carbonized in an inert atmosphere at 300°c to 700°c and then activated with carbon dioxide (flue gas) at 700°c to 1,000°c. Although steam could be used, in our experiments so far we have used only carbon dioxide as activating agent. The activation gases were in all cases preheated at the entrance to the furnace. Variables studied include the particle size of the raw material, the carbonization temperature, the activation temperature, and the activation time.

Characterization of the active carbon produced was done by the BET adsorption method, which results in the determination of the BET specific surface area (Brunauer et al., 1938). A carbon of high surface area will have a BET area larger than 1,000 m²/g, and one of medium surface area will have a BET area between 200 and 400 m²/g. Weight losses were also studied, since they can facilitate the understanding of the complex kinetics taking place during activation.

The adsorptive capacity of a typical active carbon was determined for phenol, iodine, indol, detergents, and phenozine using established standard techniques (Gomella, 1970).

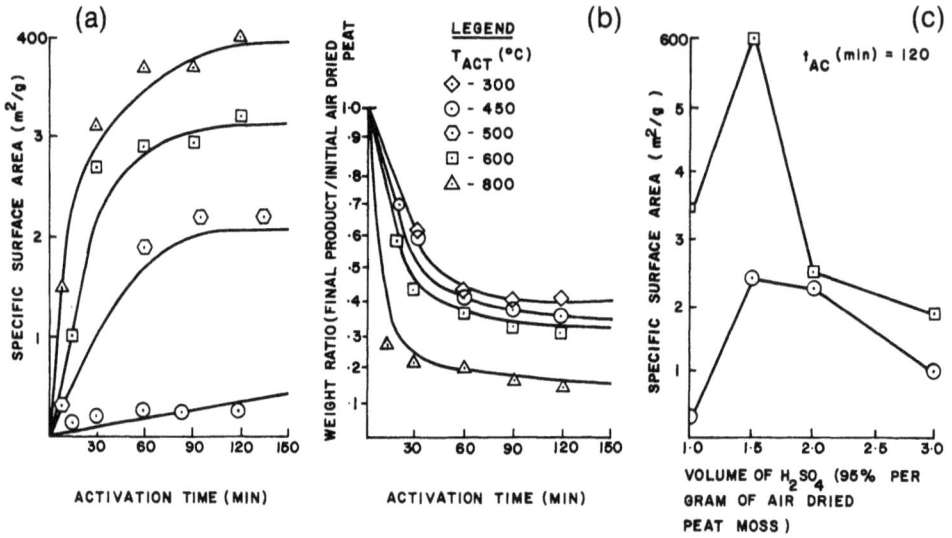

FIGURE 6
(a) Specific surface area of activated carbon after steam activation of a mixture of H_2SO_4 and peat moss (1 cc H_2SO_4 (95%)/g air-dried peat moss).
(b) Weight losses as a function of activation time and temperature for the steam activation of a mixture of H_2SO_4 and peat moss (1 cc H_2SO_4(95%)/g air-dried peat moss).
(c) Specific surface area of activated carbon after steam activation of a mixture of H_2SO_4 and peat moss as a function of the amount of H_2SO_4 added (expressed as cc H_2SO_4(95%)/g air-dried peat moss).

Details of the different experimental arrangements can be found elsewhere (Rannou, 1972; Son, 1973).

It is important to realize that the present method of active carbon characterization defines the extent of the activation in terms of weight losses as well as of surface area obtained. The latter is a more precise parameter than the adsorption indexes, but, since it is a geometric parameter, in order to give a complementary view of the selectivity of the active product, the standard indexes mentioned above have also been studied.

Results and Discussion, Chemical Activation

Activation of a Mixture of Sulphuric Acid and Peat
Addition of sulphuric acid (95% concentrated) at various weight ratios was followed by drying in an oven at 110°c for 48 hours. A black charcoal was obtained, which was subsequently activated with steam at flow rates of about 100 cc/min at 15 psig. These conditions were also employed for the mixture of peat with zinc chloride and phosphoric acid. The initial air-dried peat was a fine material of particles smaller than 100 mesh, obtained through sieving of the original material.

The specific surface areas as functions of activation time and temperature are shown in Figure 6a. It is clear that activation times of about 1.5 to 2 hours are necessary to produce the largest surface areas. No appreciable activation takes place at temperatures lower than 300°c.

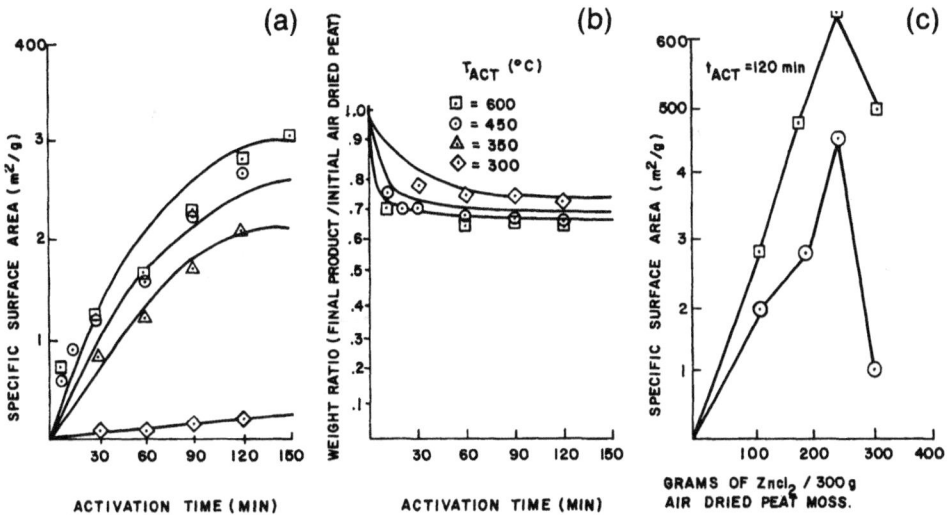

FIGURE 7
(a) Specific surface area of activated carbon after steam activation of a mixture of $ZnCl_2$ and peat moss (120 g $ZnCl_2$ in 400 cc water/300 g air-dried peat moss).
(b) Weight losses as functions of activation time and temperature for the steam activation of a mixture of $ZnCl_2$ and peat moss (120 g $ZnCl_2$ in 400 cc water/300 g air-dried peat moss).
(c) Specific surface area of activated carbon after steam activation of mixtures of $ZnCl_2$ and peat moss as a function of the amount of $ZnCl_2$ added (expressed as g $ZnCl_2$ in 400 cc water/300 g air-dried peat moss).

Weight losses based on 1 g of air-dried peat moss are shown in Figure 6b. It is significant that even at 300°c, when no activation occurs, the weight losses are considerable. This is due to evaporation of waxes and moisture still present in the initial mixture. The activation itself at temperatures between 450°c and 600°c results in small weight losses, but at 800°c the losses are high. Under the conditions shown, we can expect an approximate efficiency of about 15 to 30 per cent for the whole process (active carbon produced from initial air-dried peat).

The effect upon the final product of the amount of sulphuric acid used is shown in Figure 5c. Although the data are somewhat scattered, the best active carbon will be obtained from a mixture of 1.5 cc sulphuric acid (95%) and 1 g of air-dried peat moss. The activation time (120 min) for these experiments has been chosen to correspond to a time within the steady state plateau of the specific surface area. (This was also done for zinc chloride and phosphoric acid.)

Activation of Mixtures of Zinc Chloride and Peat
The samples were prepared by mixing solutions of zinc chloride in 400 cc of water (at different concentrations) with 300 g of air-dried peat moss. The mixtures were subsequently dried in a furnace at 110°c and activated with steam.

Figure 7a shows the specific surface area as a function of activation time and temperature. No investigations were carried out at temperatures beyond 600°c, although it is expected that both the specific surface area and the weight loss would increase as in the case of sulphuric acid.

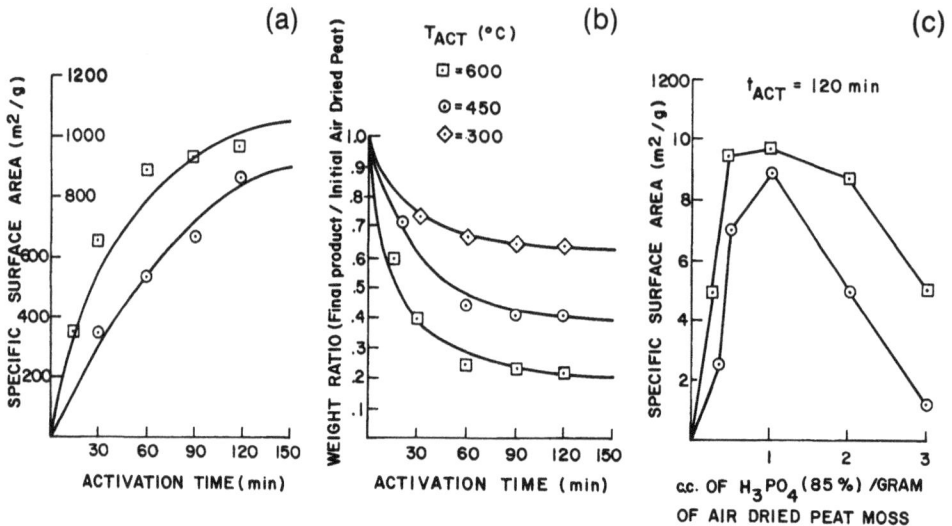

FIGURE 8

(a) Specific surface area of activated carbon after steam activation of a mixture of H_3PO_4 and peat moss (1 cc H_3PO_4(85%)/g air-dried peat moss).

(b) Weight losses as function of activation time and temperature for the steam activation of a mixture of H_3PO_4 and peat moss (1 cc H_3PO_4(85%)/g air-dried peat moss).

(c) Specific surface area of activated carbon after steam activation of mixtures of H_3PO_4 and peat moss as a function of the amount of H_3PO_4 added (expressed as cc H_3PO_4(85%)/g air-dried peat moss).

Weight losses were considerably smaller than for sulphuric acid, as illustrated in Figure 7b. This suggests that the carbon atoms present in the hydrocarbon waxes remain in the charcoal obtained with the addition of zinc chloride.

The effect of the weight ratio of zinc chloride to peat moss at two different temperatures is shown in Figure 7c. The behaviour of both curves clearly suggests that the optimum mixture corresponds to adding 250 g zinc chloride for each 300 g of air-dried peat moss.

Activation of Mixtures of Phosphoric Acid and Peat Moss

Phosphoric acid at 85 per cent concentration was used. The mixtures were dried at 110°c for 48 hours and further activated with steam.

Results shown in Figure 8a reveal that phosphoric acid is most effective in the production of active carbon. Surface areas of the order of 1000 m²/g can be easily obtained with steam activation at moderate temperatures.

Weight losses depicted in Figure 8b are relatively high, as compared with zinc chloride, but lower than those obtained using sulphuric acid. These results indicate that from 1 ton of air-dried peat moss 0.25 ton of high-quality active carbon will be obtained.

The effect of the amount of phosphoric acid added is shown in Figure 8c, where the most favourable results are seen to be obtained with addition of 0.5-1.0 cc phosphoric acid (85%) for each gram of air-dried peat moss.

Conclusions on Chemical Activation

A few points of interest have been brought out by this study:

1. No activation occurs at temperatures below 300°c.
2. Phosphoric acid seems most suitable for activation of peat moss with steam at temperatures below 600°c. The process with zinc chloride is also interesting, particularly in view of the excellent efficiency achieved in the yield of active carbon.

In each of the three processes considered there is an optimum ratio of the weight of dehydrating salt to the weight of air-dried peat moss.

Physical Activation with Carbon Dioxide

The activation process is due to the reaction

$$C(s) + CO_2(g) \longrightarrow 2CO_2(g)$$

which is highly endothermic (Walker et al., 1957). One must realize that heat must be supplied to the reactor in order to maintain a constant temperature.

Peat Carbonization
This step is carried out in the absence of air or oxidizing gases. Nitrogen, helium, or argon is used. The quality and properties of the charcoal obtained are similar for the three gases.

The carbonized product has an appreciable surface area varying from 120 to 180 m^2/g depending upon the carbonization temperature. This surface area reflects the creation of the macropore structure and, to some extent, the transition pores, as a consequence of the destruction of waxes and tars accomplished by the pyrolysis.

The effect of the carbonization temperature, T_c, on the specific surface area of the active carbon is shown in Figure 9a, where an activation temperature, T_{act}, of 700°c has been chosen. An activation time, t_{act}, of 30 min was selected. This time lies well in the range of the steady state for the formation of a given surface area. Figure 9b shows the same effect at different activation temperatures. The pattern is quite regular, and to a certain degree reproducible. The decrease of the specific surface area shows a shift towards higher T_c with increasing activation temperatures.

Since the highest surface areas are readily obtained with carbonization between 400°c and 600°c, this range seems to be the most convenient for the operation.

Weight losses during the processes indicate an efficiency of about 20-25 per cent (on the basis of active carbon produced per unit weight of initially air-dried peat moss) for the conditions shown.

Effect of Carbon Dioxide Flow Rate upon Activation
The reactor employed was a one-inch-diameter Vycor tube. Gas flow rates were varied to see whether any effect could be detected either in the final surface area or in the kinetics of the process. The results are summarized in Figure 10. Air-dried peat moss of mesh size 60/65 was chosen for these studies. The carbonization temperature was 540°c and the activation was carried out at 740°c. No significant differences in the kinetics of the process, related to the changes in surface area with time, were found. There seems to be an effect on

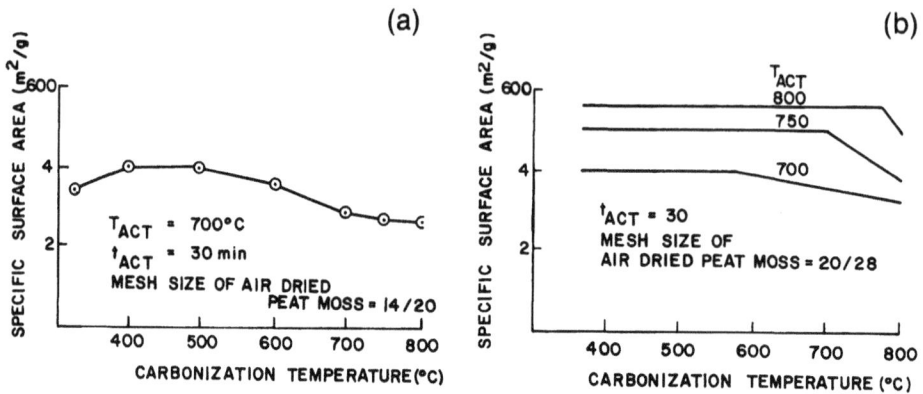

FIGURE 9
(a) Effect of the carbonization temperature on the specific surface area of the active carbon obtained.
(b) Comparative effect of the carbonization temperature on the specific surface area of the active carbon at three different activation temperatures.

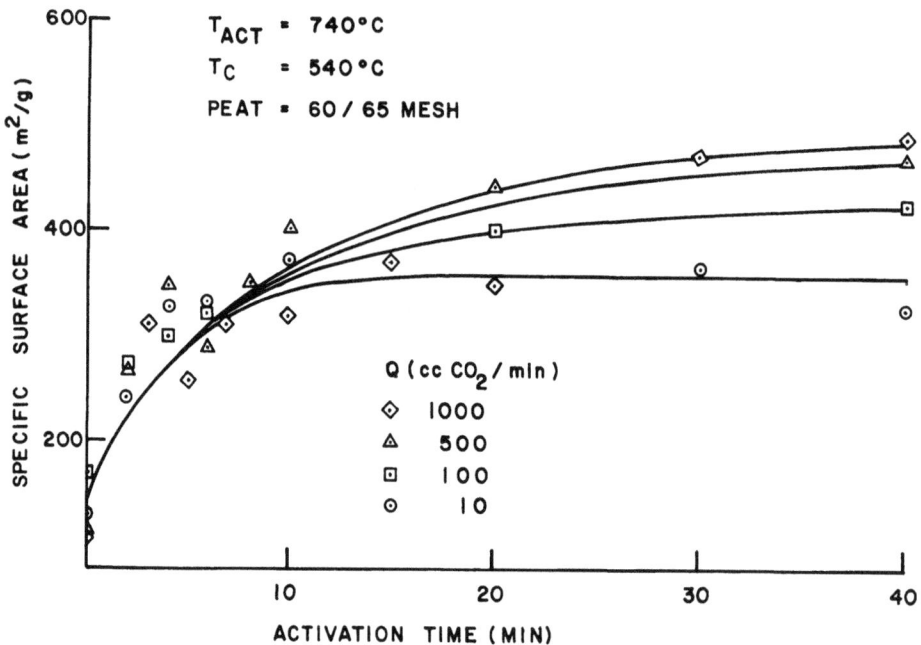

FIGURE 10   Effect of the CO_2 flow rate on the specific surface area, as a function of activation time.

the total surface area, despite the scattering of the data. A simple interpretation of this effect involves the presence of carbon monoxide as product of the reaction. Higher carbon dioxide flow rates will tend to decrease the amount of carbon monoxide present in the pores, which is in equilibrium with the carbon monoxide adsorbed. This effect will facilitate the continuation of the activation reaction.

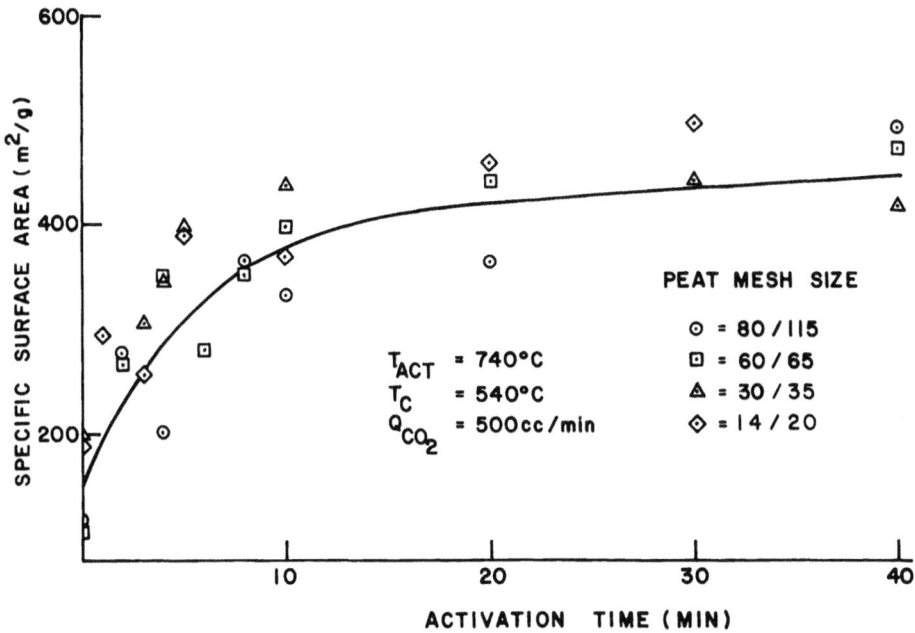

FIGURE 11   Effect of the grain size of the air-dried peat moss on the specific surface area of the active carbon.

Effect of the Grain Size on the Activation

The grain size of the initial air-dried peat moss used for the carbonization was varied from 14/20 to 80/150 mesh. The specific surface area of the activated product as a function of activation time is shown in Figure 11. There is no significant effect on either the kinetics or the total surface area. These results are consistent with those of Ergun (1956), who found no grain size effect on the kinetics of the carbon-carbon dioxide reaction.

These results suggest that the whole process of activation is controlled by pore diffusion.

Effect of the Activation Temperature on the Activation

The effect of the activation temperature on the active carbon produced is shown in Figure 12. The activation begins at 600°C. The specific surface area is a linear function of the activation temperature for these conditions. Weight losses follow increases in the specific surface area. For carbons of low surface area (200 m²/g) the efficiency of the process is about 25 per cent. For higher surface areas (600 m²/g) the weight of the active carbon produced is about 20 per cent of the original weight (based on the air-dried peat used for carbonization).

Indexes for Phenol, Iodine, Indole, Phenazine, and Detergents

These indexes were studied in order to classify the product according to established standard techniques. The exact meaning of the different indexes can be found in the literature (Gomella, 1970), but the definitions are given in the Appendix for easy reference.

The iodine number was studied as a function of activation temperature. The results are shown in Table 5. These results suggest that the active carbon has negligible activity for

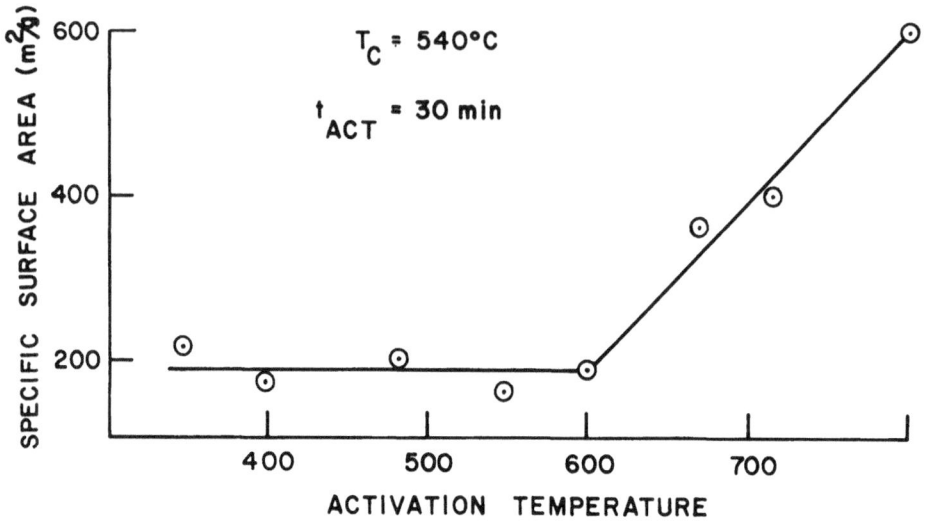

FIGURE 12   Effect of the activation temperature on the specific surface area of the active carbon.

TABLE 5
Iodine index and iodine number as a function of activation temperature

T_{act} (°C)	Iodine number	Iodine index
300	57.7	1
400	55.2	1
500	45.0	0
600	36.8	0
640	60.2	1
700	57.7	1
740	82.5	4
800	92.0	5

iodine when activated at temperatures below 700°c. There is an abrupt change beyond that temperature, and at 740°c and above the activity is appreciable.

The other indexes were also studied for an activation temperature of 740°c. Under these conditions, the classification FINAD is 44032, which indicates the following:

F.   About 0.040 g of active carbon is necessary to decrease the phenol content of 1 litre of water solution from 100 ppb to 10 ppb.
I.   About 90 g of iodine is adsorbed by 100 g of active carbon.
N.   About 0.013 g of active carbon is necessary to decrease the indole content of a 1-litre water solution from 600 ppb to 100 ppb.
A.   About 15 g of phenoxine is adsorbed by 100 g of active carbon.
D.   About 0.080 g of active carbon is necessary to decrease the sodium lauryl sulphonate content of a 1-litre water solution from 250 ppb to 25 ppb.

The active carbon prepared under the specified conditions seems to have a substantial activity for phenol, iodine, and phenozine. It is somewhat poor for detergents and has little attraction for indole.

Conclusion on Physical Activation

Physical activation seems to be plausible even at the low temperatures used for carbonization and activation. The active carbon obtained at activation temperatures higher than 700°c is most suitable in view of its large surface area and ability to adsorb phenol, iodine, and phenozine.

The carbon dioxide flow rate and the grain size of the peat moss are not critical and have little effect on the kinetics as well as on the specific surface area.

The carbonization temperature has a marked effect on the properties of the final product. From the present study we know that carbonization at 400-500°c is quite sufficient to produce a good-quality charcoal.

It should be expected that activation temperatures higher than 800°c will yield larger surface areas, but this benefit may be offset by the weight losses.

General Conclusions

From studies now available, it is indicated that active carbon can be produced from peat by either the chemical or the physical process. In both cases, a reasonable high-quality product can be obtained, and it can be varied depending on the method of preparation.

Activation times are shorter for physical activation with carbon dioxide than for any of the chemical methods. In both cases, because of the endothermic nature of the processes, heat must be supplied to the reactor.

A solid-liquid extraction (in one or various stages) should facilitate the removal of waxes and tars present in peat. This will considerably reduce the energy (heat) necessary for carbonization.

Further work is necessary in the following areas:

1. Defining a relationship between the weight loss and the specific surface area.
2. The production of metal-doped active carbon for specific applications.
3. The design and operation of a fluidized reactor for continuous activation. This area is now under study at the University of Sherbrooke.

The formation of coke and activated charcoal presents one of the greatest potential new uses for peat. This new application is technically feasible, but market studies must still be done.

PEAT MOSS AS A CONSTRUCTION MATERIAL

The current development of construction methods for housing has created a growing need for materials which satisfy the following requirements:

1. Workability and ease of installation: i.e. in sawing, nailing, glueing, and fastening or fixing.
2. Good structural properties: high tensile strength, impact resistance, compressive strength, and flexural resistance.
3. Superior insulation characteristics: thermal insulation, and desirable acoustic absorbance and transmittance performance.
4. Durability: fireproof, waterproof, and moisture resistant; corrosion and erosion resistant; and atmospherically stable.

TABLE 6
Properties of USSR Isolation Plate

Moisture content	< 15%
Water-absorbing capacity (24-hour period)	< 180%
Coefficient of heat conductivity	0.05 cal/m
Weight/volume ratio	170-220 kg/m^3
Flexural strength	> 3 kg/cm^2

5. Appearance, in the case of materials exposed to view: attractive colours and hues, and interesting textures and finishes.

There are many construction materials in present use made of wood, or wood by-products and wastes. Accordingly, peat moss should find a natural niche in the development of new construction materials.

Advantages of Peat Moss as a Source of Construction Materials

Peat moss offers a number of advantages for the production of construction materials. It is a raw material of very low cost and of comparatively constant quality, it is easily harvested and accessible, and it is available in great quantities (Simard, 1972). These advantages can only be approached by industrial wood waste materials such as bark, sawdust, and wood chips produced from scrap, which are all available in limited quantities.

In view of the low value of such waste products, transportation over even moderate distances cannot be considered and the construction of central plants for the processing of wood wastes from several sources at different locations is not practicable. Furthermore, since sawmills are generally located close to forestry operations, and since these are widely scattered and far from centres of population where markets for finished construction materials exist, such materials must of necessity be heavily penalized by transportation costs. Peat moss has the advantage that deposits of sufficient quantity occur at one location to allow the installation of large-scale manufacturing facilities at the site with little or no transportation of raw materials.

In their natural state, peat moss particles have low interparticulate cohesion, but by simple industrial processing (including drying and light pressing) it is possible to produce a material with surprising interparticulate cohesion. Some Russian plants are now using a process whereby raw peat is simply pressed and cooked to form peat slabs for the insulation of domestic housing and refrigerators (a major portion of the more recent housing is insulated with such peat slabs) (Ruel, 1972; Sukhanov, 1972). However, this material is of rather poor quality, crumbling and breaking too easily (its properties are given in Table 6). The marketing of such a product in North America would be difficult at best. To improve the interparticulate cohesion of peat moss and thus obtain a material of appropriate strength, a binder must be used.

Choice of Binder

Naturally occurring peat moss is saturated with water and may typically contain 90 per cent water. Commercially available peat moss, such as that harvested by the vacuum process, contains about 40 to 50 per cent water. Accordingly, two different approaches may be taken to select a binder. The first is to select one that will tolerate the water, the most appropriate binder by far being Portland cement. The second approach is to remove the water and to use organic resins that can be polymerized in situ.

At the University of Sherbrooke, trials carried out under Project PUDDING (Potential Uses for Discarded or Detrimental Industrial and Natural Garbage) have included experiments with both types of binder. Research in this field of endeavour was previously carried out by the Institute for Industrial Research and Standards of Dublin in Ireland (1967). From both of these preliminary research projects have emerged two new materials of construction that are currently undergoing performance evaluation of their mechanical properties and thermal and acoustic insulation characteristics. The first is *Peatcrete* (Olivier, 1971; Aitcin, 1972), which uses Portland cement as binder, and the second is *Peatwood* (Aitcin et al., 1972), made with a phenolic binder.

Choice of Peat Moss Characteristics

Several of the characteristics of the peat moss can have an appreciable effect on the construction material made therefrom. The key characteristics are considered to be the following: the type of peat moss (degree of humification, or fibrous decomposition), the particle size distribution, and the moisture content.

Effect of Binder on Processing Conditions

The fabrication process is dependent upon the type of binder used, and the following binder characteristics affect the choice of process conditions: the binding mechanism, or mode of reaction of the binder; the optimum process temperature; the optimum process pressure; and the duration of the pressing cycle.

Fabrication Process Variables

There are two general fabrication processes: the batch process and the continuous process. In preliminary investigations, only the batch process has been tried. Experimental production has been limited to plaques formed sequentially in 1 x 1 foot moulds, in a hydraulic press.

Peatcrete (Olivier, 1971; Aitcin, 1972)

The idea of using ligneous (wood-based) aggregates for making light-weight concrete is not new. There is literature dating back to 1924 that bears upon this subject. The early experiments were followed by numerous research projects throughout the world (Parker, 1947), the goal being to develop a light fibre-reinforced concrete with excellent acoustic and thermal insulating properties. In fact, light-weight mineral aggregates that can be used for the production of such concretes are rather scarce, so that frequently they must be synthesized by industrial processes to meet the demand, which is strong enough to warrant the production costs involved.

In the USA, Germany, England, France, and numerous other countries, several researchers have succeeded in producing concretes based upon sawdust or wood fibre. These projects focused on the pretreatment required by the sawdust or for fibrizing the wood, on the optimum formulation, and on the conditions of erection and curing. The great number of patents awarded testifies to the extent of research that has been carried out (Passer, 1963; Levy, 1955).

Significant Peat Moss Characteristics

The type of peat moss preferred for making construction materials is sphagnum moss of commercial quality harvested by the vacuum process. From this raw material, all oversize roots or pieces of branches, or other large extraneous matter, are scalped out on a 4-mesh screen. The screened moss is then mixed with Portland cement and water and pressed hydraulically (Hydropress). Trials have also been carried out using regulated set cement with an initial setting time of 10 to 15 minutes. Factorial design has been resorted to for the determination of optimum process parameters to yield the preferred combination of peat-crete characteristics. Evaluation of the acoustic and thermal performance of various peat-crete formulations is now underway.

Advantages and Disadvantages of Peatcrete

The principal advantages of peatcrete construction materials include:

1. They are fabricated from undried peat moss.
2. They use a low-cost binder (Portland cement costs about $0.01 per pound).
3. They are fireproof.
4. They are easily sawed and nailed.
5. They may be cast and moulded to any shape.
6. They can be produced by the continuous process.
7. Their density is low, 0.7 to 1.2 Sp Gr (45-70 lb/ft^3).

On the other hand, peatcrete is associated with the following drawbacks:

1. The moss must be screened to remove oversize detritus.
2. The mechanical strength is relatively low (of the same order as that of competitive materials). The higher the concentration of cement, the stronger peatcrete becomes, but the denser it becomes, the more it loses its insulating qualities.
3. The hardening is slow (4 to 5 hours with normal cement, and 30 to 60 minutes with high early strength cement, but the latter is twice as expensive).
4. The water resistance is moderate but could be improved by adding various chemicals (Sukhanov, 1972), which would make it more expensive.
5. The selling price for peatcrete must remain quite low to compete with other construction materials.

Until the acoustic and thermal insulation properties of peatcrete have been fully evaluated, it will be difficult to predict its commercial future. Experiments now underway are expected to yield data on its insulation properties and thereby settle the question of its commercial viability.

Peatwood (Aitcin et al., 1972)

As early as October 1969, A. Marsan of the Department of Chemical Engineering at the University of Sherbrooke experimented with the pressing of peat moss with and without the addition of synthetic resins such as methyl methacrylate, urea formaldehyde, and dextro-urea-formaldehyde. In May 1972, a research group including P.C. Aitcin, J.L. Bougamond, J.A. St. Laurent, and M. Cossette developed a process for pressing peat-moss with phenolic resins to yield a product dubbed *Peatwood*.

Choice of Binding System
The phenolic resin finally chosen was selected primarily on the basis of cost effectiveness. Liquid phenolic resins, prepared on site from elementary components, ranging in price around $0.11 per pound were considered but rejected in view of the anticipated difficulty of producing homogeneous resin-moss dispersions. Powdered resin in the price range of $0.25 to $0.30 per pound was finally chosen. It was found that such resin could be effective even at low concentrations because of the ease with which it could be blended uniformly with the moss. In addition, this resin is practically colourless and so does not affect the natural appearance of the finished peatwood. A low polymerization temperature was considered to be an additional asset for the processing of a material with very poor heat conductivity.

Peat Moss Processing
The peat moss used is first sifted to remove oversize material on a 4-mesh screen, and to remove undersize dust on a 50-mesh screen. Oversize material can be used as fuel for the process, or can be used for making coarse-textured *Peatcork* that matches natural cork closely with regard to appearance and performance. Undersize material can be used for the moulding of objects with thin sections.

The screened peat moss is then dried and blended with the powdered resin. The mixture is then ready for pressing in a heated mould.

A factorial design has been carried out to determine optimum pressure, pressing time, mould temperature, and resin formulation. The mechanical and insulating properties of various formulations of peatwood are now being evaluated.

Advantages and Disadvantages of Peatwood
The principal advantages of peatwood, as a construction material, include:

1. It has a very attractive texture, which may be different on each face of a panel owing to the segregation of the resin during the mould filling operation.
2. It has good strength.
3. It may be sawed, nailed, screwed, and glued.
4. It hardens quickly, and may be moulded at high production rates.
5. It is light in weight (40-60 lb/ft³).
6. It may be produced in a continuous process in any shape or form.

On the other hand, peatwood has the following drawbacks:

1. It requires screened and dried peat moss (unused fractions may be used for other purposes).
2. It is sensitive to moisture; droplets will form water marks during pressing.
3. The polymerization step requires heating of the entire mass which is being formed.
4. The resin system used is moderately expensive at $0.25 to $0.30 per pound, although the concentrations needed are low.

It should, however, be noted that in some trials we have succeeded in producing peatwood without the addition of synthetic resins, making use of only the naturally occurring resins and waxes in the peat moss for a binding system. Densities achieved for such peatwood were higher. A more extensive economic study and a more complete evaluation of the resulting properties will be necessary to determine if this approach is worth pursuing, and to determine the most economical process-formulation combination to achieve a desired performance level.

Research Outlook

Two new construction materials based upon peat moss are now under study. One is *Peatfoam*, an ultra-light material based on peat moss and a foamed resin system, comparable to styrofoam. The other is *Peatcork*, mentioned above. This is a synthetic cork based upon coarse peat moss fractions.

GENERAL CONCLUSION

As mentioned in the introduction, various research projects have identified promising products that might be manufactured from peat and mentioned new uses to which it might be put. Many facets of these new products or uses remain to be assessed. An important factor needed for any research and development program is the interest and participation of the industry, which we hope will keep increasing.

In this chapter we have identified primarily the new uses for peat that seem most promising on the North American continent and which have been developed recently. We are sure that peat has other potential uses.

APPENDIX: DEFINITION OF TERMS

indexes   In all cases the active carbon was crushed, sieved through 70-mesh screen (0.088 mm), and dried at 105°c for 24 hours.

phenol   The active carbon (in grams) necessary to reduce the concentration of phenol present in 1 litre of water solution from 100 ppb to 10 ppb.

iodine   The iodine (in grams), present in a N/5 solution, adsorbed by 100 g of active carbon.

indole (2,3-benzopyrol) The active carbon (in grams) necessary to reduce the concentration of indole present in 1 litre of water solution from 600 ppb to 100 ppb.

phenazine (dibenzopyrazine)   The phenazine (in grams), present in a 4 per cent water solution, adsorbed by 100 g of active carbon.

detergent   The active carbon (in grams) required to reduce the concentration of sodium lauryl sulphonate present in 1 litre of water solution from 250 ppb to 25 ppb.

REFERENCES

Aitcin, P.C. 1972. Peatcrete, a new construction material. Proc. Peat Moss in Canada Conf., Univ. Sherbrooke, pp. 304-313.

Aitcin, P.C., Bougamont, J.L. Cossette, J., and St-Laurent, J.A. 1972. What is Peatwood? Peatwood research project protocol, Univ. Sherbrooke.

Brunauer, S., Emmett, P.H., and Teller, E. 1938. J. Am. Chem. Soc. 60: 309.

Coupal, B. 1972. Use of peat moss in controlled combustion technique. Environmental Emergency Branch, Environmental Protection Service, Environment Canada, Rep. EPS 4-EE072-1.

Coupal, B., and Lalancette, J.-M. 1971. Removal and recovery of metal from polluted waters. US patent application, 23/12/71.

— 1972a. Removal and recovery of chromium from polluted waters. US patent application, 15/7/72.

— 1972b. The use of peat moss for the removal of heavy metals in the form of hydroxyde. US patent application, 15/9/72.

— 1972c. Treatment of cyanide polluted water. US patent application, 11/1/72.

D'Hennezel, F., and Coupal, B. 1972. Peat moss: a natural absorbent for oil spills. Can. Mining and Met. Bull. (Jan.).

Drucker, P.F. 1973. Insight and Innovation for the Management of Change, 3 (May/June).

Dufort, J., and Ruel, M. 1972a. Peat moss as an adsorbing agent for the removal of coloring matter. Proc. 4th Internat. Peat Congr., Otaniemi, Finland, IV: 299-310.

— 1972b. La tourbe: adsorbant de matières colorantes. Proc. Peat Moss in Canada Conf. Univ. Sherbrooke, pp. 274-89.

Ekman, E. 1972. Research in Finland on the manufacture of active carbon from peat. Proc. Peat Moss in Canada Conf., Univ. Sherbrooke, pp. 243-65.

Ekman, E., and Sandelin, R. 1971. The use of peat in combatting oil pollution. Bull. Internat. Peat Soc. 2: 19-23.

— 1973. Private communication, April 18. State National Research Centre, Finland.

Ergun, S. 1956. J. Phys. Chem. 60: 480.

Farnham, R.S., and Brown, J.L. 1972. Advanced waste water treatment using organic and inorganic materials, Part I and II. Proc. 4th Internat. Peat Congr., Otaniemi, Finland, IV: 271-98.

Fuller, H.I. 1971. The use of floating absorbents and gelling techniques for combatting oil spills on water. J. Inst. Petroleum, 57 (553): 35-43.

Gamble, D.S. 1972. Peat humic materials: a review of the chemistry. Proc. Peat Moss in Canada Conf., Univ. Sherbrooke, pp. 315-24.

Gomella, C. 1970. Techniques et sciences municipales: l'eau.

Hassla, J.W. 1963. Activated Carbon (Chem. Pub. Co., New York).

Information Canada, Ottawa. 1971. Report of the Task-Force — Operation Oil, vol. I, II, III and IV, Cat. No. T 22-2470/1,/2,/3, and 4.

Institute for Industrial Research and Standards, Dublin, Ireland. 1967. Peat Concrete Research Project. Interim Rept., March.

Jozef, F. 1972. Investigations on peat utilization for the production of activated carbon. Proc. 4th Internat. Peat Congr. pp. 185-96.

Kossman, K.H. 1953. Shawinigan Chemicals Co., Interim Rept. 3, File 22,611.

Lalancette, J.-M., and Coupal, B. 1972. Recovery of mercury from polluted water through peat treatment. Proc. 4th Internat. Peat Congr., Otaniemi, Finland, IV: 213-18.

Levy, J. 1955. Les bétons légers (Editors Eyrolles, Paris).

MacFarlane, I.C. 1969. Engineering characteristics of peat. In Muskeg Engineering Handbook (Univ. Toronto Press), pp. 78-126.

Mueller, J.C. 1972. Peat in pollution abatement. Peat Moss in Canada Conf., Univ. Sherbrooke, pp. 274-96.

Nguyen, T.C. 1973. La séparation des solides en suspension des eaux usées des usines de pâtes et papiers dans un décanteur à tubes inclinés et par une colonne de filtration utilisant la tourbe comme milieu filtrant. (Master's thesis, Chemical Engineering, Univ. Sherbrooke).

Olivier, R. 1971. Peatcrete. Eng. J. 54 (11): 25-7.

Parker, T.W. 1947. Sawdust cement and other sawdust building products. Chem. and Ind. Sept.: 593-6.

Passer, M. 1963. Ind. Eng. Chem. 55 (7): 53-8.

Rannou, B. 1972. Rapport interne, Univ. Sherbrooke.

Reilly, J., and O'Donoghue, J.R. 1941. Sci. Proc. Roy. Dublin Soc. 22 (37): 361.

Risi, J., Brunette, C.E., Spence, D., and Girard, H. 1953. Etude chimique des tourbières du Québec (Ministère des Mines, PQ).

Ruel, M. 1970. Un nouvel agent d'adsorption pour combattre la pollution. Conf. presented to Corporation of Engineers of Quebec, Sherbrooke, February.

— 1971-72. Evaluation tests regarding the use of peat moss to remove colloidal iron. Rep. I, October 1971, Rep. II, March 1972, Tech. Rept. to Quebec Cartier Mining Ltd., Port Cartier, PQ.

— 1972. Report, peat moss technical exchange mission in USSR. Unpublished observations, June.

Ruel, M., and Sirianni, A.F. 1972. Agglomeration and extraction of peat moss. US patent application 179,746, 1972.

Sanyal, S. 1973. The use of peat in fixed bed adsorption for removal of total oxygen demand in black liquor and alkyl benzene sulfonate in aqueous solution (Master's thesis, Chemical Engineering, Univ. Sherbrooke).

Silva, O.E.J. 1972. Some experiments on purification of waste waters from slaughter houses with sphagnum peat. Proc. 4th Internat. Peat Congr., Otaniemi, Finland, IV: 311-18.

Simard, A. 1972. Les tourbières au Canada; étendue totale et réserves de tourbes de mousse. Proc. Peat Moss in Canada Conf. Univ. Sherbrooke, pp. 34-42.

Smisek, M., and Cerny, S. 1970. Active Carbon: Manufacture, Properties and Applications (Elsevier).

Son, P.N. 1973. Thèse de maitrise, Univ. Sherbrooke.

Sukhanov, M.A. 1972. The use of peat as a thermal-insulating material in large panel building. Proc. 4th Internat. Peat Congr., Otaniemi, Finland, pp. 319-32.

Surakka, S., and Kamppi, A. 1971. Infiltration of waste water into peat soil. Suo, 22: 51-7.

Van, O.T., Leblanc, R., Janssen, S.M., and Ruel, M. 1971. Peat moss — a natural adsorbing agent for the treatment of polluted water. Can. Mining and Met. Bull. March: 99-104.

Walker, P.L., Rusinko, F., and Austin, L.G. 1959. Advances in Catalysis (Academic Press, New York), p. 133.

PART III

Environmental Considerations

10
Transportation

J.R. RADFORTH and A.L. BURWASH

In developing and using transportation systems in Canada, large areas of muskeg have been and will continue to be encountered. Transportation and the muskeg environment interact to the extent that transportation systems can alter the characteristics of the environment, and the nature of environmental features controls the design of transportation systems. When a transportation system is introduced to the muskeg environment there is an interface where their relationship is established.

Across this interface, permanent transportation structures like roads, railways, and airstrips can change vegetation growth and groundwater flow patterns. Off-road vehicle traffic can alter the thermal balance in the ground. Conversely, peat structure, groundwater content, and incidence of permafrost influence embankment design, selection of off-road vehicles, and installation of electrical power and communications facilities.

To optimize the design of transportation systems effectively, these interactions must be recognized and accounted for.

The relationships between terrain features, equipment capabilities, and operational requirements need to be considered in the light of environmental conservation. Although public awareness of and demand for conservation has recently increased, good conservational practice is also often beneficial to the effectiveness and economy of transportation systems.

EFFECT OF TRANSPORTATION SYSTEMS ON THE ENVIRONMENT

Any natural environment, whether it includes muskeg or not, can be considered as a system of interrelated components, including land, water, vegetation, wildlife, and the atmosphere. The interrelationships among these components are constantly changing and the components subsequently respond in various ways in the process of adapting to the changes. An obvious example of this type of activity is flooding due to heavy rainstorms which causes soil erosion and local changes in vegetational growth patterns.

When man installs a transportation system of any kind, the components of the environment also respond to this activity. These changes are just as natural as the responses to any alteration of their own interrelationships, but there is one important difference. Since the transportation system is man-made, its effect on the environment can be controlled, provided there is adequate knowledge of the effects of man-made disturbances on the natural environment. In many cases this type of knowledge is essential in order to design the transportation system in such a way that its stability and operating capability are protected.

For installations of transportation systems in an environment of which muskeg forms a portion, there are several well-known ways in which the system will affect the muskeg. Any system will contact the vegetation, water, and peat which combine to form muskeg. The nature of the effect of the transportation system on the muskeg environment depends, to a great extent, on the type of transportation system involved, that is, on whether it is a road, railway, a number of off-road vehicles, or an electrical power line.

Vegetation Disturbance

Until recently, that is within the past five years, little concern has been shown for the effect of any transportation system on vegetation. Growing concern, however, has paralleled the development of petroleum exploration activity in the Canadian and American Arctic, and the possibility of installing pipelines from those regions to the south has added fuel to the fire of controversy. Any permanent installation is likely, of course, to completely cover vegetation along the transportation system right-of-way. Most concern has been about the effects of off-road vehicles on vegetation health and survival, and again, mostly in the arctic areas.

The arctic areas which have been the major target of concern do unquestionably contain organic terrain. The thickness of the organic layer, however, is usually not great, so that many authorities question the validity of referring to it as muskeg in the usual sense. However, the plant types growing in this arctic terrain resemble, in form and structure, vegetation growing on muskeg areas farther to the south through which future transportation development will unquestionably pass.

The controversy surrounding vegetation disturbance on the tundra and in muskeg has been generated largely by the high visibility of disturbance in its initial stages and the changes in vegetation growth pattern which subsequently occurred at the site of the disturbance. Many authorities have implied that such disturbance corresponds to an ecological upheaval and therein lies the basis of the controversy. There are, as yet, no firmly established and agreed upon criteria for deciding what constitutes an actual ecological disturbance of vegetation of undesirable magnitude.

Attempts have been made to evaluate the immediate and long-term effects of operating both tracked and wheeled vehicles on tundra areas. In appraising the results of analysis of the immediate effects on the vegetation, the projected regrowth of vegetal cover in terms of health and survival of plant cover, effects on thermal properties of subsurface material, and the aesthetic quality of the appearance of the landscape are all involved.

In order to be able to discuss vegetation disturbance in comparative terms, a classification scheme is desirable. A system developed in 1970 (Bellamy-Radforth) was introduced for use in evaluating terrain disturbance due to the vehicular traffic in tests performed in the Arctic Land Use Research Programme sponsored by the Canada Department of Indian Affairs and Northern Development. Table 1 presents a description of the levels of vegetation and ground surface structure disturbance defined in this system.

As a result of this and similar test programs, it has been found that factors relating to both terrain and vehicle are influential in determining the response of vegetation to traffic. Terrain

TABLE 1
Vegetation and structure disturbance classification system

Disturbance level	Structure	Vegetation
1	Undamaged	Undamaged
2	Slightly damaged	Shrubs broken, leaves knocked off
3	Mound top scuffing/flattening	Cutting and/or flattening of all vegetation
4	Mound top destruction	Tearing and scattering of vegetation; 10% destroyed
5	Ruts start to form, less than 50% structure destroyed	25% destroyed
6	Ruts slightly deeper, more than 50% structure destroyed	50% destroyed
7	Ruts half bare	90% destroyed
8	Ruts entirely bare	100% destroyed
9	Ruts to permafrost	

factors include ground moisture content, stature and structure of vegetal cover, and depth to permafrost if present. Areas having high moisture content, when subjected to vehicular traffic in the summer, are subject to the most rapid development of vegetation disturbance. In contrast, dry shrub-covered hillsides support more rugged vegetation which is more resistant to disturbance by vehicular traffic (Bellamy, Radforth, and Radforth, 1971).

Vehicle characteristics such as weight, ground pressure, and track or tire design also influence the response of vegetation to traffic. For example, tracked vehicles that utilize a detent as a wheel guide along the centre of the tracks create an initially shallow rut after one or two vehicle passes. A shearing effect in the centre of the rut is caused by differential velocities between the detent and the side portions of the track. This shearing effect tends to hasten the progress of vegetation disturbance (Figure 1). Tracks having grouser bars can pick up chunks of peat as the tracks come in contact with the ground and scatter them behind the vehicle as it travels (Figure 2). Flat tracks minimize this effect and wheeled vehicles utilizing large low-pressure tires have been found to produce an initial disturbance one or two levels lower than tracked vehicles of similar weight.

The season of the year during which vehicular traffic takes place has an important effect on levels of vegetation disturbance sustained. In areas where permafrost predominates, the traffic early in the thaw season encounters good support on hard-frozen ground, preventing ruts from forming rapidly and avoiding disturbance to the root structure of vegetation. Later on in the season however, off-road vehicles making repeated passes over the same path can wear ruts in the ground surface displacing and damaging plant root structure and leading, in some cases, to permanent destruction of plant material (Figure 3).

In every case, vegetation disturbance is confined to the immediate location of the vehicle tracks. Vehicle traffic, therefore, does not initiate an unstable situation for plant health over a wide area extending beyond the area occupied by the tracks.

Regrowth of disturbed vegetation depends upon the amount of initial disturbance, the season of the year during which disturbance took place, and the type of habitat and geographical location of the disturbed site.

FIGURE 1   Rut formed by a detent.

FIGURE 2   The effect of grouser bars on terrain disturbance.

FIGURE 3   The effect of season on rut formation in a permafrost area: June traffic in top picture, August in bottom.

FIGURE 4   Slight thermokarst effect in a vehicle test lane.

Thermal Effects

In many areas, permafrost is a common feature of organic terrain. This feature, too, responds in various ways to the influence of transportation systems of all types. This subject is
more fully dealt with in a separate section of this publication, but a few features are worthy of
special attention at this time.

Installation of a road, railway, or pipeline, or off-road vehicle traffic, can alter the thermal
properties of the surface layers of the terrain in such a way as to affect the stability of the
permafrost beneath the surface. Either compression of the surface layer and vegetation,
thereby increasing its thermal conductivity, or removal of the surface layer exposes permanently frozen sublayers to increased insolation. If peats having a high ice content are present,
any disturbance that results in melting of this ice allows subsidence referred to as thermokarst (Figure 4) and surface ponding to occur as the ice disappears. An additional problem
can arise on slopes where removal of vegetation may allow erosion to occur, leaving gullies,
and may contribute to the melting of ice near the surface. A reduction in shear strength due to
the loss of the bonding effect of ice can threaten the stability of slopes or of foundations on
level muskeg areas.

To assist in the prediction of the extent of disturbance that can occur as a result of a given
set of engineering activities, a soil sampling and testing program can be carried out to
describe the peats present according to a recognized classification system and determine the

TABLE 2

A selection of references describing methods for freeze-thaw predictions

Graphic methods	Analytical	Finite elements or finite differences
US Corps of Engineers (1949), modifications to design curve in Brown (1964)	Neumann method in Ingersoll et al. (1954), additional information in Aldrich (1956) and Berggren (1943)	Hwang, Murray, and Brooker (1972), finite elements
	Brown and Johnston (1970)	Lachenbruch (1970). finite differences
	Aldrich and Paynter (1958)	
	Lachenbruch (1959)	Fertuck, Spyker, and Husband (1971), finite differences
	Jumikis (1955)	

water (ice) content, depth of peat, thermal properties, and similar information for the underlying strata. At present, research on ice distribution in peats has not reached the point where definite statements can be made regarding the tendencies of certain peats to contain more ice than others. There is also not enough information yet to make definite statements regarding the variation of the incidence of ice in organic or mineral soils with depth. Williams (1968) and Samson and Tordon (1969) indicate that the distribution of ice with depth varies greatly, probably depending on the deposition history and soil type.

The thermal properties of the soil, water, ice, and air system are important indicators of the depth of thaw that can be expected under any given set of conditions. In most cases, if all factors are considered, very complicated mathematical exercises are involved. Fortunately, however, it is often possible to make simplifying assumptions that allow a reasonably accurate solution to be reached. A selection of references describing methods currently in use is presented in Table 2.

The reliability of estimates obtained by any of these methods also depends upon the accuracy of the values of the thermal properties of peat which are used. At the present time, a comprehensive study of the factors affecting the thermal conductivity and specific heat of peat has not been completed, although important steps toward this goal have been taken. First, in 1949 Kersten published data for 'fibrous brown peat.' The results of these tests indicated that the thermal conductivity increases with increasing water content if the dry density is held constant. The thermal conductivity also increases with increasing dry density when the water content is held constant. Romanov (1968) reports the results of field tests to determine the thermal conductivity and specific heat values for *Sphagnum* peats. Burwash (1972) reports preliminary results indicating that the structural composition of peat as well as its water content is an important variable in determining thermal conductivity values. Woody, coarse fibrous peats have consistently higher thermal conductivities than amorphous granular or non-woody, fine fibrous peats.

Some attempts have been made to describe quantitatively the levels of terrain surface disturbance relative to different thermal properties of the soil. An example of this is an attempt made by Beatty and Gray to correlate albedo changes with the levels of disturbance proposed by Radforth and Bellamy. In this case, albedo is defined as the ratio of the amount of soil radiation (short wave) reflected by a surface to the amount of incident radiation expressed as a percentage. A few passes by a vehicle over a piece of the ground are sufficient to increase the albedo up to 3.4 per cent in some areas. This results from the change in colour of the vegetation as it is crushed by the vehicle. If enough passes are made to cause a change to dark brown, then the albedo decreases by 2.5 per cent from the undisturbed surface value. Although the albedo is also affected by the moisture content, increasing levels of surface disturbance resulting from vehicle traffic produce a corresponding decrease in albedo.

An effective preliminary step has been taken towards determining quantitatively the effects of vehicle traffic on thermal properties of terrain. In order, however, to calculate accurately the heat flux entering the soil under disturbed conditions, values for additional components of the energy budget need to be measured. Further research is also needed to determine the factors other than terrain disturbance which are responsible for variations in albedo.

Winter Roads

Off-road traffic over peatlands in the winter is simpler from a mobility viewpoint than in the summer because a frozen layer provides a firmer surface than unfrozen ground for vehicle traffic. One of the most important considerations is determining the time for starting operations so that the ground is sufficiently frozen and there is enough snow cover to minimize terrain disturbance. Vegetation along the right of way can be protected by accumulating sufficient snow and packing it with vehicles as light as a motorized toboggan, followed in some cases by application of mixed snow and water to build up an ice embankment.

This procedure can commence as soon as there is sufficient frost in the ground surface to support the weight of the vehicles used in the operation. Once sufficient material has been built up and cold temperatures have persisted to the extent that there have been more than 150 degree-days of frost, heavy traffic can begin to use the road. It is then relatively easy to prepare a smooth surface suitable for use by highway trucks for hauling heavy loads up to 40 tons.

When these procedures are followed, the vehicle traffic does not come in contact directly with the vegetation on the ground surface and this results in a minimum amount of disturbance. The road embankment tends to persist longer than the surrounding snow cover during the thaw season, however, and this can have a retarding effect on plant growth along the right-of-way for the one season. This is not found to be of great significance to long-term vegetation health in studies undertaken so far.

Drainage

The construction of permanent transportation facilities can alter the hydrological conditions of the muskeg environment. Placement of a road embankment causes consolidation of peat and a subsequent reduction in its permeability. If proper culverts are not installed in the embankment, it can also act as a dam inhibiting surface water flow in its vicinity. Compaction of peat and reduction of permeability result in a decrease of water flow through the peat in the vicinity of the road. Removal of vegetation in the vicinity of a road results in decreased evapotranspiration, which encourages ponding.

TABLE 3
Importance of terrain disturbance factors in relation to the prevailing climate

	Regrowth of vegetation	Thermal disturbance	Drainage disturbance	Secondary effects
Continuous permafrost	Slow	Highly significant if ice content is high	Significant if a large amount of ice melts	Significant
Sporadic permafrost	Moderate	Highly significant if ice content in perma-frost areas is high	Significant if primary drainage systems disturbed	Significant
Some freezing in winter	Moderate to rapid	May influence the thickness of frozen layer	Significant if primary drainage systems disturbed	Significant
Little or no freezing	Rapid	Not significant for transport systems	Significant if primary drainage systems disturbed	Significant

Changes in the water regime of a muskeg area can influence paludification (muskeg formation). If drainage of excess water from a bog is hindered so that water is dammed between a bog and a road, paludification may proceed at a faster rate. If the water level is raised and spillage occurs onto adjacent mineral terrain, trees that may be growing there will be killed. Part of the muskeg area may float, depending upon the type, and trees on an affected bog would die.

Low-lying lagg areas in confined muskeg are most easily disturbed by an over-supply or under-supply of water. Peatlands deriving their high moisture content from the presence of groundwater are more easily affected by this type of disturbance than well-drained bogs that receive most of their water supply from the atmosphere.

Drainage of muskeg areas during road construction can enhance tree growth where the peat is shallow, but generally over-drying of deeper peat deposits can inhibit vegetation growth. Additionally, drainage of muskeg areas increases their susceptibility to fire.

The importance of terrain disturbance factors in relation to the prevailing climate is summarized in Table 3.

Secondary Environmental Effects

The installation of permanent transportation systems in previously remote areas tends to have local effects on wildlife populations to some extent. An increased human population in any area tends to result in a displacement and change in the numbers of animals. Muskeg often forms an important part of the habitat of animals such as the beaver and moose, and although the presence of a transportation system itself does not appear to affect these animals to any great extent, the increased accessibility to them by human beings could affect their numbers and distribution.

When one considers the question of fire hazard, there are arguments for and against the installation of transportation systems. In remote areas by far the largest number of fires is believed to be caused by thunderstorms. Transportation systems of some kind provide accessibility to remote areas, enabling forest fires to be dealt with more rapidly. On the other hand, presence of the transportation system increases the danger of fires being started by

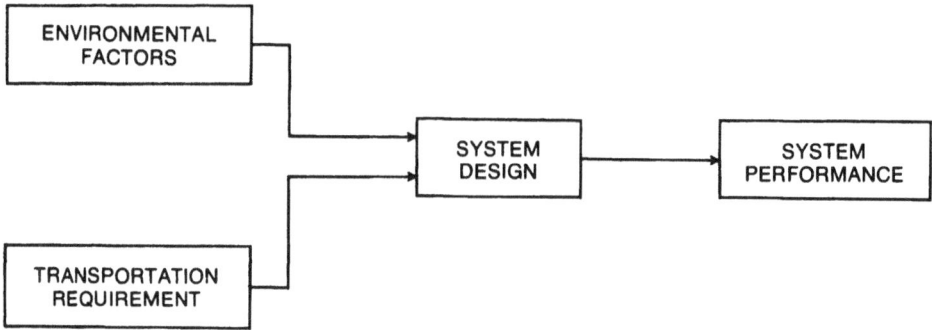

FIGURE 5   The interdependence of transportation system performance, design, environmental factors, and transportation requirements.

human activity. There is an increasing body of research dealing with the effects of fires on areas in which organic terrain abounds.

EFFECT OF THE MUSKEG ENVIRONMENT ON TRANSPORTATION SYSTEMS

When any transportation system is installed and put into operation, it not only has an effect on its environment, but it is in turn affected by that environment. A number of environmental factors influence the operation of the transportation system and the manner in which they do so is controlled largely by the design of the system.

Experience has now revealed most of the problems which arise as a result of incompatibilities of transportation systems with the muskeg environment. Road embankments settle beyond the failure point, interruption of natural drainage promotes flooding, and off-road vehicles become immobilized or, as has happened in extreme cases, have to be abandoned, to name only a few examples.

This accumulated experience constitutes a body of knowledge which will only be useful if it is applied to the design of new transportation systems. To be effective it must be applied so that the transportation system design is compatible with all the environmental features as well as the transportation requirement. This is illustrated in Figure 5, which demonstrates the dependence of system performance on the need to consider environmental factors in the system design.

The environmental factors are, of course, not the only constraints, besides the transportation requirement, which dictate the system design. Other considerations such as financing, required life-span, and fuel supply, for example, will have an important bearing on the final design of the system. For the purposes of this discussion, however, emphasis need only be placed on the impact of environmental factors.

In primary terms, any transportation system comes in contact with air, land, and water. These three elements interact with each other and combine to form the environmental features of concern here.

Muskeg, as a type of terrain, is a product of the primary environmental influences. As shown in Figure 6 these include wind, humidity, slope, precipitation, temperature, and mineral sublayer. The features of the terrain, depending on local conditions, join forces with the primary factors to form the interface with the transportation system. As the diagram

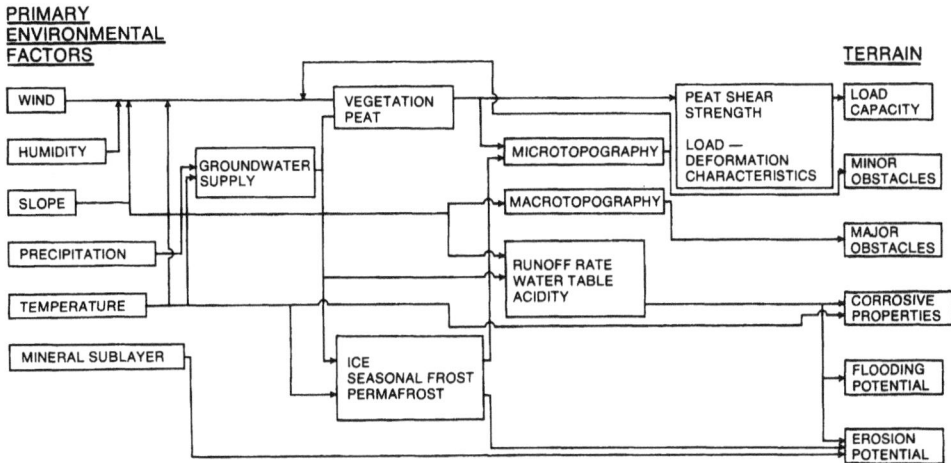

FIGURE 6 Interrelationship of environmental factors influencing transportation systems.

shows, the interface comprises terrain load-bearing capacity, minor obstacles, major obstacles, corrosive properties, flooding potential, and erosion potential.

On the transportation side of the interface, the system performance falls into a number of categories:

1. Operating effectiveness and capacity: How well does the system fulfil its requirement to move a given quantity of material, personnel, or energy from one place to another?
2. Operating efficiency: How does the value and quantity of the output of the system compare with the cost in money and energy of operating it?
3. Operating cost: Do the installation and operating costs meet the design cost constraints and compare favourably with those of alternative design?
4. System lifetime: Will the system design, operation, and maintenance program, as influenced by the environment, ensure the system's survival over its specified lifetime?
5. System maintenance: Do the mechanical design and maintenance schedule of the system recognize the full impact of the environment on the system?
6. Effect on environment: Has the system been designed not to interfere with the environment in a detrimental manner?
7. Safety: Have adequate measures been taken in the system's design to ensure the safety of persons and property likely to come in contact with it in any way?
8. Stability: Do fixed parts of the system remain in position without maintenance?

To be more specific about the nature of these performance parameters, it is necessary to consider the type of transportation system involved in a particular situation. Transport can be categorized as continuous (roads, railways, pipelines), intermittent (powerlines, communications, airfields), and transient (off-road vehicles). Each of these categories is influenced to a different degree and extent by the various environmental factors. For example, in selecting a route for the system, it is important to consider the flooding potential for roads or railways but not for off-road vehicles. Slope is critical in assessing trafficability of terrain for air cushion vehicles, but not as critical for other off-road vehicles.

TABLE 4
Transportation design checksheet for continuous transportation systems (roads, railways)

	Operating effectiveness	Efficiency	Operating cost	System lifetime	Maintenance	Effect on environment	Safety	Stability
Load capacity								
Peat depth	X					X		X
Shear strength	X							X
Load/deformation characteristic	X	X	X	X	X		X	X
Minor obstacles								
Trees						X		
Bushes						X		
Soft spots								
Hummocks/ mounds								
Lagg	X							
Ponds	X							
Streams	X					X		
Major obstacles								
Steep slopes	X	X	X		X		X	X
Lakes	X		X			X		
Rivers	X		X		X	X		X
Corrosive properties								
Moisture content								
Acidity								
Temperature								
Flooding potential								
Runoff rate	X	X	X	X	X	X	X	X
Erosion potential								
Water table	X	X	X	X	X	X	X	X
Mineral sublayer	X		X	X			X	X
Permafrost	X	X	X	X	X	X	X	X

To simplify the process of identifying the environmental factors which are critical in a particular transport operation, checksheets have been prepared (see Tables 4-6). The marks in the spaces on the checksheets correspond to combinations of system performance and terrain factors which are important to consider in the design of a transportation system.

There are three checksheets corresponding to the three different types of transportation systems grouped according to the amount of contact they have with the ground.

The ways in which these systems react to the environment are, or should be, well known to designers by now, as are the design techniques used to accommodate the reactions. The purpose of this article is only to draw attention to their existence. It is sufficient to say here that these factors must be taken into account in the design process to ensure success of a system design. A more detailed treatment of them is contained in the *Muskeg Engineering Handbook* (MacFarlane, 1969).

COMPATIBILITY OF TRANSPORT SYSTEMS TERRAIN AND THE OPERATION

In view of the interaction of transportation systems and the muskeg environment, careful

TABLE 5
Transportation design checksheet for intermittent transportation systems (transmission lines)

	Operating effective-ness	Efficiency	Operating cost	System lifetime	Mainte-nance	Effect on environ-ment	Safety	Stability
Load capacity								
Peat depth	X							
Shear strength	X				X			
Load/deforma-tion characteristic	X				X			
Minor obstacles								
Trees			X					
Bushes								
Soft spots					X			
Hummocks/mounds								
Lagg	X							
Ponds	X							
Streams	X				X			
Major obstacles								
Steep slopes	X				X			
Lakes	X							
Rivers	X							
Corrosive properties								
Moisture content	X		X	X				
Acidity	X		X	X				
Temperature	X	X	X	X				
Flooding potential								
Runoff rate	X	X						
Erosion potential								
Water table	X	X						X
Mineral sublayer	X	X						X
Permafrost	X	X						X

planning of system design and operation is an important consideration. Such planning requires the cooperation of users, vehicle manufacturers, and governments to ensure compatibility between the proposed usage and the specific terrain involved.

A primary consideration in planning new transportation facilities is whether a first-class installation is justifiable in terms of cost, expected lifetime, purpose, and maintenance and safety requirements. The decisions made on these matters will determine design specifications and construction strategy.

Selection of an appropriate route for transportation facilities is of great importance to minimize both cost and terrain disturbance while optimizing operational effectiveness. An efficient preliminary survey involves examination of airphotos of the broad area considered for transportation use. Significant terrain features can be identified and inferences can be drawn concerning the sensitivity of peatland to disturbance. After this preliminary analysis is completed, some alternative routes can be examined in greater detail from the ground and a comprehensive investigation of the most promising route can be carried out.

If avoidance of terrain disturbance is a significant factor, selection of the appropriate season for off-road travel and construction of transportation facilities is of great importance.

TABLE 6
Transportation design checksheet for transient transportation systems (off-road vehicles)

	Operating effectiveness	Efficiency	Operating cost	System lifetime	Maintenance	Effect on environment	Safety	Stability
Load capacity								
Peat depth						X	X	
Shear strength	X	X	X			X		X
Load/deformation characteristic	X	X	X					X
Minor obstacles								
Trees	X	X	X	X	X	X	X	X
Bushes	X	X			X	X		
Soft spots	X	X			X	X	X	
Hummocks/ mounds	X	X	X	X	X		X	X
Lagg	X	X						
Ponds	X	X			X		X	X
Streams	X	X	X			X	X	X
Major obstacles								
Steep slopes	X	X				X	X	X
Lakes	X						X	
Rivers	X					X	X	X
Corrosive properties								
Moisture content			X	X	X			
Acidity			X	X	X			
Temperature		X	X	X	X		X	X
Flooding potential								
Runoff rate								
Erosion potential								
Water table								
Mineral sublayer						X		
Permafrost	X	X	X			X		X

Overland travel by off-road vehicles is easier in winter when ground is frozen and there is ice cover on rivers and lakes.

In order to assess the potential disturbance of terrain due to traffic, construction, or the presence of permanent transportation facilities, it is necessary to recognize that many types of terrain, with varying sensitivities, are present in muskeg areas. Preliminary results indicate that the most sensitive areas have a high water regime in either the frozen or thawed state. Areas with a low water regime and similar land forms are still quite sensitive. The least sensitive areas are well-drained slopes or well-drained flat areas.

Compatibility of vehicles with a specific operation and the terrain is of great importance in reducing terrain disturbance. Considerations include vehicle weight, ground pressure, track or tire design, and drawbar pull-slip-weight relationships desired. Vehicles having large airbags instead of conventional tires have been found to cause less disturbance on tundra test sites (Powlan and Christopherson, 1972).

Proper route selection and foundation design is necessary to ensure the stability of permanent transportation installations.

If the factors discussed here are taken into account in transportation system planning, it is more likely that the desirable harmony between a system and the muskeg environment will be attained.

REFERENCES

Aldrich, H.P. 1956. Frost penetration below highway and airfield pavements. Highway Res. Bull. 135, Nat. Acad. Sci.

Aldrich, H.P., and Paynter, H.M. 1958. Analytical Studies of Freezing and Thawing of Soils (us Army Eng. Div., New England, Arctic Const. Frost Effects Lab., Waltham, Mass.).

Bellamy, D., Radforth, J.R., and Radforth, N.W. 1971. Terrain, traffic and tundra. Nature 231 (5303).

Berggren, W.P. 1943. Prediction of temperature distribution in frozen soils. Trans. Am. Geophys. Union, 3: 71-7.

Brown, W.G. 1964. Difficulties associated with predicting depth of freeze or thaw. Can. Geotech. J. 1 (4): 213-26.

Burwash, A.L. 1972. Thermal conductivity of peat. Proc. 4th Internat. Peat Congr., 2: 243-54.

Fertuck, L.J., Spyker, J.W., and Husband, W.H.W. 1971. Numerical estimation of ice growth as a function of air temperature, wind speed and snow cover. Trans. csme Eng. J., 54 (12).

Hwang, C.T., Murray, D.W., and Brooker, E.W. 1972. A thermal analysis for structures on permafrost. Can. Geotech. J. 9 (1): 33-46.

Ingersoll, L.R., Zobel, O.J., and Ingersoll, A.C. 1954. Heat Conduction (rev. ed., Univ. Wisconsin Press, Madison).

Jumikis, A.R. 1955. The Frost Penetration Problem in Highway Engineering (Rutgers Univ. Press, New Brunswick, NJ).

Kersten, M.S. 1949. Laboratory research for the determination of the thermal properties of soils — final report. us National Technical Information Service Document AD 712 516.

Lachenbruch, A.H. 1959. Periodic heat flow in a stratified medium with application to permafrost problems. Geol. Survey Bull. 108.3-A, Washington, DC.

— 1970. Some estimates of the thermal effects of a heated pipeline in permafrost. us Geol. Survey, Washington Circ. 632.

MacFarlane, I.C. (ed). 1969. Muskeg Engineering Handbook (Univ. Toronto Press).

Powlan, F., and Christopherson, R. 1972. Some considerations for the design and construction of pipelines in arctic muskeg. Proc. 14th Muskeg Res. Conf., pp. 27-51.

Romanov, V.V. 1968. Hydrophysics of Bogs, trans. N. Kaner (Israel Program for Scientific Translation, Jerusalem), pp. 101-31.

Samson, L., and Tordon, F. 1969. Experience with engineering site investigations in northern Quebec and northern Baffin Island. Proc. Third Can. Conf. on Permafrost. NRC Tech. Memo. 96, pp. 21-38.

us Corps of Engineers. 1949. Addendum No. 1, 1945-47, to Report on Frost Penetration 1944-45 (Corps of Engineers, us Army, New England Div.)

Williams, P.J. 1968. Ice distribution in permafrost profiles. Can. J. Earth Sci. 5 (12): 1381-6.

11
Pipelines

R.D. MEERES

Pipeline construction in areas of Canada where muskeg (MacFarlane, 1958) is encountered has generally followed closely behind the exploration for crude oil and natural gas. Although some experience was gained from prior road building and exploration activities in those areas, the unique logistics problems associated with the construction of a pipeline transmission facility necessitate different solutions than were appropriate for the other types of construction.

During the 1950s, efforts were made to construct pipelines in muskeg during the summer months when warm weather prevailed, but such facilities, constructed by conventional summer pipeline construction techniques, proved economically unattractive and environmentally undesirable.

By 1960, exploration activities were accelerating in the remote muskeg areas of north-western Alberta and northeastern British Columbia. These areas were served by a haphazard system of dirt roadways that provided only limited access to most areas being explored.

During the late 1950s and early 1960s, a totally new construction approach evolved that allowed work to be performed during the winter when freezing temperatures could be used to advantage in preparation of temporary access roads and working surfaces for execution of the work. Damage to the environment was thereby reduced while at the same time project economics was enhanced through reduced construction costs.

Methods developed for winter construction in muskeg proved so successful that they are now the accepted practice in the industry.

ENGINEERING CONSIDERATIONS

Route Selection

Economic optimization over the life of the project is the ultimate objective of the pipeline designer. Other things being equal, the straight-line route between source and destination will prove most economic. Usually, however, deviations are required from it to avoid, where possible, unfavourable topographic features, water bodies, rock outcrops, man-made struc-

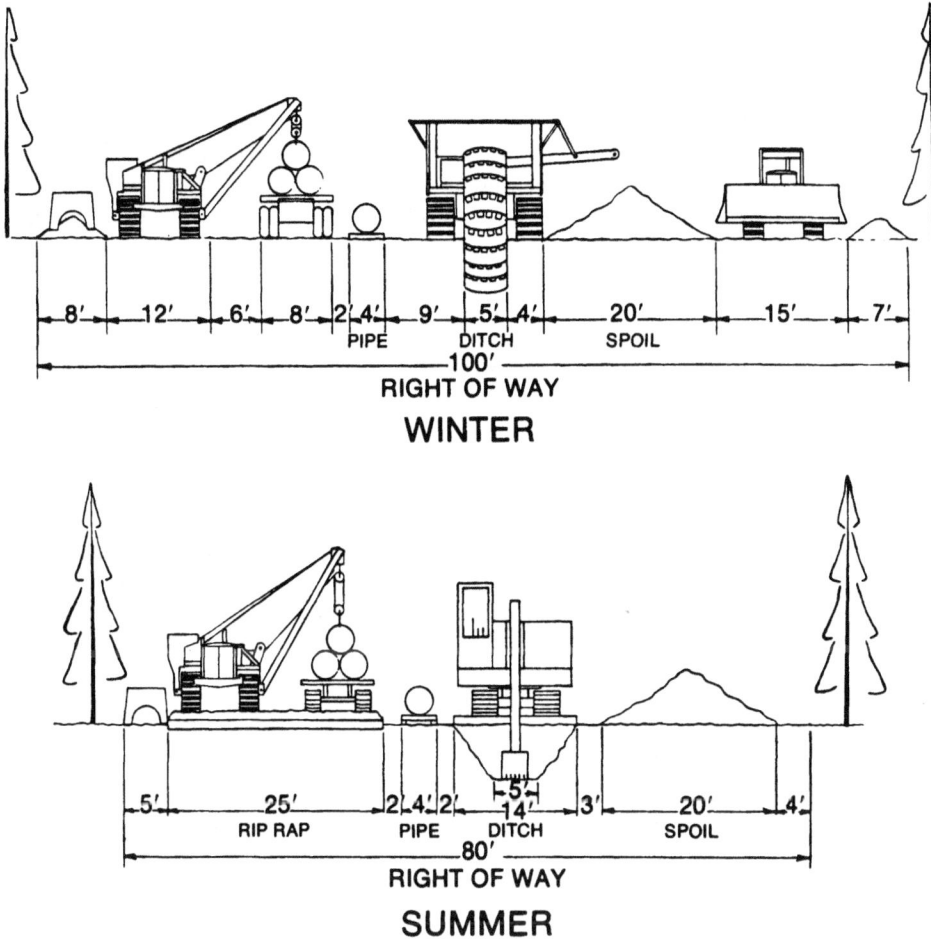

FIGURE 1   Right-of-way requirements for 36 inch O.D. pipeline in muskeg.

tures, and muskeg. The cost reduction associated with each undesirable feature avoided must be equated against the costs related to the increase in length of the line.

Access to the proposed route during both construction and subsequent maintenance of the facility must be considered in this analysis. When there are sizable differences in the lengths of alternative routes being considered, changing horsepower requirements and operating costs also have to be taken into consideration.

In areas where large-scale government topographic maps are available, these will usually serve for initial appraisal of alternative routes, and the determination of possible means of access. Further refinement can be accomplished through interpretation of aerial photographs.

The experienced pipeliner can generally complete his appraisal of field conditions along the proposed route by means of an aerial reconnaissance from light fixed-wing aircraft or helicopters with limited on-the-ground verification of aerial observations.

Right-of-way Requirements

Having established the route for the pipeline, it is then necessary to determine the width required for the right of way.

Figure 1 shows the width required for efficient construction of a 36-inch-diameter pipeline in summer and in winter. As can be seen, winter construction requires approximately 25 per cent more right-of-way width than does summer construction. Even so, the economics of winter construction justifies the expense incurred in procuring and clearing the additional right-of-way width.

Basic Design

Code requirements for pipelines in muskeg are essentially the same as for lines located in other types of soils. Current editions of Canadian Standards Association Codes (1973) are applicable.

The maximum operating pressure for big-inch pipes is limited by the codes to the pressure that will stress the steel to 80 per cent of its yield strength for oil transmission pipelines and 72 per cent for gas transmission systems. In populated areas, at road crossings, and in proximity to other man-made structures, these allowable operating stresses are further reduced.

In areas where the pipe will be exposed to extremely low ambient temperatures, special pipe must be provided with suitable low-temperature toughness characteristics.

Buoyancy Considerations

Generally speaking, pipelines are buried to a depth that will provide 30 inches of cover over the pipe in order to protect it from subsequent movement of equipment and other activities.

In areas of muskeg, the high-water table which usually prevails will cause the material backfilled around the pipe to behave like a fluid. The wall thickness of the pipe is determined in accordance with code requirements, to withstand internal hoop stresses, with a suitable safety factor. The resultant weight of the pipe generally is somewhat less than the weight of the fluid displaced by the pipe. Consequently the pipe will float out of the ditch unless special precautions are taken.

For pipelines which transport liquids, the problem is generally of a short-term duration, until such time as the pipeline is filled and put into operation. However, provision must be made to ensure that the pipeline will remain in place at the bottom of the ditch if it has to be emptied at some future time for repairs.

Pipelines intended for transportation of natural gas are faced with this buoyancy problem throughout their lifetime. There are, of course, exceptions to this, particularly in smaller pipe sizes where the wall thickness is determined by practical considerations such as the need for protection from external damage, bending limitations and weldability. In these smaller sizes, the pipe will have the required negative buoyancy naturally.

The buoyancy problem can be solved by weighting the pipeline with saddle weights, bolt-on weights, or continuous concrete coating (see Figure 2). The saddle weights are usually used in muskeg areas, and the bolt-on weights in rivers and sections where ditch conditions are such that saddle weights may fall off the pipe. Continuous concrete coating has an added advantage when used in rivers because it affords some protection to the pipe as well as providing the necessary weight.

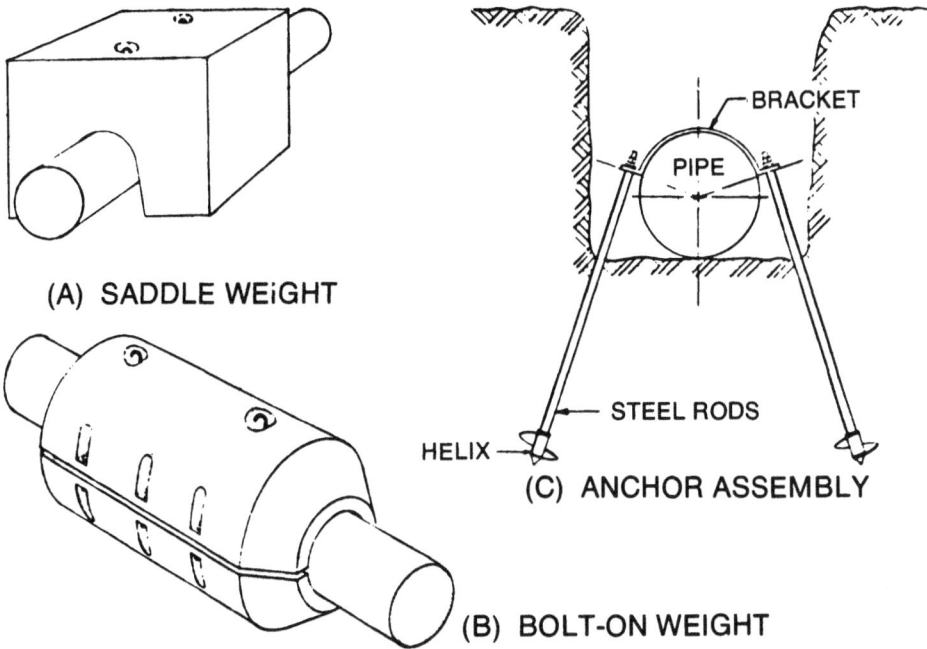

(A) SADDLE WEIGHT

(B) BOLT-ON WEIGHT

(C) ANCHOR ASSEMBLY

FIGURE 2   Typical weights and anchor.

Concrete weights are installed on the pipeline at a predetermined spacing based on the buoyancy calculations. Most of the industry assumes the density of liquid muskeg to be 80 pounds per cubic foot compared to 62.4 pounds per cubic foot for water. It is usual to specify that a 10 per cent net negative buoyancy be imparted to the pipeline, although some owning companies require as much as 25 per cent net negative buoyancy.

Typical spacing of saddle weights for various pipe sizes is shown in Table 1. This is the recommended spacing for standard-sized weights and usual wall thicknesses of pipe.

Screw anchors (Figure 2) have been used to hold down pipe in muskeg with limited success in areas where firm soils underlie shallow muskegs. There is always a certain amount of risk in their use, however, since the shear strength of the soil may change because of groundwater infiltration or saturation from the muskeg. If the soil strength becomes inadequate at any time the anchors could fail to hold, allowing the line to float to the surface.

Companies utilizing the screw type anchors usually require a 'pull' test as an assurance that the anchors are doing their job. Table 2 shows the spacing for the anchors and the pull test requirement for each helix.

A considerable amount of advance planning must be implemented in order to use anchors. Since they are not manufactured close to the areas where they would be used, they must be ordered in advance, in large enough quantities to ensure that enough are on hand for the project. Advance testing of the right of way with soil testing equipment is required to find the depth of the muskeg and strength characteristics of underlying soils, so that the anchoring requirements can be determined, and the necessary quantities purchased.

Such advanced planning and soils testing are not required to the same extent when weights are used because they can be manufactured quickly in the field close to the job site.

TABLE 1

Maximum saddle-weight spacing (in feet) *

	800†	1,000	1,600	2,000	2,800	3,400	4,000	5,600	6,800	8,000	10,000
Wall thickness (inches)	6-5/8‡	8-5/8	10-3/4	12-3/4	16	18	20	24	30	36	42
0.188	52	26	22	-	-	-	-	-	-	-	-
0.219	78	32	25	20	14	-	-	-	-	-	-
0.250	153	40	29	22	16	15	13	12	-	-	-
0.281	-	-	34	24	18	16	14	13	-	-	-
0.312	-	-	-	28	19	17	15	13	9.4	-	-
0.375	-	-	-	38	23	20	17	15	10.	7.7	6.8
0.406	-	-	-	-	-	-	18	15	10	7.9	6.9
0.469	-	-	-	-	-	-	-	17	11	8.4	7.3

* 10% negative buoyancy; specific gravity of fluid 80 pounds per cubic foot; concrete density 145 pounds per cubic foot.
† Size of weight in pounds.
‡ Pipe size in inches.

TABLE 2

Spacing requirements for anchors (in feet) for various sizes of pipe (based on 95 pound per cubic foot mud)

Pull force per screw (pounds)	Pipe diameter (inches)				
	6-5/8	16	20	30	36
13,000	-	-	-	50	39
12,000	-	-	-	47	36
11,000	-	-	-	44	33
10,000	-	-	-	41	30
9,000	-	-	-	38	27
8,000	-	-	-	35	24
7,000	50	90	90	32	21
6,000	50	90	90	29	18
5,000	50	90	76	26	15
4,000	50	90	61	23	12
3,000	50	70	46	20	9

Berm Construction

Another method has been used in a few instances to construct large-diameter pipelines through muskeg without the use of weights or anchors. The 'berming' method involves placing the pipe in a shallow ditch excavated only enough to hold the bottom one-third to one-half of the pipe. Material is then bermed over the pipeline to protect it. The maximum depth of ditch selected is based on the depth of water that would cause the pipe to float. This method eliminates the need for a large amount of ditch excavation, particulary for large pipe sizes. On a 36-inch project, the ditch need only be 1.5 feet deep instead of the usual 6 feet.

The pipe must still be buried to full depth in areas between muskegs and at locations where other rights of way are crossed in order to ensure that the line is not subjected to damage from other construction activity which may occur.

Considerable economic advantage derives from the use of this type of construction because the need to install weights to keep the pipe in place is substantially reduced.

CONSTRUCTION IN MUSKEG

Muskeg, in its natural state, characteristically has poor drainage patterns and a high water content. The surface consists of a living organic mat overlying partially decomposed, highly compressible organic material (MacFarlane, 1958).

Up to 30 per cent or more of the terrain traversed by a typical pipeline may be muskeg. By its very nature, undisturbed muskeg has a very low bearing capacity which will not support the heavy equipment required for pipeline construction.

Heavy rainfall (Table 3) experienced during the summer makes access roads to and along pipeline rights of way impassable, often for extended periods of time.

Generally speaking, the unsaturated surface vegetation acts as an almost perfect natural insulation, preventing the underlying organic material from freezing. Even in the coldest weather, this material will remain unfrozen except for a shallow layer at the surface.

This is the environment faced by contractors as the pipeline industry presses into muskeg territory. Methods devised to overcome the unique problems posed by muskeg are illustrated in the following comparison of summer and winter pipeline construction.

Right-of-way Preparation

For summer construction a roadbed must be prepared to support heavy equipment required to haul materials and construct the pipeline. Timber cut from adjacent sections of the right of way is hauled to muskeg sections, where it is placed to form a 'floating' roadbed (Figure 3). The thickness of this timber 'rip-rap' is governed by the size of construction equipment required to construct the pipeline.

Hand clearing of the right of way using power saws is often required in muskeg terrain, because of the immobility of heavy equipment prior to preparation of the working surface.

'Tree farmers' are used to skid the logs to a central location. The logs are then loaded on large-tracked vehicles for transport to muskeg sections. 'Forklifts' unload the logs from the tracked vehicles and carry them to the point where they are placed to form the roadbed (Figure 4). Large earth-moving scrapers then transport soil from the higher lands on either side of the muskeg area to be placed over the rip-rap.

Subsequent activities tend to break up the rip-rap and push it down into the muskeg (Figure 5), necessitating an on-going repair program to maintain the roadbed.

Contours in muskeg terrain are usually fairly flat and do not require extensive grading. High lands between the muskegs may, on the other hand, be fairly choppy necessitating considerable grading to prepare a suitable working surface. Conventional dozing equipment is used for that work.

The key to economic right-of-way preparation in the winter is utilization of nature's freezing action to transform the muskeg into a firm working surface. Essentially, the process developed involves a gradual compressing of the surface vegetation, which eliminates the insulating characteristics of the material, causing it to freeze when suitable ambient temperatures prevail.

TABLE 3
Temperature and precipitation data

Element and station	J	F	M	A	M	J	J	A	S	O	N	D	Year	Type of normal
ALBERTA														
Athabasca (lat. 54°43'N, long. 113°17'W, elev. 1,700 ft ASL)														
Mean daily temperature (°F)	1.5	6.2	17.9	36.6	48.8	55.3	60.5	57.9	49.4	38.8	21.4	7.6	33.5	1
Mean daily maximum temperature	12.3	18.7	30.5	49.5	64.2	71.1	76.8	74.0	64.4	51.6	31.3	17.6	46.8	1
Mean daily minimum temperature	-9.3	-6.4	5.3	23.6	33.3	39.5	44.1	41.7	34.3	25.9	11.7	-2.5	20.1	1
Mean rainfall (inches)	0.00	0.00	0.02	0.43	1.67	2.77	3.00	2.50	1.37	0.36	0.10	0.03	12.25	1
Mean snowfall	11.7	9.7	8.3	3.9	1.3	0.0	0.0	0.0	0.2	4.8	9.3	11.0	60.2	1
Mean total precipitation	1.17	0.97	0.85	0.82	1.80	2.77	3.00	2.50	1.39	0.84	1.03	1.13	18.27	1
Fort McMurray A (lat. 56°39'N, long. 111°13'W, elev. 1,213 ft ASL)														
Mean daily temperature (°F)	-6.3	1.0	15.3	34.8	48.9	55.9	61.6	58.3	48.3	36.7	16.5	0.5	31.0	7
Mean daily maximum temperature	3.8	13.2	28.4	47.8	62.8	70.0	75.5	72.1	60.4	47.2	24.9	9.6	43.0	7
Mean daily minimum temperature	-16.4	-11.3	2.1	21.7	35.0	41.7	47.7	44.4	36.1	26.2	8.1	-8.6	18.9	7
Mean rainfall (inches)	0.01	0.01	0.05	0.34	1.24	2.36	2.93	2.36	1.87	0.60	0.07	0.01	11.85	7
Mean snowfall	8.3	6.4	8.3	4.1	0.7	T	0.0	0.0	0.6	4.3	8.6	8.7	50.0	7
Mean total precipitation	0.84	0.65	0.88	0.75	1.31	2.36	2.93	2.36	1.93	1.03	0.93	0.88	16.85	7
Fort Vermilion CDA (lat. 58°23'N, long. 116°03'W, elev. 915 ft ASL)														
Mean daily temperature (°F)	-9.5	-2.9	11.7	33.1	49.7	57.0	61.7	58.2	47.4	34.2	12.5	-4.2	29.1	1
Mean daily maximum temperature	0.0	8.4	24.5	44.8	62.0	70.0	74.7	71.3	59.7	43.9	20.2	4.2	40.3	1
Mean daily minimum temperature	-18.9	-14.2	-1.1	21.4	37.3	43.9	48.6	45.1	35.0	24.5	4.7	-12.5	17.8	1
Mean rainfall (inches)	0.01	0.00	0.03	0.21	1.26	1.83	2.21	1.69	1.15	0.35	0.08	0.01	8.83	1
Mean snowfall	8.3	7.8	8.1	3.4	0.8	T	0.0	0.0	0.2	4.1	8.1	10.1	50.9	1
Mean total precipitation	0.84	0.78	0.84	0.55	1.34	1.83	2.21	1.69	1.17	0.76	0.89	1.02	13.92	1
Grande Prairie A (lat. 55°11'N, long. 118°53'W, elev. 2,190 ft ASL)														
Mean daily temperature (°F)	3.1	8.4	19.5	37.0	50.0	56.3	60.3	58.6	49.9	38.1	23.3	9.5	34.5	3
Mean daily maximum temperature	12.1	18.6	29.7	47.0	61.6	67.2	71.8	70.0	61.2	48.3	32.0	18.6	44.8	3
Mean daily minimum temperature	-5.9	-1.8	9.3	27.0	38.4	45.4	48.8	47.2	38.6	27.9	14.6	0.4	24.2	3
Mean rainfall (inches)	0.01	0.03	0.05	0.27	1.48	2.47	2.38	1.96	1.15	0.60	0.21	0.11	10.72	6
Mean snowfall	13.2	11.7	7.7	4.4	0.9	0.0	0.0	0.3	1.0	5.4	8.7	12.2	65.5	6
Mean total precipitation	1.33	1.20	0.82	0.71	1.57	2.47	2.38	1.99	1.25	1.14	1.08	1.33	17.27	6
High Prairie (lat. 55°26'N, long. 116°29'W, elev. 1,965 ft ASL)														
Mean daily temperature (°F)	2.5	8.5	20.7	37.4	50.2	56.4	61.0	57.9	49.6	39.3	21.7	9.3	34.5	2
Mean daily maximum temperature	12.0	19.3	32.5	49.8	63.5	69.4	74.1	71.1	62.1	50.6	30.4	18.2	46.1	2
Mean daily minimum temperature	-7.0	-2.3	8.8	25.0	36.9	43.3	47.8	44.6	37.0	28.0	12.9	0.3	22.9	2
Mean rainfall (inches)	0.06	0.03	0.09	0.49	1.48	2.73	2.91	2.30	1.39	0.67	0.27	0.11	12.53	2
Mean snowfall	9.1	8.4	7.3	4.4	0.3	0.0	0.0	0.4	0.7	5.6	8.7	9.8	54.7	1
Mean total precipitation	0.97	0.87	0.82	0.93	1.51	2.73	2.91	2.34	1.46	1.23	1.14	1.09	18.00	1

Keg River (lat. 57°47'N, long. 117°52'W, elev. 1,402 ft ASL)

	Jan	Feb	Mar	Apr	May	Jun	Jul	Aug	Sep	Oct	Nov	Dec	Year	
Mean daily temperature (°F)	-4.8	0.8	13.3	33.9	48.7	55.5	60.3	57.2	47.9	36.7	16.2	1.2	30.6	3
Mean daily maximum temperature	6.2	13.1	26.4	46.4	63.6	69.0	74.0	70.8	60.8	47.9	25.9	11.0	42.9	3
Mean daily minimum temperature	-15.8	-11.5	0.2	21.5	34.0	42.0	46.7	43.5	35.0	25.5	6.5	-8.5	18.3	3
Mean rainfall (inches)	T	0.00	0.02	0.26	1.59	2.01	2.46	2.05	1.40	0.45	0.09	0.02	10.35	6
Mean snowfall	7.5	7.2	7.6	4.7	T	0.0	0.0	T	0.8	4.1	9.2	9.8	50.9	6
Mean total precipitation	0.75	0.72	0.78	0.73	1.59	2.01	2.46	2.05	1.48	0.86	1.01	1.00	15.44	6

Wabasca (lat. 55°50'N, long. 113°50'W)

	Jan	Feb	Mar	Apr	May	Jun	Jul	Aug	Sep	Oct	Nov	Dec	Year	
Mean daily temperature (°F)	0.4	4.5	20.6	33.8	50.1	57.2	63.5	59.3	49.4	37.9	20.8	7.1	33.7	8
Mean daily maximum temperature	9.5	14.9	31.4	44.5	61.1	67.8	74.3	70.8	59.6	46.4	28.4	15.8	43.7	8
Mean daily minimum temperature	-8.7	-6.0	9.8	23.0	39.1	46.6	52.6	47.8	39.2	29.3	13.2	-1.6	23.7	8
Mean rainfall (inches)	0.05	0.00	0.01	0.18	1.06	2.71	2.01	2.00	1.31	0.42	0.20	0.01	9.96	8
Mean snowfall	5.1	4.6	6.6	1.7	0.1	0.0	0.0	0.0	0.4	2.1	6.3	5.7	32.6	8
Mean total precipitation	0.56	0.46	0.67	0.35	1.07	2.71	2.01	2.00	1.35	0.63	0.83	0.58	13.22	8

Whitecourt (lat. 54°08'N, long. 115°40'W, elev. 2,430 ft ASL)

	Jan	Feb	Mar	Apr	May	Jun	Jul	Aug	Sep	Oct	Nov	Dec	Year	
Mean daily temperature (°F)	5.0	10.5	21.8	36.9	48.3	54.4	59.7	56.5	48.5	38.0	21.7	8.8	34.2	3
Mean daily maximum temperature	14.8	22.8	33.5	49.2	62.0	66.9	72.9	69.2	60.9	50.0	31.7	18.5	46.1	3
Mean daily minimum temperature	-4.8	-1.8	10.1	24.6	34.6	41.9	46.5	43.8	36.0	26.0	11.7	-0.9	22.3	3
Mean rainfall (inches)	0.01	0.02	0.03	0.46	1.76	2.87	3.86	3.33	1.26	0.43	0.16	0.09	14.28	6
Mean snowfall	11.1	9.7	7.8	7.4	0.9	0.1	0.0	0.0	0.5	6.4	6.9	9.5	60.3	6
Mean total precipitation	1.12	0.99	0.81	1.20	1.85	2.88	3.86	3.33	1.31	1.07	0.85	1.04	20.31	6

BRITISH COLUMBIA

Fort Nelson A (lat. 58°50'N, long. 122°35'W, elev. 1,230 ft ASL)

	Jan	Feb	Mar	Apr	May	Jun	Jul	Aug	Sep	Oct	Nov	Dec	Year	
Mean daily temperature (°F)	-8.4	0.4	16.3	34.7	50.0	57.8	62.2	58.5	48.8	34.1	10.2	-4.8	30.0	6
Mean daily maximum temperature	-0.6	10.2	28.2	46.2	61.7	69.1	73.9	70.4	59.8	43.1	17.1	2.0	40.1	6
Mean daily minimum temperature	-16.2	-9.4	4.4	23.1	38.2	46.4	50.5	46.6	37.8	25.0	3.3	-11.6	19.8	6
Mean rainfall (inches)	0.00	0.01	0.02	0.23	1.39	2.60	2.56	1.99	1.17	0.37	0.02	T	10.36	6
Mean snowfall	9.5	10.3	10.0	5.1	1.5	T	0.0	T	1.7	6.4	12.1	11.1	67.7	6
Mean total precipitation	0.95	1.04	1.02	0.74	1.54	2.60	2.56	1.99	1.34	1.01	1.23	1.11	17.13	6

Fort St John A (lat. 56°14'N, long. 120°44'W, elev. 2,275 ft ASL)

	Jan	Feb	Mar	Apr	May	Jun	Jul	Aug	Sep	Oct	Nov	Dec	Year	
Mean daily temperature (°F)	4.2	10.6	22.1	38.0	50.6	56.5	61.1	58.8	50.8	39.7	21.6	9.0	35.3	3
Mean daily maximum temperature	11.6	18.9	30.4	47.2	61.6	66.5	71.2	67.7	60.5	47.4	28.3	16.2	44.1	3
Mean daily minimum temperature	-3.1	2.3	13.7	28.7	39.6	46.5	51.0	48.6	41.1	31.6	14.9	1.7	26.4	3
Mean rainfall (inches)	0.01	0.01	0.04	0.27	0.98	2.33	2.52	2.11	0.96	0.45	0.13	0.01	9.82	6
Mean snowfall	12.1	11.5	10.0	6.3	2.3	0.3	0.0	0.9	1.6	7.6	10.6	12.8	76.0	6
Mean total precipitation	1.22	1.16	1.04	0.90	1.21	2.36	2.52	2.20	1.12	1.21	1.19	1.29	17.42	6

FIGURE 3

FIGURE 4

FIGURE 5

FIGURE 6

TABLE 4
Equipment bearing pressures

Equipment type	Horse-power rating	Ground pressure (psi)	
Bombardier type muskeg vehicle	95	3.1	
8 ton foremost type muskeg vehicle	250	1.3	
320 type ditcher	150	8.7	
3/4 yard cable type backhoe/clam	140	6.6	
1 yard cable type crane	180	7.6	
D-6 type dozer 24 inch tracks	140	5.3	
D-6 type dozer 30 inch tracks	140	4.2	
D-7 type dozer 27 inch tracks	180	6.9	
D-8 type dozer 26 inch tracks bare	270	9.6	
D-8 type dozer 26 inch tracks with ripper	270	11.2	
D-8 type dozer 26 inch tracks with winch	270	10.9	
D-9 type dozer 27 inch tracks std bare ripper	385	13.5	
571 type sideboom	180	10.5	
572 type sideboom	180	11.1	
583 type sideboom	270	12.4	
594 type sideboom	385	13.7	
955 type loader	130	10.9	
977 type loader	190	10.7	
Model 12 type grader	125	31.0	
Model 16 type grader	225	27.5	
		Front	Rear
3/4 ton crew cab		46.3	58.7
3/4 ton pickup		42.6	54.0
2 ton truck		57.0	70.5
5 ton truck		46.0	75.0
10 ton truck		46.0	63.5
Tractor trailer (highway)			
Tractor		52.0	58.0
Low boy trailer			45.0
Oilfield tractor trailer c/w D-9 load			
Tractor		57.0	75.0
Low boy trailer			55.0

Insulating snow cover is first removed from muskeg areas by light dozers. Areas which cannot safely be crossed by the dozers are first traversed by tracked vehicles. As frost penetration progresses downwards, larger dozers work back and forth over the area (Figure 6) until frost penetration has reached the desired level. Underlying organic material is thereby compressed and caused to freeze until the right of way is capable of supporting the heaviest loads which will be involved during construction.

Only the working side of the right of way is prepared in this manner (Figure 7), the 'ditch' side receiving different treatment as noted below. In its natural unfrozen state, muskeg often cannot support a load of 2 pounds per square inch, but when the material has been frozen to a depth of 18 inches, the same muskeg is capable of supporting repeated passes of loads in excess of 60 pounds per square inch! Bearing pressures for various types and sizes of equipment used during construction are summarized in Table 4.

Clearing of timber is a machine operation in the winter (Figure 8). Special sharp cutting blades attached to caterpillar type dozers readily cut down the frozen trees. The timber is piled by machines (Figure 9) and then burned. Mechanization of the clearing operation and elimination of the rip-rap operation reduce manpower requirements substantially, while at the same time productivity increases dramatically, in comparison with the previously described summer clearing operation.

Grade preparation in the high lands is one of the few operations which is more costly in the winter than in the summer. Large dozers equipped with heavy-duty rippers are used to break up frozen soils and permit levelling of the right of way. Ideally, grading of accessible high lands should start as early in the winter as possible.

A snow 'roach' (Figure 10) is built up over the centre line of the future ditch to retard frost penetration in that area. Care is taken to avoid traversing this line with equipment, particularly in the high lands between muskegs where excavation becomes much more difficult if the soil is frozen:

Transportation of Materials and Personnel

In a summer operation, materials are hauled from the stockpile area or rail siding to the closest access point by conventional trucks. The material is then transferred to tracked vehicles or to athey wagons towed by caterpillar type tractors. The hauling equipment moves at a snail's pace over rip-rap areas. During the summer months, when heavy rainfall is experienced in many of these areas, the soils can become saturated, even in the highlands, and the road quickly becomes a quagmire as the heavily laden equipment passes over.

Tracked vehicles are usually used to transport personnel to the work site from the nearest access road, but in some cases helicopters may be required.

Transporting materials for winter pipeline projects is much easier than for a summer job. All track-type hauling equipment is eliminated, and conventional trucks (Figure 11) are used instead. These can maintain considerable speed because of the smooth travelling surface (Figure 12). Tow cats are required only on very steep hills between stretches of muskeg.

Men, fuel, and supplies are quickly transported to the work area by the use of conventional buses and trucks.

Ditch Excavation

Summer ditching in muskeg is accomplished using backhoes supported on the muskeg by 'timber mats.' Each backhoe normally works with three mats, standing on two and moving one ahead in its line of travel.

The mats are usually 8 feet by 18 feet, but vary in size according to the contractor's preference. They may be constructed of logs held together with cable or of heavy planking nailed together in a laminated fashion. Each mat is fitted with a cable loop that is 'hooked' by

FIGURE 7

FIGURE 8

FIGURE 9

FIGURE 10

FIGURE 11

FIGURE 12

FIGURE 13

FIGURE 14

FIGURE 15

FIGURE 16

FIGURE 17

the backhoe bucket and set in place.

Working off mats is a very slow operation; in muskeg it is not uncommon for a 3/4 yard backhoe to excavate less than 500 feet of ditch per day.

Conventional wheel type ditchers are used for excavation in the high land encountered between muskeg areas.

In a winter operation wheel type ditchers (Figure 13) are normally used in muskeg areas (Figure 14) as well as on high land, but where there are creek crossings and water channels, or the ground has frozen too hard, backhoes (Figure 15) are employed.

Dozers clear off the snow berm ahead of the ditchers, and the frozen surface is broken up by a ripper tractor (Figure 16). Under favourable conditions, a wheel type ditcher can trench close to 2,000 feet per day.

In the winter, ditching is scheduled to follow welding instead of leading it, so that the pipe can be lowered-in and back-filled before the spoil bank has frozen into unmanageable lumps (Figure 17).

Pipe Bending

The pipe must be bent to conform to natural contours of the land, and to the alignment of the selected route. The amount of bend required on each joint of pipe is predetermined by bending engineers. The pipe is then fed through a bending machine where the desired curvature is obtained by using a progression of short 'pulls' along the pipe. Because of the relatively flat terrain encountered in muskeg construction, limited bending is required, except over the high land encountered periodically along the route. Bending does not present

FIGURE 18

any special problems during summer construction, except for the very slow travel over rip-rap areas.

In the winter, welding precedes ditching (Figure 18) and since bending must be accomplished prior to welding it also must precede ditching. Because of the increased production of winter operations bending must proceed at a rapid pace to provide ample bending time in the hilly high land. Care must be taken to ensure that no snow or ice is stuck to the pipe, both inside and out. Any snow or ice on the inside of the pipe would impede the passage of the internal bending mandrel, and snow or ice on the external surface would make it impossible to achieve a perfect fit into the bending shoes.

Line-up and Welding of Pipe

Pipe ends are buffed down to bright metal in preparation for welding. A sideboom tractor then carries the joint ahead (Figure 19), where it is aligned and held in place by an internal, air-activated line-up clamp (Figure 20), ready for the initial welding pass. The first welding pass is made by the 'stringer bead' welders who then move on to line up the next joint. The 'hot pass' welders follow right behind with a complete second pass around the pipe. Subsequent passes required to fill and cap the weld are performed by the 'firing line' or 'back end' welders.

Welding machines for back end welders are mounted on sleds and towed by tracked vehicles for summer construction.

In the winter, when the working surface is firm, the progress of the 'pipe gang' can be remarkable. There are some special precautions which are taken in the winter, however:

FIGURE 19

FIGURE 20

snow is swabbed from the inside of each joint of pipe to allow the passage of a pipeline 'pig' after the pipe if welded into a continuous piece; pipe ends are preheated (Figure 21) to predetermined temperatures; hot pass welders follow close behind the pipe gang to ensure that the weld area does not cool between weld passes.

The weld area is again preheated ahead of the firing line welders, to maintain the weld zone at the desired temperature throughout the filler and cap passes. Welding machines are mounted on light trucks for winter operations, allowing greater mobility than is possible in summer work in these muskeg areas (Figure 22).

An asbestos insulating 'blanket' is wrapped around the joint following completion of the weld to control cooling.

Coating and Lowering-in

Once the pipe is welded, a protective coating is applied to prevent corrosion of the buried line. A number of different coating materials may be used, such as asphalt enamel, asbestos felt wrap, or any of a number of plastic tape coatings. Special machines clean and prime the pipe and apply the various coating materials selected for a particular project.

In the summer the heavy tractors required to 'cradle' the pipe during the coating operation play havoc with the rip-rap. Tow tractors are seldom idle as machines are frequently breaking through or slipping off the rip-rap. The ditch tends to close in because of its fluid nature. Clams often work around the clock to keep the ditch open ahead of the lowering-in operation.

Winter coating and lowering-in crews have a much easier time than summer crews. The firm roadbed enables the heavy equipment to move freely down the right of way. The

FIGURE 22

tendency for the ditch to slough in is substantially reduced because of the penetration of frost. Only as much ditch is opened up each day as can be reasonably coated and lowered-in that same day, in order to ensure that the ditch spoil does not freeze before it is backfilled. A cleaning machine with heavy revolving brushes cleans ice and snow from the pipe (Figure 23). This is followed by a 'train' (usually three) of direct-fired propane heaters (Figure 24) that heat the pipe to approximately 100°F in order to melt excess snow and drive off the moisture. Immediately after the heaters comes the actual coating application machine (Figure 25). This machine has revolving arms that hold huge rolls of tape and outer-wrap, and as it moves down the pipe, the arms revolve and apply the tape and outer-wrap in a spiral configuration.

Installation of Weights and Anchors

Weights and anchors each have their advantages and disadvantages. Although anchors are light and compact and may be transported in great numbers at a time, they are only effective in certain areas. The ditch should be free of water, which may be impossible to achieve if there is no means of drainage. Soil conditions must be ideal to accommodate the auger-type anchor, which is bored into the ground. The presence of rock or gravel virtually eliminates their use. Unless the ditch line of a project is tested before actual construction, the potential success of the anchors cannot be determined. Even when conditions are favourable, difficulties will be encountered necessitating last-minute decisions to switch to weights in some areas. Concrete weights are heavier and more expensive, but can be installed quickly.

In the summer placement of weights is impaired by the narrow confines of the rip-rap area,

FIGURE 23

FIGURE 24

FIGURE 25

FIGURE 26

and equipment must be operated expertly to keep it on the unstable, slippery surface (Figure 26).

In a winter operation a major expense is eliminated in the transportation and setting of weights. Large trailer trucks can transport them rapidly over good roadway directly to the place of work. Weights are off-loaded from the trucks and stock-piled on the edge of the right

FIGURE 27

FIGURE 28

of way (Figure 27). Immediately behind the lowering-in operation, another crew sets the weights on the pipe (Figure 28). Solid underfooting is afforded the equipment used for setting the weights on the pipe.

Bolt-on weights (Figure 29) are used for river and stream crossings and other areas where the welded pipe section must be carried to the final location (Figure 30).

When anchors (Figure 31) are used instead of weights, the ditch must be dewatered before the anchors are installed. The anchors are then installed using special auger type driving heads (Figure 32) that are extended to reach out over the ditch.

FIGURE 29

FIGURE 30

FIGURE 31

FIGURE 32

FIGURE 33

Tie-ins

Pipelines are welded into sections of up to a mile in length for ease of installation. A special 'tie-in' crew then joins the sections together into one continuous line, after they have been placed in the ditch.

In muskeg regions tie-ins must sometimes be performed above the ditch when the water cannot be pumped from the ditch proper, though they are generally planned for the high land between muskegs to avoid this problem.

Preheating and controlled cooling of tie-in welds is required, as for all other welds on winter construction.

Back-fill, Clean-up, and Right-of-way Restoration

Wide-tracked dozers are used for summer back-fill and clean-up in the high land between muskegs. 'Mormon boards' are used to back-fill the ditch in muskegs. The 'Mormon board' is similar to a drag line except that the bucket is replaced by a blade similar in size to a small

FIGURE 34

dozer blade, but somewhat lighter in construction. Operating from the rip-rap working surface, the machine operator 'casts' the blade over the far side of the ditch spoil and 'reels' in, pulling the spoil into the ditch and covering the pipeline.

Normally on Crown land forestry regulations permit the rip-rap to be left on the surface, provided it is in a neat fashion (Figure 33). However, when the property is privately owned, the land owner can demand its removal.

To avoid the risk of forest fires, the Department of Lands and Forests' regulations normally do not permit the disposal of timber by burning during the summer months. This operation is usually performed in early November but not before the first snowfall. Men and equipment must again be mobilized and transported to the job site. Usually caterpillar tractors, equipped with large rakes, are used in the burning operation, for piling the brush and stirring up the fires. The machines are also equipped with winches in the event that one of them requires help because of the soft terrain. Power saw operators trim the edges of the right of way.

In the winter back-fill and clean-up can be accomplished quickly and efficiently with a minimum of equipment, since the brush disposal has already been taken care of by the clearing crew. Dozers (Figure 34) may be used to back-fill the ditch. The frozen working surface permits machines to move about freely on either side of the ditch spoil, and a meticulous clean-up job can be performed (Figure 35). In muskeg areas back-fill is placed high over the ditch line to allow for subsidence when spring thaws and rains occur.

Hydrostatic Testing

Every pipeline must be tested to meet rigid requirements established by the authorities before it may be put into operation. Water is used almost exclusively as the testing medium, because of its abundance and relative safety. The pipe is cut into sections based on elevation

FIGURE 35

changes along the route and each section is then filled and pressured up to required limits. The 'squeeze' is held for the period of time specified by the authorities, usually 24 hours.

As test sections are relatively long, because of the flat nature of the terrain, sidebooms must be walked long distances over rip-rap for summer testing. Large fill pumps, squeeze pumps, test shacks, and fill pipe must be loaded on huge swamp transporters and moved to the test site. Welding equipment must be mobile to provide welders access to any part of the test section where their services may be required.

Although movement of men and equipment for winter testing is relatively simple because of the reliable roadbed afforded by the frost, other major problems do exist. The ever present danger of the water freezing in the pipeline is a constant worry to the contractor. Temperatures often plunge to 40 degrees below zero and special precautions must be taken to avoid anything detrimental to the test. Test sections are usually shortened in length to enable water to be evacuated quickly in the event of a major equipment breakdown. Large water heaters heat the water, which is then circulated through the pipe. When the required discharge temperature is achieved, the section is shut in and the squeeze put on. All exposed piping is covered over and heated to prevent freezing. Temperature recorders are inserted in the ditch line down to the pipe at regular intervals and monitored constantly by qualified personnel during the 24-hour test.

In the event that abnormally severe weather sets in and the recorders indicate a freezing condition, the pressure is relieved immediately and hot water is again circulated. After the successful completion of the test on a section, the same procedure is repeated on down the line until the entire line has been tested.

As the sections are completed, tie-in crews follow up and set the necessary valves and tie the pipeline into continuity.

Equipment Maintenance

One of the major difficulties of summer maintenance in muskeg terrain is access to equipment. Mechanics must have wide-tracked equipment to carry their tools over the rip-rap. Often on a major repair the machine must be fixed 'on the spot.' All machines are subjected to abnormal wear and tear due to the wet conditions of the terrain. Mud and water play havoc with tracks, rollers, and undercarriage of swing machines, which constantly require attention.

Cold temperatures present new problems in the winter. Severe low temperatures cause equipment components to become brittle so that breakage occurs with increasing frequency. Mechanics and welders work round the clock on winter projects; often repairs must be carried out in the field. On major repairs portable tents are provided to give protection to the maintenance personnel. If a cold snap sets in and temperatures drop drastically, equipment must be left running; otherwise hydraulic systems, lubicating systems, and cooling systems may fail to function properly when the machines are restarted and mechanical failure will result.

Summary

Table 5 shows a comparison of relative manpower requirements and Table 6 shows a comparison of relative equipment requirements, for a typical pipeline project under both summer and winter conditions. Productivity on this typical project would be 50 per cent higher in the winter than in the summer. The over-all economic saving would be in the order of 20 to 40 per cent.

ENVIRONMENTAL CONSIDERATIONS

From the foregoing description of winter and summer pipeline construction in muskeg regions, it is apparent that winter methods are environmentally superior because: (a) damage to the right-of-way is substantially reduced, (b) no rip-rap is left embedded in the muskeg (Figure 36), (c) soil erosion control is enhanced (Figure 37), (d) quality workmanship is more readily assured.

Now let us look at the over-all perspective of pipelines versus other methods of transportation.

Given the fact that oil and gas discoveries are made in muskeg regions, and that society will insist that these reserves by made available for consumption, our attention can be focused on the question of how best to transport these products from source to market. The three basic transportation modes available to us are truck, rail, and pipeline. Examining these three alternatives from the environmental point of view, it is not too difficult to see that cross-country pipelines are superior.

Railway rights-of-way must follow grade contours with limited slope and therefore meander over the countryside in a zigzag fashion, often following scenic valleys for miles to reach

TABLE 5
Comparison of manpower requirements, summer and winter,
for a 36 inch o.d. pipeline and 33 per cent muskeg right-of-way

	Manpower	
Classification	Summer	Winter
Supervision and administration	28	26
Equipment operators	102	84
Equipment mechanics	8	10
Equipment service	26	30
Truck drivers	40	48
Bus drivers	12	8
Welders and fitters	36	40
Welders' helpers	38	42
Common labour	88	94
Skilled labour	54	14
Totals	432	396

TABLE 6
Comparison of equipment requirements, summer and winter,
for a 36-inch o.d. pipeline and 33 per cent muskeg right-of-way

	Quantities	
Description	Summer	Winter
Sideboom tractors	16	15
Bulldozers	24	30
Tow tractors	12	6
Road graders	2	2
Scrapers	3	-
Tree farmers	3	-
Backhoes/clams	18	14
Wheel-type ditchers	1	4
Compressors	3	4
Quad welders	3	3
Bending machines	1	1
Line-up clamps	1	1
Line preheaters	-	3
Coating machines	1	1
Tracked vehicles	24	2
Fuel/service trucks	5	5
Tandem tractor/trailers	30	40
Welding rigs	2	24
Mechanics rigs	6	8
Small trucks	36	42
Buses	12	10
Athey wagons	4	-
Buffing rig	1	1
Water pumps	3	3
Fill pumps	1	1
Squeeze pumps	1	1
Welding machines	18	

FIGURE 36

suitable river-crossing sites. Because of the grade restrictions, deep cuts and fills are often made and the finished product is a permanent, sometimes obtrusive, monument.

Highway rights-of-way are somewhat similar in construction to railways. Lesser grade restrictions reduce the severity of cuts and fills, but the necessity for two-way traffic requires much wider road surfaces. Paving of road surfaces permanently disables the life-sustaining cycle in muskeg areas, and the operation of the huge tank trucks themselves over such highways is a significant pollutant.

Pipeline rights-of-way, on the other hand, can tolerate much steeper slopes, the primary requirement being ability to traverse the route with construction equipment. Cut and fill methods are used to a much smaller degree than in road construction to create a working surface for construction. The right of way is restored to its original contour, in most areas, upon completion of construction and the entire area is then revegetated. A year or two after construction, it is indeed often difficult or impossible to detect visually where the pipeline is located.

It is interesting to note that the number of miles of pipelines in Canada now exceeds the number of miles of railways. The average person is, however, scarcely aware of the existence of pipelines anywhere in this vast country, whereas roads, highways, and railroads are in evidence wherever we travel.

Safety aspects of transportation systems are another important environmental consideration. Pipelines provide the cleanest and safest means of transportation today. A recent report (1971) by the National Transportation Advisory Board of the United States covering a

FIGURE 37

six-year period showed that the average fatalities for each billion ton-miles of freight move-ment by various modes were: (a) federally regulated trucking, 10.90; (b) railroad freight, 2.50; (c) commercial shipping, 0.31; (d) petroleum pipeline movements, 0.011. In other words, pipeline movements were 30 times safer than commercial shipping, 250 times safer than railroad freight movements, and 1,000 times safer than regulated trucking. That is a truly outstanding safety performance.

   This safety record is no mere quirk. The design requirements for cross-country pipelines are set out in long-established codes. Recently issued Canadian codes for both oil and gas pipeline transmission systems (Canadian Standards Association, 1973) provide even more stringent requirements than those used in the past.

   Inspection during the construction of pipelines is another measure taken to ensure the highest quality. Line pipe is inspected in the mills either by x-rays or ultrasonically and each joint of pipe is then hydrostatically tested. Usually every circumferential field weld is x-rayed before the line is buried and hydrostatically tested.

   Although it is impossible to eliminate the possibility of accidents, pipeline operating companies are making great strides, on their own initiative, to minimize them. Many of our major pipelines systems are already equipped to detect sudden pressure drops and to shut down the system instantaneously until the cause of the drop can be determined and corrected.

Highly sophisticated automation of future pipelines is anticipated, but existing pipelines have not been forgotten either. Electronic devices that can be transported through the operating pipeline are being developed. These devices can detect defects in the line before failure so that they can be repaired during routine maintenance programs.

In addition to all other safety measures, the owning companies have adopted a policy of upgrading older lines by taking them out of service and resubjecting them to pressure tests. Defects located in this manner can then be repaired before the line is put back in service.

In conclusion, pipeline construction in Canada's muskeg regions is not only highly desirable from the economic viewpoint but it is also environmentally far superior to available transportation alternatives.

REFERENCES

Banister, R.K. 1967. Winter pipeline construction in western Canada. Proc. 10th Internat. Gas Conf., Hamburg, IGU/C6-67.
Canadian Standards Association. 1973. Pipeline codes. CSA Standard Z183; Oil Pipeline Transmission Systems.
Canada, Department of Transport Meteorological Branch. 1967. Temperature and Precipitation Tables, vols. 1 and 3:
Gant, W. and Meeres, R.D. 1972. Construction Problems Arctic Pipelines. Nat. Res. Council. Tech. Memo. 104, NRC 12498.
MacFarlane, I.C. (ed.). 1958. Guide to a Field Description of Muskeg (Based on the Radforth Classification System). Nat. Res. Council, Tech. Memo. 44 NRC c.4214.
— 1969. Muskeg Engineering Handbook (Univ. Toronto Press), chap. 7, pp. 234-42, and chap. 8, pp. 261-86.
National Transportation Advisory Board, United States of America. 1971. Fatality Rates for Surface Transportation 1963-1968.

12
Water Resources

D.A. GOODE, A.A. MARSAN, and J.-R. MICHAUD

INTRODUCTION

Quantitative Aspects: The Water Balance of Peatlands

Popular opinion has long held the view that peat bogs regulate river flow by gradually releasing water from the peat 'sponge' over long periods and thereby sustaining the flow of rivers during periods of drought. In Europe it is suggested that this idea originated from an utterance of Alexander von Humboldt, which perhaps explains its unquestioned acceptance. At first sight the concept seems logical on account of the obvious wetness of these areas, but does it really hold true?

Any influence which muskeg might have on water resources is obviously of considerable importance in view of its vast extent and the fact that it is the predominant landform of many northern river basins. It is therefore surprising that practically no research has been done on the water balance of peatlands in Canada. This would be true for North America as a whole, but for the notable exception of the hydrological studies of forested bogs in Minnesota carried out by the United States Forest Service and described in several publications by Bay (1968) and Boelter (1965). Most of the information on the water balance of Boreal peatlands comes from Europe and particularly from the European territory of the USSR, where intensive studies of the hydrological regime of natural peatlands have been under way for nearly 30 years.

In the absence of water balance studies on muskeg it might be possible to gain some measure of its effect on water resources by comparing the flow characteristics of rivers draining from muskeg with those of predominantly mineral ground catchments, but this requires continuous data over long periods which might not be available. So, for the present, consideration of the effect of muskeg on water resources will have to rely to a large extent on the evidence from water balance studies elsewhere in the Boreal Zone.

For any catchment the water balance equation can be written

$$P + I = E + R \pm \Delta W$$

where P is the precipitation, I is the inflow from surrounding areas (which is zero in ombrogenous peatlands), E is the evaporation, R is the runoff, and ΔW is the change in water storage, which can be a negative quantity.

There are three main questions which require an answer in considering the role of muskeg in water resources.

1. Is there a significant difference in the ratio of the individual components of the balance equation when one compares peat-covered with mineral soil catchments? For instance, is there a difference in the ratio of evaporation to runoff?
2. What is the effect of the storage characteristics of peatland on the pattern of runoff, as compared with non-peat catchments?
3. Is the type of peatland important in determining the nature of such differences?

Before considering the evidence from studies of peat catchments it might be useful to consider how far inherent physical properties of peatlands affect individual components of the water balance equation.

Qualitative Considerations

(a) The Concept of Environmental Quality
Before an attempt is made to assess the contribution of muskeg to the quality of the environment, it seems necessary to return briefly to the basic definitions, in order to establish the following discussion on firmer grounds.

An ecosystem is a more or less closed environment where the resources of the site are cycled by a biomass of plant and animal populations associated in mutually compatible processes (Dansereau, 1967). According to Dansereau (1971) the quality of any actual environment for any given species of plant or animal (including man) at any given time can only be defined in terms of the total ecological adjustment of the living population. These modes of adjustment are requirement, tolerance, and tapping power. These notions set the framework for a discussion of the quality of environment centred on man. An environment of high quality will first of all ensure the biological survival of man and, in addition, it will provide man with aesthetically and culturally satisfying surroundings which are adequate for his spiritual nature. One quality of the environment which should be mantained at all cost is diversity, for in the biota it promotes that biological stability without which chaos is forever eminent, while in the landscape it satisfies man's innate desire for interest, beauty, and peace of mind (Provost, 1972).

Then, even if the notion of environmental quality is an elusive one, there exist, for a given organism, including man, some sets of natural variables that characterize an environmental state which is best suited to the needs of that particular organism. Therefore, the quality of a natural environment, as described by the different physicochemical or biological characteristic parameters, can theoretically be assessed objectively. This is true also for changes in those characteristics, which may occur naturally or be the effect of human activities. However the value of the assessment will be no better than the state of our actual knowledge on the subject, and such objective evaluation of secondary effects is bound to change as more scientific facts become unveiled.

As expressed clearly in Provost's definition (Provost, 1972), the notion of quality of the environment is closely associated with man's evolution when the requirement stage is satisfied. The anthropocentric notion of environmental quality raises the question of human

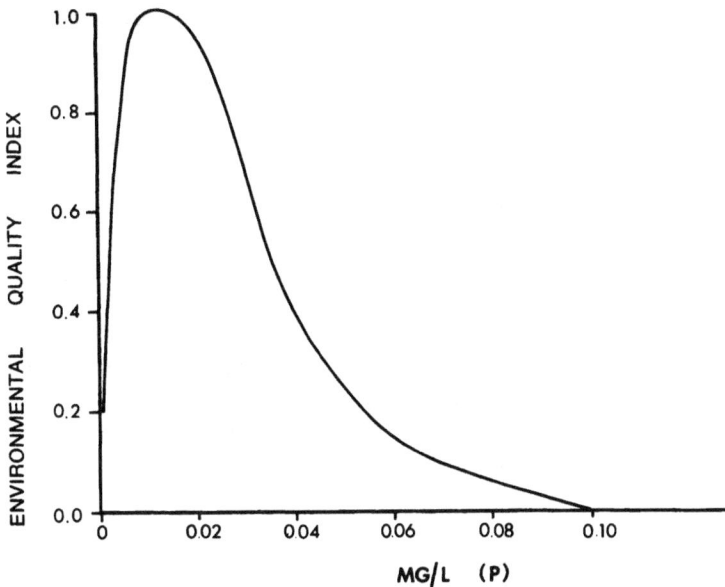

FIGURE 1   Environmental quality function for phosphates.

perception of the values of natural factors. The importance attached by individuals to changes in the environment and the perception of ecological factors are eminently variable between individuals of any one culture and between the cultures and socio-economic classes of one community. Even within the scientific community many divergences exist concerning the importance of environmental considerations in benefit-cost analysis. Given a set of environmental quality parameters that will be affected by some proposed development, an interdisciplinary group of highly qualified scientists will find it extremely difficult to agree on quantitative weighing factors that would rank the environmental parameters according to their relative importance (Battelle, 1972). The use of standards is important in administering and enforcing a desired policy, but they are not a complete tool for evaluating environmental quality. Essentially, environmental quality is not confined to a bad or good scale, but includes a range of values (Dee et al., 1973). Therefore, and this is certainly the more difficult part of the problem, values, culture, and socio-economic status will play an important role in development priorities. In the end, an environmental cost must always be paid and human ecology recognizes the interplay of education, awareness, leadership, and economic development priorities that set the cost a human community is prepared to pay.

(b) Environmental Quality Functions
The concept of environmental quality functions was recently developed by Battelle (Columbus Laboratories) as a means of assessing environmental impacts (Battelle, 1972). This system considers sets of ecological, environmental, aesthetic, and social parameters associated with environmental quality and evaluates each individual contribution on a zero-to-one normalizing scale. The zero value is applicable when the quality of the environment as

characterized by the contemplated parameter is so bad that the ecological processes that allow cycling of resources are hindered. This could be due to toxicity effects of various pollutants resulting from human activities, insufficient concentrations of essential nutrients, or overdoses of substances that are otherwise harmless or beneficial.

For example, Figure 1 give the environmental quality function of the phosphates in an aquatic environment. Sufficient work has been done on the effect of this essential nutrient in lakes to establish the general value function illustrated by the figure (McKee and Wolf, 1963). No primary productivity is possible in a water body devoid of phosphorus; consequently, the environmental quality index (EQI) drops rapidly to zero as the concentration of inorganic phosphate approaches zero. Above 0.07 mg/l (as P) excessive growth of algae can develop and this value is considered as the maximum permissible value (Curry and Wilson, 1955). This value has been given an index of 1.0, considering that it will permit high productivity without impeding on other environmental processes essential to animal life. Above this value the quality index falls rapidly to zero. Thus, environment quality indices can be determined corresponding to the different values assumed by environmental parameters in various situations.

THE WATER BALANCE OF PEATLANDS

The Influence of the 'Active Layer' on the Hydrological Regime of Peatlands

(a) Nature of the Active Layer *
The concept of an active layer at the surface of raised bogs, within which most moisture exchange takes place, has been proposed by Ivanov (1953a) and Romanov (1953, 1961). This active layer includes the living vegetation (as long as it is sufficiently dense to form pores for the capillary movement of water) together with the underlying layers of partially decomposed vegetation. The lower boundary is not clearly defined, but is considered to be the level above which considerable changes in water conditions and degree of decomposition occur, whereas below it these are negligible. Fluctuations of the water table within the active layer cause inverse fluctuations in aeration; consequently decomposition of plant material proceeds at a fairly rapid rate and so the active layer coincides with the peat-forming layer. The main feature of this layer is the pronounced increase in decomposition of organic material with depth. This, together with the pressure of successive layers, results in a substantial decrease in porosity towards the base of the active layer.

Variations in water conditions within the active layer are primarily dependent on this change in physical structure. Rapid changes in permeability with depth within the active layer have been widely demonstrated, e.g. by Ivanov (1953a), Vorobiev (1963), and Romanov (1961) in Russia, and Boelter (1965) and Korpijaakko and Radforth (1972) in North America. The hydraulic conductivity close to the surface is often thousands of times greater than at the base of the active layer. Consequently there is a very marked decrease in lateral subsurface flow as the water table falls towards the base of the active layer. Although interflow can occur in substantial amounts within the active layer, water is unable to move downwards through the relatively impermeable inert layers. Ivanov (1953a) considers that

* The term 'active layer' as used in peat hydrology should not be confused with the 'active layer' of permafrost.

lateral flow within the inert horizons constitutes only about 1 per cent of the total runoff from a bog massif. He cites values of hydraulic conductivity below the active layer in raised bog as low as 1 x 10^{-5} cm/sec. Boelter (1965) gives similar figures for highly decomposed peat (0.75 x 10^{-5} cm/sec)and points out that this is lower than many glacial till soils. These peats have many small pores which are not easily drained and which produce specific yields of only 0.10 to 0.15 cm^3 per cm^3. In contrast Boelter gives a figure exeeding 3,810 x 10^{-5} cm/sec for undecomposed moss peat at horizons close to the surface of the active layer.

The relation between the hydraulic conductivity and degree of humification is shown in Figure 2, which summarizes the results of over 2,000 measurements of permeability in different peat types (Baden and Eggelsmann, 1963). This does not include peat in its least humified state, which would be considerably more permeable. A similar relationship is shown in Figure 3 (Korpijaakko and Radforth, 1972), which includes the values for unhumified material. In both cases the relationship is exponential. In raised bogs Romanov (1961) has shown that the percentage volume of dry matter increases with depth throughout the active layer. Since there is a linear relationship between this percentage and the degree of humification on the von Post scale (Korpijaako and Radforth, 1972), Figures 2 and 3 provide an indication of the change in hydraulic conductivity with depth. (The influence of depth alone is relatively insignificant and in *Sphagnum* peat, where the humification is greater than 5, it is unimportant. This distinction need not concern us here.) The essential fact is that the hydraulic conductivity decreases exponentially with depth in the active layer of raised bogs. If the increase in degree of humification with depth is irregular, with abrupt changes from unhumified to strongly humified conditions, then the effect on the hydraulic conductivity will be dramatic. Such discontinuities have been described in the case of blanket bog (Goode, 1970), raised bog (Godwin and Conway, 1939), and wooded bog (Boelter, 1972a). They can be expected wherever renewed bog growth occurs after drier phases or drainage. It is also worth mentioning here that frozen subsurface layers will have the same effect.

Ivanov (1953b) took account of different types of peat growth in his suggestion (based on filtration flume experiments) that the vertical variation of hydraulic conductivity is adequately expressed by the equation

$$K_z = A/(z + 1)^m$$

where K is the hydraulic conductivity (cm/sec) at depth z (cm) from the surface, and A and m are constants which are applied for various elements of bog microtopography (the limits of z being 0 and 5 for any one value of A or m). Representative samples of A and m are given below:

	A	m
Pine-dominated bog with *Sphagnum* and dwarf shrubs	41,700	3.8
Sphagnum bogs with dwarf pine and dwarf shrubs in centre of raised bog	3,670	2.38
Pools with *Sphagnum* and Eriophorum	2,290	2.67

The physical structure of the active layer and its effect on moisture characteristics are discussed at length in Romanov (1961), Novikov (1972), Doodge (1972), and Verny and Boelter (1972). Although this concept has been mainly developed and applied to domed ombrogenous bogs, it is clear that similar conditions prevail in the ridges of minerotrophic

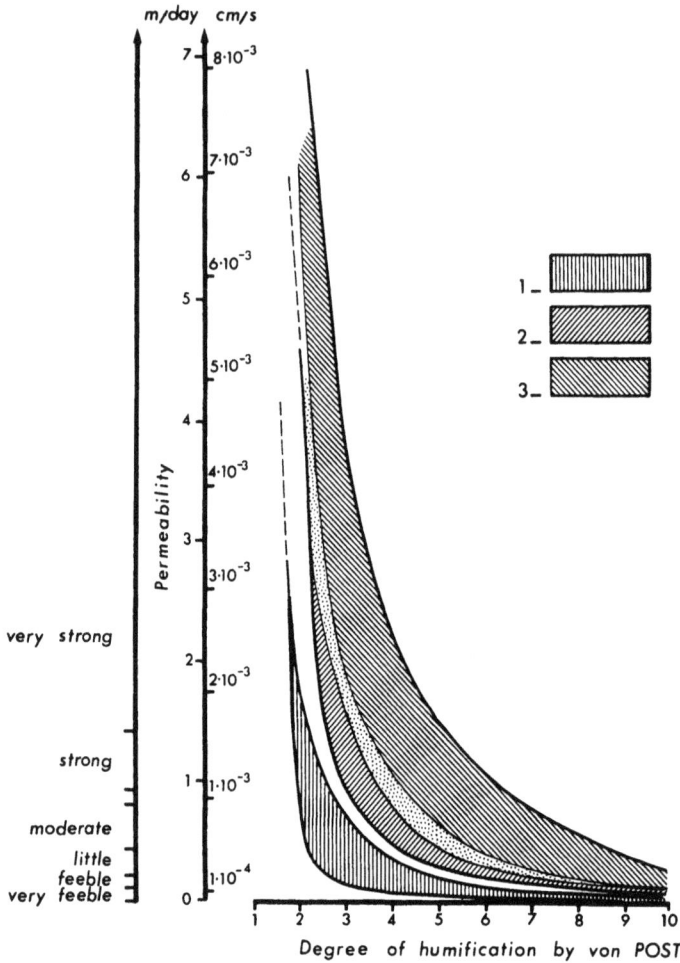

FIGURE 2   Relationship between permeability and degree of humification for the main kinds of peat: 1, *Sphagnum* peat; 2, Bryales-sedge-peat; 3, Phragmites and Carex-peat (after Baden and Eggelsmann).

peatlands such as string-bogs or aapa-moors, and in the surface layers of blanket bog. It is important to distinguish between these situations and topogenous fen peats consisting predominantly of cyperaceous plants and wood material. The hydrophysical properties of such peats have not been investigated to the same extent as for bog peats and it is not known if a comparable active layer exists.

(b) Relation between the Active Layer, Water Table, and Runoff

Runoff is progressively reduced as the bog water table falls within the active layer. This relationship has been demonstrated particularly well by Romanov (1961) for raised bogs near Leningrad (Figure 4). In this case the base of the active layer lies at about 25 cm depth. Results from a raised bog in Scotland (Robertson et al., 1963) accord with those of Romanov

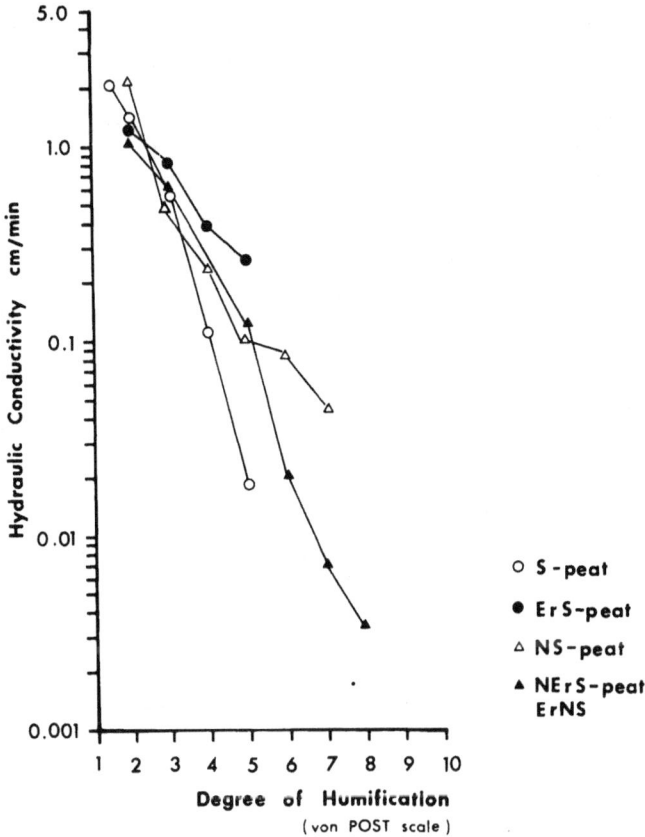

FIGURE 3   Hydraulic conductivity of different kinds of *Sphagnum* peat in vertical direction vs. von Post's degree of humification: s = *Sphagnum*, ER = Eriophorum, N = shrubs (after Korpijaakko and Radforth, 1972).

(1953), with a critical depth at 25 cm below which runoff is considerably reduced. Pronounced reduction in runoff when the water table falls below a critical level has been widely reported from various other types of peatland. In blanket bog this level seems to be relatively close to the surface. Chapman (1965) showed a major reduction in runoff at water levels of 8 cm depth, and total absence of runoff at 20 cm (Figure 5). By comparing the water level recession during periods of maximum and minimum evaporation conditions Goode (1970) has shown that similar conditions exist on blanket bog with pool and ridge microtopography in Scotland. In this case the critical level was at a depth of only 5 to 10 cm, depending on site factors, and it corresponded with a pronounced change in humification. A close relationship between runoff rate and height of the water table has been demonstrated in several small forested bogs in Minnesota (Bay, 1968). This showed an exponential increase in runoff with rising water table (Figure 6). Runoff always ceased at approximately the same water table position in each bog and further lowering of the water level was accomplished only by evapotranspiration.

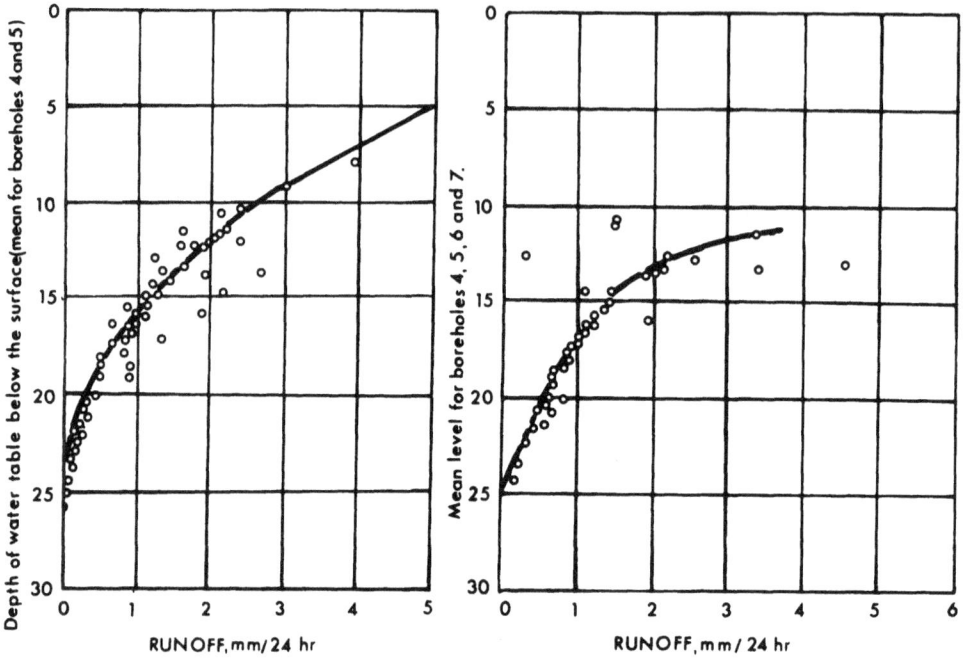

FIGURE 4   Runoff curves relative to the water table (after Romanov, 1961). Central part of the Lammin Suo Massif at the left and convex part at the right.

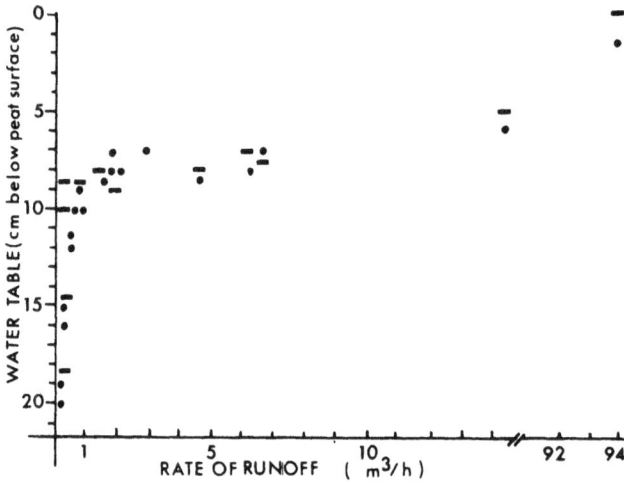

FIGURE 5   The relationship between the height of water table and the rate of runoff for catchment area 2: –,water table at site 1; •, water table at site 2 (after Chapman, 1965).

Essentially then the results from these various types of peatland are closely comparable in the exponential relationship between water level and runoff and in the presence of a critical water level at which runoff ceases. In the raised bogs investigated by Ivanov (1957) the water level at which runoff ceased was always below the level at which flow recommenced following precipitation. The difference varied from 1 to 10 cm in different bogs. Artificial drainage of bogs provides additional evidence of the limiting effect on runoff caused by low hydraulic conductivity in the subsurface layers. Boelter (1972a) demonstrates that water table drawdown associated with a deep ditch across a forested bog in Minnesota is ineffective below the base of the active layer at horizontal distances greater than 5 metres from the ditch. He gives figures for hydraulic conductivity within the surface layer (5.0×10^{-2} cm/sec) and subsurface layer (2.2×10^{-5} cm/sec); see Figure 7. This is a good example of the effect of a discontinuity in humification referred to earlier.

In most bogs the water table periodically falls below the critical level referred to above, as a result of the loss of water by evapotranspiration. The duration of such occurrences depends on the climatic regime and, to a small extent, on the physical characteristics of the peat. The proportion of time during which the water table lies below this critical level is relatively small in oceanic blanket bog compared with continental raised bog. This is referred to again in discussing water balance.

Understanding the influence of the active layer on the water level within different elements of bog microtopography is vital to the interpretation of runoff from peatlands of various types. Examination of the water level changes within a complex of ridges and pools on gently sloping blanket bog (Goode, 1970) shows that the vertical fluctuation is greater within the active layer of the ridges than in the adjacent pools over the same time period (Figure 8). Below the active layer the rate of fall is much the same beneath ridge or hollow. The physical properties of the ridge are therefore important in determining the rate of runoff, and it can be expected that the ridges of minerotrophic string bogs will behave in much the same manner. The relationship between water level and runoff has been used by Romanov (1961) and others to calculate total runoff from water-table data. This is explained further in the next section, but it is important to note here that a knowledge of the ratio of areas occupied by positive and negative relief elements (ridges and pools), together with their storage capacity, is essential to the calculation.

The Water Balance of Natural Peatlands

It was mentioned previously that a great deal of work on the water balance of Boreal peatlands has been done in the USSR. This is summarized in two books by Romanov (1961, 1962) which have been translated into English and in a paper by Novikov et al. (1972). Most of the investigations outside the USSR have used conventional techniques for measuring or predicting the individual components of water balance and are generally restricted to small catchments. Similar catchment studies in the USSR have established empirical relationships between storage capacity, water level, and runoff which have led to the development of techniques for calculating runoff from extensive peatland catchments solely from climatic and water-level data. This type of approach would obviously be of great value if applied to the extensive tracts of unconfined muskeg, but first it would be necessary to define empirically the relation between the individual parameters mentioned above for various types of peatland microtopography.

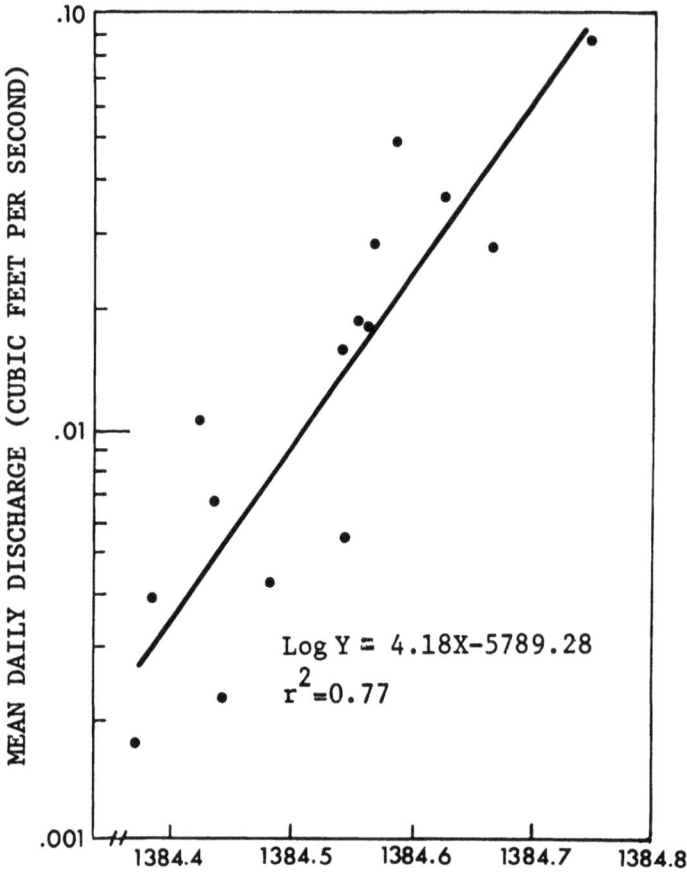

FIGURE 6   Relationship between the mean daily discharge and the bog water table elevation for bog S-2 (after Bay, 1968).

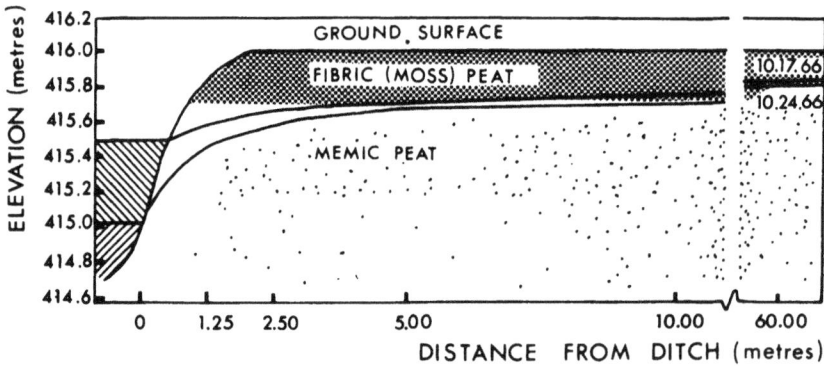

FIGURE 7   The effect of a water level drop in the ditch on bog water table drawdown (after Boelter, 1972).

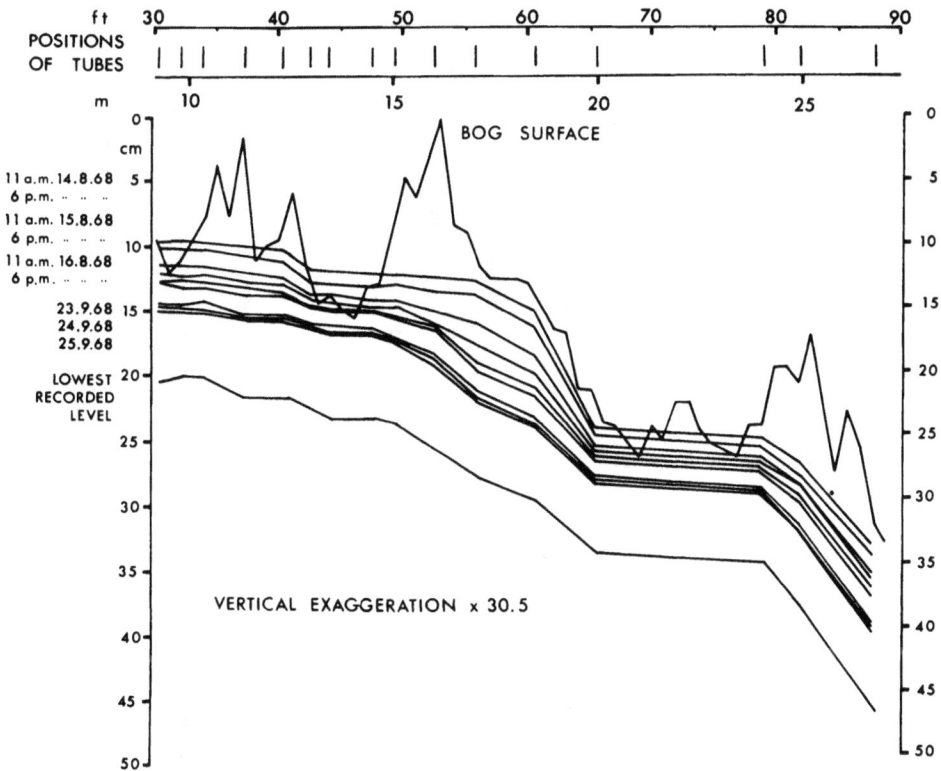

FIGURE 8   Profile showing successive fall in water levels during two periods within pool and hummock bog (after Goode, 1970).

So there are two fundamentally different ways of approaching the question of peatland water balance. Consideration of each of these will provide a better understanding of the possible effect of muskeg on water resources.

(a) Catchment Studies

There are comparatively few complete studies of water balance in which all parameters have been measured directly (Figure 9). Precipitation and runoff are most often measured directly, whereas evaporation has often been calculated from potential evapotranspiration, or by other means, using meteorological data. This is acceptable since it is known that there is a linear relationship between evaporation and radiation balance in raised bogs (Romanov, 1962), and good correlations between observed evapotranspiration in lysimeters and potential evaporation have been described by Virta (1966). It has also been clearly demonstrated that evaporation decreases with increasing depth of water table. Romanov (1962) (Figure 10) indicates an abrupt decrease in specific evaporation (i.e. evaporation per unit absorbed energy) at certain water levels which he interprets as being the point when the capillary fringe loses contact with the rooting zone. This corresponds to a water table depth of 45 cm below dwarf shrubs and 25 cm below *Sphagnum* hollows in a raised bog. Any values given for evaporation must therefore be related to depth of water table. Another factor affecting

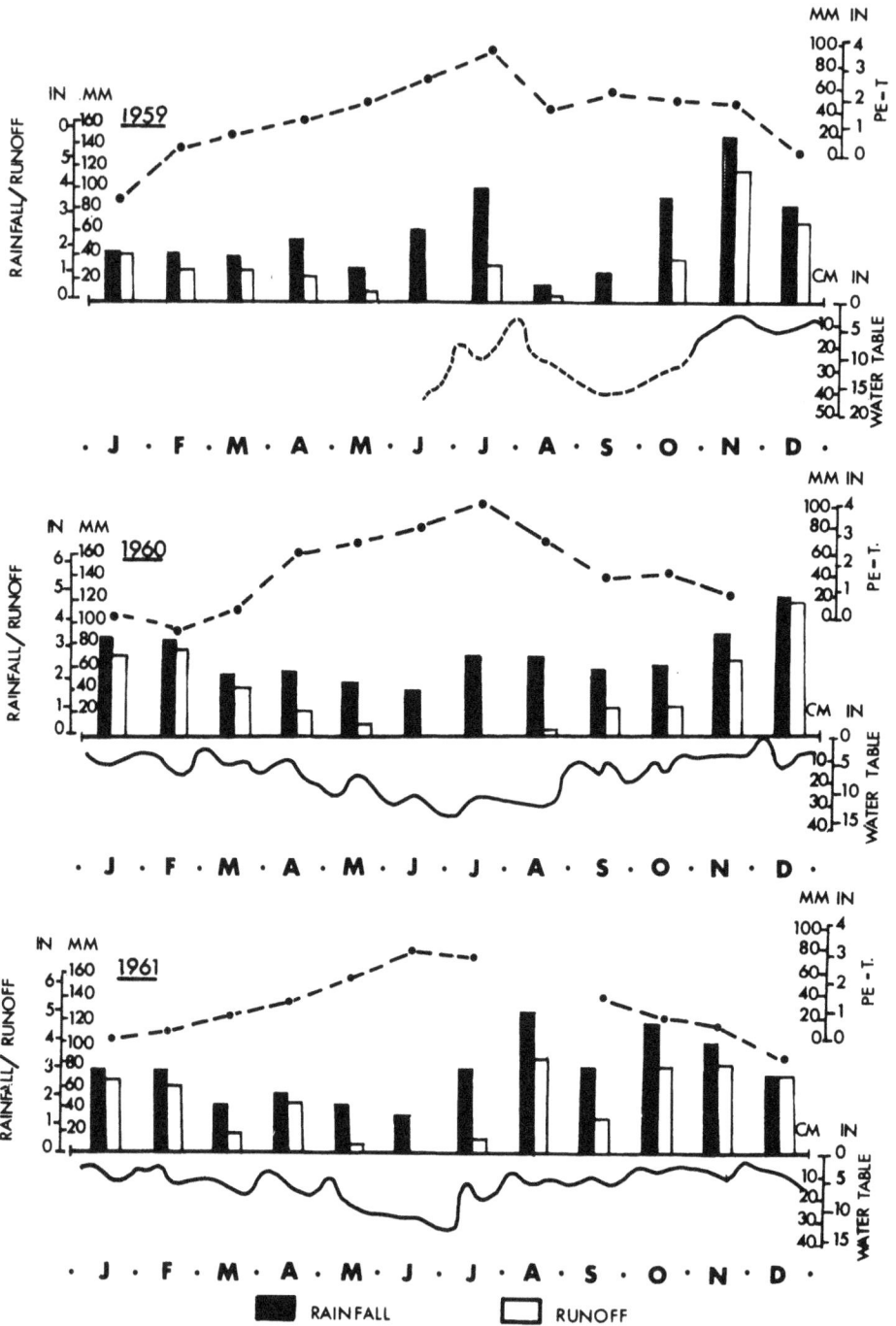

FIGURE 9   Rainfall, runoff, evapotranspiration, and groundwater table relationships, in an oceanic raised bog (after Robertson et al., 1968).

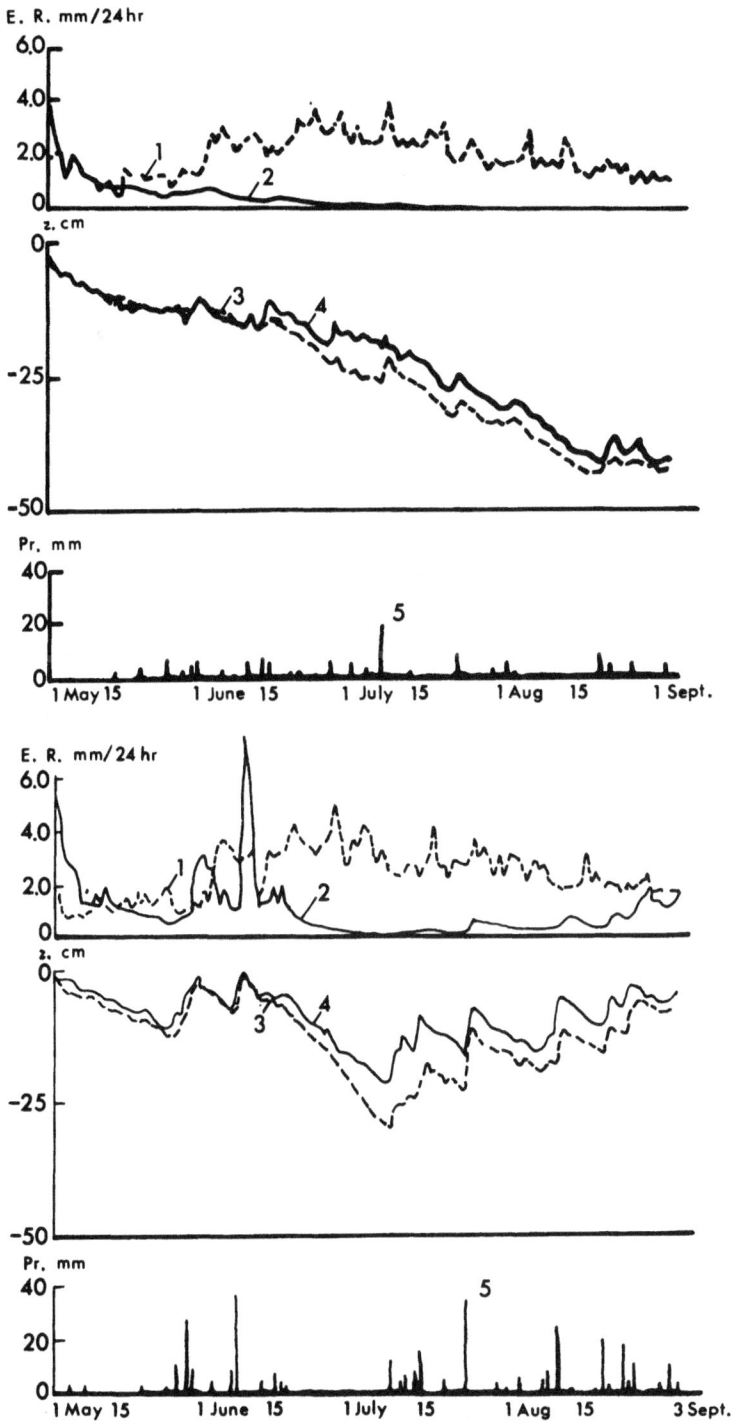

FIGURE 10   Course of water balance items for the Lammin-Suo bog massif: 1, evaporation; 2, runoff; 3, water table (calculated); 4, water table (observed); 5, precipitation (after Romanov, 1961). Top: for 1950 (dry year). Bottom: for 1953 (wet year).

evaporation is the seasonal growth stage of vegetation, a relationship which has also been closely investigated by Romanov (1962).

The relationship between water-table depth and runoff has already been discussed, but we have not yet considered the storage capacity, which in many ways is the most important single element in the water balance so far as bogs are concerned. Changes in storage include both the saturated zone below the water table and the unsaturated zone above. Porosity in the upper parts of the active layer is so great that precipitation will immediately cause a rise in water table in that zone. At deeper levels the relative response of moisture within the unsaturated and saturated zones might be expected to vary according to changes in bulk density. However, it has been found that the two components of storage do in fact behave in a similar manner in response to given amounts of precipitation or evaporation. This being so, a change in water level can be used as a measure of change in storage, so long as it is related to the bulk density or porosity of the peat concerned. The amount of water loss when the water table is lowered by a given amount is referred to as the water yield coefficient or storage coefficient, which is expressed as a volume ratio for any given peat type. Since porosity changes with depth this is an important factor which must be taken into account in using changes of water level to indicate storage changes.

A second problem of storage is the large difference in storage coefficient between the different elements of microtopography. This is well displayed in Figure 8. Microtopography also affects calculations of storage in that the depth of water table varies because of the occurrence of local hummocks and hollows, and so a mean depth is required for whole bog landscapes rather than the individual elements. Virta (1966) describes a method by which the change in storage is related to a mean 'water stage' which represents the gradient of the bog surface irrespective of surface microtopography. The difference between the actual water level and the mean water stage provides a measure of the storage for a larger area than would otherwise be possible, and this method is important in computing water balance in the absence of runoff data.

We can now proceed from the individual components of water balance to the behaviour of individual peatland catchments. It is important to realize that the evidence concerning the water balance of any particular catchment depends on the seasonal characteristics of climate in that area and cannot be applied to other climatic zones. For this reason examples have been chosen from areas with different climatic regimes to illustrate the broad features of the water balance equation.

The work of Bay and others in northern Minnesota (Bay, 1969) provides a good example of the water balance under a continental climatic regime, though it must be realized that this area lies close to the southern limit of Boreal peatlands and therefore has higher potential evapotranspiration than areas farther to the north. The catchments in this case were small forested watersheds less than 120 acres in extent, each of which contained a central area of bog varying from 8 to 20 acres. Some of the bogs were 'perched' ombrogenous bogs and others were minerotrophic, i.e. receiving water from the surrounding areas of the watershed in addition to precipitation.

In this region the precipitation is largely restricted to spring and summer, with high potential evaporation in summer and several months during which the ground is continuously frozen in winter. Most of the annual water yield from the perched bogs occurred in spring (between spring thaw and the beginning of June) as a result of the low evaporation conditions, snow melt, and spring rains. The water table was therefore highest at this time of year. During the summer (1 June to 1 September) the water yield was substantially less (25 per cent of the annual yield) even though 42 per cent of annual precipitation occurred during this period. In effect this means that evapotranspiration was sufficient to maintain the water level at a depth sufficiently great for runoff to be substantially reduced (due to the lower permeability of the deeper horizons). At times runoff entirely ceased, indicating that the water table fell below the active layer. During fall (a period with negligible precipitation and declining evapotranspiration) runoff was extremely low, ceasing entirely with the onset of frozen ground conditions.

The general pattern described above is very much in line with the water balance studies in the continental regions of the USSR. The water balance of a raised bog in Scotland (Robertson et al. 1968) provides an interesting comparison since this is an area of oceanic climate in which precipitation is fairly evenly spread throughout the year. Summer potential evapotranspiration is substantially less than in continental regions and periods of freezing are relatively short-lived. The general features of water balance are shown in Figure 9. In this case it was found that runoff occurred throughout most of the year but was considerably reduced in midsummer since the amount of runoff as a percentage of precipitation over short periods was strongly influenced by evapotranspiration conditions. Runoff ceased during short periods in the summer. This always corresponded to periods when the water level fell below 25 cm. Such conditions might be produced either by short periods of particularly high evapotranspiration or by extended periods without precipitation which would allow the water table to reach the same level under lower evapotranspiration conditions. Goode (1970) has suggested that the length of dry periods is just as important as the actual values of rainfall or evapotranspiration in determining the periods of minimum runoff from oceanic bogs. The relationship between periodicity of rainfall and phases of runoff is well demonstrated by Baden and Eggelsmann (1968) for a suboceanic raised bog in north Germany. They have shown that for much of the summer the runoff is negligible but that only one day of heavy rain is needed to restore the water table to a height at which runoff occurs, with the result that the summer period is characterized by long periods during which little or no runoff occurs, interspersed with brief periods of intense runoff. It is thought that this phenomenon is probably due in part to the low storage coefficient of the deeper layers together with the fact that evapotranspiration becomes less effective at depth.

So in the case of continental bogs it can be said that their regulation of river flow is likely to be restricted to periods in the summer when there will be an effective storage capacity, but that the length of time over which this water is released can only be very short because of the rapid drawdown of water table by evaporation. Such bogs are not effective in storing the bulk of annual precipitation since most of this is lost very rapidly in the spring when there is a considerable surplus over the storage capacity.

For oceanic bogs the pattern is rather different in that rainfall and runoff occur throughout the year but with less runoff in proportion to rainfall in the summer because of the higher evapotranspiration. For much of the year the water table fluctuates within the active zone or close to it, so that total recharge is frequent. The water table only falls to the depth at which runoff completely ceases during short periods of high evapotranspiration or as a result of long, dry periods at other times. A great deal depends on the frequency of ranfall in such a regime, rather than on the total amount. This being so, bogs in an oceanic climatic regime affect river flow in two ways. For much of the year runoff will occur rapidly and produce frequent spates, but at other times, particularly in summer, there may be periods of several weeks during which runoff is virtually non-existent. There is in fact a direct response to weather conditions and virtually no storage except for brief periods in the summer.

Most of the catchment studies referred to have applied to bogs. Few studies have been made of fen or minerotrophic peatland because of the difficulties of measuring inflow. Bay (1968) has demonstrated that the groundwater affects the water table and runoff from minerotrophic bogs. Exceptionally high water levels in the groundwater maintained the bog water table at a high steady level through the summer in a very wet year. On the other hand Bavina (1972) has demonstrated that high evaporation from fen areas can reduce the ground water supply to rivers during summer periods. She compared the water balance of a fen with that of a river basin which was only partially peat covered and found that under normal conditions runoff was approximately equivalent but in particularly dry years water passing into the fen was lost as evaporation and consequently the river flow was reduced by the negative water exchange of the fen area. Much research is needed to establish a better understanding of the effect which fen areas have on drainage basins.

(b) Runoff Prediction
The fact that there is a direct relation between the depth of water table and runoff for a given type of peatland has resulted in the prediction of runoff from extensive tracts of peatland solely from meteorological and water-table data (Romanov, 1961) and similarly Novikov (1964) has suggested that the water level can be computed directly from meteorological data as long as the storage coefficient is known. Figure 10 shows the results of this type of approach for raised bogs near Leningrad (Romanov, 1961). The only requirements are knowledge of water-level changes, values of the water yield coefficient for specific types of microtopography, and meteorological data. The value of this type of approach lies in the fact that the water balance can be investigated in extensive areas of peatland where it would be difficult if not impossible to measure runoff directly — as in vast areas of unconfined muskeg. However, it does require that the storage coefficient be established for individual peat types and the storage characteristics would need to be assessed for all types of surface microtopography. Once this is done it is necessary to assess the relative proportions of different elements in any one hydrologic system, relate these to a mean water stage, and measure changes in water level. A further possibility is that aerial photography (using infrared film) could lead to the direct assessment of water storage conditions. Such an approach might have considerable potential in assessing the role of muskeg in water supply.

In the meantime it is essential that an assessment be made of the water balance by standard catchment studies in order to discover precisely how storage capacity and hydrophysics affect runoff from extensive tracts of minerotrophic muskeg. The European work cannot contribute much to this subject. Recent views concerning the role of peatlands in catchment control and regulation of river flow (Vidal, 1968; Eggelsmann, 1971; Gordon, 1972; Heikurainen, 1972) take their evidence largely from ombrotrophic bogs. It has been shown during the course of this discussion that the evidence from these bogs indicates that peat is far from being a moderating influence, and it is reasonable to assume that other types of peatland such as extensive minerotrophic string bogs will behave in much the same manner. We must await the results of direct catchment studies in order to discover whether this is in fact the case.

PEAT DEPOSITS AND WATER RESOURCES: A DISCUSSION FROM AN ENVIRON-
MENTAL STANDPOINT

Under this heading, it is intended to discuss the more important parameters that influence the
quality of the water resources associated with peatlands. Virgin peatlands will be considered
first, in the light of environmental quality functions when applicable, with the general
objective of establishing an environmental base-line for these areas. This will being in
perspective the effects of bog drainage and reclamation on the water resources.

Virgin Peatlands

Accumulating evidence indicates that the ionic balance and cation content of peatland
waters in relation to water source and hydrotopography are major factors influencing
vegetation and ultimately peatland evolution and, depending on morphology and climate,
peat bogs will differ in their hydrological characteristics; runoff and ground water conditions
are primarily a function of climate, that is, rainfall, temperature, wind, evapotranspiration.
Other important variables are the topography of the region, its vegetation, and the physical
conditions of the soil (Vidal, 1968). Deevy (1958) reported that one reason accounting for the
lack of nutrients in bog water is that the local rock is usually granite, which contributes
almost no minerals to the water flowing through it. According to Bavina (1973) the type of
hydrological regime of swamps is determined by three principal features: (1) the condition of
water feed of the swamps, (2) the surface relief of the swamps that determines runoff, and (3)
the vegetation cover and its distribution over the swamp.

Doodge (1972) distinguished between bogs and fens, on the basis of the type of water
supply. According to this author bogs are fed mainly on meteoric water, whereas fens are
supplied with either surface or underground water, which is richer in nutrients.

Heinselman (1970) made a particularly complete synthesis of these views and proposed a
classification of peatland types, based on the following criteria:
1. Location with respect to mineral soil margin lands, islands, substratum relief, and water
   sources.
2. Topography of the peat surface.
3. Water level and water movement.
4. Chemical properties of water supply and runoff.
5. Botanical origin, decomposition, and thickness of peat.
6. Natural vegetation.

Heinselman (1970) is in close agreement with Doodge (1972) and Sjors (1961) when, on the
basis of these criteria but with special reference to ionic concentration, he describes the two
following diametrically opposed types of peatlands:
1. *Minerotrophic peatland (fens)*, which are supplied at least partly by water percolating
from mineral soils. These swamps support a relatively rich vegetation: grasses, sedge herbs,
and tall shrubs. Minerotrophic peatlands are oriented parallel to slopes, occupy long slopes,
exhibit concave cross-sections, and usually terminate at outlets. The water flow patterns are
confluent and this is reflected by the vegetation patterns.
2. *Ombrotrophic peatland (bogs)*, the water supply of which is atmospheric precipitation.
Ombrotrophic peatlands occupy water table divides and have divergent or even radiating
water flow patterns. These are also reflected by the vegetation pattern. Peat surfaces are

TABLE 1   Peatland types

	Location	Topography of the peat surface	Water levels	Water movement
I. Minerotrophic swamp (rich)	Downslope of mineral soil islands and substratum ridges	Often concave Peat surface slopes (8-20 ft/mile)	At or above the peat surface	Strong flow of moderately mineral-waters out of the peatland
II. Weakly mine-rotrophic swamp (poor swamp)	Downslope of type 1, often crossing most of peatland	Slopes gently and slightly concave Peat surface slopes (2-6 ft/mile)	At or above the surface	Sluggish
III. String bog and patterned bog	Downslope to-ward drainage outlets	Slopes (3-6 ft/mile) concave	High (6-18 in. free water)	Sluggish
IV. Forest island and fen complex	Downslope from or sur-rounded (type 2)	Sloping and con-cave, slopes (4-5 ft/mile)	High	
V. Transitional forested bog	Downslope from 6 or 7, upslope type 1	Convex in at least one cross-section sloping toward the margin (8-15 ft/mile)	Below the surface	Rapid but moving water is visible only at high water
VI. Semi-ombrotro-phic bog (semi raised bog)	Surrounded by type 2 of up-slope from type 5	Semiconvex slopes, slight to moderate (4-10 ft/mile)	Just below the surface	Little standing water
VII. Omprotrophil bog (raised bog)	Occurred toward the lower margin	Totally convex slopes, slight (2-6 ft/mile)	Near the sur-face most of the season and therefore fully convex	
VIII. Raised bog drains		3-5 ft/mile	Above much of the bog surface except in very dry weather	Barely discernible

Ref: Heinselman (1970).

| Chemical properties | | | Thickness | | |
pH	Ca(ppm)	Mg	(feet)	Vegetation	Water feeding
6.0-6.5 neutral	15-28 (high)	High	1-6	Rich swamp forest	Strong, from margin or other mineral
4.5-6.0 Slightly to moderately acid	4-12 Inter.	Inter-mediate	10-25	Poor swamp forest	Waters derived from 2 also from drainage from type 6 and 7 and precipitation
5.0-6.6 moderately acid	4-10 inter.		10-25	Cedar string-bog or fen complex	Waters received from 2, ultimate sources are a mixture of waters from mineral soils, precipitation, runoff from type 7
	Similar type 3		15-28	Poor swamp forest within islands	
3.2-3.8 strongly acid	5-7 inter.		2-7	Black spruce feathermoss forest	Peatland receives a mixture of waters from ombrotrophic bogs, direct precipitation, and some richer minero-trophic sites
3.2-3.8 very acid	1.5-3.5 low	Low	3-20 or more	Sphagnum-black spruce leatherleaf bog forest	Inflow of waters from one or two sides but a tendency for precipita-tion waters to drain away in two or three directions causing mineral depletions
3.2-3.7	0.6-2.3 low	Almost absent	3-30	Sphagnum-black spruce leatherleaf bog forest on crests and upper slopes, and sphagnum-leatherleaf - Kalmia spruce heath on lower slopes	Water table divides causing dependence on precipitation for metallic cations (thus the term ombrotrophic, meaning rain-fed)
3.6-4.5	2.0-4.7	High	8.20	Open, depauperate, poor swamp forest	

convex in cross-section in one or more directions. The water table is convex and the water discharges into at least two separate local watersheds. This convex configuration prevents inflow of nutrient-rich water from mineral soils. Consequently, the water is highly deficient in nutrient salts, causing low metal ion saturation in the peat and a low pH. There results a very poor vegetation, mainly of shrubs and perennials of arctic barren and cold steppes.

Actual peatlands may belong to one type or the other, or many will be a composite of the two, depending primarily on the relative importance of precipitation and surface and ground water in their water supply. Heinselman (1970) has thus defined eight peatland types, described in synoptic form in Table 1. This classification is in reasonable agreement with Jasmin and Heeny (1960) and Bogomolov and Kats (1972).

The parameters pertaining to the water resources which appear on Table 1 will now be discussed.

Water Quality Parameters

(a) Water and Nutrient Movement
As the hydraulic conductivity of a soil increases, so does the water circulation; this in turn promotes greater aeration and a better distribution of dissolved solids. This situation enhances primary productivity and vegetation diversity (Armstrong and Boatman, 1967; Ingram, 1967). Considerations of peat hydrology discussed above have firmly established that the water movement takes place in an active top layer characterized by a very high porosity.

The most important effect of groundwater movement in bogs is the control of the chemical quality of water in marshes. According to Eisenlohr (1972) the slow lateral water movement is particularly important because it allows seepage outflow that prevents the water from turning brackish. In soils with a high content of organic materials and especially in acid peat soils, the essential nutrients (Ca, K, N, P) are very movable and easy to leach out (Huntze and Eggelsmann, 1972). This in turn will have an effect on the vegetation, which will colonize the areas richer in nutrients (water tracks). The relief of the peat surface itself will reflect past interactions between vegetation and water chemistry (Heinselman, 1970) and influences the flow of surface water.

The oxygen concentration of the mire water increases with increasing flow rate, approaching saturation at rates over 1 cm/sec. Sparling (1966) used a model approach to show that at speeds less than 0.3 to 0.4 cm/sec the diffusion of oxygen into the bog water would be similar to that into a stationary pool and because of the respiration of roots and microorganisms in the peat the oxygen would tend to be depleted. Above 0.4 cm/sec, the turbulence in the water seems sufficient to keep a suitable oxygen concentration.

Since the hydraulic conductivity is a function of the degree of decomposition of the peat material and since it was established that runoff (water movement) is exponentially related to the height of the water table within the active layer, a value function has been drawn that reflects these phenomena (Figure 11). Peat material in the active layer, characterized usually by a von Post number equal to one, was accordingly given an environmental quality index of 1.0. As the degree of decomposition increases, the hydraulic conductivity decreases and limits are thereby set on the aeration and the water and nutrient movement; the environmental quality index falls off rapidly to negligible values.

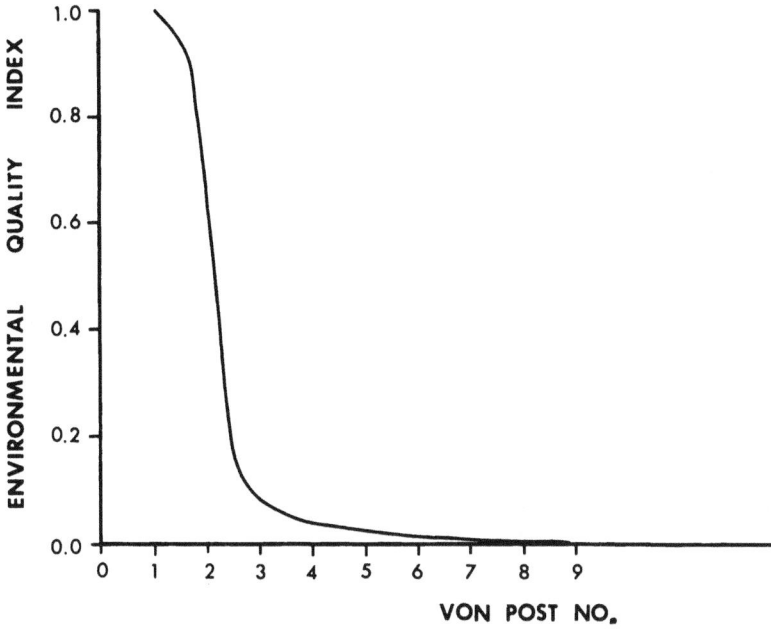

FIGURE 11   Environmental quality function: water and nutrient movement vs. von Post index.

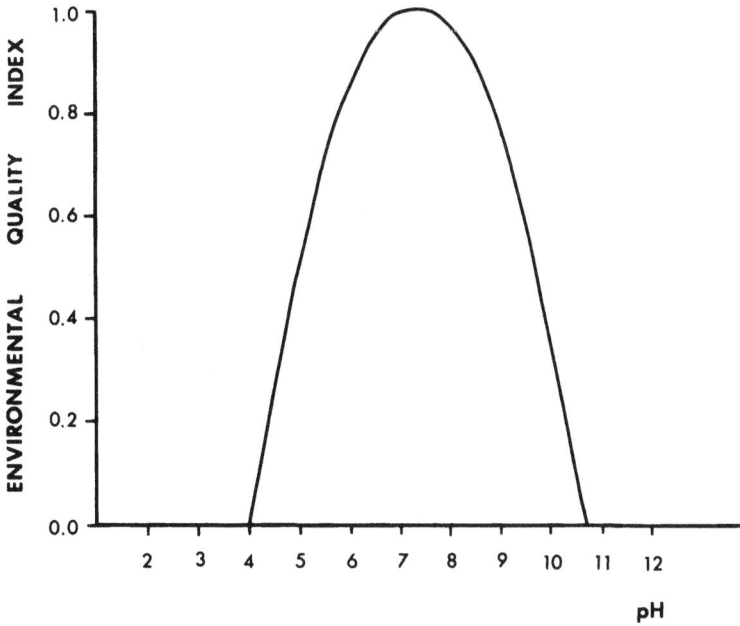

FIGURE 12   Environmental quality function: pH fresh water organism.

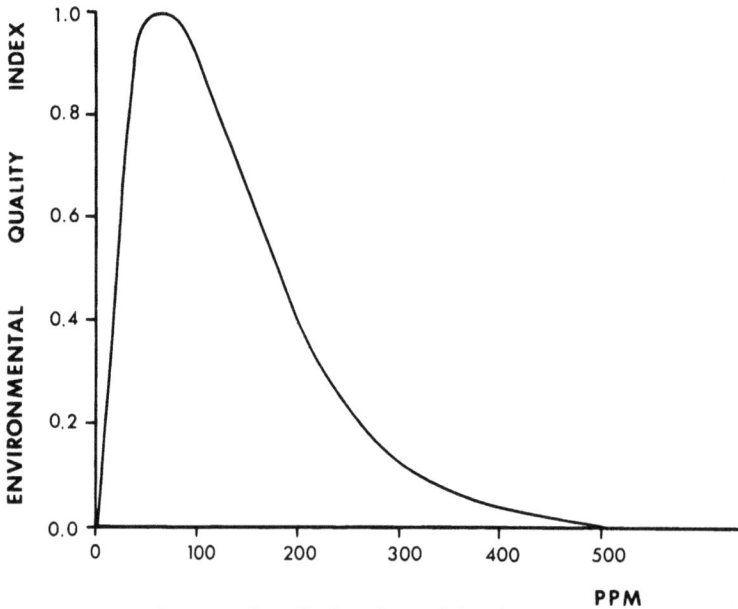

FIGURE 13   Environmental quality function: calcium ion.

(b) Concentration of Hydrogen Ion (pH)
It has been observed that productive natural fresh water has a pH ranging from 6.5 to 8.5
EIFAC, 1969). In the United States, only 5 per cent of the waters that support a good fish fauna
have pH values lower than 7.6, and in 95 per cent the pH is less than 8.3. Direct lethal effects
are not produced within a range of 5 to 9.5. These guidelines, as well as minimum, maximum,
and recommended pH values reported by McKee and Wolf (1963), were built into the
environmental quality function proposed by Battelle (1972) and reproduced in Figure 12.

In ombrotrophic bogs characteristic pH values, ranging between 3.2 and 3.8, are too low
and the quality of the environment index is zero for most fauna species except perhaps
specialized types of organisms. However, because of the very small dissolved solids concentration, the buffering capacity of muskeg water is poor and can be altered very easily
(Didier-Dufour et al., 1972; Prince, 1972).

It should be remembered also that the hydrogen ion concentration is intimately related to
the concentration of other substances and, for example, will control the degree of dissociation of weak acids and bases. Since undissociated molecules are frequently more toxic than
their dissociation products, the pH becomes a very important parameter in the evaluation of
the toxicity of effluents. This is particularly significant for bog waters of low pH, and each
case should be investigated with particular attention (Doudoroff and Katz, 1950). To give a
specific example, hydrogen sulphide is extremely toxic to fish at low pH values because the
toxicity is related to the undissociated molecule (Jones, 1969).

(c) Calcium
Calcium is an important ion because the human body daily requires about 2 grams of it to
keep in good health. The literature survey conducted by McKee and Wolf (1963) indicates

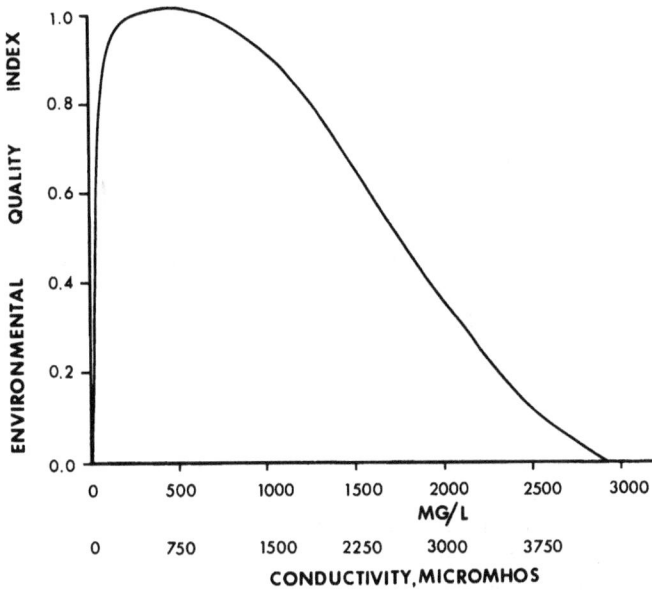

FIGURE 14   Environmental quality function: total dissolved solids and conductivity.

that the optimum concentration recommended by various authors ranges between 50 and 75 mg/l. A value of 200 mg/l is considered excessive.

For fish and other aquatic fauna, calcium is important because it seems to antagonize the toxic effect of heavy metals in concentrations around 50 mg/l, although it may become toxic in concentrations varying between 300 and 1,000 mg/l. In the United States, 95 per cent of the waters supporting a good mixed fish fauna have a concentration of calcium ion less than 52 mg/l, and 5 per cent have a concentration less than 15 mg/l. The environmental quality functions shown in Figure 13 were constructed according to these figures. When we compare the environmental quality functions for calcium with actual representative concentrations of this metal in bog water, we see that the quality is at best poor, even for minerotrophic peatlands (the calcium concentration ranges from 15 to 30 ppm). For ombrotrophic bogs, the quality is near zero, since the representative calcium ion concentration falls below 5 ppm.

(d) Total Dissolved Solids and Conductivity

A very good review of this subject has been made by McKee and Wolf (1963), and a water quality function has been suggested by Battelle (1972) and reproduced in Figure 14. This function integrates the following general findings:

1. Dissolved solids may influence the toxicity of heavy metals because of an antagonistic effect of hardness metals. Chromate, copper, cyanides, zinc, and many other substances are generally more toxic in distilled water than in hard water.

2. On the basis of taste thresholds, domestic water supplies should not contain more than 500 mg/l of total dissolved solids and the International Standards of WHO (1958) set the excessive limit at 1,500 mg/l. Waters containing more than 4,000 mg/l are generally unfit for human use.

3. Irrigation water should contain between 150 and 1,500 mg/l of total dissolved solids; waters containing over 2,000 mg/l of dissolved solids are marginal for irrigation use.

4. For inland water fish and fauna, it has been found in the United States that 95 per cent of the inland waters supporting a good mixed fish fauna contain less than 400 mg/l of total dissolved solids.

Since the total dissolved solids afford a toxicity blanket to the effect of heavy metals, this parameter should be established when studying the water chemistry of peatlands. However, the concentration of dissolved solids is low, certainly inferior to 100 ppm. For example, Schmeidl et al. (1970) report conductivity values ranging from 36 to 70 mmhos, over the April-December period, for the water of an upland bog. These values do indicate a very low concentration of total dissolved solids. Therefore, the aquatic environment of a peatland is poor, also, in this respect.

BIOLOGICAL INDICATORS OF WATER RESOURCE QUALITY

A far greater number of environmental quality functions could be established on the basis of pertinent data found in the literature. These will be useful in the assessment of base-line quality of a water resource and of the impact following some man-made or natural change. Their usefulness is also appreciated for the identification of severely limiting parameters. However, the extremely complicated interactions of the parameters set a limit on the significance of individual parameters, and biological indicators are sought that integrate the whole gamut of environmental characteristics into some general index of environmental quality. The latter could subsequently be at least loosely correlated with a limited number of important physical or chemical parameters.

Species diversity is such an indicator. Species diversity tends to be low in ecosystems subjected to strong physicochemical limiting factors and high in biologically controlled ecosystems. It is an expression of the possibility of constructing feedback systems. Higher diversity means longer foodchains and more cases of symbiosis and greater possibilities of feedback control, which reduces oscillations and hence increases stability (Odum, 1971). Therefore some measure of species diversity should reflect the degree of stress imposed upon biological processes and give a measure of the general quality of a natural or man-made environment.

Jeglum (1971) and Heinselman (1970) have carried out very thorough inventories of peatland environment and have attempted to relate data on vegetation to classes of pH, depth of water level, waterflow pattern, water chemistry, and topography. Jeglum (1971) identified the pH and depth of water level as limiting factors and recognized plant indicators for the various classes of pH and water level. Their data have been rearranged by plotting the frequency of occurrence of plant species as a function of pH and depth of water level (Figures 15 and 16). Although there is obviously no simple relationship between the occurrence of a species and these parameters, the figures demonstrate nonetheless that low pH values (below say 6.0) and above-surface water levels are detrimental to plant diversity. There are, however, clear patterns showing optimal ranges of pH (6.0-6.9) and water levels (20-38 cm below the surface). This affords some positive indication of the beneficial impact of bog drainage. The scatter of data is explained in terms of nutrient availability, which is a compound factor involving water movement and concentration of dissolved solids, topography, climate, etc. This substantiates the statement that, from an ecological standpoint, a crude measure of diversity is a better indicator of the quality of the underlying water resource than the aggregation of numerous physicochemical components with unknown interactions.

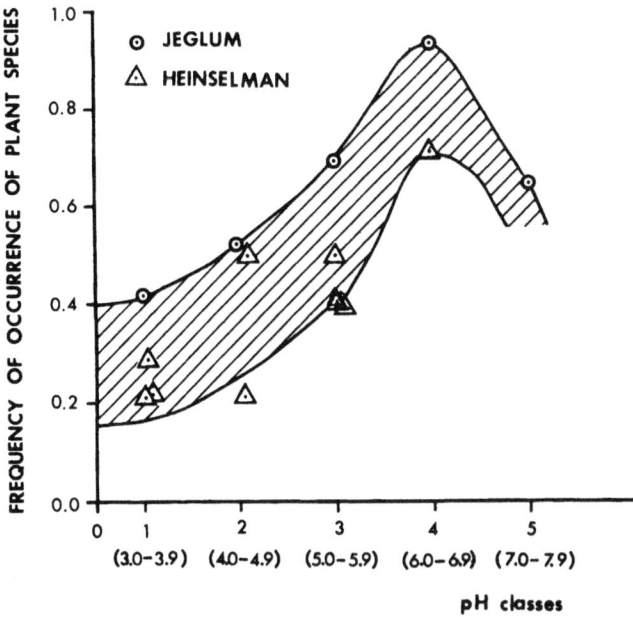

FIGURE 15   Environmental quality function (diversity); parameter: pH.

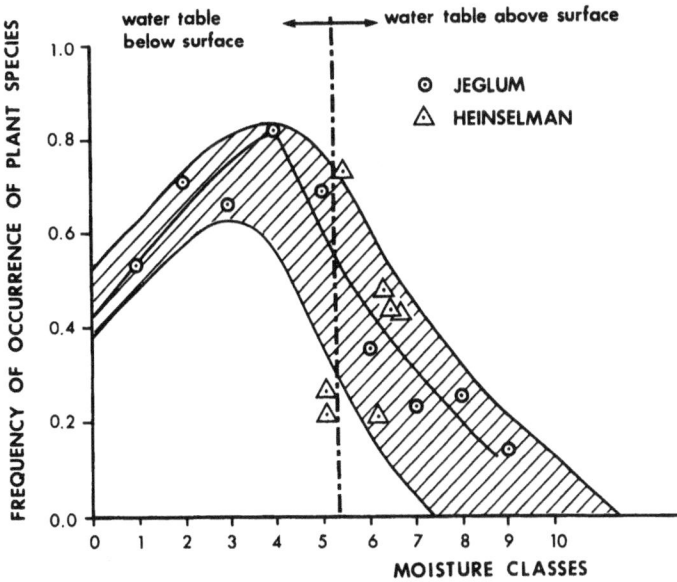

FIGURE 16   Environmental quality function (diversity); parameter: water level.

IMPACT OF BOG DRAINAGE AND RECLAMATION

Objectives of Drainage and Reclamation Works

According to Vidal (1963) bog reclamation not only provides soil improvement for the purpose of agricultural production and management, but also allows controlled consumption of the natural water reserves as well as microclimatic betterment. Agricultural production will nevertheless depend on adequate methods of drainage to ensure the proper moisture, air, and nutrient regime for different crops and highly effective use of agricultural machines (Bogomolov and Kats, 1972). Vidal (1963) is categorical when he says: 'Bog reclamations are integral improvements that are absolutely positive and valuable from the point of view of water balance, landscape ecology and political economy.'

   This positive impact will only be effected if adequate knowledge of the hydrological regime occurring in marshy lands is available. As emphasized by Eggelsmann (1972a) successful reclamation methods and engineering processes depend on: (1) concrete natural conditions of the moisture regime, (2) processes of formation of individual components of water balance, and (3) characteristics of seasonal and yearly distribution and combination of intake and discharge elements of the total water resources of the territory.

   Ivitsky (1972) has set drainage and reclamations objectives that underline modern thinking about multiple uses of natural resources. Planning and management of marshy lands should aim at:

1. Reasonable and adequate utilization of water resources of the entire catchment area, taking into account not only future water needs, but prospective ones as well.
2. Maintenance of regulated rivers and water regime of the entire catchment area.
3. The possibility of controlling the water regime and relevant air, heat, and nutrient regimes in soil.
4. The extension of peat soil life.
5. The construction of durable, effective drainage-irrigation systems.
6. Alongside drainage activities, reclamation activities should be carried out in order to provide for other priorities, for example, water transport, industrial and agricultural water supply, water power, and fishery and recreation zones.

The following section will review the short and long term effects of drainage and reclamation measures on the environment of marshy lands.

Effects of Drainage and Reclamation

Drainage and reclamation works have a beneficial effect on marshlands because they increase their storage capacity, they regularize the flow over a longer period of time, and they increase the minimum flow. As noted by Bogomolov and Kats (1972), a drainage network helps to create in the body of the peat bed, above groundwater, an absorbing capacity that retains a certain volume of spring runoff and consequently attenuates discharge fluctuations. Schmeidl et al. (1970) have observed on the basis of long-term monitoring that the drainage of a peatland increases the hydraulic conductivity coefficient by a factor of 4 approximately. This contrasts with undrained peatland, where pores are almost fully filled with water and the free volume is negligible (Lundin, 1972). Improved moisture-holding capacity accounts for the observation that runoff from an undrained area starts to flow sooner than that from a drained area (Burke, 1963). Consequently, flow characteristics are much more uniform from the drained areas and runoff hydrographs do not show the sharp peaks exhibited by un-drained areas (Heikurainen, 1972). This has been confirmed by Eggelsmann (1972) by simultaneous measurement of rainfall, groundwater, and runoff after heavy or lasting rain and after melting snow.

In many cases, a low water flow rate was almost doubled following drainage operations (Bulavko, 1971). Klueva (1972) has established, on the basis of an analysis of sixteen river basins, that after reclamation minimum summer daily runoff increased on the average from 30 to 150 per cent in all the basins studied. For the winter period, this increase was observed only for eight basins. Considering the fact that minimum flow occurs very often at the most crucial period in the utilization of the water resources of rivers, drainage may have a considerable beneficial impact (Burke, 1972; Bulavko and Drozp, 1972).

It has been observed, however, that drainage may increase runoff because of the settling of the peat near drainage ditches: the surface of reclaimed lands acquires a specific mesorelief consisting of transverse slopes and regularly connected depressions (Bulavko and Drozp, 1972). According to Klueva (1972) this increase in runoff occurs in summer and autumn periods. This is probably due to the drainage method, which may greatly affect the nature and the extent of changes in the runoff, as emphasized by Bogomolov and Kats (1972).

When the marshes have been drained, the groundwater level may drop 1 to 1.5 metres and natural moisture-loving vegetation is destroyed, with the result that evaporation may be decreased by 40 to 50 per cent (Bulavko, 1971). According to Bulavko, when marshes are reclaimed and planted with agricultural crops, evaporation will increase slightly but it will remain 10 to 15 per cent below the original value. In the same way intensive agricultural utilization of reclaimed lands will stabilize runoff; evaporation will increase somewhat and annual runoff will approach its original or a slightly lower value (Bulavko and Drozp, 1972). These points have been confirmed by Zubets and Murashko (1972), who have shown that the streamflow rates become more stable with an increase in agricultural use of reclaimed lands. Lundin (1972) has also shown that agricultural use of peatland results in an increase in free volume which improves the capacity of the soil to hold excess moisture in the period of intense nutrient requirement by plants.

Apart from the beneficial effects outlined above, in peat soil drainage triggers physical effects which never end, and which may be detrimental. For example, Eggelsmann (1972 b) mentions that the duration and intensity of drainage have an important influence on the peat permeability, which decreases very quickly immediately after drainage. This loss of per-meability is due to a reduction in the proportion of macropore to micropore space, as the water ebbs out. This also explains why, after drainage, the bulk density (mass of oven-dry material per unit bulk volume) and the consolidation of the peat soil increase. In the same way, subsidence increases with drainage. All these physical effects have been observed in every drained peat area of western Europe and Canada, Scandinavia, and Yugoslavia. As described by Van der Molen (1972) this subsidence of peat soils is due to oxidation of organic matter, shrinkage of the top-soil due to drying, and compaction of the subsoil due to an increase in the load. After drainage the settling of all peat layers is also observed, and the heat capacity seems to be improved.

Drainage of peatland for forestry is a way of increasing timber production (Meustonen and Seuna, 1972), but it will also increase annual runoff by lowering the evapotranspiration. According to Heikurainen (1972) forest drainage will increase the annual runoff by 29 per cent.

Measures to Minimize Environmental Impacts

Generally speaking, it can be stated that the hydrological regime of soil is changed radically by providing a drainage system and, as emphasized by Kovalchuk et al. (1972), local improvements are accompanied by an alteration in the moisture regime for the whole area. These changes manifest themselves in two stages. First, changes in the water regime take place directly in the reclaimed area, and then changes in hydrology and water regime take place in the lands adjacent to this area and on river catchment areas as a whole (Zubets and Murashko, 1972). Thus it is important in the planning stages of drainage networks to widen the range of studies to include the following, as expressed by Eggelsmann (1972a):

1. Relation and quantitative characteristics of formation of intake and discharge elements of the water balance.
2. Moisture transfer in the unsaturated material of the aeration zone.
3. Moisture exchange with groundwater table.
4. Moisture exchange in the layer of the atmosphere immediately above the soil.

Moreover, the optimal water regime of drainage conditions required for better growth of crops in the drained area should be established (Ivitsky, 1972). This can be done by analysing the observations of groundwater level variations and moisture content of the soil, taking into account the economic characteristics of different drainage systems (Ziverts et al., 1972). As suggested by Ivitsky (1972), reclamation projects should embrace not only the drained areas of the catchment but also its entire area. As complementary studies, hydrogeological, mining, geological, and soil observations should be made during the exploitation of drainage systems. These studies and monitoring will prevent drainage on too large a scale which might lead to detrimental irreversible hydrological changes such as irreversible drying.

FINAL CONSIDERATIONS AND CONCLUSIONS

'Water interacts dynamically with the other components of the ecosystem, and "clean water" results only when the producers (plants), consumers (animals), and decomposers (bacteria, fungi, etc.) are capable of performing their functions in the ecosystems. If any part of the ecosystem is disturbed severely — that is, if any one of the essential environmental quality parameter becomes severely limiting — the aquatic ecosystem cannot operate efficiently to cycle material and to break down added wastes. Biological purification is inhibited and the water thereby may become unsuitable for man's use. The suitability of water to support fish and other aquatic life is an indication that the ecosystem is operating properly. Quality levels which protect sensitive fish species and other aquatic life are normally adequate to protect the ecosystem as a whole and to permit certain use of water for man' (Prince, 1972).

This lengthy citation is useful because it summarizes much of the story of bog-associated water resources, when investigated locally. Nutrient-deficient water, more or less stagnating on an impermeable substrate in a temperate, wet climate, will lead to bog formation when the rate of inflow of water is greater than the rate of evaporation. Unoxidized organic matter will accumulate, absorbing further the dissolved matter, and a low-pH aquatic environment will develop which is inappropriate for the decomposers that would release the nutrients and allow them to be recycled in the ecosystem.

This aspect, considered alone, is sufficient to give support to Bulavko's claim that 'in their natural conditions, marshes and marsh ridden areas represent an unfavorable element in all respect. They make areas unproductive, hinder transport ... and despite the large water reserves contained in undrained marshes, only a very small proportion of it helps to feed the river' (Bulavko, 1971).

On a global scale, however, the role of vast expanses of peat bogs is very difficult to appraise indeed. Since the evaporation from a water-logged marsh is often 20 to 40 per cent more than from a free water surface (Bulavko, 1971), and given the fact that they accentuate rather than hinder the fluctuation in water flow after heavy rainfalls and snow melt, large expanses of peatland certainly have an effect on regional climate.

River runoff is very strongly cyclic in Canada and, according to Dickie (1971), high spring outflow is responsible for initiating the breakup of ice in sea estuaries. This creates large ice-free areas that enlarge still further and at an increasing rate because of the upwelling of warmer deep water and wind action. It is well documented that the drainage of peatlands would effect some degree of regulation of the water regime. What would be the over-all effect of a drainage program aiming at the reclamation of the vast expanses of marsh-ridden areas of North America and the USSR? Preliminary investigations by Dickie (1971) have shown that regulation of the St Lawrence River has resulted in a reduction in ice cover and a general warming up in the gulf; however, this author expects the reverse to be true for systems draining into Hudson Bay. What would be the combined effect on the regional climate of the reduced evaporation and the dampening of the natural water regime fluctuation that would follow such a drainage program? This is a question that cannot be answered now but, we submit, it deserves close investigation.

Finally, a question may be raised that seems far from water resources but that may cause an indirect important long-term effect. Deevy (1958) has suggested that the very large peat deposits in the world may be a controlling factor in the concentration of carbon dioxide in the atmosphere, according to the following equation:

$$CO_2 + (H_2O)n \rightleftarrows (CH_2O)n + O_2$$

Drainage or reclamation of peatland would initiate a slow but irreversible process of oxidation of the carbohydrate and other organic matter stored in the peat. This author estimates that approximately 400 billion tons of carbon dioxide could eventually be released in the atmosphere, an amount equivalent to 'one sixth of the amount actually present in the atmosphere ... and the whole reserve of carbon in land plant and animal is only 15 times as much' (Deevy, 1958). Even if the amount of water released by this slow oxidation may not be significant compared with the huge quantities of free water at the surface of the earth, the increased percentage of carbon dioxide in the atmosphere would absorb more of the energy of the sun. This could lead, through a general warming up of the earth's surface, to glacier recess and to raising of the surface levels of water.

More research is needed on global environmental effects of man's activities and greater effort should be devoted to global environmental monitoring. Given their geographic importance, international monitoring programs should include the effect of the drainage of peatlands.

ACKNOWLEDGMENTS

The coordinator of this review paper (A.A.M.) wishes to extend his most sincere gratitude to the Muskeg Subcommittee of the National Research Council of Canada for the opportunity to present this paper: it was a great opportunity indeed for learning.

Many people have collaborated on this work; this paper would not have been what it is without the help of its co-authors: Dr David A. Goode, who contributed the section on the water balance of peatlands; and my assistant, Jean-René Michaud, who had the difficult task of abstracting the pertinent literature. Mr F. Lunny, of the Bord Na Mona (Eire), was of immense help in locating for us the unpublished proceedings of the Minsk Conference on Marsh-Ridden Areas.

I also thank very sincerely my staff for their excellent work in preparing this manuscript.

REFERENCES

Armstrong, W., and Boatman, D.J. 1967, Some field observations relating the growth of bog plants to condition of soil aeration. J. Ecol. 55: 101-9.
Baden, W., and Eggelsmann, R. 1963. Zur Durchlas-sigkeit der Moorboden. Z. Kulturtech. u. Flurberein, 4: 226-54.
— 1968. The hydrologic budget of the highbogs in the Atlantic region. Proc. 3rd Internat. Peat Congr., Quebec, pp. 206-11.
Battelle. 1972. Environmental Evaluation System for Water Resource Planning (Bureau of Reclamation, us Dept. of the Interior).
Bavina, L.G. 1972. Water balance of swamps in the Forest Zone of the European region of the u.s.s.r. UNESCO Internat. Symp. on Hydrology of Marsh-Ridden Areas, Minsk, July (to be published).
Bay, R.R. 1968. The hydrology of several peat deposits in northern Minnesota, u.s.a. Proc. 3rd Internat. Peat Congr., Quebec, pp. 212-18.
— 1969. Runoff from small peatland watersheds. J. Hydrol. 9: 90-102.
Boelter, P.H. 1965. Hydraulic conductivity of peat. Soil Sci. 100: 227-31.
— 1972a. Water table drawdown around an open ditch in organic soils. J. Hydrol. 15: 329-40.
— 1972b. Methods of analysing hydrological characteristics of organic soils in marsh-ridden areas. UNESCO Internat. Symp. on Hydrology of Marsh-Ridden Areas, Minsk, July (to be published).
Bogomolov, G.V., and Kats, D.M. 1972. Methods and results of hydrology, soils, vegetation in marsh-ridden areas of the temperate zone. UNESCO Internat. Symp. on Hydrology of Marsh-Ridden Areas, Minsk, July (to be published).
Bulavko, A.G. 1971. The hydrology of marshes and marsh-ridden lands. Nature and Resources, 7 (NI): 12-15.
Bulavko, A.G., and Drozp, V.V. 1972. Bog reclamation and its effects on the water balance

of river basins. UNESCO Internat. Symp. on Hydrology of Marsh-Ridden Areas, Minsk, July (to be published).

Burke, W. 1968. The drainage of blanket-peat at Glenamoy. Trans 2nd Internat. Peat Congr., Leningrad (HMSO, Edinburgh).

— 1972. Aspects of the hydrology of blanket peat in Ireland. UNESCO Internat. Symp. on Hydrology of Marsh-Ridden Areas, Minsk, July (to be published).

Chapman, S.B. 1965. The ecology of Coom Rigg Moss, Northumberland. III. Some water relations of the bog system. J. Ecol. 53: 371-84.

Curry, J.J., and Wilson, S.L. 1955. Effect of sewage-borne phosphorous on algae. Sewage and Ind. Waste, 27: 1262.

Dansereau, P. 1967. The post-conservation period. A new synthesis of environmental science. Cranbrook Inst. Sci. News Letter, 37 (4): 42-9.

— 1971. Dimensions of environmental quality. Sarracenia, 14, March (ed. Virginia Weadock, Univ. du Québec à Montréal).

Dee, N., et al. 1973. An environmental evaluation system for water resource planning. Water Resources Res. (in press).

Deevy, E.S. 1958. Bogs. Sci. Am. 199 (4): 115-21.

Dickie, L.W. 1971. Personal communications. Fisheries Res. Bd. Canada, Marine Ecol. Lab., Bedford Inst., Dartmouth.

Didier-Dufour, et al. 1972. Essai de normalisation de la qualité biochimique de l'eau du lac Solitaire (unpublished document).

Doodge, J. 1972. The water balance of bogs and fens. UNESCO Internat. Symp. on Hydrology of Marsh-Ridden Areas, Minsk, July (to be published).

Doudoroff, P., and Katz, M. 1950. Critical review of literature on the toxicity of industrial wastes and toxic components to fish. Sewage and Ind. Wastes, 22: 1432.

Eggelsmann, R. 1971. Uber den hydrologischen Einfluss der Moore. Telma, 1: 37-48.

— 1972 a. The water balance of lowland areas in northwest coastal region of the FRO. UNESCO Internat. Symp. on Hydrology of Marsh-Ridden Areas, Minsk, July (to be published).

— 1972b. Physical effects of drainage in peat soils of the temperate zone and their forecasting. UNESCO Internat. Symp. on Hydrology of Marsh-Ridden Areas, Minsk, July (to be published).

EIFAC 1969. Water quality criteria for European freshwater fish — extreme pH values and inland fisheries. J. Water Res. 3: 593-611.

Eisenlohr, W.S. 1972. Hydrology of marshy ponds in coast of Missouri. UNESCO Internat. Symp. on Hydrology of Marsh-Ridden Areas, Minsk, July (to be published).

Ferda, J., and Pasak, V. 1969. Hydrologic and Climatic Functions of Czechoslovak Peat Bogs (Monograph, Zbraslav).

Godwin, H., and Conway, V.M. 1939. The ecology of a raised bog near Tregaron, Cardiganshire. J. Ecol. 17: 313-63.

Goode, D.A. 1970. Ecological studies on the Silver Flowe Nature Reserve (Ph.D. thesis, Univ. Hull).

Gordon, M. 1972. Die Bedeutung der Moore zur Erhaltung einer hydrologischen Stabilitat. Telma, 2: 149-50.

Heikurainen, L. 1972. Hydrological changes caused by forest drainage. UNESCO Internat. Symp. on Hydrology of Marsh-Ridden Areas, Minsk, July (to be published).

Heinselman, M.L. 1970. Landscape evolution, peatland types and the environment in the Lake Agassiz Peatlands Natural Area, Minnesota. Ecol. Monogr. 40 (2): 235-61.

Huntze, H. and Eggelsmann, R. 1972. The agricultural influence of water eutrophication by nutrient leaching in lowlands areas. UNESCO Internat. Symp. on Hydrology of Marsh-Ridden Areas, Minsk, July (to be published).

Ingram, H.A.P. 1967. Problems of hydrology and plant distribution in mires. J. Ecol. 55: 711-24.

Ivanov, K.E. 1953a. Gidrologiya Bolot [Hydrology of Bogs] (Gidrometeoizdat, Leningrad).

— 1953b. Experimental studies of the water conductivity of the upper layer of peat deposits of raised bogs [in Russian]. Trudy GGI 39 (93).

— 1957. Osnovy Gidrologii Bolot Lesnoi Zony [Principles of Bog Hydrology for the Forest Zone] (Gidrometeoizdat, Leningrad).

Ivitsky, A.I. 1972. Hydrological fundamentals of bog drainage. UNESCO Internat. Symp. on Hydrology of Marsh-Ridden Areas, Minsk, July (to be published).

Jasmin, J.J., and Heeny, H.B. 1960. Reclaiming Aird Domw for Agricultural Use (Dept. Agriculture, Canada Pub. 1089).

Jeglum, J.K. 1971. Plant indicators of pH and water level in peatlands at Candle Lake, Saskatchewan. Can. J. Bot. 49: 1661-76.

Jones, J.R.E. 1969. Fish and River Pollution (Butterworths, London).

Klueva, K. 1972. Drainage reclamation effect on hydrologic regime of rivers in Byelorussia. UNESCO Internat. Symp. on Hydrology of Marsh-Ridden Areas, Minsk, July (to be published).

Korpijaakko, E.O., and Radforth, N.W. 1972. Studies on the hydraulic conductivity of peat. Proc. 4th Internat. Peat Congr., Otaniemi, 3: 323-33.

Kovalchuck, V.F., et al. 1972. Geological and hydrogeological conditions of marshy areas of primoné and freezing rate of soils. UNESCO Internat. Symp. on Hydrology of Marsh-Ridden Areas, Minsk, July (to be published).

Lundin, K. 1972. On moisture accumulation in peatlands under drainage. UNESCO Internat. Symp. on Hydrology of Marsh-Ridden Areas, Minsk, July (to be published).

McKee, J.D., and Wolf, H.W. 1963. Water Quality Criteria (The Resource Agency of California, State Water Resources Control Board, Pub. 3-A, 2nd ed.).

Meustonen, S.E., and Seuna, P. 1972. Influence of forest draining on the hydrology of an open bog in Finland. UNESCO Internat. Symp. on Hydrology of Marsh-Ridden Areas, Minsk, July (to be published).

Novikov, S.M. 1964. Computation of the water-level regime of undrained upland swamps from meteorological data. Trudy GGI, 112. Translated in Soviet Hydrology: Selected Papers (Am. Geophys. Union), 1: 3-22.

— 1972. Methodological principles of investigation of swamps on stations and expeditions. UNESCO Internat. Symp. on Hydrology of Marsh-Ridden Areas, Minsk, July (to be published).

Novikov, S.M., Ivanov, K.E., and Kuprianov, V.V. 1972. Hydrological study of swamps related to their management. Proc. 4th UNESCO Internat. Peat Congr., Otaniemi, 3: 335-45.

Odum, E.P. 1971. Fundamentals of Ecology (3rd ed., W.B. Saunders Co., Philadelphia).

Prince, A.T. 1972. Guidelines for water quality objectives and standards — a preliminary report. Inland Water Branch, Dept. of the Environment, Tech. Bull. 67.

Provost, M.W. 1972. Environmental quality and the control of biting flies. Proc. Symp. on Biting Fly Control and Environmental Quality, Univ. Alberta, Edmonton, May 16-18, ed. Anne Hudson (Dept. Entomology, Univ. Alberta).

Robertson, R.A., Nicholson, I.A., and Hughes, R. 1968. Runoff studies on a peat catchment. Trans. 2nd Peat Congr., Leningrad, 1963 (HMSO, Edinburgh), 1: 161-6.

Romanov, V.V. 1953. Metody Opredeleniya Zapasa Vody Deyatel'nom Slow i Raschet Elementov Vodnogo Balansa Bolot [Methods for the Determination of Water Reserves in the Active Layer and Computation of the Items of the Water Balance of Bogs]. Trudy GGI, 39 (93).

— 1961. Gidrofizika Bolot (Gidrometeoizdat, Leningrad). English translation: Hydrophysics of Bogs (Israel Program for Sci. Translations, Jerusalem, 1968).

— 1962. Isparenie's Bolot Evropeiskoi Territorii SSRR (Hydrometeorological Publishing House, Leningrad). English translation: Evaporation from Bogs in the European Territory of the USSR (Israel Program for Sci. Translations, Jerusalem, 1968).

Schmeidl, H., Schuck, M., and Wanke, R. 1970. Wasserhaushalt und Klima einer Kultievierten und unberuhrten Hockmooreflache am Alpenrand (bauwesen, Verlag Wasserund Boden Hamburg, Munchen).

Sjors, H. 1961. Surface patterns in boreal peatland. Endeavour, 20: 217-24.

Sparling, J.H. 1966. Studies of the relationship between water movement and water chemistry in mires. Can. J. Bot. 44: 747.

Van der Molen, W.H. 1972. Subsidence of peat soils after drainage. UNESCO Internat. Symp. on Hydrology of Marsh-Ridden Areas, Minsk, July (to be published).

Verny, E.S., and Boelter, D.H. 1972. The influence of bogs on the distribution of stream flow from small bog upland watersheds. UNESCO Internat. Symp. on Hydrology of Marsh-Ridden Areas, Minsk, July (to be published).

Vidal, H. 1968. Importance of northwestern and central European bogs as regulators of water balance and climate. Trans. 2nd Internat. Peat Congr., Leningrad, 1963 (HMSO, Edinburgh, 1968), 1: 167-8.

Virta, J. 1966. Measurement of evapotranspiration and computation of water budget in treeless peatlands in the natural state. Comm. Phys.-Math. 32: 1-70.

Vorobiev, P.K. 1963. Investigations of water yield of low-lying swamps of western Siberia. Trudy GGI, 105: 45-79. English translation in Soviet Hydrology: Selected Papers 3 (1963): 226-52.

WHO 1958. International Standards for Drinking Water (Geneva).

Ziverts, A.A., Rieksts, I.A., and Skinkis, T.N. 1972. Investigation of runoff from offtake rivers and subsurface drained soils in the Latvian S.S.R. UNESCO Internat. Symp. on Hydrology of Marsh-Ridden Areas, Minsk, July (to be published).

Zubets, V.M., and Murashko, M.G. 1972. Transformation of the water regime over marsh-ridden areas in the temperate zone by modern reclamation technique and forecast of its effects on hydrometeorological conditions on the environment. UNESCO Internat. Symp. on Hydrology of Marsh-Ridden Areas, Minsk, July (to be published).

13
Waste Disposal

A.E. FEE, N.A. LAWRENCE, and A.C. MOFFATT

DEFINITION AND SCOPE

Muskeg research handbooks have dealt with the problems of muskeg, regardless of its geographical location. In dealing with environmental considerations, and with waste disposal in particular, it would be futile to cover the topic without limiting discussion to the northern Canadian areas, which is the geographical area most people associate with 'muskeg'. To limit the scope of this chapter further the subject is defined as relating to the waste problems generated by the activities of man, and for simplicity the subjects of mine tailing wastes and oil spills will be ignored.

It has only been in recent years that the populations in our northern areas have been drawn together in communities by the pressures of health and school services, by low-rental housing programs and welfare. Parallel with this have been the establishment of industrial towns such as Clinton Creek and Faro, and the growth of governmental-industrial towns and cities, such as Inuvik, Yellowknife, and Whitehorse. With the exception of Inuvik, almost all non-industrial communities in the Far North are located on ground originally occupied by founding organizations, such as the Hudson's Bay Company, as a trading post. The site of Inuvik, as an exception, was chosen after an exhaustive engineering study.

The organization of communities, by necessity, leads to the organization of the disposal of liquid and solid wastes produced by man's activities and interests. The state of the art of household waste disposal over this vast area ranges from orthodox southern systems, with modifications to adapt to northern conditions, to elaborate and costly above-ground 'utilidor' systems as installed in the original Inuvik townsite, and to primitive 'honey bucket' systems adopted as standard practice in many small northern communities.

WASTE-WATER COLLECTION

Existing Settlements

In relatively recent years, added emphasis has been placed on the provision of suitable waste-water collection and disposal systems for existing Canadian arctic settlements in order to ensure the protection of their raw water sources, to control infectious diseases, to minimize odours, and to improve the aesthetic aspects of the communities. The solutions reached have been determined by the small number of users, budget limitations, and design by a myriad of professional personnel. The results have tended to develop highly specialized solutions uniquely related to the conditions in each settlement.

The size of these existing communities, the highly varied conditions of the terrain, the scattered locations of dwellings, and the varied cultural levels of the inhabitants, all create complications in achieving acceptable solutions for formalized and economical waste-water collection systems. The small size of the settlements coupled with the 'custom-built' solutions required, has tended to defeat or negate coordinated efforts by professionals in the related disciplines to seek methods that could be experimented with and gradually developed into specialized solutions for specialized problems in arctic areas.

Added to these problems are the difficulties of easy application of the tried and true waste-water collection systems that have been developed for the more temperate southern areas of Canada, and which can be seriously affected by high ice content subsurface conditions, severe temperatures, short treatment seasons, and lack of trained personnel for maintenance.

Although a considerable number of systems servicing existing communities in arctic areas of Canada are available for analysis, each system relates to the specific restrictions imposed by the site and very few common design and operating parameters can be observed.

New Communities

The provision of adequate waste-water collection systems for new communities in arctic areas can be handled successfully through the application of existing physical planning and engineering criteria. This would result in the establishment of transportation and utility corridors and building sites specifically related to the terrain and subsurface characteristics, and would provide for both a protected raw water source and an effective collection, storage, and treatment system for waste-water disposal.

The selection of the most appropriate waste-water collection system for a new community must be based on all available previous experience in arctic areas. This requires full recognition of the unique problems that exist relative to the selection of systems and materials, availability of maintenance personnel and procedures, ease of procurement of repair equipment and parts, and an appreciation of the severity of the consequences for an isolated community if the system fails. The professional involved should have design and operating experience with arctic conditions as well as a thorough knowledge of the site, the likely inhabitants, and the potential lifetime of the community. Selection of equipment and materials that are easily maintained by relatively unskilled personnel is essential.

Sufficient knowledge is available to permit the development of well-planned communities for arctic conditions. A lack of judgment in applying this knowledge, particularly when there are budget and time limitations, can lead to difficulties.

General Characteristics of Collection Systems for Water-Borne Wastes

The essential characteristic of waste-water collection systems in arctic areas must be simplicity. Every effort must be made to reduce the mechanical portion of the system to a minimum, and to avoid the need for highly trained maintenance personnel and delays in

Table 1

Information on communities of the Northwest Territories

Community	Population	Lat.(N)	Long.(W)	Source	Treatment	Piped - All year	Piped - Summer (Waterpoints)	Trucked Homes	Waterpoints	Trucked Ice	Individual	Disposal	Piped	Truck (Holding Tanks)	Truck (Honey Bags)	Individual	Disposal	Trucked	Individual	Fire Fighting
						WATER SUPPLY						SEWERAGE					GARBAGE			
Aklavik	675	68°12'	135°00'	Lake and River	Plant		++	+	+			Dump		+	+		Dump	+		
Arctic Bay	200	73°02'	85°11'	Lake					+	+	+	Dump			+	+	Dump	+	+	
Arctic Red River	100	67°27'	133°46'	Lake					+			Dump			+		Dump	+		
Baker Lake	900	64°18'	96°03'	Lake				+	+			Dump			+		Dump	+		
Broughton	120	67°35'	63°50'	Stream				++		+	+	Bay			+	+	Dump	+		
Cambridge Bay	420	69°03'	105°50'	Lake	chlor.	+		+	++			Bay			+		Bay	+		
Cape Dorset	480	64°14'	76°32'	Lake			+	+		+	+	Bay			+		Dump	+		
Chesterfield Inlet	250	63°21'	90°42'	Lake		+		+				Bay			+		Dump	+		
Clyde	120 (+250)	70°25'	68°30'	Lake						++	++	Dump			+	++	Dump		++	
Colville Lake	80	~ 66°	127°	Lake													At Large			
Coppermine	500	67°50'	115°05'	River			+		+			Bay			+		Bay			
Coral Harbour	115 (+175)	64°08'	83°10'	Lake		+				+		Bay			+		Bay	+		
Discovery	275	63°11'	113°51'	Lake				+						+						
Eskimo Point	470	61°07'	94°03'	Lake		+	+				++	Dump			+	++	Dump	+		
Fort Franklin	410	65°11'	123°06'	Lake	chlor.	+		+	+	++	++	Lagoon & Dump			++	++	Dump	+		
Fort Good Hope	400	66°50'	128°38'	Lake and River	chlor.				++			Dump					Dump	+		
Fort Liard	225	60°15'	123°28'	Wells					+			Dump			+		Dump	+		
Ft. McPherson	850	67°27'	134°53'	Lake	Plant	+		++		++	++	Lagoon & Dump	+			+	Dump	+		
Fort Norman	250	64°54'	125°34'	River								Dump					Dump			
Ft. Providence	450	61°21'	117°39'	River	Plant							Dump					Dump			
Ft. Resolution	600	61°10'	113°40'	Lake	Plant			+		+	+	River & Dump	++			+	Dump		+	
Fort Simpson	700	61°52'	121°23'	River	Plant	++			+			Lagoon	++				Dump			
Fort Smith	2,000	60°00'	111°53'	River	Plant	++						Bay	++				Dump			
Frobisher Bay	2,100	63°44'	68°28'	Lake					+			Bay					Dump			
Gjoa Haven	100	68°38'	95°57'	River and Lake						+	++	Dump & Bay		+	++	++	Dump & Bay	++	+	
Grise Fiord	70	76°25'	83°01'	Stream				+		+++	++	Dump		+		+	Dump	++		
Hall Beach	65	68°41'	82°17'	Lake			+			+	+	Lagoon				+	Dump		+	
Hay River	3,500	60°51'	115°43'	Lake		+		+				Dump & Bay	+				Dump	++	+	
Holman	180	70°43'	117°43'	Pond						+	+	Dump				++	Dump	++		
Igloolik	530	69°24'	81°49'	Lake				++		++	++	Dump		++	++	+	Dump	++		
Inuvik	3,500	69°22'	133°43'	River and Lake	Plant	+		++				Lagoon & Dump	+			+	Dump			
Jean Marie River	50	61°32'	120°38'	Wells and River								Dump					Dump		++	
Lake Harbour	90 (+150)	62°51'	69°53'	Pond						+	+	Bay & Dump		++		+	Dump		++	
Mary River–Milne Inlet	80	71°19'	79°21'	River and Wells													Bay & Dump	+	+	
Nahanni Butte	75	61°05'	123°23'	Creek		+				+	+	Privies				+	Dump		++	
Norman Wells	250	65°17'	126°51'		Plant							Septic Tanks and Lagoons				+	Dump		++	
Padloping		67°03'	62°45'																	
Pangnirtung	320 (+300)	66°08'	65°44'	River					+		+	Bay & Dump			+	+	Bay & Dump	+	+	

Table 1 (continued)
Information on communities of the Northwest Territories

Community	Population	Lat.(N)	Long.(W)	Source	Treatment	Piped - All year	Piped - Summer (Waterpoints)	Trucked Homes	Waterpoints	Trucked Ice	Individual	Disposal (Sewerage)	Piped	Truck (Holding Tanks)	Truck (Honey Bags)	Individual	Disposal (Garbage)	Trucked	Individual	Fire Fighting
Pelly Bay	75	68°53'	89°51'	Lake	Plant	+		+			+	Dump	+		+	+	Dump	+	+	
Pine Point	550 (?)	61°01'	114°15'	Wells								Lagoon					Dump	+		
Pond Inlet	120 (+180)	72°41'	78°00'	Lake				+				Dump			+		Dump	+	+	
Rae	1,300	62°50'	116°03'	Lake				+			+	Dump, Privies & Septic Tank			+					
Rankin Inlet	550	62°45'	92°10'	Lake		+			+			Septic Tank & Dump	+		+		Dump	+	+	
Repulse Bay	200	66°	86°	Lake				+				Bay & Dump	+		+		Bay & Dump	+		
Resolute	300 (+200)	74°41'	94°54'	Lake				+							+					
Sachs Harbour	95	71°59'	124°44'	Stream, Lake, Ice						+	+	Bay		+		+	Bay			
Snowdrift	150	62°23'	110°47'	Lake							+	Privies				+	Bay		+	
Spence Bay	125	69°32'	93°31'	Lake							+	Bay				+	Dump	+	+	
Trout Lake	42	60°26'	121°15'	Well								Privies					Dump	+	+	
Tuktoyaktuk	666 (+120)	69°27'	133°02'	Lake & Sea	chlor.		+					Dump			+		Dump	+		
Tungsten	125	62°15'	128°	River		+		+	+			Septic Tank	+	+			Dump			
Whale Cove	150	62°09'	92°35'	Lake				+				Dump		+	+	+	Dump	+	+	
Wrigley	130	63°16'	123°37'	River	Plant	+					+	Privies				+	Dump	+		
Yellowknife	6,000	62°28'	114°27'	Lake	Plant	+						Lagoon	+				Dump	+		
TOTAL	33,518																			

obtaining replacement parts. The high cost of energy and the consequences of an interruption in the system must be appreciated.

There are four types of waste-water collection systems currently in use in Canadian arctic areas: (1) underground piped systems, (2) above-ground piped systems (utilidors), (3) pump-out tanks, (4) honey buckets.

(a) Underground Piped Systems

All the design criteria applying to underground piping systems in temperate climates apply equally to systems planned in the northern areas. These include a sufficient gradient to produce self-cleansing velocities, the necessity for access at regular intervals, and economical trench depths. The normal methods of attaining these are applicable, including pumping stations to overcome problems of grade and drop manholes to avoid excessive grade. In addition to these, provisions for coping with arctic conditions must be incorporated. The prime items to be considered are:

1. *Ground stability.* Very simply, if the ground in which the pipe is to be laid is 'stable when thawed' an underground piping system is feasible. Heat transfer from the liquid in the pipes will tend to upset thermal regimes, and thawing will occur. If the thawed material is unstable, the pipes will be unsupported or will heave, leading to destruction of the system. Pipes can be installed in low-ice-content gravels or soils.

2. *Ability to trench.* Soils which are stable when thawed may still pose severe problems in trenching. The use of specialized equipment and blasting or jackhammering of frozen ground adds appreciably to the cost. Trenching costs of $25.00 per running foot are common.

3. *Protection from freezing.* The contents of the pipe must be prevented from freezing. If the flow is adequate the latent heat of the waste water will prevent freezing. In many parts of a system normal flow is spasmodic and freezing temperatures may exist. In such cases, the system design will have to include an artificial flow from the water source, together with effective insulation of those parts of the system in order to reduce the heat transfer. The greatest advantage of an underground system lies in the relatively high temperatures existing at normal pipe depths, compared to minimum air temperatures. For example: at a depth of 8 feet the ground temperature at Inuvik is approximately 17°F (or -10°C). This poses a far less difficult problem to overcome than exposure to air temperature, which frequently drops to -46°F (or 82 degrees below freezing). If Inuvik did not have a subsurface with a high ice content, which is unstable when thawed, an underground system at that latitude would be feasible. Faro, with only slightly higher ground temperatures but with low-ice-content sand and gravels, does have a buried system. The recent innovation of extruded non-absorbent foamed insulations, capable of withstanding burial conditions, has extended the range of conditions under which underground piped systems will operate. Insulated manholes assist in preventing freezing.

4. *Provision of contingency items for the operator.* The operator must be provided with a procedure to be applied in emergencies which gives time for correction of problems. These could include duplicate equipment and pumps or a second power supply, thawing machines or steamers, a ready supply of calcium chloride, a warm-water supply, or a means of draining all systems. The lack of direct view of and access to the problem area (which may well be buried under feet of snow and hard-frozen ground) is the major problem with buried systems. This major disadvantage is also its greatest advantage: that is, it is below the surface and thus offers no impedance to surface traffic or construction.

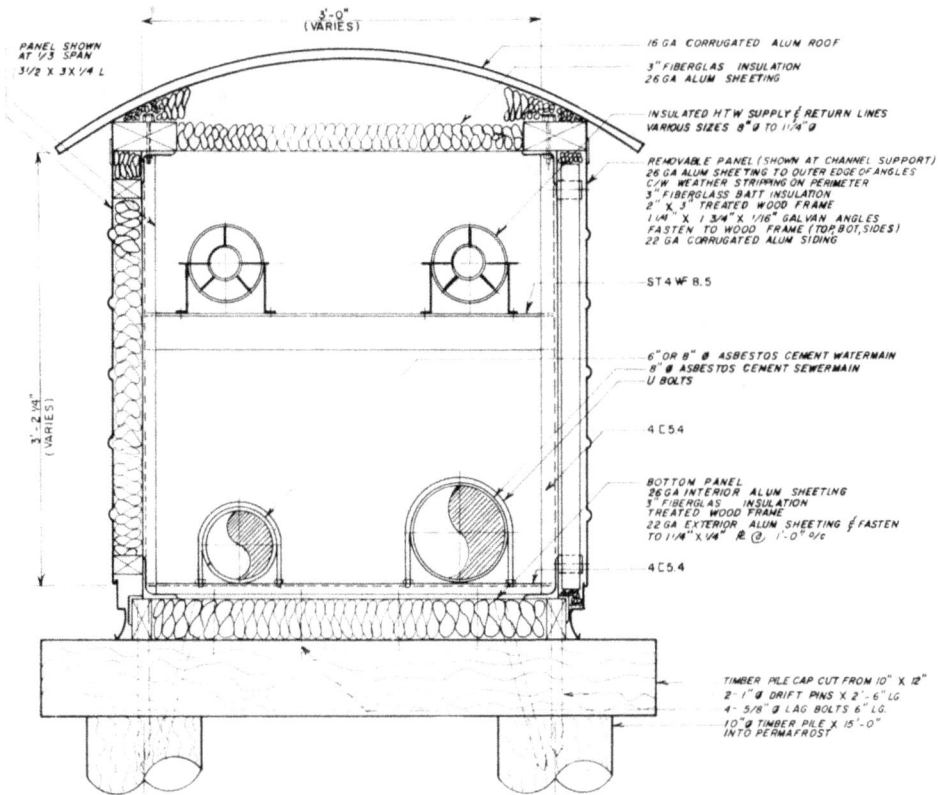

Original Utilidor at Inuvik (Type 1)

(b) Utilidors

In areas of difficult ground conditions such as high-ice-content permafrost, rock, or swamp, the restrictions of site conditions, economics, and operating procedures may dictate that the sewer mains be installed above the ground surface. Then the necessary gradients to provide a combination of gravity-flow characteristics and self-cleansing velocities in the sewer system are achieved by use of artificial supports above the ground level combined with pumping stations or drop manholes as required.

For above-ground construction, the same hydraulic design criteria are applicable as for buried conduits. Normally sewer, water, steam, and electric lines are planned for installation in a utility corridor involving a common heated and insulated space called a 'utilidor'.

If the subsurface conditions are stable, the utilidor can take the form of a box tunnel installed in a shallow trench with the cover of the utilidor utilized as a sidewalk. In those situations where the subsurface conditions will not permit even a shallow trench type installation, the utilidor must be installed completely above the ground surface, supported on mud-sills or piles.

The shape of the utilidor and the exterior cladding can vary to suit the economics of material supply and fabrication provided the necessary space is provided in the interior for the placement, support, installation, maintenance, and inspection of the conduits involved

8"∅ ASBESTOS CEMENT SEWER
1 V4"∅ POLYETHYLENE PIPE
SNAKED IN 1' IN 100'
4"∅ ASBESTOS CEMENT WATERMAIN
LOOSE VERMICULITE MASONRY FILL INS.
1/4" HALFROUND CAULK BEAD
3/4" FIR PLYWOOD
1/8" X 1" STEEL STRAP

1'-2 1/2"

1'-3/4"

1'-4"

1'-0"

2"X 3" SPRUCE
3"X 4"X 3' CEDAR @ 8'-0" %
3/8" LAG SCREW

PLYWOOD SHIMS
4"X 4"X 3'-6" CEDAR
@ 8'-0" %

GRAVEL BED

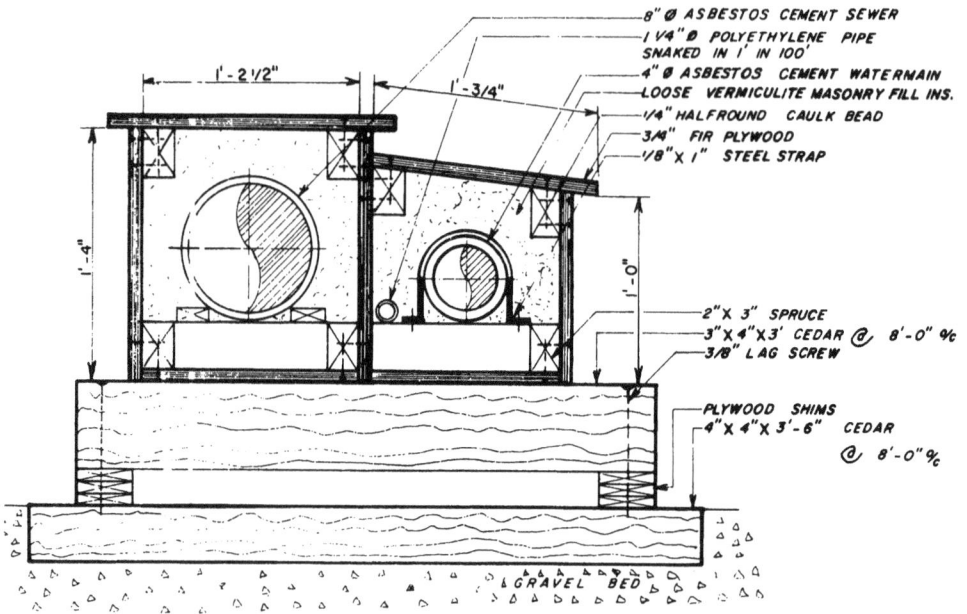

Plywood Box on Fill (Type 2)

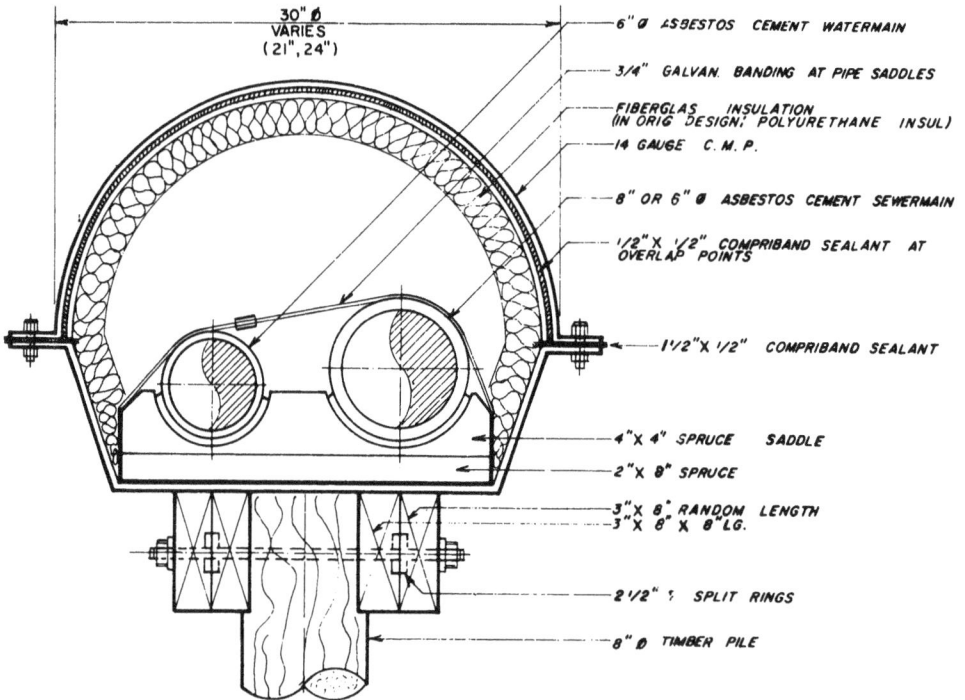

30" ∅
VARIES
(21", 24")

6"∅ ASBESTOS CEMENT WATERMAIN

3/4" GALVAN. BANDING AT PIPE SADDLES

FIBERGLAS INSULATION
(IN ORIG DESIGN; POLYURETHANE INSUL)

14 GAUGE C.M.P.

8" OR 6"∅ ASBESTOS CEMENT SEWERMAIN

1/2" X 1/2" COMPRIBAND SEALANT AT
OVERLAP POINTS

1 1/2" X 1/2" COMPRIBAND SEALANT

4"X 4" SPRUCE SADDLE

2"X 8" SPRUCE

3"X 8" RANDOM LENGTH
3"X 8" X 8'LG.

2 1/2" SPLIT RINGS

8"∅ TIMBER PILE

Corrugated Metal Pipe (Type 3)

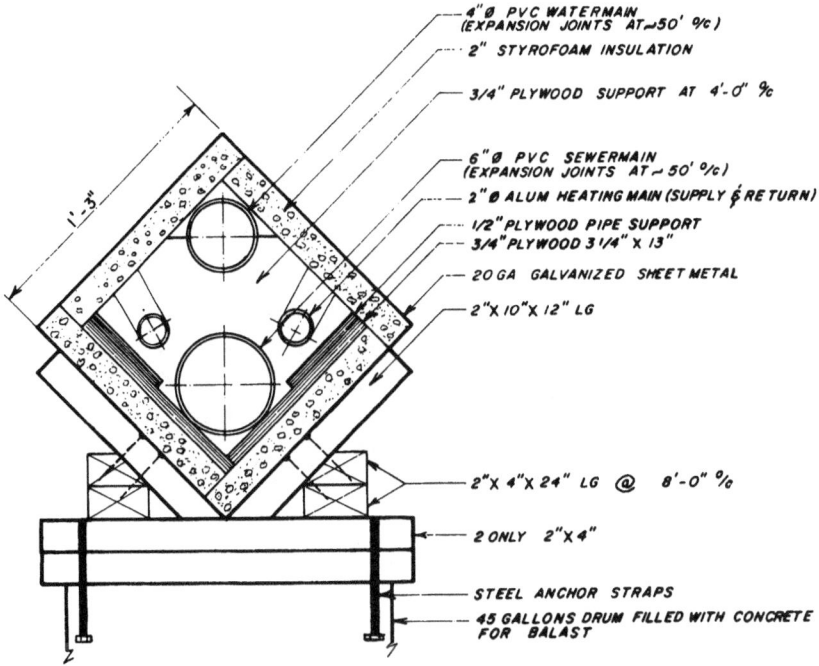

4" Ø PVC WATERMAIN
(EXPANSION JOINTS AT ~50' %c)

2" STYROFOAM INSULATION

3/4" PLYWOOD SUPPORT AT 4'-0" %c

6" Ø PVC SEWERMAIN
(EXPANSION JOINTS AT ~50' %c)

2" Ø ALUM HEATING MAIN (SUPPLY & RETURN)

1/2" PLYWOOD PIPE SUPPORT

3/4" PLYWOOD 3 1/4" X 13"

20 GA GALVANIZED SHEET METAL

2" X 10" X 12" LG

2" X 4" X 24" LG @   8'-0" %c

2 ONLY 2" X 4"

STEEL ANCHOR STRAPS

45 GALLONS DRUM FILLED WITH CONCRETE
FOR BALAST

1'-3"

Econodor (Type 4)

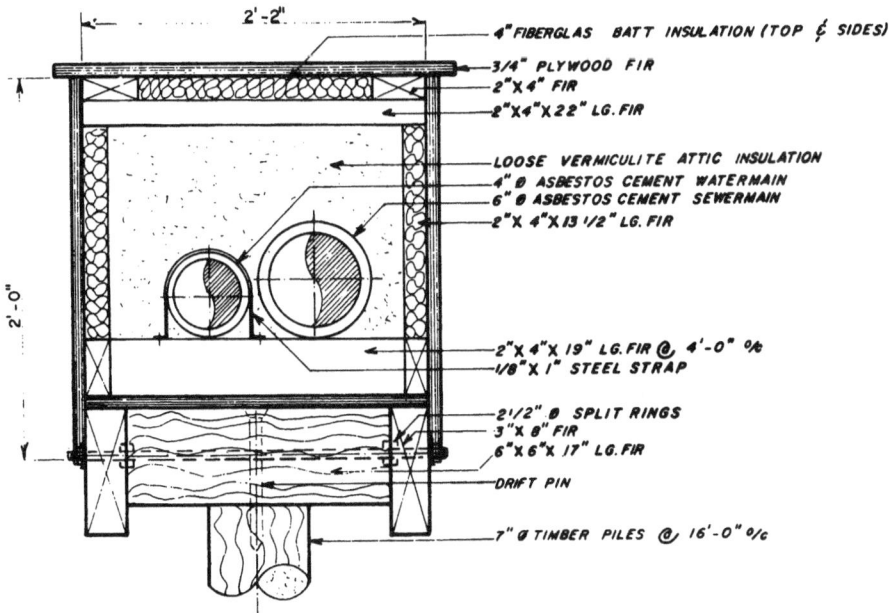

2'-2"

4" FIBERGLAS BATT INSULATION (TOP & SIDES)

3/4" PLYWOOD FIR

2" X 4" FIR

2" X 4" X 22" LG. FIR

LOOSE VERMICULITE ATTIC INSULATION

4" Ø ASBESTOS CEMENT WATERMAIN

6" Ø ASBESTOS CEMENT SEWERMAIN

2" X 4" X 13 1/2" LG. FIR

2" X 4" X 19" LG. FIR @ 4'-0" %c

1/8" X 1" STEEL STRAP

2 1/2" Ø SPLIT RINGS

3" X 8" FIR

6" X 6" X 17" LG. FIR

DRIFT PIN

7" Ø TIMBER PILES @ 16'-0" %c

2'-0"

Plywood Box on Piles (Type 5)

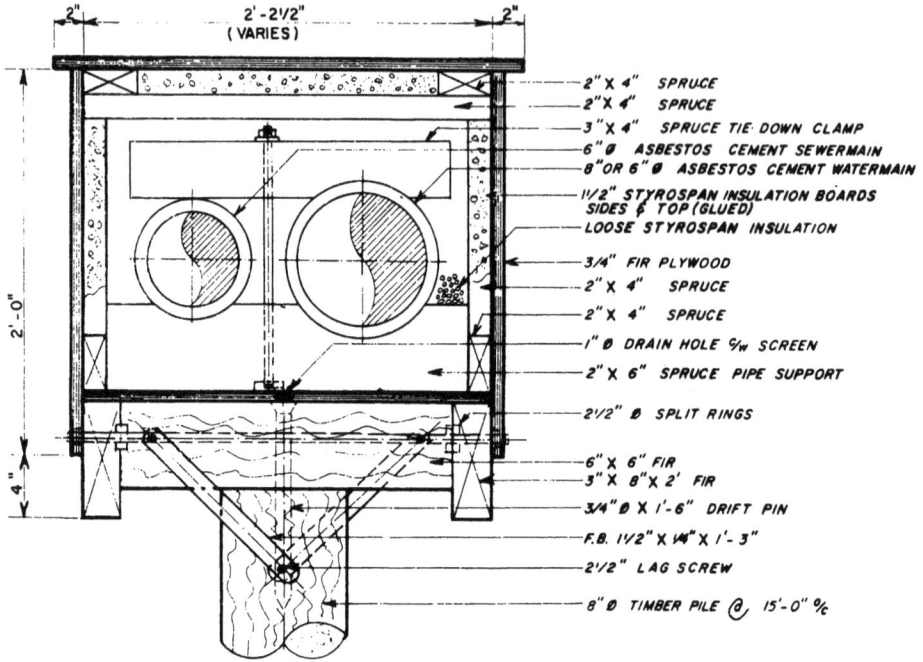

2" X 4" SPRUCE
2" X 4" SPRUCE
3" X 4" SPRUCE TIE·DOWN CLAMP
6" Ø ASBESTOS CEMENT SEWERMAIN
8" OR 6" Ø ASBESTOS CEMENT WATERMAIN
1 1/2" STYROSPAN INSULATION BOARDS
SIDES & TOP (GLUED)
LOOSE STYROSPAN INSULATION
3/4" FIR PLYWOOD
2" X 4" SPRUCE
2" X 4" SPRUCE
1" Ø DRAIN HOLE C/W SCREEN
2" X 6" SPRUCE PIPE SUPPORT
2 1/2" Ø SPLIT RINGS
6" X 6" FIR
3" X 8" X 2' FIR
3/4" Ø X 1'-6" DRIFT PIN
F.B. 1 1/2" X 1/4" X 1'- 3"
2 1/2" LAG SCREW
8" Ø TIMBER PILE @ 15'-0" %

Plywood Box on Piles (Type 6)

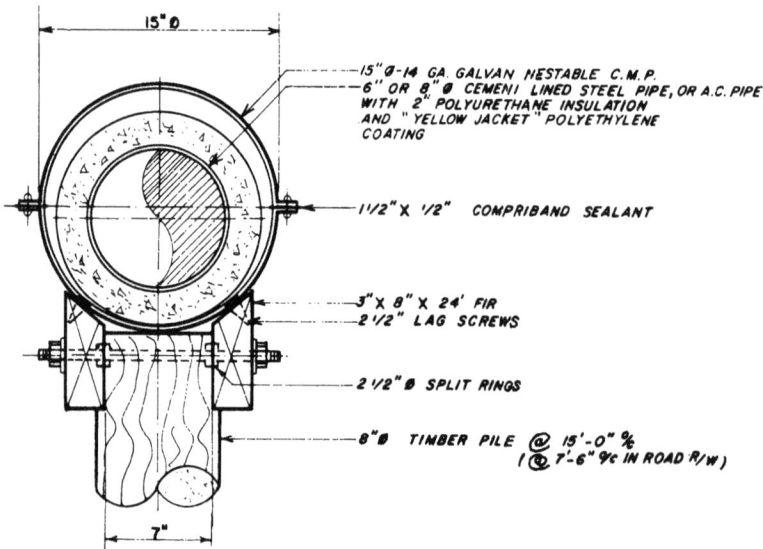

15" Ø-14 GA. GALVAN. NESTABLE C.M.P.
6" OR 8" Ø CEMENT LINED STEEL PIPE, OR A.C. PIPE
WITH 2" POLYURETHANE INSULATION
AND "YELLOW JACKET" POLYETHYLENE
COATING

1 1/2" X 1/2" COMPRIBAND SEALANT

3" X 8" X 24' FIR
2 1/2" LAG SCREWS

2 1/2" Ø SPLIT RINGS

8" Ø TIMBER PILE @ 15'-0" %
(@ 7'-6" %c IN ROAD R/W)

Single pipe in C.M.P. (Type 7)

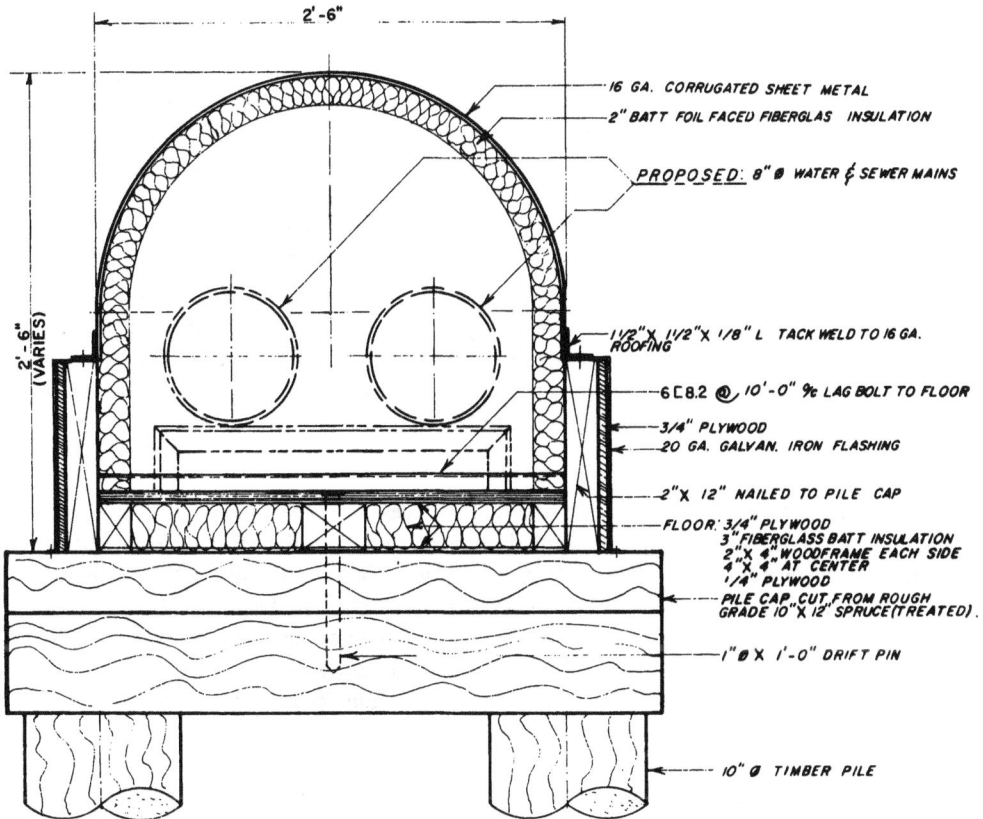

Round C.M.P. Top Box (Type 8)

and for the protecting insulation. Sketches of various utilidor types in use in Inuvik are shown above.

For an above-ground installation, the utilidor is subjected to extreme weather conditions resulting from variations in air temperatures and wind velocities. Thus care must be taken in the selection of materials and method of fabrication to avoid any possibility of freezing the water-carrying conduits. A single penetration point resulting in a short section of frozen line can immobilize the entire system.

The above-ground location also subjects the utilidor to the continuing possibility of damage from personnel and vehicles, and thus the accidental disruption of all services to the community. In addition, this exposure creates both a physical and visual 'pollution' of the urban environment.

Depending upon the mix of conduits included in the utilidor, various methods can be used to provide heat and circulation to the systems with a maximum of interaction to achieve a common benefit. Recirculating water systems, electric heating tapes, and steam-heated pipes are common methods of providing temperature control within utilidors. Discharge points are provided to drain the system for maintenance or emergency shutdowns.

Care must be taken in the selection of conduit materials, joints, connecting materials, and methods and ease of maintenance for each utility included in the utilidor. Insulations capable of maintaining their effectiveness even when wetted from leaks are recommended. Detection of leaks may be hampered by the long hours of darkness. Emergency maintenance requirements can be made more difficult by the need to disassemble an effectively sealed utilidor to reach the source of the problem and by the severe weather conditions that may exist in exposed areas.

In many arctic communities the selection of an above-ground utilidor is imperative. In such cases, study of existing utilidor systems can provide useful information for the development of effective installations. However, the above-ground location results in visual and physical interference with community activity, is vulnerable to damage, and has a high cost of initial installation and continuing maintenance, so that the use of utilidors requires greater than normal care in the planning of buildings and the entire community layout. Costs ranging from $70.00 to $200.00 per foot are to be expected.

(c) Pump-out Tank

For certain existing settlements and possibly the occasional new settlement, it will not be practical and economical to install a piped sewer system, and therefore other methods of collection and disposal must be used. One system in use in a number of existing settlements consists of holding tanks to retain the water-borne wastes close to the points of origin. These are normally installed under the building or immediately adjacent to it. Such tanks are pumped out periodically by means of tank trucks and the waste-water is transported to a disposal site. Both the holding tanks and the tank trucks must be protected to avoid freezing. This type of system has the advantages of low capital and reasonable freedom from contamination within the household. Its disadvantage is in its operating costs and limited capability for expansion. Contracts to operate such systems range from 2 to 4 cents per gallon of waste disposed of. Spillage adjacent to buildings and along travelled ways is a common occurrence.

When site and budget conditions dictate the use of pump-out tanks, this type of system can provide an effective method for the collection and transportation of water-borne wastes. It is acceptable only if the capital costs of a piped system are beyond the capabilities of the community.

(d) Honey Bucket Systems

There are and will continue to be many smaller settlements for which the most practical system involves the use of small containers to hold human wastes for short periods prior to collection and disposal by means of a community-operated collection system. This type of system is instituted in areas where conditions dictate that only the simplest methods be instituted. Plastic bag linings of toilet pails have become standard items. The plastic bags are collected frequently under community-arranged contracts, and transported to a disposal site. Sanitary conditions and aesthetics leave a good deal to be desired, the main objection being the necessity to handle the bags by hand, plus their vulnerability to breakage and spillage.

This type of system requires an all-weather, year-round road system, personnel to operate the vehicle and collect the wastes, and a suitable site and method for disposal of highly concentrated human wastes. Disposal becomes a solid wastes problem with all its ramifications.

Industrial Camps

Collection systems for waste water from industrial camps are relatively easy to provide because the buildings are compact and interconnected, they are occupied around the clock, and they have adequate heated spaces. The collection system is generally included as an integral part of the site development plan, and skilled maintenance personnel are available.

If the terrain and subsurface conditions will permit a buried waste-water collection system, a pumping station can be used both as a retention facility as well as to pump the wastes periodically to the treatment site. If site conditions dictate above-ground installations, then holding tanks can be used with either piped systems or tank truck discharge. Less concern about aesthetics is experienced in industrial camps than in other communities. Single workers are involved, the hours of work are long, the leisure hours short, and the personnel are mobile. Disposal methods are regulated in order to prevent contamination of water supplies or other environmental damage.

Industrial camps in arctic areas will soon be required by regulation to provide secondary treatment of all domestic wastes prior to discharging the effluent to the environment. Industry is generally prepared to meet these requirements.

WASTE-WATER TREATMENT AND DISPOSAL

The Effect of Temperature

In any inhabited area care must be taken in the disposal of human wastes so as to avoid the spreading of sewage-borne, disease-producing organisms. In temperate climates, sewage-borne pathogens are subject to natural die-away or scavenging by microbial predators when discharged to the environment. The situation is different in the Arctic. Microbiologists have found that some pathogens are preserved under freezing conditions. Freezing is also a common means of virus preservation. Therefore in actic regions concern is necessary for the protection not only of the present population but of future populations as well. On this basis it can be argued that, contrary to former opinion, greater care is needed to protect the health of the inhabitants in cold areas than in temperate climates.

Sewage-borne pathogens are spread in many ways, including by direct handling of wastes and contamination of water supplies. In the primitive means of waste disposal used in many northern communities there are frequently many potential ways that disease can spread. It is also known that fox, dog, seal, walrus, whale, and polar bear can harbour parasites dangerous to man. In view of this, care must be taken in the disposal of wastes in regions where animals, birds, or marine life can scavenge.

The cold temperatures which inhibit the natural self-purification of wastes in the environment unfortunately also can seriously affect the efficiencies of conventional waste treatment systems. Most treatment systems are biological, relying on microbial action to break down sewage solids. As the temperature in the treatment units becomes lower, biological activity decreases until as the temperature nears 32°F it almost ceases.

Waste-Water Treatment Methods

In general, waste treatment processes provide for removal of organic matter and suspended solids from the waste and disinfection of the waste. Treatment processes are conventionally classified as:

1. *Primary,* with 35 to 45 per cent Biochemical Oxygen Demand (BOD): reductions and removal of the major portion of the suspended solids
2. *Secondary,* removal of most of the suspended, colloidal, and dissolved solids with up to 95 per cent BOD reductions
3. *Tertiary,* advanced treatment of the secondary effluent, often involving filtration and removal of nutrients by chemical means.

Separate treatment units are used for destruction and removal of pathogens and coliform bacteria. Most conventional processes are either aerobic or anaerobic, depending on whether or not they are carried out in the presence of free oxygen. The effect of low temperature on biological systems generally results in less damage to the functioning of aerobic systems than anaerobic systems. Aerobic systems are generally less affected by cold temperatures than anaerobic systems. For both, however, limits of satisfactory operation lie well above the normal air temperatures. This necessitates enclosure and/or heating for satisfactory operation.

(a) Septic Tanks

Septic tanks are commonly used in the Arctic to serve 25 to 100 persons or even more. Usually they are enclosed in a building which is heated. An improvement of septic tank design called the 'Sewage Digestion Tank' was developed in Sweden. This unit was developed after considerable research, and BOD reductions of 70 to 80 per cent are claimed. The unit, encased in a steel tank, has five separate compartments through which the sewage flows in series. Present units being marketed are sized for 5, 10, 20, and 50 persons.

(b) Activated Sludge

This process involves violent mixing of air and raw settled sewage, with bacterial colonies being utilized to convert non-settlable substances, in finely divided, colloidal, and dissolved form, into stable sludge which can be removed. The required oxygen is supplied in the form of fine air bubbles with high liquid-gas interface ratios allowing efficient transfer of oxygen into solutions. Retention is normally eight hours. The advantages of this process are possible BOD reductions in excess of 90 per cent and freedom from offensive odours. The disadvantages include sensitivity to changes in flow, necessity for constant skilled attendance, vulnerability of mechanical equipment to cold weather, and difficulty in dewatering and disposing of the large volume of sludge produced. Temperature plays an important part in this process. Operation under low temperatures significantly reduces the bacterial population, causes greater accumulation of solids, and decreases the BOD and COD (chemical oxygen demand) removals.

   Treatment and disposal of the sludge are normally carried out in separate tanks under either aerobic or anaerobic conditions. This complicates the process considerably.

(c) Extended Aeration

Extended aeration is a modification of the activated sludge process. Treatment is effected by oxidation of the wastes in the presence of previously treated sludge. Aeration and violent mixing are provided by bubbling diffused air from electric-motor-driven blowers through the wastes. Retention in the aeration tank is usually 24 hours, with no primary settling tank being provided. Following aeration the treated wastes are held in a final settling tank for about four hours, with the settled sludge being returned to the aeration tank. With continuous recycling the sludge is exposed to active aerobic conditions for extended periods of time. To prevent excessive sludge build-up in the aeration tank, sludge is periodically removed and disposed of in storage pits or on land.

The chief advantages of the extended aeration method over the activated sludge method for small municipalities are its lower initial cost, its simplicity of operation, the relative absence of odours, the very small quantities of sludge to be disposed of, and the resistance to upset by shock loads inherent in the longer aeration period. Disadvantages include the larger-sized aeration tanks required to retain the sewage for the longer period, the tendency to produce turbid effluents and greater growth of filamentous organisms under low-temperature conditions, and the time (up to two to three months) required for adequate sludge build-up in the aeration tank before full effectiveness is reached.

(d) Trickling Filtration

This process involves spraying or trickling settled sewage continuously over a rock or synthetic bed. A bacterial growth develops on the medium which is effective in removing BOD. Performance under low-temperature conditions indicates that the filters are more seriously affected by temperature than are activated sludge units; therefore, covering of filters in cold climates is considered a necessity. The use of synthetic media could be one advantage in arctic application of trickling filters. Disadvantages are the need for primary treatment, final sedimentation, and regular sludge disposal.

(e) Waste-Water Lagoons

Waste-water lagoons, or stabilization ponds, are being used successfully in northern climates. Operating depths are normally 4 to 5 feet in summer and 6 to 8 feet in winter. The process involves the activity of algae and bacteria which effect a reduction in the organic matter in sewage and leave a relatively inoffensive liquid. Oxygen supplied by surface aeration is greatly aided by oxygen release of the algae through photosynthesis.

Under winter conditions, with ice cover, biological activity virtually ceases and the settling out of solids produces only the equivalent of primary treatment. However, during the northern summer, even though it is short, the sunlight reaches the surface of the pond for long periods of time each day and a high rate of algae growth results. As a consequence high dissolved oxygen levels occur and treatment comparable to secondary treatment is achieved.

Normally, complete retention of the wastes throughout the winter is practised. Where receiving streams have large flows during the spring flood the stored wastes are often discharged at time of break-up. However, if the receiving stream or lake is small, good practice requires that the winter's accumulation of wastes be held for treatment over most of the summer and released gradually prior to the next freeze-up.

Lagoons are only practicable where topography and soils are suitable. A significant area of relatively flat land is required (approximately 2 to 4 acres per 100 persons served, depending on the mode of operation). Suitable impervious soils for dyke construction must also be available, or alternatively an existing slough or lake can be converted into a lagoon.

The principal advantages of lagoons are the relatively minimal amount of operation required and the low capital cost which can be achieved under favourable site conditions. The disadvantages arise from sludge build-up adjacent to the inlets because of the slow bacterial action. Odours can occur when the ice first comes off a lagoon in the spring. For this reason it is good practice to provide a separation of at least half a mile from the nearest residence in a community.

(f) Aerated Lagoons

An aerated lagoon is a basin of significant depth (usually 10 to 20 feet) in which oxygenation and mixing is accomplished by mechanical or diffused aeration units and by induced surface aeration. The efficiency of the process is highly dependent on temperature. Detention time generally varies from 15 to 25 days, with BOD removal varying from 50 to 80 per cent. Where improved effluent quality is required effluent from an aerated lagoon can be further treated in a conventional stabilization pond.

Final Disposal of Waste Water

(a) Subsurface Drainage

Subsurface drainage systems are generally used for small systems in southern areas with good effect, with soil bacteria providing a polishing action on the effluent. In the Arctic frozen soils will not absorb the effluent, so practice is to construct above-ground discharge over the edge of river banks or similar features. The degree of treatment obtained by this process is thus minimal.

(b) Discharge to Streams or Lakes

This method takes advantage of the dilution available and the self-purification capabilities of the receiving waters. Rates of biochemical stabilization and bacteria die-away are markedly lower at near freezing temperatures than at higher ones. The re-aeration in ice-covered streams may be negligible. In view of the retarding effect of low temperatures on bacterial die-away, the need for disinfection is greater in arctic regions than in temperate climates.

The introduction of nutrients from the sewage into almost sterile arctic lakes has been observed to increase biological activity throughout the life chain. The effects, in the opinion of some fish biologists, may be beneficial.

(c) Discharge to Sea

Discharge to the sea takes advantage of the tremendous dilution available in order to disperse the sewage and thus minimize or eliminate its effect on the receiving waters and its natural biota. Attention must be paid to proper design of the outfall. Since fresh water and sewage are less dense than salt water, the sewage will concentrate on the surface. The purification activity of salt water is less than that of fresh water. The oxygen content is also lower by about 20 per cent and the rate of reoxygenation is in proportion. Problems created by arctic conditions involve the construction of an adequate outfall and thermal and physical protection of the outfall line from ice damage. Except for small installations, sea outfalls are not recommended.

Prospects for the Future

Considering the increased awareness of the importance of protecting the northern environment which now exists, it may be expected that in the next several years there will be a great deal of activity in sewage treatment in northern communities. Treatment facilities will be constructed in communities which now have no such facilities, and existing systems in other population centres will be upgraded to provide improved treatment.

The methods used will vary widely. For collection systems, increased use of better insulation and other construction material and the development of vacuum systems, pressure systems, and recirculation systems offer hope for more economical construction. Selection of the best treatment method for a given community can only be made after the most careful study of all factors involved, including the location for the disposal of effluent, the degree of treatment required, the nature and location of available sites for construction of disposal facilities, the existing and future populations to be served, the calibre of operating personnel available, and the limits of funds for capital and operating costs.

A guess as to the possible directions future waste treatment efforts in the north might take would include the following:

1. *Self-contained household treatment units.* Research and development on individual household treatment units may be expected to be emphasized. Such units, which will probably be based on physical-chemical treatment combined with incineration of solids and optional reuse of liquid portions, could eliminate the high costs involved in constructing sewage collection systems in permafrost areas. The problem of disposing of bath and laundry water must still be solved even if treatment and disposal of toilet wastes can be accomplished by some internal recirculation system.

2. *Lagoons.* Stabilization ponds and aerated lagoons will continue to be used where conditions are such as to make their use appropriate.

3. *Extended aeration plants.* Enclosed prefabricated extended aeration plants will probably be more widely used. Ease of operation, good treatment efficiency, and adaptability for small flows make this system attractive.

4. *Physical-chemical treatment.* For mobile construction camps where 'packaging' of treatment units is essential and where a high degree of treatment is required, physical-chemical treatment will be considered. This method provides for chemical clarification (normally using lime, alum, or ferric chloride), filtration on dual or multi-media filters, and absorption of soluble organics by carbon columns. Unlike conventional waste treatment methods, the physical-chemical method does not depend on biological activity. It has the advantage of being capable of being put into operation quickly, and for this reason will find application in large camps having relatively short occupancy. It has the capability of permitting recirculation of treated waste water for some limited use such as laundry or toilet flushing.

5. *Reverse osmosis.* At locations where waste-water volumes are small and water is scarce, treatment systems may incorporate reverse osmosis units (which use semi-impermeable membranes to provide, essentially, a molecular sieve) to produce high-quality reclaimed water for reuse.

6. *Nuclear power plants.* Temperature is the principal variant distinguishing northern sanitation practices from those in other regions. Considerable waste heat energy is available from nuclear power plants. The use of excess or waste energy from future nuclear power plants may constitute an important potential advance in the field of waste treatment in northern communities.

7. *Incineration.* In areas having cheap fuel the complete destruction of all human wastes by incineration is practical. Furnaces must be provided with high-temperature secondary burners or the accompanying odours will nullify all advantages.

SOLID WASTES DISPOSAL

General

Included in the category of solid wastes are the items normally associated with the topic in southern areas, i.e. paper, wrappings, tin cans, boxes, bottles, rags, and discarded household items. It also includes old car bodies, mobile toboggans, canoes and motors, 45-gallon oil drums, etc. In many arctic communities it also consists of the 'honey bags,' as noted previously.

In earlier times combustible paper and boxes would never find their way into the waste disposal system: they were too valuable as fuel. The government-sponsored program of providing oil-burning furnaces has eliminated any method of burning paper and wood within the houses provided in the low-rental housing program in the Northwest Territories, so that they have now become the major solid waste item.

The approach to collection is orthodox. Containers, usually 45-gallon fuel drums with the tops removed, are used as a disposal receptacle. Community vehicles collect the contents from individual homes or from neighbourhood stands and transport them to the point of disposal. To some degree plastic garbage bags are being introduced.

Problems of Disposal

The major favourable factor in disposal is the relative isolation of most communities. Northern conditions discourage the development of suburbs or homes outside the central area of the communities. There is normally no problem in finding an area suitably located, in so far as isolation is concerned, to dispose of wastes. Almost invariably the problem of access is the ruling factor.

The use of a disposal site on a continuous basis requires an all-weather road for access. Each community has had to solve this problem individually. Where problems of conflict between dumps and sites for other community activities have occurred, it is because a dump site has been selected beside an existing access way for expediency rather than expending funds to construct access to a more ideal disposal site.

The twin problems of low average annual temperature and the presence of permafrost complicate the solid waste disposal problem. A basic need in organic waste disposal on land is the continuance of the biological process wherein organics are reduced to humus. With continuous low temperatures and permafrost the process is inhibited, and organic garbage buried below the 'active layer' becomes preserved for posterity. Several examples of this have recently been reported from Alaska, where wartime armed forces' sanitary land fills have been experimentally re-excavated and the contents found virtually unchanged.

Normally the warm days of a short summer are sufficient to promote biological degradation if organic wastes are left exposed, but the disadvantages of permitting community wastes to remain exposed on the ground surface must also be considered. First, paper and other debris may be dispersed over a wide area by wind. Secondly, the dump forms a feeding ground for ravens, rodents, or stray dogs, and these animals can return to the town limits.

The problems of covering such wastes even lightly are difficult in permafrost areas primarily because of the lack of suitable earth for covering. Occasionally a community may be blessed with a location providing dry gravels along an esker ridge or creek, but this is the exception rather than the rule.

As communities develop and grow, the sheer volume of waste material becomes a problem. In Inuvik, where access to suitable sites is difficult, low average annual temperatures are encountered, and high-ice-content permafrost is prevalent, the problem is increased by the extension of the industrial-commercial area right to the dump site. To add insult to injury, the present public concern over the strewing of solid wastes over the Arctic has resulted in the transportation of many tons of solid wastes for hundreds of miles by water and by air for disposal on the Inuvik dump. Fairbanks, Alaska, has been similarly blessed with air back-haul of tons of waste from the Prudhoe Bay area.

As communities develop, more sophisticated solid waste disposal methods must be instituted. Although simplicity in methods is desirable, more complex procedures are warranted to overcome the effects of the slow natural degradation.

1. Separation of wastes by type should be instituted in order to permit obnoxious wastes and combustible items to be dealt with as a separate item and to permit recycling of metals at some future date.
2. Incineration of wastes will avoid some problems. This will be mandatory in industrial camps and will be very attractive in any communities being served with natural gas.
3. Experimentation with shredding, compaction, and recycling should be carried out with a view to prevention of nuisance, to protect the environment, and to conserve materials.

The primary solution involves 'concerned management,' rather than an infusion of complicated processes.

Industrial Solid Wastes

The entry of industrial, petroleum, and exploration firms into the North has quickly led to the setting of a more rigid standard on industrial activities than on existing communities. As these firms tend toward greater mobility, and as their activities spread over a large area, government guidelines for waste disposal are essential. These are as yet in the draft stages of assembly. Industry has been quick to cooperate and recommended waste disposal practices have been put into effect prior to the issuance of the guidelines.

Exploration crews are equipped with mobile incinerators. Ashes and tin cans are buried. Special crews have been assigned to inspect each camp site under summer conditions and ensure that no wastes are left unburied and that camp sites are left clean enough to pass inspection by government teams.

Base camps of a more permanent nature have instituted waste disposal practices on an individual basis. Liquid wastes are generally directed to pits or lagoons. Combustible solid wastes are incinerated with ultimate disposal of ashes, tin cans, and bottles by burial. Larger mobile camps will soon be required to provide portable secondary treatment plants, the physical-chemical process being favoured for this.

Solid wastes of a more durable nature are still a problem. Wrecked vehicles, oil drums, and construction materials are being accumulated in various areas and no decision is being made as to their ultimate disposal. A considerable quantity of such material was returned to the town dump at Inuvik. Provision for controlled central dumps is expected to be a requirement in the government guidelines.

Government Guidelines

As previously noted, the government guidelines are still (as of May 1973) in the draft stages. They are particularly aimed at the larger construction program expected to develop because of pipeline or other industrial activity in the North.

Indications are that the guidelines adopted in Alaska will apply. The main provisions are as follows:

1. Prior approval is required for construction affecting environmental health.
2. Area-wide systems approaches shall be considered. This precludes each company or subcontractor establishing his own solid waste materials dump. It means that a common water supply shall be established and protected.
3. Responsibility for development and maintenance of environmental quality shall be assigned to specific persons in the companies involved. This individual may well have the most unpopular job in the area, as he will have to bear the brunt of all criticism from all sources as regards effectiveness of procedures involving not only disposal of waste but damage to tundra, drainage, spills, etc.
4. Occupancy of any site so temporary that it is not deemed feasible to provide complete facilities for handling waste does not preclude the necessity for following a completely safe and approved practice.
5. Burial of waste organic material will not be permitted where average soil temperatures of 26°F or less exist. This precludes the practices of either augering or blasting a hole in the permafrost and burying organic waste. This does get it out of sight but also preserves it for posterity. In many areas where successive waves of crews occupy the same area such practices can only result in unacceptable conditions for the latecomers.
6. Sewage stabilization ponds are not acceptable in Alaska as a principal method of sewage treatment at sites characterized by annual average soil temperatures less than 26°F or daily mean air temperatures of 32°F for 200 or more days of the year.

There is some disagreement between health authorities on the last provision. The Alaskan authorities do not consider that under normal circumstances such a facility may be occupied for a sufficient time by construction crews to establish and maintain a proper biological action, even if such action could get established in the temperature zone noted. The other view is that even if the ponds effect only a separation of settlable solids from suspended solids, as long as they retain the sewage a sufficient length of time to effect a die-off of bacteria, they have done a job. In general, the Canadian outlook appears to be that a ruling will be made for each situation on its own merits within the established guidelines. The objective of secondary treatment will be sought.

SOCIAL AND ENVIRONMENTAL PROBLEMS

The original inhabitants of the North were not concerned with waste disposal. Their nomadic habits permitted them to move and occupy fresh ground as campsites became befouled. They accumulated very little in the manner of solid waste material.

Few people realize how recently many of these people settled into fixed communities. The resulting lack of concern for debris and waste, within the newly developed housing areas, is thus a traditional attitude. Concerned effort and the provision of tools and incentive are being used to change these attitudes. Schools and health officials are hoping that education into the effects of poor sanitary conditions will be the prime factor in coping with the problem.

The very low density of population in the North precludes lasting serious effects on the ecology by municipal waste disposal methods. Nevertheless concern for future effects is resulting in the establishment of procedures by communities and industry that will ensure that this situation endures. Such concern must be supported at all levels.

NORTHERN EUROPEAN AND RUSSIAN PRACTICE

Methods of collection and disposal of human wastes vary throughout the world according to tradition, economics, culture, aesthetics, and environmental standards. Therefore, a comparison of the sewage collection and disposal practices used in other arctic areas of the world with Canadian practices may not be too beneficial. Certainly the terrain, subsurface, and weather conditions in the Canadian Arctic are as severe as, if not more severe than, in other Arctic areas.

Methods of collection of waste water are similar in all areas, with a greater preference given to one type or another depending upon the size of the community; the dual use of tunnels as utility conduits and pedestrian walkways is favoured in northern Russian cities.

The basic methods of sewage treatment are similar in all arctic areas, although variations exist depending upon the degree of protection required for raw water supply, availability of year-round discharge to ice-free oceans or receiving river waters, use of adjacent waters for recreational purposes, and the period of suitable treatment weather.

CONCLUSION

Arctic waste disposal methods vary from primitive to advanced. A very large and costly program is being formulated to improve conditions throughout the Northwest Territories. Guidelines and regulations for industrial activity are being developed and these are being supported by industry. These programs should be encouraged and supported by all.

14
Wildlife, Conservation, and Recreation

R.D. MUIR

Muskeg exists at all latitudes in Canada. It has been considered by many as a nearly continuous band of impenetrable wet terrain stretching across Canada north of the more heavily populated areas, breached only by the Rocky Mountains. But the phenomenon reflects more than one image. The biogeography of muskeg in Canada must reconcile apparently contradictory evidence. Certain plants and animals typical of muskeg appear 'out-of-place' if surrounded by very different types of landscape. Thus, pockets of spruce muskeg are found in southern Ontario at latitudes 42° or 43° north, surrounded by an eastern deciduous forest. Conversely in muskeg near the mouth of the Mackenzie River, NWT, at latitudes 67° or 68° north (within the Arctic Circle), muskrats and birds and plants of southern affinity are found. On northern Ellesmere Island and Axel Heiberg Island, green areas of muskeg in vast stretches of polar desert attract muskoxen, shorebirds, and waterfowl.

These apparent anomalies are explained by post-glacial climatic change, edaphic influences, ecological succession, and local climatic effects. Plant communities change relatively slowly and reflect long-term climatic change or persistent local modification of climate. Birds and mammals, being mobile, respond very quickly to environmental change and are therefore a very sensitive indicator of environmental stability.

Conservation of muskeg environment is based on maintenance of desirable qualities, limitation of undesirable change, and the human perception of value in the natural order. For many, the frontier of civilization is where sizable areas of muskeg and rough unsettled land are first encountered and is therefore often the closest environment offering an alternative to the pressures of society and civilization. Recreation values of muskeg and wilderness stem from the relatively recent awareness of the need for this alternative.

WILDLIFE

The definition of muskeg as organic terrain encompasses a variety of landscape types,* based on vegetation and preservation of plant remains.

---

* Some plant ecologists and phytogeographers contend that there is no such thing as a typical or type plant community that can be given a name, that possesses constant characteristics wherever found, and that converges from different beginnings on a deterministic 'objective' or condition. The view is

Cattail marshes of southern Canada are typified by those of Lake St Clair, Point Pelee, Rondeau, Long Point, Hamilton, Toronto, Bay of Quinte, and numerous smaller inland pocket marshes throughout southern Ontario. They qualify as muskeg in this discussion because they build progressively upon accumulations of former plant life with varying degrees of preservation. Cattails (*Typha* sp.) are invasive, quickly occupying stagnant aquatic habitat wherever near-by stands are available to provide seed, if the area is left undisturbed and the water is not too deep. They colonize ditches and persistent standing water, even on agricultural lands. Well-developed cattail stands can arise in less than five years following inundation of the terrain. They usually signify eutrophication arising from near-by development.

Fully developed cattail marshes in Canada occur in southern Ontario and are found eastward into the Maritime provinces, westward on the Prairies, among the Cordilleran Mountains, and northward into the boreal forest.

Roadside examples are good places to observe marsh wildlife. It is often possible to sit quietly in a parked car and observe Red-winged Blackbird, Muskrat, American Bittern, and other marsh birds among the cattails in a water-filled ditch or corner of a field. Large cattail marshes covering many acres are also the home for other species: Coot, Gallinule, Black Tern, rails, grebes, Marsh Hawk, marsh wrens, and Least Bittern. Ducks and geese typically do not nest among the cattails but take their brood to water, often in cattail marshes, immediately after hatching, for food and safety. Other less obvious forms of wildlife inhabit cattail marshes: Snapping Turtle and Painted Turtle, several species of snakes, the Bullfrog and other frogs, peepers, and treefrogs. In the larger cattail marshes connected to the Great Lakes, Carp and Bass are found.

Cattail marshes have low commercial value in themselves, but the habitat they provide attracts or supports commercially or recreationally valuable wildlife. High populations of Muskrat can be maintained, their houses of dead cattail stalks looking like miniature beaver lodges. They are dotted throughout suitable marshes, sometimes many to the acre. Muskrat are subject to fluctuations in numbers induced by water-level changes and other factors. Extreme levels, either high or low, are detrimental. Marshes controlled by levels of the Great Lakes are classic cases. As the water of Lakes Erie and Ontario pass through cyclic highs and lows, marshes at Point Pelee, Rondeau, and elsewhere show synchronous Muskrat fluctuations. 'Pocket' cattail marshes, not affected by Great Lakes levels, are subject to more erratic water levels. An unusually dry or wet season, the spring runoff, or even a few days' heavy rain can flood these marshes causing distress in the Muskrat population. At such times, one may see muskrats wandering in unusual places throughout the countryside. Many are killed on highways, though this may also happen during the time when young rats are leaving the home range and dispersing in search of unoccupied territory.

Muskrat trapping has been a thriving livelihood for trappers wherever rats occurred, and in recent years the importance of rat trapping has been maintained as fur prices increase. The

---

advanced that species distribution and other characteristics are a matrix or 'continuum' in space and time, the result of many complex factors and significantly different in different parts of the range.

An opposing and older view holds that once ecological succession begins, environmental factors and the plant community interact to 'direct' the community along a predetermined series of stages towards a 'fated' or predetermined mature community. It is proposed that a finite number of mature type communities (climaxes) will eventually represent the interactions of succession.

For purposes of this discussion, it is convenient, though not necessarily accurate, to name community types so that generalized statements can be discussed in the space available.

value of Muskrat fur taken annually in all parts of Canada has been as high as 11 million dollars and over the past fifty years has never fallen below 1.1 million dollars. During 1971-72 the value of muskrats taken in Canada was 3.1 million dollars.

Under natural conditions, Mink preys upon the Muskrat. As the numbers of muskrats fluctuate, Mink numbers usually vary in a similar fashion but peak numbers of Mink typically occur after rat numbers have passed their peak and are declining. This well-known relationship, where there is a close or exclusive dependence on one prey species, is referred to as 'predator lag.' Mink are also highly prized for their fur. Even though ranched Mink are increasingly important (because of mutant strains exhibiting fashionable shades of fur), wild Mink pelts remain in steady demand. Otter, a larger member of the same (weasel) family, also frequent wetlands. Mink and Otter furs add another 1.3 million dollars to the total value of wild fur originating on organic and other wet terrain.

Waterfowl are possibly the best-known resource of cattail marshes. In spring, their flocks are elegantly feathered in anticipation of the breeding season as they pass through, stopping briefly for food, rest, or improvement in the weather. In autumn, mixed flocks of waterfowl including both adults and young of the year migrate southward through marshes and muskeg areas. (In former times, wild rice attracted waterfowl in great numbers in eastern Canada, but now naturally occurring wild rice is uncommon.) Ducks and geese use cattail marshes during migration and are taken in great numbers during the fall hunting season. Currently the annual bag of ducks legally taken in Canada is approximately three and one-half million birds. The monetary value of this annual harvest in the national economic context is impossible to determine.

Cattail-spruce mixed muskeg is found, as the name implies, where elements of both southern cattail marshes and the more northern spruce-*Sphagnum* bog grow in the same drainage unit. Typically the bog is saucer-shaped, with deeper open water in the centre, a ring of cattails at the water's edge, and sedge, shrubs, and spruce trees on progressively higher ground grading into a dense spruce forest. This intermediate type of muskeg occurs in many places along the transition zone between the northern coniferous forests and more southerly parklands, prairies, and deciduous forests. Examples are easily seen from public roads in the Haliburton district and parts of the Trans-Canada highway in Ontario, in Riding Mountain National Park, Manitoba, in Prince Albert National Park, Saskatchewan, and other places.

Wildlife usually occupies suitable habitat wherever it occurs. It is therefore not surprising that in mixed muskeg typically southern species occupy the cattail areas and boreal species occupy areas covered by *Sphagnum,* shrubs, spruce, and Tamarack. Thus, the Red-winged Blackbird, if present in the general region, and the Canada or Gray Jay may be found together in mixed muskegs. Mink, Otter, Muskrat, and Beaver may also use these areas.

Among the lesser plant species in mixed muskeg, Pitcher Plant and bog orchids, Labrador Tea, cranberry, and Cotton Grass occupy the 'spruce bog' portion of the muskeg and Arrowhead and Pickerel Weed may be found among the cattails or southern component. The Yellow Pond Lily and the White Water Lily occur in both components.

Typical *spruce-Sphagnum* muskeg occurs throughout the coniferous forest zone, ranging in size from small wet pockets of organic terrain a few yards or so in diameter to vast stretches referred to as unconfined muskeg, covering many square miles. Open water may or may not occur. Typically, spruce and often Larch (Tamarack) trees are associated with this type of muskeg. These may form a closed forest if organic accumulation has proceeded far enough to obliterate open water. Shrubs such as Labrador Tea and Leatherleaf may cover unforested areas. *Sphagnum* and other mosses are usually present and often sedges and grasses.

Few species of wildlife make the spruce muskeg their home range. Many species pass through, however, and obtain food, shelter, or the answer to some specific need such as a nest site. A detailed account of wildlife in the spruce muskeg environment is not possible here, but a few generalizations appear valid. Caribou, both Woodland and Barrenland, pass through muskeg when it lies across their route of travel during migration or local movements. Moose sometimes feed on the aquatic vegetation of muskeg lakes. White-tailed Deer normally do not occur in areas where muskeg forms a significant part of the landscape. None of the herbivorous animals eat spruce or significant amounts of moss. Caribou of both species do, however, actively seek out arboreal lichens, the so-called 'beard moss,' which is often found festooned on old or dead spruce trees. During the months when the snow is deep, arboreal lichens may be the main reliable food supply available to caribou wintering among the trees. Bears and the medium-sized mammals, such as wolves, foxes, raccoons, hare, and lynx, find little of interest in muskeg though they pass through during their travels.

Beaver and Muskrat may use open-water portions of muskeg for lodges and push-ups, typically seeking food elsewhere. Beaver living in muskeg lake lodges may travel considerable distances along streams or overland to stands of aspen, their preferred food, which grows on higher, drier land.

The ecological role of Beaver may be pivotal. Muskeg has developed in post-glacial time where poorly drained land or shallow standing water has permitted an aquatic plant succession to begin. Species participating are those whose distribution ranges are appropriate to the local climate. Additionally, any event creating new occurrences of poor drainage or shallow standing water may initiate muskeg development at any time. The most frequent such natural event is the arrival of beaver followed by the establishment of a dam and pond. Previously dry land is inundated or rendered soggy, and trees die and fall down. Aquatic plants become established around the edges in shallow water. The muskeg succession has begun, and tends to be self-perpetuating. The beaver may be removed or may leave when food species of plants are exhausted, but the muskeg succession continues. A large cattail marsh or an open, wet 'beaver meadow' of sedge, shrubs, spruce, and Tamarack develops. In this way beaver are the agents of sudden ecological change. Over a single twenty-four hour period they may establish residence, begin dam construction, and initiate a new plant succession. This condition may not have arisen on the site in the thousands of years since deglaciation and would not arise without such intervention.

In the 1971-72 trapping season, the value of Beaver pelts taken in Canada was 6.4 million dollars.

Small mammals such as mice, voles, lemmings, and Red Squirrel may exist in rather low numbers in some types of muskeg, especially if tree seeds are plentiful. Among the birds, Spruce Grouse associate closely with mature muskeg, eating coniferous and other buds and nesting on the ground among the trees. A number of small insectivorous birds, especially warblers, nest in coniferous tree tops, obtaining much of their food in the coniferous canopy. Of the raptors, the Sharp-shinned Hawk, Pigeon Hawk (Merlin), and the cavity nesters (Hawk Owl and Saw-whet Owl) may reside in muskeg forest if suitable cavity sites are available in large trees. Small-cavity nesters such as nuthatches, creepers, and both the Downy and Hairy Woodpecker are also found in suitable muskeg forests. The Olive-backed or Swainson's Thrush nests at low levels in dense muskeg forest, and the Gray-cheeked Thrush nests along the northern fringe of the muskeg forest, sometimes in isolated clumps of stunted spruce beyond the main continental tree line. The piercing alcoholic call of the Olive-sided Flycatcher rings throughout the muskeg forest, part of the musical backdrop of bird song shared with the warblers, the White-throated Sparrow, and the Canada or Gray Jay.

Open (non-treed) *muskeg* lying north of the continental tree line is characterized by more shrubs and sedges as trees and other southerly species drop from the flora, excluded by an increasingly rigorous climate and the presence of permafrost.

Vast tracts of open muskeg extend in some regions as far as the eye can see, relieved only by small rocky rises and other minor terrain features. The condition is found across much of northern Canada, from the Tuktoyaktuk peninsula across mainland Mackenzie district, much of Keewatin district, and southward around Hudson Bay, behind Cape Henrietta Maria and taking in part of the James Bay coast down to where the tree line meets the bay. Across northern Quebec and touching parts of Labrador, open muskeg is again encountered, but in that rougher landscape it occurs in smaller expanses. On Newfoundland, much open muskeg occurs, though the rough landscape reduces the size of continuous tracts of muskeg. Good examples of open muskeg are visible from the roads in Terra Nova National Park in Newfoundland. The road to the Ochre Hills Outlook passes through a small area of open muskeg and a larger expanse, known as the Gross Bog, is visible from the road to Terra Nova Village. The road northward along the west coast of Newfoundland passes numerous open areas of muskeg.

Wildlife on open muskeg is represented by fewer species and they are more truly arctic. Caribou and Muskox are the only ungulates. The Arctic Fox, Arctic Hare, Wolf, lemmings, ptarmigans, colonial nesting geese, and shorebirds are encountered. Only a few reminders of the forested muskeg persist, such as the Red Fox and Wolverine, though a few others such as the Canada or Gray Jay, Gray-cheeked Thrush, or Tree Sparrow may be found in the last clumps of stunted spruce separated by miles of open muskeg from their nearest continuous forest cover.

Surprisingly, in a landscape where climatic suppression of the biota is evident and the lakes and streams appear so unpromising, fish are found in fair numbers. The main requirement appears to be lakes deep enough that they do not freeze to the bottom in winter and rivers which do not either dry up or freeze solid. Lake Trout, Char, Grayling, Whitefish, and Pike are found in appropriate aquatic habitat throughout the region of open muskeg. Record-sized 'trophy' fish, especially Lake Trout, are taken from Great Bear Lake, the Henik Lakes, and Chantrey Inlet on the Arctic coast and elsewhere.

Open muskeg is encountered on the Arctic islands almost as far north as land extends. An extensive lowland muskeg bisects Bathurst Island between Bracebridge and Goodsir inlets. It is characterized by a generally wet, broad, flat lowland with numerous lakes, ponds, pools, and puddles set in a mossy, wet, sedge muskeg. The Bathurst Island lowland forms an oasis of practically all wildlife species found in the High Arctic islands. The Muskox, Peary Caribou, Greater Snow Goose, Eider, shorebirds, foxes, the Wolf, Snowy Owl, and many other species breed in or close to the muskeg and obtain food directly or indirectly from the relatively high vegetative productivity of these areas.

Many other smaller areas of muskeg occur throughout the Arctic islands. An extensive sedge muskeg lies north of Stanwell Fletcher Lake on Somerset Island. Another lies on the western perimeter of the Bjorne Peninsula, Ellesmere Island. Sizable mossy muskeg lies within sight of the glacier at the head of Middle Fiord on Western Axel Heiberg Island. At the head of Admiralty Inlet on Baffin Island a sedge muskeg meadow is used by the Greater Snow Goose.

Smaller pockets of muskeg are encountered in many other places. Those of the Fosheim Peninsula of Ellesmere Island are becoming well known as industrial and scientific parties examine the area for research and resource potential. Muskoxen regularly feed on the lush green muskeg meadows in summer and greater snow geese take their broods there for food and protection immediately after hatching. Some of these muskeg areas are characterized by frost polygon pools. The raised edges are vegetated, but the flooded low centres form square and polygonal pools in regular checkerboard patterns.

A large muskeg pond at the south end of Eastwind Lake has been the nesting site of successive pairs of red-throated loons for over fifteen years. The nest, probably replaced or repaired frequently, lies on the same square foot of muskeg edge each year. Muskoxen regularly feed on the sedges around the edge of this pond. Near by an exposed soil profile caused by a collapsed ice lens showed that organic accumulation amounted to about ten inches of peat. Near the north coast of Ellesmere Island and on the west coast of Axel Heiberg Island at the head of Strand Fiord, early stages of string bog formation show that the main muskeg processes operate at and beyond latitude 80° north, whenever suitable conditions exist.

A number of muskeg areas in Canada are important to wildlife, apart from other values which they may possess. For instance, the only known breeding area for the Whooping Crane*lies within a single continuous muskeg unit in Wood Buffalo National Park. The area is a nearly impenetrable network of small, shallow lakes and ponds, separated by narrow strips of muskeg supporting a thin fringe of spruce trees. It is an unusual combination of limestone bedrock, marl-bottomed ponds with mildly alkaline waters, acid bog vegetation, and a high production of aquatic invertebrates. Habitat conditions in this strange secluded area appear to be not only adequate but possibly necessary for the breeding success of the Whooping Crane, whose nests are usually large flat-topped piles of dead vegetation with a small depression for the eggs, located in shallow water along the edge of the ponds. The Sass River, which drains the region, flows at a lower level than that of the lakes, constituting an implied threat to the stability of the muskeg complex and therefore to the Whooping Crane. Should natural events or negligence initiate changes, a sequence of habitat alterations could doom the species.

Waterfowl and Muskrat use other wet areas, including a muskeg complex at the Peace-Athabasca Delta in the southeastern quarter of the park. The hydrology of the area is extremely complex, and changes brought about by construction and filling of the W.A.C. Bennett Dam in British Columbia have caused massive ecological change detrimental to most wildlife living downstream. Low water strikes at wildlife both directly and through the changes it causes in food-producing areas.

---

*Grus americana L., a large, white, heron-like bird, larger than the Great Blue Heron or the dusky Sandhill Crane. Cranes, however, fly with necks fully extended, whereas herons fly with necks folded back against the body. The Whooping Crane was reduced to little more than a score of individuals in the 1940s but has slowly increased to about fifty in 1969. The entire known wild population winters in Texas. Approximately eighteen whooping cranes are held in captivity.

At Old Crow in Yukon Territory, an extensive, lake-dotted muskeg lies astride the Porcupine River, which flows 140 feet below the level of the muskeg flats. Conditions favour heavy production of Muskrat and waterfowl. Moose and Caribou (of the Porcupine herd) use the area also, but it does not supply their total requirements. Notably lacking is suitable winter range.

The Tuktoyaktuk peninsula, part of the Mackenzie lowland complex, is a vast rolling muskeg of both wet and relatively dry types. Caribou formerly inhabited the area but disappeared during the early part of this century. A herd of domesticated Reindeer, obtained from Asia, was driven into the area to provide a reliable economic base for resident Eskimo. Though the sociological aspect of the project was not very successful, a remnant of the Reindeer herd still inhabits the peninsula. Ducks, geese, and swans are found there also, utilizing both the coastal waters and the inland lakes and ponds.

Near the lower end of the Saskatchewan River, near Cumberland House in the Province of Saskatchewan and eastward toward The Pas in Manitoba, a complex of wetlands involving large areas of muskeg originally provided important habitat for waterfowl, Muskrat, Beaver, and Moose. In recent decades, habitat management through water-level control has been aimed at increasing wildlife productivity. A number of projects have been implemented and more are planned by the federal and provincial governments and by Ducks Unlimited. An area of 400,000 acres is involved. Dikes, ditches, and about forty control structures are planned to maintain optimum water levels. Diversions from the Saskatchewan River have been made, and the Saskeram River too has been dammed for many years as part of a pilot project by Ducks Unlimited.

At Delta, Manitoba, a research station and wetlands management project has been operated for many years by Ducks Unlimited. The area is a vast, open cattail muskeg, heavily used by waterfowl for breeding and during migration. *Phragmites,* a very tall (up to 10 feet) member of the grass family, is spreading at the expense of cattails, possibly reducing the value of the area for waterfowl.

Lesser areas of muskeg important to wildlife occur in the western cordilleran mountains. On the tops and sides, in the passes and broad intermontane valleys, among the coastal ranges, and on the Pacific islands, patches and pockets of muskeg occur under a variety of circumstances. All are characterized by a wet flora rich in sedge, moss, ericaceous shrubs, and spruce. At or near the tree line, from 6,000 to 8,000 feet above sea level, muskeg develops under suitable conditions. Part of Opabin Pass in Yoho National Park, British Columbia, at an altitude of 7,250 feet, also contains a small muskeg complex of moss- and sedge-filled water courses, associated with small ponds and pools. On Mount Revelstoke and around the Jade Lakes in Mount Revelstoke National Park, BC, and across the river valley in the Monashee Mountains, muskeg conditions are associated with ponds and pools. They have given rise to peat accumulations of 1.3 metres, the bottom layer of which was dated by the carbon-14 method at 5,490 years BP, plus or minus 140 years. Very little wildlife is directly associated with this type of terrain, though evidence of mountain sheep, Elk, Caribou, Grizzly Bear, foxes, marmots, and bats (species undetermined) has been recorded on site. Of the birds, the Canada or Gray Jay, Clark's Nutcracker, ptarmigans, the Golden Eagle, and hummingbirds have been recorded at 6,000 feet above sea level.

The broad valley bottoms between the individual ranges of mountains often contain stretches of marsh and muskeg at places where the gradient is low and the water is slowed or held back in lakes and ponds. One of the most accessible such areas is the Vermillion Lakes complex at Banff, Alberta, visible from the Trans-Canada highway. Conifers, shrubs, sedges, and grasses line the edges of a series of lakes, ponds, and waterways. Moose, Beaver, and Muskrat feed on aquatic vegetation. Ducks, geese, and shorebirds rest and feed on the lakes during migration. Grebes sometimes nest on platforms of dead vegetation floating in the shallow lakes and osprey have nested in tall trees beside the water. Many other similar areas are found among the various ranges of the Rocky Mountains and other ranges of the Cordillera.

Up and down the western coastal mainland, on Vancouver Island, the Queen Charlotte Islands, and elsewhere, muskeg conditions are found. Some on relatively high ground result directly from heavy rainfall along the rim of the Pacific Ocean; others farther inland arose from blocked drainage on low-lying land.

Virtually all muskeg occurrences share common features across Canada, modified by local conditions and the plant and wildlife species. So many aspects of the general biology of muskeg are essentially alike or so obviously derived from common conditions and circumstances that for practical purposes at least, it appears reasonable to refer to all occurrences as simply 'muskeg.' As a derived communal complex, its ecological relations display more convergent characteristics of familiarity than factors of divergence and strangeness.

Thus muskeg, shown previously to express the common characteristics of organic terrain from southernmost to northernmost Canada, is also found to express these characteristics from Newfoundland to Vancouver Island.

CONSERVATION

Conservation, by definition, has two meanings, avoiding waste and preserving original conditions. Conservation during commercial utilization of muskeg concerns the former and is a matter of economic efficiency. Rationale for the conservation of natural conditions is based on scientific and sociological concern, and is dealt with in the section on 'recreation.'

Until about thirty years ago little thought was given to the conservation of either muskeg or other wetlands. The north seemed to have an inexhaustible amount of muskeg. It was viewed as the main barrier to the northward spread of civilization and was considered impossible to convert to meaningful use. In the south, where muskeg exists chiefly in the confined condition, some commercial extraction of peat has taken place. A few agricultural operations based on well-humified peat have succeeded. Possibly the best-known enterprise is an ongoing market garden project at Bradford Marsh north of Toronto. On Point Pelee in southwestern Ontario, near the west end of Lake Erie, draining and use of the north part of the marsh for agriculture necessitated building a large dike across the point to retain the waters of the vast cattail marsh which constitutes three-quarters of Point Pelee National Park. This dike, constructed to preserve natural conditions in the national park as required under the National Parks Act, is one of a very few significant efforts made in Canada to conserve muskeg in its natural condition.

Muskeg constitutes a self-maintaining set of ecosystems when left undisturbed. Preservation of organic materials is based on an anaerobic, usually acid regime in wet, usually cool circumstances. Living plants act as a thermal shield and add to the accumulation when they die. Any change which removes the surface growth allows oxygen to penetrate the organic accumulation, causing decay. Conservation of muskeg is concerned mostly with maintaining the physical integrity of the terrain.

Water relations in an undisturbed muskeg are the result of long-term checks and balances controlling plant growth, accumulation, and preservation. Maintaining natural water levels (including seasonal fluctuations) is probably the most important single factor in conservation of muskeg. Too high a water level kills much of the vegetation and exposes wildlife to danger. Too low a level exposes buried organic material to oxidation and decomposition.

The quality of the water entering a muskeg is also important to its conservation. Though the organic soils of muskeg have a very high capacity to exchange ions and thereby buffer the chemistry of incoming water, this capacity is not unlimited. In recent years tests have been conducted using organic soils as ion filters to remove undesirable ions from mine waste and other waters.

In some places it is industrially convenient to dump mine waste into near-by muskeg. At Pine Point on Great Slave Lake, a complex of open pits and extraction and concentration plants results in large amounts of waste rock slurry. This is pumped some distance and discharged onto muskeg. The coniferous forest, locally overwhelmed by the silt-laden waters, is dead and partly buried. The effluent will percolate through considerable muskeg terrain before reaching the shore of Great Slave Lake. The long-term implications of this practice are not fully understood.

An additional factor in the conservation question is the recent awareness of the diminishing availability of fossil fuels. Under these circumstances reserves of combustible carbon assume new importance. Muskeg throughout Canada contains many millions of tons of peat, much of which has fuel potential.

Conservation of muskeg terrain appears at present most important in localities where: (a) important ecological factors such as the occurrence of muskeg-dependent rare and endangered species are involved; (b) renewable natural resources such as fur-bearing animals, food animals, or forest resources are concerned; (c) sociological factors such as recreation are concerned; (d) muskeg is an uncommon landscape.

Since about 1945 increasing public interest has focused on preserving unique, rare, or ecologically significant occurrences of natural habitat of all types. Particular attention is paid to those close to civilization, which are most immediately threatened and are of interest to the growing numbers of natural history enthusiasts. Pockets of muskeg are often the only undeveloped land left in otherwise solidly built-up areas. In spite of difficulties in draining or removing the organic soils, the commercial value of these remnants is high. Private individuals, conservation clubs, and environmental agencies find that previously worthless 'swamp' is now very highly priced and difficult to acquire. As this is written, Rattray Marsh, a 55-acre remnant of a shoreline marsh west of Toronto, is the subject of a dispute over development or preservation. The dispute involves the acquisition price, whether it should be 1.1 or 3 million dollars.

Within sight of the city of Ottawa, the Mer Bleu, a large muskeg of spruce and *Sphagnum*, has been the focal point of conservation discussions in recent years. During the Second World War parts of it were used as a practice bombing range. Other parts have been raked by fires over the years. The Mer Bleu contains relict populations of plants and insects. In 1963 a pair of hawk owls successfully raised young along the edges of Mer Bleu, constituting a southern breeding record for the species. During the preceding winter, numerous sightings of this species had been recorded in the area, and the breeding pair undoubtedly remained to nest because the Mer Bleu habitat closely resembled that of the northern coniferous forests and muskeg where the species normally breeds. Efforts over the years to have the Mer Bleu placed in preservation status have not yet been entirely successful, largely because of lack of sufficient funds for acquisition.

Apart from the foregoing, unaltered natural terrain of all kinds including muskeg is in itself a resource, and part of the heritage of society. There is a general ethical requirement to maintain natural conditions and qualities everywhere, unless it can be shown that defined and minimized alteration is in the best long-term interests of the responsible society.

RECREATION

Let us begin with the dictionary definition of the word 'recreation' and of another word with which we shall be concerned in this section, 'integration':

Recreation: refreshment of the strength and spirits after toil; forming anew.
Integration: to make entire.

Recreation is an increasingly important part of the lives of many people. Three social factors are mainly responsible. Relatively few people now work in jobs that provide healthy outdoor exercise. Most people work a five-day week, leaving two days each week in addition to annual holidays for discretionary activity. Finally, our increasing affluence and the growing awareness that civilization with all its attractions is becoming oppressive have affected our recreational activity. An alternative rooted in the natural world is essential for physical and mental well-being. The increasingly divisive effects of crowded society are being offset by the 'integrating' effect of recreational outdoor experience.

Often the first encounter with muskeg is when it is sighted while fishing, hunting, canoe tripping, or hiking. Sometimes the encounter is unplanned and goes unappreciated. Muskeg typically is difficult underfoot, insect-infested,* impenetrable, too hot, too cold, or unattractive for some other reason. For some it is a trial of character and fortitude, but others marvel at the greenness, the sponginess, the wetness or the sombre conifers, the strange plants, and the solitude.

Hunters fantasize the struggle for food. Canoeists imagine voyageurs against the wilderness. Artists interpret the stark landscape on a single plane. Naturalists ponder the array of specialized plants. Engineers shudder at the cost of construction. Developers search for a glimmer of profit. Sociologists mourn the existence of land unfit for people. Trappers wait until cold and snow make travel easy and profitable.

In Point Pelee National Park a boardwalk takes visitors half a mile into a cattail wilderness. In Yoho National Park, British Columbia, a trail and boardwalk lead through, around, and across a muskeg and beaver pond. In Terra Nova National Park, Newfoundland, a nature trail leads visitors through a spruce and shrub muskeg. As awareness of the total environment increases, all natural landscapes including muskeg are being examined minutely for new information, understanding, and experience. The search for beginnings, identity, and destiny is no longer confined to the ivory tower.

The mossy proto-muskegs of the polar desert of the highest Arctic islands are present-day examples of how it all probably began. The pocket coniferous muskegs of southern Ontario

---

* As this text was being written, a pocket electronic instrument was announced which may repel biting flies. If it works, a major discomfort of recreational use and enjoyment of muskeg may disappear.

hold in their preserved organic material the history of our southern landscape during post-glacial time. Environmental destiny can only be guessed at by comparing these relic areas of muskeg with their present-day surroundings and projecting the trend. To do so suggests a warmer future, but there are hints of a pause or even a regression in recent years. Whatever the future holds, it is likely that muskeg will never again be overlooked as valueless.

MUSKEG-RELATED LIVING SPECIES

Plants

Spruce   *Picea mariana*
Tamarack (Larch)   *Larix laricina*
Cattail   *Typha latifolia, Typha angustifolia*
Sphagnum (peat moss)   *Sphagnum* sp.

Sedges

Scirpus sp.
Eriophorum sp.
Kobresia sp.
Carex sp.

Shrubs

Labrador Tea   *Ledum groenlandicum*
Leatherleaf   *Chamaedaphne* sp.
Laurel   *Kalmia* sp.
Bog Rosemary   *Andromeda* sp.
Phyllodoce sp.
Cassiope sp.
Vaccinium sp.

Arboreal Lichens (beard moss)

Alectoria sp.
Usnea sp.
Evernia sp.
Arrowhead   *Sagittaria* sp.
Pickerel Weed   *Pontederia* sp.
Yellow Pond Lily   *Nuphar* sp.
White Water Lily   *Nymphaea* sp.
Pitcher Plant   *Sarracenia purpurea*

Orchids

Showy Lady's Slipper   *Cypripedium reginae*
Rein Orchis   *Habenaria* sp.
Beard Flower   *Pogonia* sp.

Grass Pink   *Calopogon* sp.
Swamp Pink   *Arethusa* sp.
Calypso   *Calypso* sp.

Birds

Common Loon   *Gavia immer*
Red-throated Loon   *Gavia stellata*
Horned Grebe   *Podiceps auritus*
Pied-billed Grebe   *Podilymbus podiceps*
Great Blue Heron   *Ardea herodias*
American Bittern   *Botaurus lentiginosus*
Least Bittern   *Ixobrychus exilis*
Canada Goose   *Branta canadensis*
Greater Snow Goose   *Chen caerulescens*
Mallard Duck   *Anas platyrhyncos*
Black Duck   *Anas rubripes*
Pintail Duck   *Anas acuta*
Common Eider   *Somateria mollissima*
King Eider   *Somateria spectabilis*
Sharp-shinned Hawk   *Accipiter striatus*
Golden Eagle   *Aquila crysaetos*
Bald Eagle   *Haliaeetus leucocephalus*
Marsh Hawk   *Circus cyaneus*
Pigeon Hawk   *Falco columbarius*
Spruce Grouse   *Canachites canadensis*
Willow Ptarmigan   *Lagopus lagopus*
Rock Ptarmigan   *Lagopus mutus*
White-tailed Ptarmigan   *Lagopus leucrurus*
Whooping Crane   *Grus americana*
Sandhill Crane   *Grus canadensis*
Virginia Rail   *Rallus limicola*
Sora Rail   *Porzana carolina*
Coot   *Fulica americana*
Gallinule   *Gallinula chloropus*
Black Tern   *Chlidonias niger*
Hawk Owl   *Surnia ulula*
Snowy Owl   *Nyctea scandiaca*
Barred Owl   *Strix varia*
Great Gray Owl   *Strix nebulosa*
Saw-whet Owl   *Aegolius acadicus*
Rufous Hummingbird   *Selasphorus rufus*
Hairy Woodpecker   *Dendrocopos villosus*
Downy Woodpecker   *Dendrocopos pubescens*
Canada or Gray Jay   *Perisoreus canadensis*
Black-capped Chickadee   *Parus atricapillus*

Clark's Nutcracker   *Nucifraga columbiana*
Red-breasted Nuthatch   *Sitta canadensis*
Brown Creeper   *Certhia familiaris*
Winter Wren   *Troglodytes troglodytes*
Long-billed Marsh Wren   *Telmatodytes palustris*
Short-billed Marsh Wren   *Cistothorus platensis*
Swainson's or Olive-backed Thrush   *Hylochicla ustulata*
Gray-cheeked Thrush   *Hylochicla minima*
Yellow Warbler   *Dendroica petechia*
Magnolia Warbler   *Denroica magnolia*
Myrtle Warbler   *Dendroica coronata*
Blackburnian Warbler   *Dendroica fusca*
Bay-breasted Warbler   *Dendroica castanea*
Yellow-throated Warbler   *Geothlypis trichas*
Red-winged Blackbird   *Agelaius phoeniceus*
Slate-colored Junco   *Junco hyemalis*
White-throated Sparrow   *Zonotrichia albicollis*

Mammals

Muskrat   *Ondatra zibethicus*
Muskox   *Ovibos moschatus*
Beaver   *Castor canadensis*
Mink   *Mustela vison*
Otter   *Lutra canadensis*
Varying Hare   *Lepus americanus*
Caribou   *Rangifer tarandus*
Marmot   *Marmota caligata*
Red Squirrel   *Tamiasciurus hudsonicus*
White-footed Mouse   *Peromyscus leucopus*
Deer Mouse   *Peromyscus maniculatus*
Red-backed Vole   *Clethrionomys gapperi*
Meadow Vole   *Microtus pennsylvanicus*
Brown Lemming   *Lemmus trimucronatus*
Bog Lemming   *Synaptomys* sp.
Collared Lemming   *Dicrostonyx groenlandicus*
Gray Wolf   *Canis lupus*
Arctic Fox   *Alopex lagopus*
Red Fox   *Vulpes vulpes*
Grizzly Bear   *Ursus horribilis*
Black Bear   *Ursus americanus*
Raccoon   *Procyon lotor*
Wolverine   *Gulo luscus*
Moose   *Alces alces*

Fish

Carp *Cyprinus carpio*
Bass *Micropterus* sp.
Lake Trout *Salvelinus namaycush*
Arctic Char *Salvelinus alpinus*
Grayling *Thymallus arcticus*
Whitefish *Coregonus clupeaformis*
Pike *Esox lucius*

Amphibians and Reptiles

Snapping Turtle *Cheledra serpentina*
Painted Turtle *Chrysemys marginata*
Bullfrog *Rana catesbeiana*
Spring Peeper *Hyla crucifer*
Chorus Frog *Pseudacris* sp.

FURTHER READING

Plants

Fernald, M.L. Gray's Manual of Botany

Mammals

Banfield, A.W.F. 1974. The Mammals of Canada (Univ. Toronto Press)
Cahalane, V.H. Mammals of North America
Errington, P.L. 1963. Muskrat Populations (Iowa State Univ. Press)
Peterson, R.L. 1966. The Mammals of Eastern Canada (Oxford Univ. Press, Toronto)

Birds

Godfrey, W.E. 1966. The Birds of Canada (National Museum of Canada)
Bent, A.C. Life Histories of North American Birds of Prey, Vols. 1 and 2 (Dover Publications)
Linduska, J.P. (ed.). 1964. Waterfowl Tomorrow (US Dept. of Interior)

Fish

Scott, W.B., and Crossman, E.J. 1969. Checklist of Canadian Freshwater Fishes with Keys for Identification (Royal Ontario Museum Misc. Pub.)

Reptiles and Amphibians

Conant, R. Field Guide to Reptiles and Amphibians of Eastern North America
— Field Guide to Reptiles and Amphibians of Western North America

General

Dirschl, H.J., and Coupland, Robert T. 1972. Vegetation Patterns and Site Relationships in the Saskatchewan River Delta (National Research Council publication)

Dirschl, H.J. 1972. Geobotanical Processes in the Saskatchewan River Delta (National Research Council publication)

Nieman, D.J., and Dirschl, H.J. 1969, 1970. Waterfowl Populations on the Peace-Athabasca Delta (Canadian Wildlife Service Occasional Paper 17)

Dirschl, H.J., and Goodman, A.S. 1967. Wildlife in the Saskatchewan River Delta (joint report to Saskatchewan River Delta Development Committee)

MacKay, J. Ross. 1963. The Mackenzie River Delta, NWT (Memoir 8, Geog. Branch, Dept. Energy, Mines and Resources, Ottawa)

The Wisconsin Deglaciation of Canada. Arctic and Alpine Research, 5/3 (1973), Part 1 (symposium)

Leverin, H.A. 1946. Peat Moss Deposits in Canada (Canada Dept. of Mines and Resources No. 817)

Errington, P.L. 1957. Of Men and Marshes (Macmillan Co., New York)

Dunbar, M.J. 1968. Ecological Development in Polar Regions (Prentice-Hall, Englewood Cliffs, NJ)

Project Mar (the conservation and management of temperate marshes, bogs, and other wetlands). 1962. Vol. 1, Proceedings MAR Conference IUCN, ICBP and IWRB.

A List of Terms and Definitions

W. STANEK

At the 21st meeting of the Subcommittee on Muskeg held on 24 May 1974 in Ottawa, the motions were carried to add a glossary to the muskeg environment handbook and to give me the responsibility for coordinating its compilation. From 1970 to 1974, as a member of the IUFRO (International Union of Forestry Research Organizations) Working Group S1.05.1 'Afforestation of Peatland and Extremely Wet Soils,' I was responsible for Project 1, Peatland Terminology, which the leader of the Working Group, Dr L. Heikurainen, Professor of Peatland Forestry at the University of Helsinki, originated in 1964 with a basic list of terms. In 1970 I revised his version, reviewed pertinent literature, and listed the definitions, sources, and most important synonyms. In 1973 the combined IUFRO - IPS (International Peat Society) Committee on Peatland Terminology in Glasgow accepted it as the basis for compilation of an international terminology of peatlands.

The glossary at hand is the third, greatly abbreviated revision of the original material. Its use in this volume has been sanctioned by the IUFRO Working Group leader. In preparing the definitions I have aimed at conciseness and preservation of the meaning of the first users or apparent originators. Where two or more definitions are documented I have prepared a composite definition based on points common to all of them and on current Canadian usage. The list of references is by no means complete. Some source material was unavailable and there may be other material of which I am unaware. In addition to the references listed in the bibliography I would like to recommend the following sources of terms and definitions:

American Society of Agricultural Engineers. 1967. Glossary of Soil and Water Terms (Special Publ. SP-04-67; St. Joseph, Michigan).
Bick, W., Robertson, R. Allan, Schneider, R., and Schneider, S. 1973. Glossary for Bog and Peat (Torfforschung GmbH, Bad Zwischenahn).
Carpenter, J.R. 1938. An Ecological Glossary (Hafner Publ. Co., New York).
Plaisance, G. 1969. Dictionnaire des forêts (Marseille, France).
Soil Science Society of America. 1965. Glossary of soil science terms. Proc. Soil Sci. Soc. Am. 29: 330-51.

At this point it should be noted that the glossary definitions may differ from the meanings attached to the terms by the authors of various chapters of this book. In such instances the glossary definitions are not intended to replace the meanings in context or the definitions given by the chapter authors. The purpose is to provide additional information, and this should be looked upon as an invitation to all to assess critically the use of peatland terms with the aim of achieving an unbiased interpretation in the future.

The compilation of the terminology would not have been possible without the aid of Miss Susann Collver and Messrs G.T. Atkinson, V. Jansons, and T. Silc. I would like to express my appreciation to my colleagues Dr J.K. Jeglum and Mr V.F. Haavisto; to Dr R. Brown, Dr. P.J. Rennie, and Dr M.J. Ruel; and to Mr G.P. Williams, Research Advisor of the Muskeg Subcommittee, and Dr R.M. Strang, RLVR, for reviewing the manuscript and making valuable suggestions.

TERMS AND DEFINITIONS

aapa bog   Combined Finnish-English term for 'string bog.'

aapa fen   Combined Finnish-English term for 'string fen.'

aapa moor   Combined Finnish-German term for 'string peatland.'

acidophilous   Refers to the response shown by organisms adapted to life in an acid medium (Dansereau, 1957).

aefja   (A) horizon with an algal cover (Kubiena, 1953).

airform pattern   Pattern based on the arrangement of shapes apparent on aerial photographs taken at a particular altitude. It is characteristic of significant terrain entities and their spatial relationships and thus useful in the application of aerial interpretation. (Terms and definitions of airform patterns are given in Radforth [1958, 1964] and MacFarlane [1969].)

alder swamp   Wet, eutrophic site, often with muck soils. The vegetation is dominated by alders (*Alnus* spp.). The nutrient-poorer alder swamps with abundant black spruce and poorly growing alder shrubs have been separated as mesotrophic alder swamp. These swamps often occur at bog borders influenced by seepage water (Damman, 1964). See also *swamp*. Synonym: wood fen (Kubiena, 1953).

allochthonous peat   Peat of sedimentary origin. This peat is formed from the remains of plants brought in (mainly by water) from outside the site of deposition (Waksman, 1942; Farnham, 1968).

anaerobic   Having no molecular oxygen in the environment.

anmoor   See *muck*.

autochthonous peat (Greek *autos*, 'self'; *chthon*, 'earth')   Designation for peat which has been formed in situ (Waksman, 1942; Farnham, 1968).

basin bog   Bog which has built up to the water level in a confined depression or a large kettle as in a lake or a river channel. The surface of the peat is horizontal, gently sloping, or slightly concave (Jessen, 1949; Dyal, 1965; Pollett, 1968; Tarnocai, 1970). See also *bog lake, bog, closed pond, raised bog.*

basin fen   Fen occupying a defined basin (as in basin bog), where flooding occurs only for a short period as a result of spring runoff or seepage; however, water is retained for a long period and fen-type vegetation develops (Anon., 1972a). Synonyms: *fen, retention fen, soligenous peatland, topogenous peatland.*

basin marsh   Marsh occupying a confined basin (ice block depression, kettle, glacial or sink basin). The water source may be runoff, overflow, surface drainage, or groundwater discharge (Anon., 1972a).

basin swamp   Swamp developed in a confined basin or kettle whose water source is dependent upon seepage or surface runoff (anon., 1972a).

blanket bog   Term used in the British Isles for bog covering undulating semi-uplands. Blanket bogs develop in cool, temperate regions under a maritime rainfall at lower elevations or in low-lying land, on hills under high rainfall and low temperatures. They sometimes cover the whole landscape — the hills, slopes, and valleys. Usually the peat is fibrous and ombrogenous and seldom more than 6 feet deep (Conway, 1954; Ratcliffe, 1964; Damman, 1965; Pollett, 1967, 1968). Terrain covering bogs (Osvald, 1925) or maritime raised bogs (Potonié, in Osvald, 1925).

bog   Ombrotrophic peatland, wet, extremely nutrient-poor, acid with a vegetation in which *Sphagnum* species play a very important role, tree cover less than 25 per cent (Du Rietz, 1949; Hustich, 1949a; Moss, 1953; Heinselman, 1963; Radforth, 1964; Ratcliffe, 1964; Zoltai et al., 1974). Damman (1964) differentiated between oligotrophic and mesotrophic bogs. Several landscape types of bog have been differentiated: raised, blanket, basin, string, continental, island, palsa, maritime, bowl, sinkhole, floating, shore, flat, polygonal peat plateau, peat plateau, peat mound, etc.

bog forest   Forest growing on peat in bogs (Hustich, 1949a, b) and nutrient-poorer than swamp. Similar to continental forest raised bog (Kulczynski, 1949).

bog iron   Impure iron deposits (limonite) that develop in bogs or swamps by the chemical or biochemical oxidation of iron carried in solution (Anon., 1972b).

bog lake   Area of open water, commonly surrounded either wholly or in part by true bog margins, possessing peat deposits around the margin, in the bottom, or both; usually with a false bottom composed largely of very finely divided, flocculent vegetable matter; containing considerable amounts of colloidal materials and so constituted generally that in time it may become completely occupied by bog vegetation (Welch, 1935).

bog pool   See *bog lake*.

bog ridge (Drury, 1956)   Synonymous with *string*.

brown moss peat   Peat composed of brown mosses, mainly *Amblystegium, Paludella*, and *Hypnum* (Cajander, 1913).

carex peat   See *sedge peat*.

closed pond   Pond filled with organic material often made up of living plants (Radforth, 1964).

collapse scar   Depression adjacent to the slumping edge of a palsa, peat plateau, or peat island, usually water-saturated, without permafrost, round in outline, treeless, minerotrophic. The thawing permafrost edge appears as a steep bank with leaning, mostly dead trees (Zoltai, 1971).

continental bog   Bogs in continental climate, similar to raised bogs but less raised and covered in most cases with open, coniferous tree stands, the *Sphagna* (mainly *S. angustifolium*) hidden by dense shrubs, i.e., *Ledum palustre* and *Chamaedaphne calyculata*. Described by Osvald (1925) in eastern Europe as meadow raised bogs, ridge raised bogs, and karst raised bogs.

coprogenous earth   A material composed of plant debris and faecal pellets less than a few tenths of a millimetre in diameter and having dry colour values less than 5. It has slightly viscous water suspensions, is slightly plastic but not sticky, and shrinks upon drying to form clods that are difficult to rewet. It has few or no plant fragments recognizable to the naked eye. The cation exchange capacity is less that 240 meq per 100 g soil (Anon., 1973a).

cupola   Dome-shaped portion of an ombrophilous bog dominated by *Sphagnum* and flanked on the fringe by a lagg (Bellamy, 1966).

decomposition   See *humification* and *von Post humification scale*.

duff   See *raw humus*.

dwarf shrub bog   Nutrient-poor, relatively dry bog covered with ericaceous dwarf shrub and *Sphagnum* species (Damman, 1964).

dy   Subaqueous, muddy, acid humus horizon on top of parent material (AC-soils), biologically extraordinarily inert, occurring at the bottom of brown waters which consist to a great extent of an amorphous precipitation of humus gels (Kubiena, 1953). Is poorer in nutrients than gyttja and is characterized by a high carbon:nitrogen ratio (Anon. 1974).

dystrophic (Greek *dys*, 'dis-, ill-'; *trophos*, 'nourishment')   Designation for humus and soil formation with particularly unfavourable biological relations tolerated only by certain organisms (Kubiena, 1953); very poor both floristically and faunistically (Dansereau, 1957); term used to denote high concentrations of humic acid in water (Odum, 1959).

eccentric bog   Bog having peat masses elevated near the margin, sloping unilaterally, sometimes forming a fan or saddle shape, occurring on lowlands and uplands (Anon., 1972a).

emergent fen   See *marsh*.

eutrophic (Greek *eu*, 'well'; *trophos*, 'nourishment')   Designation for soils with high nutrient content and high biological activity (Kubiena, 1953; Pogrebniak, 1944), applied to a bog composed of plants growing in hard waters which are rich in nutrients (Barry, 1954).

fen   Meadowlike, sedge-rich peatland on minerotrophic sites, better nutritionally and less acidic than a bog. *Sphagnum* species are subordinate or absent, whereas *Campylium polygamun*, *C. stellatum*, *Scorpidium scorpioides*, and *Drepanocladus* species are abundant (Waksman, 1942; Du Rietz, 1949; Tansley, 1949; Sjörs, 1959). Damman (1964) recognized two main types: eutrophic fen, usually on sites with nutrient-rich telluric water (*Sphagnum* species are absent and green sedges predominate), and mesotrophic fen, in which moderately nutrient-poor, greyish-green sedges are predominant and *Sphagnum* species occur frequently. Zoltai et al. (1974) proposed the following types of fen: string, seepage, net, floating, shore, draw, horizontal, pond, collapse, palsa, spring, and slope.

fen-soaks   Natural drains in bogs, with minerotrophic water often present in bogs where springs come forth (Du Rietz, 1954).

fibrous peat   Composed mainly of the remains of sedges (*Carex*) and similar plants (Rigg, 1958).

filling-in   See *lake filling-in process*.

flark   Usually elongated, wet, and muddy depressions in string peatlands; may be several hundred metres in length. On slopes flarks are narrow, only a few metres wide. On horizontal peatland they may be a hundred or more metres wide. The flark's long axis is always perpendicular to the direction of the contours. In a well-developed string bog these have a mud bottom with *Drepanocladus* and *Scorpidium* mosses and emergent sedges. The term is applied to any water-filled or sedgy area between strings (Cajander,

1913; Hustich, 1955; Sjörs, 1961a, 1963; Pollett, 1968). Synonyms: rimpi, flachette (in Newfoundland), mare, hollow.

flowe   Scottish name given to areas of sloping bog land (Ratcliffe and Walker, 1958).

flushing   Lateral movement of water, usually charged with mineral ions, through peat.

fluvial marsh   See *marsh*.

gyttja   Swedish folk name (von Post, 1862). Subaqueous humus form, active, muddy, predominantly coprogenic, grey-brown to blackish sediment, rich in organisms occurring in waters sufficiently rich in nutrients and oxygen, containing great quantities of organic food (partly in excess) (Kubiena, 1953).

heath   Expanse of ground dominated by shrubs of the heath family (Ericaceae), specifically heather (*Erica, Calluna*) species on sandy or mainly mineral soil. The ground may or may not be peat covered; if it is, it can be termed a bog (Wenner, 1947; Heinselman, 1963).

hochmoor   A German term. See description under *raised bog*. Properties of hochmoor peat given by Fleischer (1913) and Brüne (1948) are as follows: bulk density, 90-120 g/1,000 cm^3; specific gravity, 1.61-1.66 g/cm^3; nitrogen content, 1.20; calcium oxide content, 0.35; phosphate content, 0.10; potassium content, 0.05 (as a percentage of the dry weight).

humification   The extent of decomposition or 'huminosity' (Waksman, 1942) and a process by which organic matter decomposes to form humus (Anon., 1972b). See *von Post humification scale*.

humin   The fraction of the soil organic matter that is not dissolved when the soil is treated with dilute alkali (Anon., 1972b).

hummock   Microtopographic feature which refers to an elevated, comparatively dry area on raised bog, surrounded by relatively wet depressions. The hummocks are composed principally of hummock-forming *Sphagna*, such as *Sphagnum fuscum, S. medium, S. imbricatum*, and *S. flavicomans*. In normally growing raised bogs the depressions are occupied by *S. cuspidatum*. The hummocks, in some areas, may be a peat core topped with ericaceous plants, lichens, and small bush (Cajander, 1913; Auer, 1930; Sjörs, 1963; Radforth, 1964; Tanttu, 1915). Smaller than peat mound. Growth of hummocks is explained as 'lenticular regeneration of peat' (Sernander and von Post, in Kulczynski, 1949).

humus   The fraction of the soil organic matter that remains after most of the added plant and animal residues have decomposed. It is usually dark coloured. Humus is also used in a broader sense to designate the humus forms referred to as forest humus. They include principally mor, moder, and mull (Anon., 1972b).

hydromorphic soil   A general term for soils that develop under conditions of poor drainage in marshes, swamps, seepage areas, or flats (Anon., 1972b).

hygric   Refers to the quality of an organism adapted to wet conditions or to the site that provides such conditions (Dansereau, 1957). See *moisture regimes*.

lacustrine bog   Transitional stage in which some mineral water is still a major influence in the development of the bog (Gorham and Pearsall, 1956).

lagg   Swedish term denoting the zone where the water collects at the margin of a peatland, near the mineral ground of the surrounding site. The water is relatively rich in bases and supports a eutrophic type of vegetation, with the communities resembling those of a fen (Osvald, 1923; Godwin and Conway, 1939; Du Rietz, 1954; Sjörs, 1961a). It is formed as a result of drainage of water from a convex bog, and is similar to moat (Drury, 1956).

Synonyms: marginal fen (Conway, 1949); ditch (Allington, 1961); slough (Lewis et al., 1928); fen soaks or seepages (Sjörs, 1961b).

lake filling-in process Filling in of lakes, etc., by vegetation (Stoeckeler, 1949); synonymous with terrestrialization (Weber, 1908b).

lowland swamp Swamp occupying a large, poorly drained tract of lowlands, where the source of water is due to groundwater flow, seepage, or runoff; often located on glacial lake beds, or level till plains (Anon., 1972a).

mare See *flark*.

maritime bog Term used by Potonié (in Osvald, 1925) for terrain covering bogs (blanket bogs) which develop in areas of cold and very wet climate.

marl Loose, earthy deposit of calcium or magnesium carbonate, believed to have accumulated in freshwater basins fed by mineral water springs.

marsh An open flat or depressional area, covered by less than 25 per cent woody cover with standing or slowly moving water and subject to seasonal flooding. It is characterized by unconsolidated graminoid mats which are frequently interspersed with open water, or by a closed canopy of grasses, sedges, or reeds (shore vegetation). Usually there is little peat accumulation. The substrate can vary from usually shallow, well-decomposed peat to mineral soils. Waters are weakly acid or alkaline (Moss, 1953; Baldwin, 1958; Heinselman, 1963; Damman, 1964; Adams and Zoltai, 1969; Anon., 1972a). Zoltai et al. (1974) differentiated the following types of marshes: *estuarine marsh*, in river estuaries or adjoining bays where tidal flats, numerous channels, and pools are inundated by fresh, brackish, or salt waters; *coastal marsh*, on marine terraces remote from estuaries, or in embayments or lagoons behind barrier beaches, where there is periodic inundation by tidal, brackish, and salt waters; *fluvial marsh*, occupying water courses or floodplains; *lentic marsh*, occupying lake shores or bays of flowage lakes; *catchment marsh*, occupying topographically defined basins; *seepage marsh*, not in topographically defined basins, usually at low elevation or at the base of slopes.

meadow marsh Similar to fen but part of the marsh formation which is typified by generally shallow peats and high fluctuations of water table.

mesotrophic See *nutrient classes*.

mesotrophic fen See *fen*.

minerotrophic Nourished by mineral water; refers to peatlands which receive nutrients from mineral groundwater in addition to precipitation by flowing or percolating groundwater indicating that nutrients are brought to the peat by water that has previously extracted them from a mineral soil (Du Rietz, 1954; Sjörs, 1961a) as is the case in topogenous fens, soligenous mires, springs, flushes (Ratcliffe, 1964), thin blanket peats, fen hummocks, and mud bottoms, sloping fens, marshes (Pollett, 1972), and peatlands which occur in bands of varying widths around upper slopes and lateral margins, downslope from mineral soil islands and substratum ridges and along lower peatland margins that have a strong outflow of moderately mineral-rich waters. Waters are nearly neutral, high in calcium and magnesium (Heinselman, 1963).

mire Wet, spongy earth; bog; soft or deep mud, slush, etc. (Anon., 1967) occasionally used in the sense of peatland.

moat Area of open water several yards across, but in most bogs a broad zone of coarse sedges growing in deep water and closely resembling a lagg that is actively expanding by thawing of the surrounding frozen alluvium. It is associated with discontinuous permafrost in peatlands with palsas. A moat is wider than the marginal channel of most American bogs, and is formed by the break in vegetation resulting from the rise of the floating vegetation mat when flooded in the spring (Drury, 1956).

moisture regimes   Actually physiological moisture regimes which affect the biogeo-coenoses over long periods of time. They are expressed in the total effect and only seldom can be related to a single observation of physical properties, such as absolute moisture content. A genetic soil type would perhaps best express the soil-moisture regimes; however, the relation between vegetation responses and soil type is still poorly defined. Von Kruedener and Becker (1941) and Klika et al. (1954) used five classes: hydric (very wet), hygric (wet), mesohygric (moist), mesic (fresh), and xeric (dry). In North America three regimes were proposed: aqueous, aquic, and moist unsaturated regimes, and several moisture subclasses (Anon., 1973a). Descriptive conditions and associated peat landforms of several moisture subclasses as applied to organic soil families are given below:

aqueous: free surface water - wetlands, marsh, floating fen collapse scars;

peraquic: saturated for very long periods, very poorly drained - flat fens, patterned fens, spring fens, swamps;

aquic: saturated for moderately long periods, poorly drained - blanket bogs, transitional bogs;

subaquic: saturated for short periods, imperfectly drained - domed bogs, plateaus;

perhumid: moist with no significant seasonal deficit, imperfectly to moderately well drained - frozen plateaus, frozen palsas, frozen peat polygons;

humid: moist with no significant seasonal deficit, moderately well drained - drained peatlands, folisol.

moor   German term for peatland, applied to any area of deep peat whether acid or alkaline. In England the word is applied to high-lying country covered with mainly ericaceous dwarf shrubs. It is often used to refer to land having any of the oxyphilous (acid-loving) communities (Tansley, 1949).

mor   See *raw humus.*

moss   Scottish for *bog.*

moss peat   See *peat moss.*

muck   Generally a mineral-rich, gyttja-like, well-decomposed organic material, dark in colour and accumulated under conditions of imperfect drainage, containing 50-80 per cent ash, often used as manure (Harshberger, 1909; McCook and Harmer, 1925; Ziegler, 1946; Brüne, 1948; Kubiena, 1953; Anon., 1967; Ford-Robertson, 1971).

muck soil   An organic soil consisting of highly decomposed material. Mucky peat and peaty muck are terms used to describe increasing stages of decomposition between peat and muck (Anon., 1974).

muskeg   North American term frequently employed for peatland. The word muskeg is of Indian (Algonquin) origin and applied in ordinary speech to natural and undisturbed areas covered more or less with *Sphagnum* mosses, tussocky sedges, and an open growth of scrubby trees (Lewis and Dowding, 1926; Raup, 1935; Dachnowski-Stokes, 1941; Radforth, 1952; Brown, 1970; Porsild, 1937; Hustich, 1949a, b; Ritchie, 1958).

muskeg complex   See *peatland complex.*

myr   Swedish for *peatland.*

Niedermoor, Niederungsmoor (Weber, 1908; Brüne, 1948)   German term for minerotrophic peatlands. It is typical of these peat formations that they consist throughout of nutrient-rich peat types. They could originate only on nutrient-rich mineral soil or when mineralized water seeps in. In young or recent moraine landscape they originate from filling in of lakes. They contain, therefore, in the deposits, layers of gyttja many metres thick on top of which many demanding plant societies subsequently develop luxuriantly. Properties of 'niedermoor' peat (Brüne 1948) are as follows: bulk density, 250 g/1,000 cm^3, specific gravity, up to 2.52 g/cm^3; nitrogen content, 2.50; calcium oxide content, 4.00; phosphate content, 0.25; potassium content, 0.10 (as a percentage of the dry weight). All peat soils with at least 2.5 per cent calcium oxide are niedermoor peat (Fleischer, 1913; Brüne, 1948). Synonyms: swamp, fen, minerotrophic peatland, rheophilous peat-bog, valley bog (Kulczynski, 1949), etc.

nutrient classes   The designations for a vegetation site with regard to its nutrient status, i.e., the availability to the plants of nutrients, including oxygen and water, and the radiant energy of the sun, under prevailing environmental conditions. The nutrient status determination by measuring the involved factors is complex and difficult. In its place descriptions of the vegetation stands, plant indicators, site index, humus form of the soil profile, etc., have been used (von Kruedener and Becker, 1941; Hills, 1952; Kubiena, 1953).

1) Poor (oligotrophic) (Greek *oligos*, 'little'; *trophos*, 'nourishment'): Sites with low nutrient availability and relatively low biological activity, generally formed on base-deficient parent rocks (Kubiena, 1953). Designation for peatlands (muskegs) formed of plants in soft waters poor to extremely poor in nutrients. An example of this type is a raised bog (Barry, 1954; Sjörs, 1961a; Damman, 1964). Vegetation: stands are of low synusia (usually only one tree and one moss layer). Trees are stunted and the shrub layer is poorly developed. Organic material is little decomposed. *Sphagnum* peat is characteristic.

2) Moderately poor (suboligotrophic): Vegetation stands are of low synusia (usually one tree, one moss, and one shrub layer). Trees reach merchantable size. Organic matter decomposition is slow.

3) Less than mediocre (submesotrophic): Vegetation stands are of intermediate synusia (usually one tree, one moss or herb, and one shrub layer). Trees are of good pulpwood size. Organic matter decomposition is slow, but more organic matter is being utilized by organisms than in the moderately poor nutrient class.

4) Mediocre (mesotrophic): Vegetation stands are of well-expressed synusia (usually two tree layers, one shrub, one herb, and one moss layer). Organic matter decomposition is slow. Formation of raw humus and some mull takes place. The mineral soil indicates processes of gleization and podzolization.

5) Better than mediocre (permesotrophic): Vegetation stands are of well-expressed synusia (two tree layers, one to two shrub layers, one herb layer, and poorly developed moss and lichen layers). The tree stands are of good timber quality. Organic matter decomposition is good, formation of a thin mull layer takes place. The mineral soil horizons can indicate gleization.

6) Moderately rich (subeutrophic): On peatland only in locations that are not water-logged, usually mixed stands, producing very good timber quality. Vegetation stands are of well-expressed synusia (four to five layers). The vegetation is less luxuriant than in the rich nutrient class. Decomposition and incorporation of organic matter in the A horizon are good and formation of duff mull or mull takes place.

7) Rich (eutrophic): not occurring on undrained peatlands, stands of well-expressed

synusia (two to three tree layers and one shrub layer, usually in openings, and one thick and luxuriant herb layer, weakly developed mosses). Organic matter decomposes well. Clay-humus complex can develop. Earthworms are found.

oligotrophic   See *nutrient classes.*

ombrogenous   Produced by rain, refers to muskeg areas which receive nutrients from precipitation (von Post and Granlund, 1926; Du Rietz, 1949). See also *ombrotrophic.*

ombrotrophic   Nourished by rain, refers to muskeg areas dependent on nutrients from precipitation (Du Rietz, 1949). Waters are acid, low in calcium, with magnesium almost absent (Heinselman, 1963); found in raised muskeg (bog). See also *ombrogenous.*

organic matter   More or less decomposed material of the soil derived from organic sources, usually from plant remains. The term covers matter in all stages of decay (Ziegler, 1946).

organic order of soils   In Canada (Anon., 1973a, 1974), an order of soils that have developed principally from organic deposits. The majority of organic soils are saturated for most of the year, or are artificially drained, but some of them are not usually saturated for more than a few days. They contain 30 per cent or more of organic matter and: 1) if the surface layer consists of fibric* organic material and the bulk density is less than 0.1 (with or without a mesic* or humic* cultivated surface layer less than 15 cm [6 inches] thick), the organic material must extend to a depth of at least 60 cm (24 inches); 2) if the surface layer consists of organic material with a bulk density of 0.1 or more, the organic material must extend to a depth of at least 40 cm (16 inches); 3) if bedrock occurs at a depth shallower than stated in 1 or 2 above, the organic material must extend to a depth of at least 10 cm (4 inches). The organic soils are classified on the basis of duration of the period of saturation; decomposition; type of plant fibres; presence of mineral, frozen, water, or rock layers or strata, into the following great groups (subgroups are not shown here):

Fibrisol is saturated for most of the year. The soils have a dominantly fibric* middle tier, or middle and surface tiers if a terric,* lithic,* hydric,* or cryic* contact occurs in the middle tier.

Mesisol is saturated for most of the year. The soils have a dominantly mesic* middle tier or middle and surface tiers if a terric,* lithic,* hydric,* or cryic* contact occurs in the middle tier.

Humisol is saturated for most of the year. The soils have a dominantly humic* middle tier, or middle and surface tiers if a terric,* lithic,* hydric,* or cryic* contact occurs in the middle tier.

Folisol is not usually saturated for more than a few days. The soils are composed of leaf, litter, twigs, and mosses (mor or raw humus). A lithic* layer occurs at a depth of less than 160 cm, or the organic layers rest on fragmental material with interstices filled with organic material.

organic terrain   Term sometimes used in the sense of muskeg.

palsa   Peat-covered mound with a permafrost core; usually ombrotrophic; generally much less than 100 metres across and from one to several metres high. Palsas may be treeless or may contain a few stunted tamaracks or black spruce (wooded palsa). Palsa growth is due to the build-up of segregated ice mainly in the mineral soil (Kihlman, 1890; Anufriew, 1922; Gorodkov, 1926; Wenner, 1947; Lundquist, 1951; Sjörs, 1961a; Forsgren, 1968; Zoltai, 1971; Zoltai and Tarnocai, 1971; Crampton, 1973).

---

* Related to the organic soil material; fibric = least decomposed; mesic = moderately decomposed; humic = most highly decomposed (refers to contact of the organic soil to: permafrost (cyric); water (hydric); bedrock (lithic); unconsolidated mineral soil (terric)).

palsa bog   See *palsa peatland*.

palsa peatland   Peatland complex of the discontinuous permafrost region, with palsas protuberant in the adjacent peatland without permafrost, associated with collapse scars and moats. Palsas are usually ombrotrophic, the surroundings minerotrophic.

paludification   Process of peat formation of which the characteristic feature is anaeroby caused by waterlogging (Auer, 1930; von Post, 1937). Synonym: swamping.

palynology   Study of pollen, spores, and other microfossils. For example, pollen analysis is one aspect of palynology which concerns itself with a record of fossil pollens in bogs (Dansereau, 1957).

patterned bog   See *string bog*.

patterned fen   See *string fen*.

patterned muskeg   See *string peatland*. Synonym: patterned peatland.

peat   Organic soil of peatlands, exclusive of plant cover, consisting largely of organic residues accumulated as a result of incomplete decomposition of the dead plant constituents (incomplete decomposition as a result mainly of the prevailing anaerobic conditions associated with waterlogging) (semiterrestrial humus form, Kubiena, 1953). The physical and chemical properties of the peat are influenced by the nature of the plants from which it has originated, by the moisture relations during and following its formation and accumulation, by the geomorphological position and climatic factors. It must have an organic matter content of not less than 20 per cent of the dry weight. When peat is used as fuel, more than 50 per cent of its dry weight must be combustible. Peat with 20-50 per cent of organic matter content is called muck (Ford-Robertson, 1971). Peat consists largely of carbon, hydrogen, and oxygen and varying amounts of nitrogen, sulphur, and ash. Various animal residues are admixed. Upon drying, well-decomposed peats shrink considerably, changing into loose fragments or into hard, frequently fibrous clumps, breaking apart and forming sharp edges. Upon wetting, the air-dry substance of peat swells, the degree of swelling depending on the constituent plant residues, on the state and perhaps even the type of peat formation and on the pressure to which it was subjected during the continuous contact with water; it never gives a structural mass which would resemble soil even when fully softened. Only in little-decomposed peat are the plant residues recognizable by the naked eye (Johnson, 1859; Weber, 1903; Cajander, 1913; Dachnowski, 1919, 1920; Puchner, 1920; Auer, 1930; Waksman, 1942; Brüne, 1943; Farnham and Finney, 1965; Olenin et al., 1972). (For information on classification of peat soils see MacFarlane [1958], Anon. [1973a, 1973b, 1974], and *organic order of soils*.)

peat bog   Synonymous with bog.

peat deposit   Peatland with a well-developed peat layer. It includes peatland with water and vegetation as well as dried-up peatlands where a peat bed is the only remaining characteristic of the former vegetation (Pjavchenko, 1968).

peat moss   The term is mainly a trade name for peat sold for horticultural uses and refers to peats composed generally of *Sphagnum*, e.g., *cymbifolia* peat and *acutifolia* peat, which are not humified or are only slightly humified. It also includes peats which have a high percentage of other constituents such as *Carex*-moss peat, wood-moss peat, and moss-*Carex* peat (Leverin, 1946). It is used chiefly as a mulch or seedbed, or for acidification (Anon., 1967). Synonyms: raw moss, peat, *Sphagnum* moss, moss peat.

peat mound   Natural elevation of peat, larger than a hummock, usually ombrotrophic

(Kihlman, 1890; Anufriew, 1922; Gorodkov, 1926; Wenner, 1947; Sjörs, 1961a). With permafrost core it is called a palsa.

peat plateau Same morphology as palsa, but covers much larger areas (often over 1 km²) and seldom exceeds 1 m in height; surrounded by unfrozen, waterlogged peatland (Svensson, 1962; Sjörs, 1963; Tyrtikov, 1966; Brown, 1968). Replaces palsa of lower latitude. In the continuous permafrost zone it shows frost polygons (Sjörs, 1961a).

peatland In nature applies to every unit of peat-forming vegetation on peat and includes all peat originating from that vegetation. The peat layer must be 45 cm thick when undrained or more than 30 cm thick when drained, the ash content not more than 80 per cent (Weber, 1903; Cajander, 1913; Auer, 1930; Waksman, 1942; Brüne, 1948; Du Rietz, 1954; Drury, 1956; Hare, 1959; Heinselman, 1963; Ehrlich, 1965; Farnham and Finney, 1965; Brown, 1970). According to the nutrient levels and amount of tree cover, the most frequently used terms are: ombrotrophic and nutrient poor; quaking, floating, and raised bog; strings, palsas, mounds, and plateaus of peatland complexes; treed muskeg (black spruce muskeg); minerotrophic, floating, and spring fen; fen pools and flarks of peatland complexes; rich fen, fen, and treed swamps (alder, cedar, and tamarack swamp). Synonym: *muskeg*.

peatland complex Huge peatland areas originating from small primary peatlands which grew or fused by horizontal expansion. They are composed of many peatland types, sometimes interdependent (Cajander, 1913). Example: string bogs.

peatland type Unit of peatland usually classified by its vegetation.

permafrost Perennially frozen ground whose mean temperature remains below 0°c continually for one or more years (Lindsay and Odynsky, 1965; Brown, 1966). A continuous stratum of permafrost is found where the annual mean temperature is below about 23°F (Anon., 1968). The degree of disturbance or instability which will result from the thawing of permanently frozen ground is a function of the moisture (ice) content: the more moisture present the greater the disturbance or loss of stability (Mackay, 1971).

permesotrophic See *nutrient classes*.

plateau bog Elevated peat plateau varying from 0.5 to 2 metres high, with a relatively flat surface. The elevated ombrotrophic surface is due to peat accumulation or permafrost lenses. The plateaus cover an area of several hectares, and they are often tear-shaped, and surrounded by minerotrophic peat. The plateaus are usually forested (Tarnocai, 1970).

pocosins On the lowest and geologically youngest marine terraces of the coastal plain of the southeastern United States are found areas (often thousands of acres) of swamps and marshes called pocosins. They typically have an overstorey of pond pine and dense understorey of evergreen, ericaceous shrubs. Although the topography appears to be extraordinarily flat and featureless, pocosins are commonly slightly dome-shaped, and sometimes have lakes in their higher central portions. Organic soils are the rule and at the pocosin's salt water margin the underlying mineral soil may actually be below sea level. Mineral surface soils may also be present where fire has burned away the peat or muck layer (Stubbs, 1962; Anon., 1967).

polygon Patterned ground caused by permafrost with recognizable trenches or cracks along the polygonal circumference; in peat mainly with high centres, or low centres and ridges along the trenches (Washburn, 1956; Drew and Tedrow, 1962; Svensson, 1962).

pothole muskeg Peatland with many ponds and small lakes often closely grouped with narrow strips of land separating one from the other (Coombs, 1954); develops in flat terrain in contrast to strings and flarks on sloping terrain.

quagmire Floating bog (quaking bog) which quakes or yields underfoot (Buell and Buell, 1941; Drury, 1956; Hanson, 1962). A stage in hydrarch succession resulting in pond-filling, *Sphagnum* mats tend to proliferate. Similar to island bog and floating fen (Adams and Zoltai, 1969). The term is also used for soft, wet, miry land, which yields under the foot (Anon., 1967).

quaking bog See *quagmire*.

raised bog Nutrient-poor (oligotrophic) peatland which has grown above its site of origin (sometimes several metres) whose centre is higher than the margins and whose surface is convex. Growth is by *Sphagnum* proliferation and deposition of peat, 'lenticular regeneration' (Kulczynski, 1949), water being supplied chiefly by rainfall (ombrotrophic). There are usually one or several very acid hummocks on its surface and wet depressions or ponds draining through soaks toward the lagg sometimes completely surrounding the bog (Osvald, 1925; Du Rietz, 1954; Drury 1956; Sjörs, 1959; Ratcliffe, 1964; Heinselman, 1970). Among the woody plants occurring on virgin bogs in northwestern Europe are only stunted *Betula pubescens* and *Pinus sylvestris* and in the Alpine foothills *Pinus montana*. In northern North America *Picea mariana* is the main woody plant. Raised bog peats are nutrient-poor and their calcium oxide content does not exceed 0.5 per cent (Brüne, 1948). Synonyms: raised muskeg, domed bog, high bog, raised peatland, continental raised bog, etc.

raw humus Terrestrial humus form, characterized by the preponderance of structurally well-preserved, little-decomposed, and comminuted plant residues and always high acidity leading to the development of acid humus soils. Raw humus is formed in a cool and humid climate on poor acid soils, very deficient in calcium, magnesium, bases, nutrients, and mineral colloidal material. It is the most unfavourable form of terrestrial humus formation. In forest, the raw humus profile is characterized by a thick litter layer (förna) in which the litter of several years has accumulated owing to slow decomposition. Next to it the fermentation layer (F-layer) is always very strongly developed but the humic substance layer (H-layer) is scarcely perceptible or completely missing in very unfavourable raw humus formations. Synonyms: mor (Kubiena, 1953), duff (Rommel, 1931, in Kubiena, 1953).

retention fen Horizontal fen in which floodwater is retained for as much as a whole year (Radforth, 1962).

rheophilous Referring to peat formed of plants which grow under the influence of mobile groundwater (Kulczynski, 1949; Bellamy, 1966).

ridge See *string* and *string muskeg*.

rimpi See *flark*.

sapropel Subaqueous humus form, fetid slime (Kubiena, 1953).

sedge peat Peat composed primarily of the stalks, leaves, rhizomes, and roots of sedges (*Carex* spp.) (Kubiena 1953). In some instances so termed because it contains 50 per cent sedge by volume (Auer, 1930). Synonyms: *Carex-peat*, reed peat.

sedge-reed and *sedge-shrub peats* (*non-forest*) Peat composed of sedges, reeds, grasses; often laminated, stringy, matted, or felty; usually not too decomposed; may contain stems and roots of woody shrubs as inclusions; no large logs or pieces of wood from trees. It originates in sedge fens and reed fens, reed sedge and cattail marshes, and sedge-shrub carr (shrub-covered wetlands). It may occur as a single layer from surface to substratum, as a second layer overlying aquatic peats as a basal layer, or in multilayer sequences (Heinselman, 1970).

sedimentary peat   Used by Rigg (1958) as synonymous with gyttja.

seepages   These consist of ladderlike rows of small, shallow, narrow pools, or flarks in a steplike arrangement. The pools parallel the contours and are at right angles to the slope (Sjörs, 1963). See *flark*.

sinkhole   Microdepression in muskegs, wet, with peat well decomposed in the depression, possibly because of influence of mineralized groundwater. Sinkholes occur frequently in swamps.

sloping bog   Equivalent to a hanging bog; a bog occupying a slope which is receiving mineralized water. The internal lateral movement of water is restricted by the accumulation of peat and bog vegetation dominates. Common in more humid areas and found in the Precambrian Shield, and mountains (Adams and Zoltai, 1969).

sloughs   Shallow lakes commonly surrounded by reed swamp, fed by underground springs (Lewis et al., 1928; Moss, 1953).

smallpox muskeg   In muskegs where former lake and pond beds are free of water and characteristically saucerlike in shape. The former rims have a good growth of small trees and bushes which produce a pock-marked effect (Coombs, 1954).

soaks   Broad troughs leading waters away from raised muskegs (Du Rietz, 1954; Heinselman, 1963).

soligenous   Referring to peatlands with water percolating through them and carrying minerals into the peatland from outside sources (von Post and Granlund, 1926; Sjörs, 1948; Drury, 1956; Heinselman, 1963; Ratcliffe, 1964).

Sphagnum-peat   Peat which is little decomposed, raw, consisting mainly of *Sphagnum* spp. with admixed *Eriphorum* spp., *Carex* spp., *Andromeda glaucophylla, Ledum groenlandicum, Vaccinium oxycoccus,* and *Empetrum nigrum,* etc. (Auer, 1930; Leverin, 1946).

spring fen   Convex or sloping fen formed directly over springs or discharge areas, often forming flushes; may be characterized by small spring-fed pools. The peat is usually moderately to well decomposed (Tarnocai, 1970). Synonym: soligenous muskeg.

spruce bog   See *swamp, muskeg.*

spruce islands   Elevated areas of peat with cores of permafrost, covered with black spruce, which may gradually collapse from the edges. They occur at the southern margin of continuous permafrost, where climatic conditions for tree growth are still good (Sjörs, 1959, 1961a, b; Zoltai, 1971). See also *palsa.* The term is also applied to spruce islands in peatlands south of the permafrost occurrence (Heinselman, 1963). See also *wooded island fen.*

string   In string muskeg, the elevated, better-drained portion supporting mosses, sedges, brush, or trees; narrow, usually with its long axis across the slope; may form into net patterns (Drury, 1956; Hamelin, 1957). Synonyms: ridge, rib lanière.

string bog   Predominantly ombrotrophic, oligotrophic peatland complex of strings and flarks. See also *string peatland.*

string fen   Predominantly minerotrophic peatland complex of strings and flarks. See also *string peatland.*

string muskeg   See *string peatland.*

string peatland   Gently sloping peatland complex consisting of strings (ridges, ribs, lanières) up to 2 m high, alternating with flarks (rimpi, flashette, mare, hollow), usually wet or water-covered, troughlike depressions, oriented across the slope of the peatland and perpendicular to the water movement. These may be parallel or occur in a webbed, sinuous, or netlike pattern. In flat watersheds, with no slope, the flarks become enlarged to pools of irregular size and shape. The strings are frequently ombrotrophic, and support mainly *Carex* and *Sphagnum*, but also spruce, larch, dwarf birch, willow, ericaceous shrubs. The intervening flarks are frequently minerotrophic, and contain open, shallow water or *Carex, Drepanocladus-Calliergon* communities. Depending on nutrient content, the predominantly minerotrophic complex is called string fen and the predominantly ombrotrophic, string bog. Its typical form occurs near the tree line (Cajander, 1913; Tanttu, 1915; Auer, 1920; Troll, 1944; Drury, 1956; Hamelin, 1957; Tedrow and Cantlon, 1958; Hare, 1959; Sjörs, 1959; Ritchie, 1960; Allington, 1961; Kalela, 1962; Heinselman, 1963, 1965, 1970; Radforth, 1964; Pollet, 1968). Synonyms: patterned peatland; patterned or string muskeg, bog, fen; aapa bog or fen.

subeutrophic See *nutrient classes.*

submesotrophic See *nutrient classes.*

suboligotrophic See *nutrient classes.*

swamp Type of wet forested minerotrophic peatland rich in herbs, frequently with dominant tree species reaching pulpwood sizes at least (tamarack, white spruce, balsam fir, cedar, etc.). Black spruce, if present, is faster growing than in bogs. A more luxuriant moss cover is dominated by brown mosses (Hustich, 1955; Heinselman, 1963; Jeglum, 1972). Swamps include 'Carr,' a term for shrub-covered wetlands, related to fens and marshes (Kubiena, 1953); and wetlands having tree and/or shrub layers occupying more than 25 per cent of the area (Radforth, 1964). Although most swamps are level this is not a necessary condition, for hillside swamps are by no means uncommon, owing to a constant supply of percolating groundwater which maintains the swampy condition (sloping fen) (Tarnocai, 1970). Lake basins are occasionally filled with vegetation and sediment, thus becoming swamp. Swamps may be formed on the floodplain of rivers as well as on their deltas. They are characteristic of the flat, ill-drained areas of the Atlantic Coastal Plain. Coastal salt-water swamps may develop in the zone between high and low tides or extend up river estuaries (Anon., 1968). See also *alder swamp.* Zoltai et al. (1974) proposed: alluvial, lakeside, peat margin, catchment, and seepage swamps.

swamping See *paludification.*

telluric water Originating from the earth, carrying dissolved mineral nutrients, minerotrophic.

terrestrialization See *lake filling-in process.*

topogenous Produced by relief, as a peatland complex determined by topography; indicates that the source of water for a peatland is the regional water table in a depression that predated peat formation (von Post and Granlund, 1926; Allington, 1961; Heinselman, 1963; Ratcliffe, 1964).

transition peatland or *muskeg* Peatland with vegetation and physical and chemical properties of the peat intermediate between oligotrophic and minerotrophic peatlands (bogs and fens) (Brüne, 1948; Heinselman, 1970), poor fen (Sjörs, 1948).

turf Layer of matted earth formed by grass and plant roots; peat, especially as material for fuel (Anon., 1967); where thin, or so weathered as to be unfit for fuel, the term is applied (Skertchley, 1877).

übergangsmoor See *transition peatland* for description. Properties of the peat given by
  Fleischer (1913) in Brüne (1948): bulk density, 180 g/1,000 cm³; specific gravity, 1.65-1.78
  g/cm³; nitrogen content, 2.00; calcium oxide content, 1.00; phosphate content, 0.20;
  potassium content, 0.10 (as a percentage of the dry weight).

upland peat Peat on slopes and undulating uplands. It has no particular water table.

valley bog Term first referred to by Tansley (1949) and Kulczynski (1949). It refers to
  peatland complexes where ombrogenous bog cannot develop (Newbould and Gorham,
  1956). Synonyms: minerotrophic peatland, swamp.

von Post humification scale Scale describing peat moss in varying stages of decomposition
  ranging from H_1, which is completely uncoverted, to H_{10}, which is completely converted
  (von Post, 1922).

H_1: completely unhumified and dy-free peat; upon pressing in the hand, gives off only
  colourless, clear water.

H_2: almost completely unhumified, dy-free peat; upon pressing, gives off almost clear but
  yellow-brown water.

H_3: little humified and little dy-containing peat; upon pressing, gives off distinctly turbid
  water but the residue is not mushy.

H_4: poorly humified or some dy-containing peat; upon pressing, gives off strongly turbid
  water. The residue is somewhat mushy.

H_5: peat partially humified or with considerable dy content. The plant remains are
  recognizable but not distinct. Upon pressing, some of the substance passes between
  the fingers, together with mucky water. The residue in the hand is strongly mushy.

H_6: peat partially humified or with considerable dy content. The plant remains are not
  distinct. Upon pressing, one third (at the most) of the peat passes between the fingers.
  The residue is strongly mushy but the plant residue stands out more distinctly than in
  the unpressed peat.

H_7: peat quite well humified or with considerable dy content in which much of the plant
  remains can still be seen. Upon pressing, about half of the peat passes between the
  fingers. If water separates it is soupy and very dark in colour.

H_8: peat well humified or with considerable dy content. The plant remains are not
  recognizable. Upon pressing, about two-thirds of the peat passes between the fingers.
  If it gives off water at all, it is soupy. The remains consist mainly of more resistant root
  fibres, etc.

H_9: peat very well humified or nearly completely dylike in which hardly any plant remains
  are apparent. Upon pressing, nearly all of the peat passes between the fingers like a
  homogeneous mush.

H_{10}: peat completely humified or completely dylike in which no plant remains are
  apparent. Upon pressing, all the peat passes between fingers.

water track Term for vegetation types marking the path of mineral-influenced waters
  through a peatland. Water tracks contrast sharply on air photos with the surrounding bog
  forests or muskegs (Sjörs, 1948; Heinselman, 1963).

water track fen Component of patterned fen occupying a concave tract of peatland which
  marks the path of subsurface mineral water flow. The peat is usually well decomposed.
  Tamarack and shrubs may be present (Heinselman, 1963).

water track swamp Swamp occupying a concave drainage track across peatlands. It often
  occurs downslope from mineral-soil islands. The peat type is well-decomposed, mucky

black peat. It is usually a moderately productive site with tamarack, white cedar, and black spruce, and an understorey of alder. Sometimes the vegetation consists of pure alder or less productive tamarack forest (Heinselman, 1963).

waterlogged   Saturated with water (Anon., 1972b).

wetland   General term, broader than peatland or muskeg, to name any poorly drained tract whatever its vegetational cover or soil (Sjörs, 1948).

wooded island fen   Form of patterned fen with interspersed linear and teardrop-shaped islands of forest on peat. The islands have long axes with tails pointing downslope. The islands may be very small or more than a mile across, and usually consist of peats with bog vegetation and black spruce-tamarack forest (Anon., 1972a). See also *spruce islands*.

wooded palsa   See *palsa*.

woody peat   Peat containing many woody remains of trees and shrubs.

REFERENCES

Adams, G.D., and Zoltai, S.C. 1969. Proposed open water and wetland classification. In Guidelines for Biophysical Land Classification, ed. D.S. Lacate. Dept. Fish. For., Ottawa, Publ. 1264, pp. 23-41.

Allington, K.R. 1961. The bogs of central Labrador-Ungava; an examination of their physical characteristics. Geog. Ann. (Stockholm) 43/3-4: 404-17.

Anon. 1967. The Random House Dictionary of the English Language (Random House, New York).

— 1968. Van Nostrand's Scientific Encyclopedia (D. Van Nostrand Co. (Can.), Ltd., Toronto).

— 1972a. Contributions to a meeting of the Subcommittee on Organic Terrain Classification held April 11-12, 1972, G.D. Adams, Chairman (Migratory Bird Res. Cent., Univ. Saskatchewan, Saskatoon).

— 1972b. Glossary of terms in soil science. Can. Dept. Agric. Publ. 1459.

— 1973a. Revised system of soil classification for Canada - provisional collection of official and tentative definitions for use by Canadian pedologists. Can. Dept. Agric., Soil Res. Inst., Ottawa, Misc. Publ. Spec. Rept.

— 1973b. Soil Taxonomy: A Basic System of Soil Classification for Making and Interpreting Soil Surveys (USDA Soil Cons. Serv., Washington; preliminary, abridged text).

— 1974. The system of soil classification for Canada. Can. Dept. Agric., Soil Res. Inst., Ottawa, Publ. 1455, rev. 1974.

Anufriew, G.I. 1922. Moore der Halbinsel Kola. Arbeiten der Bodenkundl. Bot. Abt. Geog. Inst., Petrograd, Leaf 1.

Auer, V. 1920. Über die Entstehung der Stränge auf den Torfmooren. Acta Forest. Fenn. 12: 23-145.

— 1930. Peat bogs in southeastern Canada. Can. Dept. Mines, Geol. Surv. Mem. 162.

Baldwin, W.K.W. 1958. Plants of the clay belt of northern Ontario and Quebec. Nat. Mus. Can. Bull. 156.

Barry, T.A. 1954. Some considerations affecting the classification of the bogs of Ireland. Irish For. 11/2: 48-64.

Bellamy, D.J. 1966. Peat and its importance. Discovery 27/6: 12-16.

Brown, R.J.E. 1966. Permafrost, climafrost and the muskeg H factor. Proc. 11th Muskeg

Res. Conf., Nat. Res. Counc. Can., Assoc. Comm. Geotech. Res., Tech. Mem. 87: 159-78.

— 1968. Permafrost investigations in northern Ontario and northeastern Manitoba. NRC, Div. Building Research, Ottawa, Tech. Pap. 291.

— 1970. Occurrence of permafrost in Canadian peatlands. Proc. 3rd Internat. Peat Congr., Quebec, pp. 174-81.

Brüne, F. 1948. Die Praxis der Moor- und Heidekultur (Paul Parey, Berlin and Hamburg).

Buell, M.F., and Buell, H.F. 1941. Surface level fluctuation in Cedar Creek Bog, Minnesota. Ecol. 22: 317-21.

Cajander. A.K. 1913. Studien über die Moore Finnlands. Acta Forest. Fenn. 2.

Conway, V.M. 1949. The bogs of central Minnesota. Ecol. Monogr. 19: 173-206.

— 1954. Stratigraphy and pollen analysis of southern Pennine blanket peats. J. Ecol. 42/1: 117-47.

Coombs, D.B, 1954. The physiographic subdivisions of the Hudson Bay Lowlands south of 60 degrees north. Geog. Bull. 6: 1-16.

Crampton, C.B. 1973. The distribution and possible genesis of some organic terrain patterns in the southern MacKenzie River Valley. Can. J. Earth Sci.

Dachnowski, A.P. 1919. Quality and value of important types of peat material. US Dept. Agric., Bur Plant Ind. Bull. 802: 1-40.

— 1920. Peat deposits in the United States and their classification. Soil Sci. 10: 453-65.

Dachnowski-Stokes, A.P. 1941. Peat resources in Alaska. US Dept. Agric. Bull. 769.

Damman, A.W.H. 1964. Key to the Carex species of Newfoundland by vegetative characteristics. Dept. Forest. Publ. 1017.

— 1965. The distribution patterns of northern and southern elements in the flora of Newfoundland. Rhodora 67/772: 363-92.

Dansereau, P. 1957. Biogeography: An Ecological Perspective (Ronald Press, New York).

Day, J.H. 1968. The classification of organic soils in Canada. Proc. 3rd Internat. Peat Congr., Quebec, pp. 80-4.

Drew, J.V., and Tedrow, J.C.F. 1962. Arctic soil classification and patterned ground. Arctic 15: 109-16.

Drury, W.H.,Jr. 1956. Bog flats and physiographic processes in the upper Kuskokwin River region, Alaska. Contrib. Gray Herb., Harvard Univ. 178.

Du Rietz, G.E. 1949. Main units and main limits in Swedish mire vegetation (in Swedish, English summary). Svensk. Bot. Tidskr. 43: 274-309.

— 1954. Die Mineralbodenwasserzeigergrenze als Grundlage einer natürlichen Zweigliederung der nord- und mitteleuropäischen Moore. Vegetation 5-6: 571-85.

Farnham, R.S., and Finney, H.R. 1965. Classification and properties of organic soils. Adv. Agron. 17: 115-62.

Fleischer, M. 1913. Die Anlage und die Bewirtschaftung von Moorwiesen und Moorwiden (Paul Parey, Berlin).

Ford-Robertson, F.C. 1971. Terminology of Forest Science, Technology Practice and Products (Soc. Am. Forest., Washington, D.C.).

Forsgren, B. 1968. Studies of palsas in Finland, Norway and Sweden, 1964-1966. Biul. Peryglac. 17: 117-23.

Godwin, H., and Conway, V.M. 1939. The ecology of raised bog near Tregaron, Cardiganshire. J. Ecol. 27: 313-63.

Gorham, E. and Pearsall, W.H. 1956. Acidity, specific conductivity and calcium content of some bogs and fen waters in northern Britain. J. Ecol. 44: 129-41.

Gorodkov. B.N. 1926. Polar ural in the upper reaches of the River Sob. Trav. Mus. Bot. Acad. Sci. USSR 19: 1-74.

Hamelin, L.E. 1957. Les tourbières réticulées du Québec-Labrador subartique: interprétation morpho-climatique. Cahiers Geog. Québec 2: 87-106.

Hanson, H.C. 1962. Dictionary of Ecology (Peter Owen Ltd., London).

Hare, F.K. 1959. A photo-reconnaissance survey of Labrador-Ungava. Can. Dept. Mines Tech. Surv., Geog. Branch Mem. 6: 1-64.

Harshberger, J.W. 1909. Bogs, their nature and origin. Plant World 12: 34-41, 53-61.

Heinselman, M.L. 1963. Forest sites, bog processes and peatland types in the Glacial Lake Agassiz region, Minnesota. Ecol. Monogr. 33: 327-72.

— 1965. String bogs and other patterned organic terrain near Seney, Upper Michigan. Ecol. 46/1-2: 185-8.

— 1970. Landscape evolution, peatland types and the environment in the Lake Agassiz peatland natural area, Minesota. Ecol. Monogr. 40/2: 235-61.

Hills, G.A. 1952. The classification and evaluation of sites for forestry. Ont. Dept. Lands Forests, Res. Div. Rept. 24.

Hustich, I. 1949a. On the forest geography of the Labrador Peninsula. A preliminary synthesis. Acta. Geog. 10/2: 1-63.

— 1949b. Phytogeographical regions of Labrador. Arctic 2: 36-42.

— 1955. Forest-botanical notes from the Moose River area, Ontario, Canada, Acta Geog. 13/2: 3-50.

Jeglum, J.K. 1972. Boreal forest wetlands near Candle Lake, central Saskatchewan. I. Vegetation. The Musk-Ox 11: 41-58.

Jessen, K. 1949. Studies in later Quaternary deposits and flora history of Ireland. Proc. Roy. Irish Acad. 52/6: 85-290.

Johnson, S.W. 1859. Essays on Peat Muck and Commercial Manures (Brown and Gross, Hartford).

Kalela, A. 1962. Notes on the forest and peatland vegetation in the Canadian clay belt region and adjacent areas. I. Comm. Inst. Forest. Fenn. 55/33: 1-14.

Kihlman, A.O. 1890. Pflanzenbiologische Studien aus Russisch-Lappland. Acta Soc. Fauna Flora Fenn. VI/3.

Klika, J., Novák, V., Gregor, A. 1954. Praktikum Fytocenologie, Ecologie, Klimatologie a Půdoznalství [Practicum of Phytocoenology, Ecology, Climatology and Soil Science]. (Naklad. Česk. Akad. Věd, Praha).

Kubiena, W.L. 1953. The Soils of Europe (Thomas Murby & Co., London).

Kulczynski, S. 1949. Peat bogs of Polesie. Mém. Acad. Polon. Sci. Lettres, Série 15, Cracovie.

Leverin, H.A. 1946. Peat deposits in Canada. Ottawa Dept. Mines Resources Publ. 817.

Lewis, F.J., and Dowding, E.S. 1926. The vegetation and retrogressive changes of peat areas (muskegs) in central Alberta. J. Ecol. 14/2: 317-41.

Lewis, F.J., Dowding, E.S., and Moss, E.H. 1928. The vegetation of Alberta. II. The swamp, moor and bog forest vegetation of central Alberta. J. Ecol. 16/1: 19-70.

Lindsay, J.D., and Odynsky, W. 1965. Permafrost in organic soils of northern Alberta. Can. J. Soil Sci. 45: 265-9.

Lundquist, G. 1951. En palsmyr sudost om Kebnekaise (English abstract). Stockholm, Geol. Foren, Forhandl. 73/2: 209-25.

MacFarlane, Ivan C. 1958. Guide to a field description of muskeg. NRC Can. Assoc. Comm. Soil Snow Mech., Tech. Mem. 44.

— (ed.). 1969. Muskeg Engineering Handbook (Univ. Toronto Press, Toronto).

Mackay, J.R. 1971. Ground ice in the active layer and in the top portion of permafrost. In Proc. Seminar on the Permafrost Active Layer, ed. R.J.E. Brown. NRC, ACGR Tech. Mem. 96 142-17XX162.

McCook, M.M. and Harmer, P.M. 1925. The muck soils of Michigan. Mich. Agric. Exptl. Stn. Spec. Bull. 136.

Moss, E.H. 1953. Marsh and bog vegetation in northwestern Alberta. Can. J. Bot. 31: 448-70.

Newbould, P.J. and Gorham, E. 1956. Acidity and conductivity measurements on some plant communities of the New Forest Valley bogs. J. Ecol. 44: 118-28.

Odum, E.P. 1959. Fundamentals of Ecology (W.B. Saunders Co., Philadelphia and London).

Olenin, A.S., Neistadt, M.I., and Tyuremnov, S.N. 1972. On the principles of classification of peat species and deposits in the USSR. Proc. 4th Internat. Peat Congr., Otaniemi, Finland 1: 41-7.

Osvald, H. 1923. Die Vegetation des Hochmoores Komosse. Svenska Växtsoc. Sällsk. Handl. 1, Uppsulala.

— 1925. Die hochmoortypen Europas. Veröff. Geobot. Inst. Rübel Zürich 3: 707-23.

Pjavchenko, N.I. 1968. Basic terminology in bog science. Trans. 2nd Internat. Peat Congr., Leningrad, 1963, 1: 111-14.

Pogrebniak, P.S. 1944. Fundamentals in Forest Typology (Techniceskoi i Promyshlonnoi Literatury pri SNK-USSR Kiev).

Pollett, F.C. 1967. Certain Ecological Aspects of Selected Bogs in Newfoundland (M.Sc. thesis, Memorial Univ., Newfoundland).

— 1968. Glossary of peatland terminology. Dept. Forest. Rur. Dev., St. John's, Nfld. Intern. Rept. N-5.

— 1972. Classification of peatlands in Newfoundland. Proc. 4th Internat. Peat Congr., Helsinki, 1: 101-10.

Porsild, A.E. 1937. Flora of the Northwest Territories. In Canada's Western Northland (Can. Dept. Mines Resources, Lands, Parks, Forests Branch), pp. 130-41.

Puchner, H. 1920. Der Torf (F. Enke, Stuttgart).

Radforth, N.W. 1952. Suggested classification of muskeg for the engineer. Eng. J. 35/11: 1-12.

— 1958. Organic terrain organization from the air (altitudes 1,000 to 5,000 feet). Handb. 2, Dept. Nat. Def. Can., DR 124.

— 1962. Air photo interpretation of organic terrain for engineering purposes. Trans. Symp. Photo Interpretation. Int. Arch. Photogramm. Delft, Holland, 14: 507-13.

— 1964. Prerequisite for design of engineering works on organic terrain — a symposium. Part II - Definitions and terminology. NRC Can., Assoc. Comm. Soil Snow Mech., Tech. Mem. 81: 24-35.

— 1969. Airphoto interpretation of muskeg. Muskeg Engineering Handbook, ed. I.C. MacFarlane (Univ. Toronto Press), pp. 53-77.

Ratcliffe, D.A. 1964. Mires and bogs. In The Vegetation of Scotland, ed. J.H.L. Burnett (Oliver and Boyd, London), pp. 426-78.

Ratcliffe, D.A., and Walker, D. 1958. The Silver Flowe, Galloway, Scotland. J. Ecol. 46: 407-45.

Raup, H.M. 1935. Botanical investigations in Wood Buffalo Park. Nat. Mus. Can. Bull. 14.

Rigg, G.B. 1958. Peat resources of Washington. Wash. Dept. Conserv., Div. Mines Geol., Bull. 44.

Ritchie, J.C. 1958. A vegetation map from the southern spruce forest zone of Manitoba. Geog. Bull. 12: 39-46.

— 1960. The vegetation of northern Manitoba. V. Establishing the major zonation. Arctic 13/4: 211-29.

Sjörs, H. 1948. Myrvegetation i bergslagen. Acta Phytogeog. Suec. 21 (English summary): 277-99.

— 1959. Bogs and fens in the Hudson's Bay lowlands. Arctic 12/1: 1-19.

— 1961a. Surface patterns in boreal peatland. Endeavour 20/80: 217-24.

— 1961b. Forest and peatland at Hawley Lake, northern Ontario. Nat. Mus. Can. Bull., 1971: 1-31.

— 1963. Bogs and fens on Attawapiskat River, northern Ontario. Nat. Mus. Can. Bull. 186: 45-133.

Skertchley, S.B.J. 1877. The geology of the fenland. Mem. Geol. Survey England and Wales (London), pp. 145-51, 172-4. (Referred to in Waksman, 1942)

Stanek, W. 1966. Occurrence, Growth and Relative Value of Lodgepole Pine and Engelmann Spruce in the Interior of British Columbia (Ph.D. thesis, Univ. British Columbia).

Stoeckeler, E.G. 1949. Identification and evaluation of Alaskan vegetation from air photos with reference to soil moisture and permafrost conditions. A preliminary paper (Dept. Army Corps. Eng., St. Paul Dist.).

Stubbs, J. 1962. A challenging opportunity in forest management: wetland forests. Forest. Farmer 21/11: 6-7, 10-13.

Svensson, H. 1962. Tundra polygons. Norg. Geol. Unders. Arsbok. pp. 298-327.

Tansley, A.G. 1949. The British Islands and Their Vegetation (2 vols.; Cambridge Univ. Press).

Tanttu, A. 1915. Über die Entstehung der Bülten und Stränge der Moore. Acta Forest. Fenn. 4/1.

Tarnocai, C. 1970. Classification of Peat Landforms in Manitoba (Can. Dept. Agric. Res. Stn., Pedol. Unit. Winnipeg).

Tedrow, J.C.F., and Cantlon, J.E. 1948. Concepts of soil formation and classification in arctic regions. Arctic 11/3: 166-79.

Troll, C. 1944. Strukturboden, Solifluktion und Frostklimate der Erde. Geol. Rundchau 34: 545-695.

Tyrtikov, A.P. 1966. Formirovanije i razvitije krupnobugristych torfjanikov v severnoy taÿge zapadnoy sibire (Formation and development of large hummocky peat bogs in the northern taiga of western Siberia). Merzlotnyye issledovaniya (Permafrost Investigations) 4: 144-54.

von Kruedener, A., and Becker, A. 1941. Atlas standortkennzeichnender Pflanzen (Atlas of plant indicators) (Wiking Verlag, Berlin).

von Post, Hampus. 1862. Studier öfver nutidens koprogena jordbildninger, gyttja, dy och mylla R. Sv. Vet. Ak. Handb. 4(1). In Waksman (1942).

von Post, L. 1922. Sveriges geologiska undersöknings torvinventering och nagra av dess hittills vunna resultat. Bilago Svenska Mosskultur Fören. Tiaskr. (1): 1-25.

— 1937. The geographical survey of Irish bogs. Irish. Nat. J. 6: 210-27.

von Post, L. and Granlund, E. 1926. Södra sveriges torvtillgangar. Sveriges Geol. Unders. Ser. C. No. 335 Arsbok 19/2.

Waksman, S.A. 1942. The peats of New Jersey and their utilization. Dept. Cons. Dev., New Jersey. Bull. 55A.

Washburn, A.L. 1956. Classification of patterned ground and review of suggested origins. Geol. Soc. Am. Bull. 66: 823-66.

Weber, C.A. 1903. Über Torf, Humus und Moor. Adhandl. Naturw. Ver. Bremen 17: 465-84.

— 1908a. Die Entwicklung der Moorkultur in den letzten 25 Jahren. (Die wichtigsten Humus und Torfarten und ihre Beteiligung an dem Aufbau norddeutscher Moore). Festschrift zur Feier des 25 Jahrigen Bestehens des Vereins zur Förderung der Moorkultur im Deutschen Reich (Berlin), pp. 80-101.

— 1908b. Aufbau und Vegetation der Moore. Norddeutschlands. Englers Botanisches Jahrbuch, Beibe. 90, Leipzig (referred to in Kulczynski, 1949).

Welch, S. 1935. Limnology (McGraw-Hill Book Co. Inc., New York).

Wenner, C.G. 1947. Pollen diagrams from Labrador. Geog. Ann. (Stockholm) 29/3-4: 137-374.

Ziegler, C.M. 1946. Field Manual of Soil Engineering (Mich. State Hwy. Dept., Lansing, Mich.).

Zoltai, S.C. 1971. Southern limit of permafrost features in peat landforms, Manitoba and Saskatchewan. Geol. Assoc. Can. Spec. Pap. 9: 305-10.

Zoltai, S.C., Pollett, F.C., Jeglum, J.K., and Adams, G.D. 1974. Developing a wetland classification for Canada. Proc. 4th North Amer. For. Soil Conf. Quebec, Aug. 1973.

Zoltai, S.C., and Tarnocai, C. 1971. Properties of wooded palsa in northern Manitoba. Arctic Alp. Res. 3/2: 115-29.

Index

organisms, disease-producing, 343
oxbows, 42
oxidation: resistance to, 85; of peat, 327
oxides, of metal, 228
oxygen concentration in peat, 99, 318

palaeobotany, 13
palsa: collapse of, 155; definitions, 45, 47, 375;
    formation of, 152–5; mature, 155; occurrence
    of, 35, 41, 43, 152–5; relation to permafrost,
    150–5
palsa bog, 376
palsa peatland, definition of, 376
palsa muskeg, definition of, 47
paludification: definition of, 376; and deglaciation,
    10–11; and drainage, 257; in peatland survey,
    66–7; synonyms, 48
palynology, 13, 20–25, 376
paraffin, in peat, 85
paraquat, 217
parasites and waste disposal, 343
pasture, 107, 208, 211
pathogen, in sewage, 343
patterned bog, 376
patterned fen, 376
patterned muskeg, 376
patterns, airform, see Airform pattern
peat: as an absorbent, 65, 222–6; accumulation,
    15, 17, 211 (see also Paludification); botanical
    characteristics of, 66, 70–72; chemical proper-
    ties of, 84–99, 128–9; classification of, 68–73,
    184, 381; composition of, 83, 84–90, 116, 231;
    correlation of physical and chemical properties
    of, 109–20; cutting, 208; decomposition of, 15,
    82, 100, 119, 370; definition of, 17, 47, 376; de-
    gree of humification, see Humification, degree
    of; depth, 27, 31, 66, 67, 71, 74, 80, 82, 143, 255,
    303; drying, 107, 208, 228; engineering proper-
    ties of, 64, 67, 83; field testing, 68; as a fossil
    record, 20, 27; genetic features, 67; nutrient
    status of, 37–40, 91–3, 216–17, 316–17; physical
    properties of, 67, 99–109, 126–7; rewetting, 107;
    sampling, 67; as a soil, 211–12; thermal charac-
    teristics of, 83, 141, 150, 161, 254–6; volume of,
    63, 74
peat, types of: allochthonous, 368; autochthon-
    ous, 368; brown ‘moss, 368; calcareous, 44;
    carex, 369; fen, 304; fibrous, 370; frozen, 67;
    sedge, 378; sedge reed, 378; sedge shrub, 378;
    sedimentary, 379; Sphagnum, 379; upland, 381;
    woody, 382
peat, uses of: as construction material, 239–44;
    industrial, 221–46; as a soil additive, 208, 211
peat bog, 45, 376
peat carbonization, 235
peat compost, 212
peat deposit, 20, 376
peat granules, 225
peat litter, 212
peat mining, 67, 80

peat moss, 5, 25, 47, 65, 72, 373, 376
peat mound, 47, 376
peat plateau: definition of, 47, 377; relation to
    permafrost, 152–5
peat profile, 67, 70, 74, 75
peat science, 6, 189
peat spreader, 225
peatcork, 243–4
peatcrete, 241
peatfoam, 244
peatland: classification of, 16, 213, 315–18; defini-
    tion of, 182, 377; potential productivity of, 213;
    use of, 65, 211–20; see also Muskeg
peatland complex, 377
peatland type, 16, 377; minerotrophic, 314–17;
    ombrogenic, 16; ombrotrophic, 315–17; palsa,
    376; soligenic, 16; soligenous, 368; string, 379;
    topogenic, 16; topogenous, 368; transition, 380;
    treed, 74
peatwood, 242–4
pedological processes, 82
permafrost, 5, 67, 139; continuous, 32, 148–50;
    definition of, 377; development of, 150–2; dis-
    tribution of, 32, 34–6, 148–50; discontinuous,
    32, 34, 36, 148; effect on construction, 337; ef-
    fect on waste disposal, 348; remote sensing of,
    155–9; southern limit of, 148, 149; surface fea-
    tures of, 37, 152–5; thermal aspects of, 150–2;
    variation by physiographic region, 160–1
permafrost table, 149
permeability of peat, 12, 67, 80, 222; determina-
    tion of, 70, 104–7; effect of loading on, 256;
    effect of draining on, 325; see also Hydraulic
    conductivity
permeability apparatus, 70
permesotrophic, 374, 377
petroleum, 349; exploration, 250, 349
pH: of peat, 12, 16, 40, 67, 92, 95–7, 119; of water,
    320, 322; see also Peat, chemical properties of,
    Acids in peat
phenol: absorption by peat, 222; adsorption by
    activated carbon, 231
phenolic binder, 241
phenozine, 231
phosphate, 92, 222
phosphoric acid, 232
phosphorus, in peat, 37, 91; in fertilizer, 217
photography: aerial, see Aerial photography; in
    muskeg survey, 76–7
physical properties of peat, 67, 99–109, 126–7;
    correlation with chemical properties, 109–19
phyto-biocoenoses, 31
phytosociological analyses, 67
phytosociological units, 44
pipeline: automation, 298; berm construction,
    268–9; buoyancy, 266–8; cooling systems, 294;
    design, 266; effect on environment, 5, 10, 29,
    130, 250, 254, 259, 294–8; horsepower require-
    ments, 265; operating cost, 259, 265; operating
    efficiency, 259; right-of-way requirement, 266;

www.ingramcontent.com/pod-product-compliance
Lightning Source LLC
Chambersburg PA
CBHW051749200326

41597CB00025B/4489